STEP 7 in 7 Steps

A Practical Guide to Implementing S7-300/S7-400 Programmable Logic Controllers

This book is dedicated to my friend,
Russell Briskey

STEP 7 in 7 Steps

A Practical Guide to Implementing S7-300/S7-400 Programmable Logic Controllers

First Edition

C.T. Jones

Printed in the United States

ISBN 1-889101-03-6

This book is sold as is, without warranty of any kind, either express or implied, respecting the contents of this book, including but not limited to implied warranties for the book's quality, performance, merchantability, or fitness for any particular purpose. This book is not intended to be, nor should it be, used as authority for design, implementation, or installation of any control systems. Such tasks should only be undertaken by trained professionals using currently accepted design and engineering practices, and the guidance of required user manuals.

Trademarks:

ET-200, S7-300, S7-400, HiGraph, HMI, SIMATIC, SIMATIC HMI, SIMATIC NET, SINEC, SOFTNET, STEP, and WinCC, are Siemens registered trademarks. Other product names mentioned in this book may be trademarks or registered trade marks of their respective companies, and is the sole property of their respective owners and should be treated as such. Furthermore, use of a term in this book should not be regarded as affecting the validity of any trademark or service mark.

With the exception of screen captures, this book was done entirely in Microsoft Word. Screen captures in this book were created using Full Shot 6.02, Inbit Inc., Mountain View, CA.

About the Author

C.T. Jones, received his Bachelor of Science in Electrical Engineering from Howard University, and has made his career in the Industrial Automation Industry for over twenty-five years. During this time, he has served both as a PLC user and in marketing, applications engineering, and technical training roles for several automation vendors. It was in the role of Control Systems Engineer, for Procter and Gamble, where his fascination with PLCs and their endless possibilities in automation, drew him into this field, which he has pursued since.

Early on as a Controls Engineer, he became acutely aware of the unique position of end-users, who have the tough balancing-act of keeping the plant operational and staying abreast of automation technologies. As a result, he authored two books on programmable controllers, including the first PLC book "*Programmable Logic Controllers Concepts and Applications*," and "*Programmable Logic Controllers: The Complete Guide to the Technology*." While serving eight years as applications engineer and Technical Instructor for SIMATIC equipment, he devoted much of his time to developing step-by-step application notes and other educational materials for users.

For several years now, C.T. has been a SIMATIC consultant and freelance technical writer. In this role, he has worked with software companies in the development of user documentation, training and tutorial courseware, and in product development involving integration with STEP 7. He also provides applications consulting and onsite training for end users.

Acknowledgement

Garon Hart, Graphic Designer and friend, has worked with me in the development of books since I started writing in 1984. Writing a book takes a lot of time and involves many costs, along the way, for which writers usually cannot offer compensation. Garon, nonetheless, has always been very generous with his time, skills, and knowledge. To put it simply, he has always helped make it possible for me to produce a new book. Thanks Garon!

I also extend thanks to Claude Kouakou, software developer and friend. During the development of this book, I encountered many problems related to early choices I had made for the book's layout and title. Claude, who during the time of the books development was writing code to integrate STEP 7 with another software package, was always my sounding board. He always had an uncanny way of making simple, that which appeared to be insurmountable. Thanks Claude!

About This Book

This book is organized in seven practical areas associated with getting the job done efficiently. It is a task-oriented guide to configuring, programming, deploying, troubleshooting, and maintaining S7-300/S7-400 PLCs and SIMATIC Networks. Each of the seven task areas is introduced with a brief tutorial that is followed by a number of actual tasks. Each task is presented in an identical two-page layout. On the left-hand page, a task is briefly described under the headings — **Basic Concept**, **Essential Elements**, and **Application Tips**. On the right-hand page, the task is presented in a step-by-step format.

At the time this book was developed, STEP 7 was at version 5.2 and 5.3. Its usefulness, however, should serve well into the future of new releases. This is true since the tutorial discussions at the beginning of each chapter and for each task example, are based on architectural and operational features and concepts, inherent to S7-300 and S7-400 PLCs and SIMATIC Networks.

Step 1 Getting Started

Chapter 1 introduces you to the S7-300/S7-400 hardware structure and the STEP 7 basic and optional software tools. You learn to install the software and configure preferences, to configure the programming workstation and establish S7 Online connection.

Step 2 Working with Hardware Configurations

Chapter 2 introduces the concept of STEP 7 projects and libraries. The SIMATIC Manager is introduced along the basic STEP 7 objects and hierarchy. The Standard library is reviewed, and finally several example tasks of working with projects and libraries are presented.

Step 3 Working with Hardware Configurations

Chapter 3 introduces basic principles of using the Hardware Configuration Tool to configure S7-300/S7-400 hardware. After reviewing basic navigation, several example tasks of typical arrangements of the S7-300/S7-400 are covered, including modular, compact, and intelligent DP slaves. Example tasks also include defining parameters for CPUs, CPs, SMs, and IMs. Finally, addressing is described in detail for both digital and analog I/O.

Step 4 Working with STEP 7 Programs and Data

Chapter 4 introduces the basic principles of STEP 7 programming including design principles, block types, order of processing, data types and formats, addressing memory, and using the LAD/FBD/STL editor. Each instruction category and instruction of LAD/FBD is described, and examples of programming block types and every instruction is covered.

Step 5 Managing Online Interactions with S7 CPUs

Chapter 5, will show you how to access the CPU online with or without a project, and how to work with standard online operations, such as compare online/off-line programs, upload/download program, clear memory, compress memory, switch CPU operating modes, and access password protected CPUs.

Step 6 Working with Monitoring and Diagnostic Tools

Chapter 6 expands CPU online operations to include monitoring and diagnostic tools. You will learn to evaluate the S7 Program and data, using *Debug Monitor, Monitor/Modify Variables, and Forcing tools*. In addition, you will learn to diagnose problems in the program, the hardware, and in the process, using the various STEP diagnostic tools.

Step 7 Working with SIMATIC NET Networks

Chapter 7 introduces the basic principles of configuring SIMATIC NET networks, using the NetPro Network Configuration Tool. After reviewing navigation of the configuration tool, example tasks of MPI, Profibus, and Ethernet networks from the top-down are covered. You will learn how to install and configure subnets, communications processors, address and attach stations to the network, and to download network configurations.

Preface

It is easy to say that this new book is for anyone involved with S7-300/S7-400 and SIMATIC NET networks, but in fact, this is the case. If you are a controls engineer, responsible for design and programming, a maintenance technician, an applications engineer, or sales representative, you'll find that this book provides much useful information. The book offers significant coverage of configuration concepts, addressing memory and I/O, programming in STEP 7, defining network configurations, as well as a guide to using the tools for monitoring the program, and diagnosing problems in the hardware, software, or in the process.

If you are a controls or applications engineer, your work will likely involve both configuring the hardware and programming for S7-300/S7-400 and SIMATIC networks. You'll find examples of the most typical configurations, as well as an explanation for various module parameters. In most cases, you'll find that the default parameters are generally usable, but you will learn how to obtain different operational behavior in the CPU and other modules. The Chapter on *Working with STEP 7 Programs and Data* is a great reference for learning about program design, data types and formats, instruction sets, and how to write code for Functions (FCs) and Function Blocks (FBs). You'll also learn the basics for structuring a program and calling blocks.

If you are responsible for maintaining an S7-300 or S7-400 system, then you should certainly read chapters one and two to gain a basic understanding of S7 hardware and STEP 7 software — especially navigating the SIMATIC Manager. Chapters 5 and 6 offer many of the tasks that will likely serve your needs, and in Chapter 3, you can gain an understanding of how to configure new hardware and address the I/O system. While writing programs and configuring networks may not be your main concern, these topics are presented in an easily understood format. So, don't hesitate to use Chapters 4 and 7, to broaden your knowledge of STEP 7 program operation, the instruction sets, and how your plant network is configured.

Much consideration went into the content and the design for this book. I have often heard users speak of their difficulty in quickly getting their arms around the S7 implementation and in determining where to look for pertinent information; so, my first goal was to determine the tasks that as a user, you would eventually need to perform. Then, to present each of these in a manner that would cover the basic concepts of the task, the essential elements involved, and then to present the task in a brief and easily understood fashion. The idea was to gather and explain in one place, topics essential to allowing you to get right to the task.

Finally, how you should consider using this book is largely linked to its design. Not every task has been included; however, you will likely find what you are looking for. Should you read the book from cover-to-cover? You may if you'd like, but that is not necessary. If you are a new user, the tutorial at the beginning of each chapter will be a good start. As your project calls for a specific task, turn to the appropriate chapter to get a quick introduction. Step-by-step procedures for several common tasks are provided in every chapter. You can also use the index to quickly find and use a specific task, and the Appendices to find other useful information.

Contents

STEP 1: GETTING STARTED WITH STEP 7 1

S7-300/S7-400 PROGRAMMABLE CONTROLLER STRUCTURE 2

The Components 2

THE STEP 7 PROJECT MANAGEMENT SYSTEM 4

Standard Tools and Utilities 4

Optional Language Tools 6

COMMENTS ON GETTING STARTED WITH STEP 7 11

Checklist: Getting Started with STEP 7 11

INSTALLING THE STEP 7 SOFTWARE 12

INSTALLING AND REMOVING STEP 7 AUTHORIZATIONS 14

OPENING SIMATIC SOFTWARE COMPONENTS 16

CONFIGURING A SIMATIC WORKSTATION 18

DEFINING STORAGE PATH PREFERENCES 20

DEFINING LANGUAGE PREFERENCES 22

DEFINING ARCHIVING PREFERENCES 24

ESTABLISHING A DIRECT CPU PROGRAMMING CONNECTION 26

SETTING THE PG/PC INTERFACE FOR PC ADAPTER 28

STEP 2: WORKING WITH STEP 7 PROJECTS AND LIBRARIES 31

THE SIMATIC MANAGER 32

A Project and Library Manager 32

SIMATIC Manager Menu and Toolbar 35

Project Windows 36

COMMENTS ON WORKING WITH STEP 7 PROJECTS AND LIBRARIES 37

Checklist: Working with STEP 7 Projects and Libraries 37

NAVIGATING THE STEP 7 PROJECT STRUCTURE 38

NAVIGATING SIMATIC MANAGER MENUS AND TOOLBAR 40

FINDING STEP 7 PROJECTS AND LIBRARIES 42

CREATING A PROJECT USING THE NEW PROJECT WIZARD 44

CREATING A PROJECT USING THE 'NEW' COMMAND 46

ADDING A NEW STATION TO A PROJECT 48

ARCHIVING A PROJECT OR LIBRARY 50

RETRIEVING AN ARCHIVED PROJECT OR LIBRARY 52

STEP 3: WORKING WITH HARDWARE CONFIGURATIONS....................................55

S7-300/S7-400 HARDWARE COMPONENTS OVERVIEW 56
Racks 56
Interface Modules (IM) 57
Power Supply (PS) 58
Central Processing Unit (CPU) 58
Signal Modules (SM) 59
Function Modules (FM) 59
Communications Processors (CP) 59
Multi-Point Programming Interface (MPI) 59
THE HARDWARE CONFIGURATION TOOL 60
Configuring the Hardware 60
Menus and Toolbar 60
Hardware Catalog Window 61
Station Window – Rack Arrangement Pane 62
Station Window – Configuration Tables Pane 62
COMMENTS ON WORKING WITH HARDWARE CONFIGURATIONS 63
Checklist: Working with Hardware Configurations 63
NAVIGATING THE HARDWARE CATALOG 64
NAVIGATING THE HARDWARE CONFIGURATION MENU AND TOOLBAR 66
VIEWING STATION CONFIGURATION DETAILS 68
BUILDING A STATION CONFIGURATION 70
DOWNLOADING A STATION CONFIGURATION 72
UPLOADING A STATION CONFIGURATION 74
ASSIGNING SYMBOLIC ADDRESS TO INPUT/OUTPUT MODULES 76
CONFIGURING AN S7-300 CENTRAL RACK 78
CONFIGURING S7-300 LOCAL I/O EXPANSION 80
CONFIGURING S7-300 SINGLE-TIER I/O EXPANSION 82
CONFIGURING AN S7-300 AS DP MASTER 84
CONFIGURING AND ATTACHING MODULAR DP SLAVES 86
CONFIGURING AND ATTACHING COMPACT DP SLAVES 88
CONFIGURING THE S7-31X-2 DP AS AN INTELLIGENT DP SLAVE 90
CONFIGURING THE CP-342-5 AS AN INTELLIGENT DP SLAVE 92
CONFIGURING THE ET-200X (BM 147/CPU) AS AN INTELLIGENT DP SLAVE 94
CONFIGURING MASTER-INTELLIGENT SLAVE DATA EXCHANGE AREA 96
CONFIGURING DATA EXCHANGE FOR CP 342-5 INTELLIGENT SLAVE 98
CONFIGURING S7-300 CPU GENERAL PROPERTIES 100
CONFIGURING S7-300 CPU START-UP PROPERTIES 102
CONFIGURING S7-300 CPU CYCLE AND CLOCK MEMORY 104
CONFIGURING S7-300 CPU RETENTIVE MEMORY 106

Configuring S7-300 CPU Local Memory ... 108
Configuring S7-300 CPU Diagnostics and Clock Properties ... 110
Configuring S7-300 CPU Access Protection ... 112
Configuring S7-300 CPU Interrupt Properties ... 114
Configuring S7-300 CPU Time-of-Day Interrupts ... 116
Configuring S7-300 CPU Cyclic Interrupts ... 118
Configuring S7-300 Digital Input Module Properties ... 120
Configuring S7-300 Interrupt Input Properties ... 122
Configuring S7-300 Digital Output Module Properties ... 124
Configuring S7-300 Digital I/O Addresses ... 126
Configuring S7-300 Analog Input Module Properties ... 128
Configuring S7-300 Analog Input Signal Parameters ... 130
Configuring S7-300 Analog Output Module Properties ... 132
Configuring S7-300 Analog I/O Addresses ... 134
Configuring an S7-400 Central Rack ... 136
Configuring an S7-400 Multi-Computing Central Rack ... 138
Configuring the S7-400 as a DP Master ... 140
Configuring S7-400 Local I/O Expansion ... 142
Configuring S7-400 Remote I/O Expansion ... 144
Configuring S7-400 CPU General Properties ... 146
Configuring S7-400 CPU Start-Up Properties ... 148
Configuring S7-400 CPU Cycle and Clock Memory ... 150
Configuring S7-400 CPU Retentive Memory ... 152
Configuring S7-400 CPU Local Memory ... 154
Configuring S7-400 CPU Diagnostics and Clock Properties ... 156
Configuring S7-400 CPU Access Protection ... 158
Configuring S7-400 CPU Interrupt Properties ... 160
Configuring S7-400 CPU Time-of-Day Interrupts ... 162
Configuring S7-400 CPU Cyclic Interrupts ... 164
Configuring S7-400 Digital Input Module Properties ... 166
Configuring S7-400 Interrupt Input Properties ... 168
Configuring S7-400 Digital Output Module Properties ... 170
Configuring S7-400 Digital I/O Addresses ... 172
Configuring S7-400 Analog Input Module Properties ... 174
Configuring S7-400 Analog Input Signal Parameters ... 176
Configuring S7-400 Analog Output Module Properties ... 178
Configuring S7-400 Analog I/O Addresses ... 180
Configuring I/O Modules for Multi-Computing Operation ... 182

Step 4: Working With STEP 7 Programs and Data 185

Introduction to STEP 7 Programming Principles 186
Program Design Strategy 186
Developing the Control Logic 187
STEP 7 Block Types 188
System Blocks (SFB, SFC, SDB) 188
Organization Blocks (OB) 188
Function Blocks (FB) 190
Functions (FC) 191
Data Blocks (DB) 192
S7 Program Processing 194
The Normal CPU Cycle 194
Interrupting the Normal CPU Cycle 194
Addressing S7 Memory Areas 196
Input Memory (I) 196
Output Memory (Q) 196
Bit Memory (M) 197
Peripheral Memory (PI/PQ) 197
Timer Memory (T) 198
Counter Memory (C) 198
Local Memory (L) 198
Summary of S7 Memory Addressing 199
S7 Data Types and Formats 200
Elementary Data Types 200
Complex Data Types 206
Parameter Data Types 208
Overview of the LAD/FBD/STL Editor 209
Introduction 209
Menus and Toolbar 209
Program Elements/Call Structure Window 210
Block Window 211
Details Window 211
STEP 7 Instruction Set Overview 212
Bit Logic Instructions – Basic Operations 212
Bit Logic Instructions – Special Operations 213
Counter Instructions 214
Timer Instructions 215
Conversion Instructions 216
Integer and Real Arithmetic Instructions 218
Compare Instructions 220

- Program Flow Control Instructions ... 221
- Status Bit Instructions ... 222
- Word Logic Instructions ... 223
- Shift-Rotate and Move Instructions ... 224

COMMENTS ON WORKING WITH STEP 7 PROGRAMS AND DATA ... 225

- Checklist: Working with STEP 7 Programs and Data ... 225

VIEWING AND EDITING SYMBOLIC ADDRESSES ... 226

CREATING A DATA BLOCK ... 228

EDITING A DATA BLOCK ... 230

GENERATING A NEW CODE BLOCK ... 232

NAVIGATING THE LAD/FBD/STL EDITOR ... 234

OPENING, EDITING, AND SAVING A BLOCK ... 236

DOCUMENTING A CODE BLOCK ... 238

PROGRAMMING AN FC WITHOUT FORMAL PARAMETERS ... 240

CALLING AN FC WITHOUT FORMAL PARAMETERS ... 242

PROGRAMMING AN FC WITH FORMAL PARAMETERS ... 244

CALLING AN FC WITH FORMAL PARAMETERS ... 246

PROGRAMMING AN FB WITHOUT FORMAL PARAMETERS ... 248

CALLING AN FB WITHOUT FORMAL PARAMETERS ... 250

PROGRAMMING AN FB WITH FORMAL PARAMETERS ... 252

CALLING AN FB WITH FORMAL PARAMETERS ... 254

PROGRAMMING BASIC BIT-LOGIC OPERATIONS ... 256

PROGRAMMING SET-RESET OPERATIONS ... 258

PROGRAMMING EDGE EVALUATION OPERATIONS ... 260

PROGRAMMING COUNTER OPERATIONS ... 262

PROGRAMMING TIMER OPERATIONS ... 264

PROGRAMMING CONVERSION OPERATIONS ... 266

PROGRAMMING COMPARE OPERATIONS ... 268

PROGRAMMING INTEGER ARITHMETIC OPERATIONS ... 270

PROGRAMMING REAL ARITHMETIC OPERATIONS ... 272

PROGRAMMING TRIGONOMETRIC AND OTHER MATH FUNCTIONS ... 274

PROGRAMMING JUMP, LABEL, AND RETURN OPERATIONS ... 276

PROGRAMMING MCR OPERATIONS ... 278

PROGRAMMING WORD LOGIC OPERATIONS ... 280

PROGRAMMING SHIFT AND ROTATE OPERATIONS ... 282

PROGRAMMING STATUS BIT OPERATIONS ... 284

PROGRAMMING THE MOVE OPERATION TO READ AND WRITE DATA ... 286

ACCESSING DATA IN A DATA BLOCK ... 288

PROGRAMMING ORGANIZATION BLOCK 1 (OB 1) ... 290

STEP 5: MANAGING ONLINE INTERACTION WITH S7 CPUS 293

ESTABLISHING ONLINE CONNECTIONS USING STEP 7 294

The Standard Physical Connection 294

Using the Online Project Window 294

Using the Accessible Nodes Window 295

STANDARD ONLINE OPERATIONS WITH STEP 7 295

Downloading the User Program to the CPU 296

Uploading the User Program from the CPU 296

Comparing Online/Offline Programs 296

Comparing Offline Programs — Path 1/Path 2 297

Providing CPU Access Protection 297

Accessing CPU Information and Operating Characteristics 298

COMMENTS ON MANAGING ONLINE OPERATIONS WITH S7 CPUS 299

Checklist: Managing Online Operations with S7 CPUs 299

ACCESSING ONLINE OPERATIONS WITHOUT A PROJECT 300

VIEWING CPU RESOURCES AND PERFORMANCE DATA 302

VIEWING AND CHANGING CPU OPERATING MODES 304

RESETTING MEMORY FROM THE SIMATIC MANAGER 306

ACCESSING A PASSWORD PROTECTED CPU 308

SETTING THE CPU DATE AND TIME 310

COMPRESSING CPU MEMORY 312

DOWNLOADING THE S7 PROGRAM OR SELECTED BLOCKS 314

UPLOADING THE S7 PROGRAM OR SELECTED BLOCKS 316

COMPARING OFFLINE/ONLINE PROGRAMS 318

COMPARING TWO PROGRAMS (PATH 1/PATH 2) 320

STEP 6: WORKING WITH MONITORING AND DIAGNOSTIC TOOLS 323

TOOLS FOR MONITORING PROGRAMS AND DATA 324

Using the Debug Monitor to Evaluate the Program Status 324

Using Variable Tables 324

Monitoring and Modifying Variables 324

Forcing I/O and Memory Variables 324

USING STEP 7 PROGRAM REFERENCE DATA 324

Assignments List 325

Program Structure 325

Program Cross Reference Data 325

Unused Symbols List 325

Addresses without Symbols 326

DETERMINING THE CAUSE OF CPU STOP 326

Using the CPU Diagnostic Buffer 326

Diagnosing S7 Hardware Online 326

Diagnosing Program-Related Faults Using the CPU Stacks 326

COMMENTS ON WORKING WITH MONITORING AND DIAGNOSTIC TOOLS ... 327
Checklist: Working with Monitoring and Diagnostic Tools ... 327
MONITORING PROGRAM STATUS WITH A PROJECT ... 328
MONITORING PROGRAM STATUS WITHOUT A PROJECT ... 330
CREATING AND EDITING VARIABLE TABLES (VATS) ... 332
MONITORING AND MODIFYING VARIABLES ... 334
FORCING I/O AND MEMORY VARIABLES ... 336
USING THE HARDWARE DIAGNOSTICS UTILITY ... 338
USING THE CPU DIAGNOSTIC BUFFER ... 340
USING THE CPU STACKS TO DIAGNOSE PROGRAM FAULTS ... 342
GENERATING AND DISPLAYING PROGRAM REFERENCE DATA ... 344

STEP 7: WORKING WITH SIMATIC NET NETWORKS ... 347

INTRODUCTION TO SIMATIC NET NETWORKS ... 348
Networks and Subnets ... 348
COMMUNICATIONS CONNECTIONS AND SERVICES ... 350
Communication Connections ... 350
Communication Services ... 351
PROFIBUS COMMUNICATIONS PROCESSORS ... 357
Profibus CPs for S7-300 and S7-400 Stations ... 357
Profibus CPs for PC and PG/PC Stations ... 358
ETHERNET COMMUNICATIONS PROCESSORS ... 359
Ethernet CPs for S7-300 and S7-400 Stations ... 359
Ethernet CPs for PC and PG/PC Stations ... 360
THE NETWORK CONFIGURATION TOOL ... 362
Configuring the Network ... 362
Menu and Toolbar ... 362
Network Components Catalog Window ... 363
Configuration Window – Network Layout ... 364
Configuration Window – Connection Table ... 364
Comments on Working with SIMATIC NET Networks ... 365
Checklist: Working with SIMATIC NET Networks ... 365
BUILDING A NETWORK CONFIGURATION USING NETPRO ... 366
DOWNLOADING A NETWORK CONFIGURATION USING NETPRO ... 368
ADDING AND CONFIGURING AN MPI SUBNET ... 370
BUILDING AN MPI NETWORK WITH PEER S7-300/S7-400 STATIONS ... 372
CONFIGURING A PROGRAMMING STATION (PG/PC) ON MPI ... 374
CONFIGURING GLOBAL DATA COMMUNICATIONS ON MPI ... 376
ADDING AND CONFIGURING A PROFIBUS SUBNET ... 378
INSTALLING, CONFIGURING, AND ATTACHING AN S7-300 PROFIBUS CP ... 380
BUILDING A PROJECT WITH S7-300 PEER PROFIBUS STATIONS ... 382
INSTALLING, CONFIGURING, AND ATTACHING AN S7-400 PROFIBUS CP ... 384
BUILDING A PROJECT WITH S7-400 PEER PROFIBUS STATIONS ... 386
CONFIGURING A PROGRAMMING STATION (PG/PC) ON PROFIBUS ... 388
CONFIGURING PROFIBUS COMMUNICATIONS CONNECTIONS ... 390
ADDING AND CONFIGURING AN ETHERNET SUBNET ... 392

INSTALLING, CONFIGURING, AND ATTACHING AN S7-300 ETHERNET CP 394
BUILDING A PROJECT WITH S7-300 PEER ETHERNET STATIONS 396
INSTALLING, CONFIGURING, AND ATTACHING AN S7-400 ETHERNET CP 398
BUILDING A PROJECT WITH S7-400 PEER ETHERNET STATIONS 400
CONFIGURING A PROGRAMMING STATION (PG/PC) ON ETHERNET 402
CONFIGURING ETHERNET COMMUNICATIONS CONNECTIONS 404

APPENDICES 407

A STANDARD LIBRARY — ORGANIZATION BLOCKS 408
B STANDARD LIBRARY — SYSTEM BLOCKS (SFBS, SFCS) 410
C ASCII CHARACTER CHART 414
D EXAMPLES OF MODIFYING/FORCING I/O AND MEMORY VARIABLES 416
E CONDITION CODES CC0 AND CC1 AS RESULT BITS 418
F ANALOG INPUT/OUTPUT DIGITAL REPRESENTATION 419
G COMMON ABBREVIATIONS AND ACRONYMS 420

GLOSSARY 421

INDEX 43

Step 1

Getting Started with STEP 7

Objectives

- Introduce S7-300/S7-400 PLC Structure
- Introduce STEP 7 Basic Package
- Install and Remove STEP 7 Authorization
- Introduce Standard and Optional Tools
- Configure the STEP 7 Workstation
- Customize STEP 7 Operating Preferences
- Establish Direct CPU Programming Connection via PC Adapter
- Use the PG/PC Interface Utility for Setting or Switching the STEP 7 Online Interface

S7-300/S7-400 Programmable Controller Structure

The S7-300/S7-400 family of controllers consists of a range of PLCs, covering a span of requirements from simple to very complex. Though different in size and in overall capabilities, these controllers are alike in operational characteristics, memory organization, data structure, addressing, programming languages and instruction sets. Programs created for an S7-300 are ported to an S7-400 with little or no changes. They share the STEP 7 programming facilities, support the same local area networks (LANs), and integrate easily with common systems like operator panels (OPs), HMIs, and third-party I/O systems.

The Components

The S7-300 and S7-400 systems are both modular in nature, each built-up from the same basic types of components. These components, listed below, are briefly described later in Chapter 4. The programming device provides the common user interface for creating, storing, troubleshooting, and managing control programs. Each of these component types represent the various parts from which your S7-300/S7-400 system will be configured.

Table 1-1. S7-300 and S7-400 Basic Components.

Component	Abbrev.	Brief Description
Racks	-	Mounting base for installing various user-selected modules.
Power Supply	PS	Supplies internal operating voltages to racks and modules.
Central Processor	CPU	Stores and processes the user control program and data.
Signal Modules	SM	Digital/Analog I/O interfaces to field sensors and actuators.
Function Modules	FM	Intelligent modules that execute control tasks independent of the CPU (e.g., PID, stepper-positioning, servo-positioning).
Communication Processors	CP	Used to establish networking among S7 PLCs and other stations or point-to-point serial links to other devices.
Interface Modules	IM	Used to make various local and remote interconnections between the S7-300 and S7-400 central and expansion racks.
Programming Device	PG/PC	PG is a PC pre-configured with facilities for developing and managing STEP 7/STEP 5 programs; PC is a user-configured PC programming system.
Multi-Point Interface	MPI	Low-performance network and multi-point programming interface to components with MPI port [e.g., CPUs, CPs, FMs, operator panels (OPs)].
Distributed I/O	DP	I/O subsystems (or DP-slaves), connected to a DP-master according to Profibus DP standard EN 50170 Volume 2.

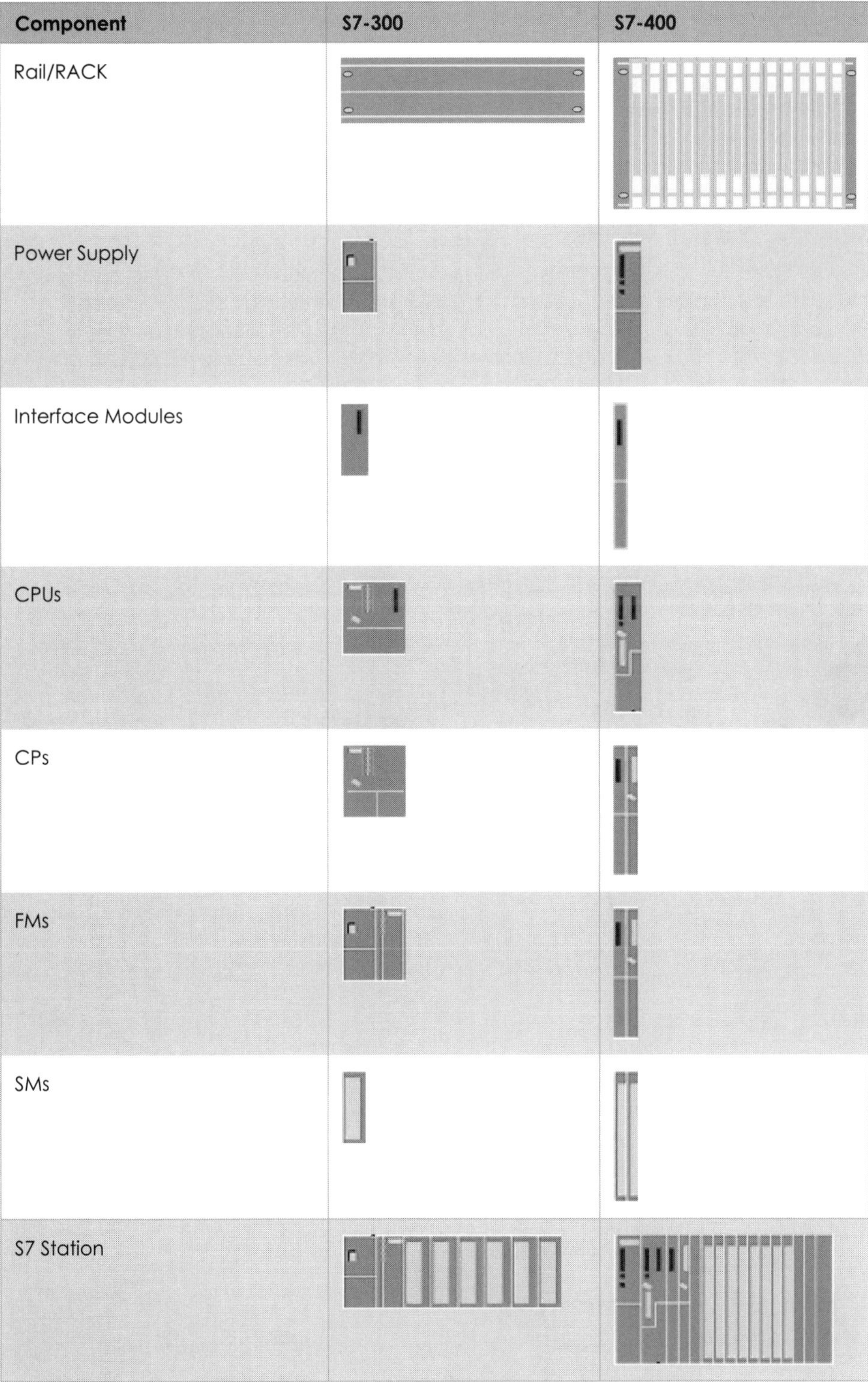

Component	S7-300	S7-400
Rail/RACK		
Power Supply		
Interface Modules		
CPUs		
CPs		
FMs		
SMs		
S7 Station		

Figure 1-1. S7-300 and S7-400 Programmable Controller Structure.

The STEP 7 Project Management System

STEP 7 is comprised of many components that make it the basis for development, management, and maintenance of SIMATIC automation systems. These systems include PLCs, HMIs, numerical controllers, electric drives, networks, and others. Furthermore, STEP 7 provides many standard and optional tools including program development languages, utilities for program conversion, monitoring and diagnostics, hardware configuration, network configuration, remote access service, and many other operations.

Standard Tools and Utilities

Standard software tools and utilities are immediately available for use, after the installation of the STEP 7 Basic package. Many of these tools, which are part of the STEP 7 Windows Program Group, may be launched by proceeding from the Start Menu button, selecting STEP 7, and finally clicking on the desired tool as shown above. Other standard tools are directly available from the SIMATIC Manager, as will be introduced later.

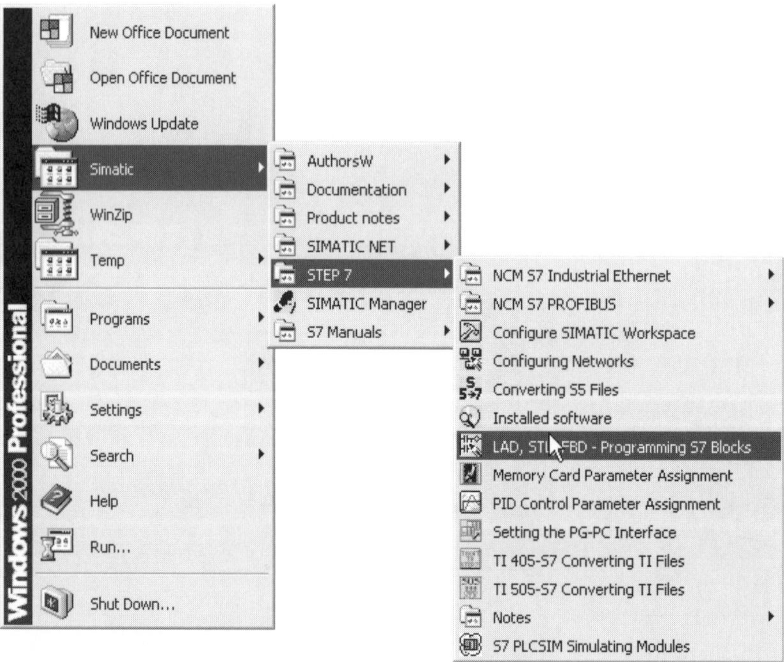

Figure 1-2. SIMATIC Start Menu with Standard Software Tools.

SIMATIC Manager

The SIMATIC Manager is the top-level tool for managing SIMATIC automation projects. Other standard software components, for example the hardware and network configuration tools and optional language editors, once installed, may be launched from the SIMATIC Manager.

Configuring SIMATIC Workspace

The SIMATIC Workspace Utility is used to set operating parameters required for working with STEP 7 in a multi-user configuration. Networking parameters such as protocols, addressing, and single- or multi-user environment are defined.

Configuring Networks

The *NetPro* configuration tool allows graphical configuration of the MPI, Profibus, and Industrial Ethernet subnets. The tool depicts network stations, communications interfaces, and physical connections; it also supports setting of module and network parameters and the creation of communications connections between partners.

Converting S5 Files

The STEP 5 to STEP 7 Conversion Utility allows existing STEP 5 programs to be converted to equivalent STEP 7 code. The purpose of the utility is to convert as much of the instruction set of existing STEP 5 programs as possible to STEP 7. The conversion result is presented as Statement List (STL) instructions.

LAD/FBD/STL Programming Editor

LAD/FBD/STL is the standard programming tool for S7-300/S7-400 CPUs. This 3-in-1 editor allows total development in any of the three language representations or a combination of the languages. LAD represents *Ladder Diagram*; FBD, or *Function Block Diagram*, is a graphic language that uses Boolean gate logic representation; and STL represents *Statement List, a text-based assembler-like language*.

Memory Card Parameter Assignment

The Memory Card Parameter Assignment Utility is used to set parameters of optional S7 memory cards. The user determines what EPROM or Flash file drivers are used and what LPT port to use for connecting external EPROM programming devices.

PID Control Parameter Assignment

This utility provides templates for defining PID control parameters for use with S7-300/S7-400 standard PID functions (FB41 and FB42), for continuous and step controllers respectively. User-defined instance data blocks serve as the data interface for turning controller functions on or off and for defining control values.

Setting the PG-PC Interface

This utility is used to select the interface parameter set to establish an S7 online connection, and for installing or removing communications adapters and protocols.

TI-405 Converting Files

The TI-405 Conversion Utility allows SIMATIC TI405 program code to be converted to equivalent STEP 7 programs or blocks. The utility produces one or more STEP 7 Statement List (STL) text files that are subsequently edited and/or compiled.

TI-505 Converting Files

This conversion utility allows SIMATIC TI505 programs to be converted to equivalent STEP 7 programs or blocks. The utility produces one or more STEP 7 Statement List (STL) text files that are subsequently edited and/or compiled. SIMATIC TI Special Function Programs (SFPGMS) require the SCL optional STEP 7 language package.

Optional Language Tools

Optional tools for STEP 7 support code development using either graphic elements or text-based source files. To accommodate typical programmable control, the languages include support for functions such as I/O access, binary signal manipulation, timers, counters, as well as standard functions for communications and data manipulation. In either case, the results may be combined with code developed using the standard editor (LAD/FBD/STL).

An advantage of these optional tools is that they offer language choices that may be matched to user skills and preference, and that may be selected to better suit code development based on the operations and requirements of the application. These tools, which must be installed after STEP 7, are generally accessible from the STEP 7 Program Menu and from within the SIMATIC Manager.

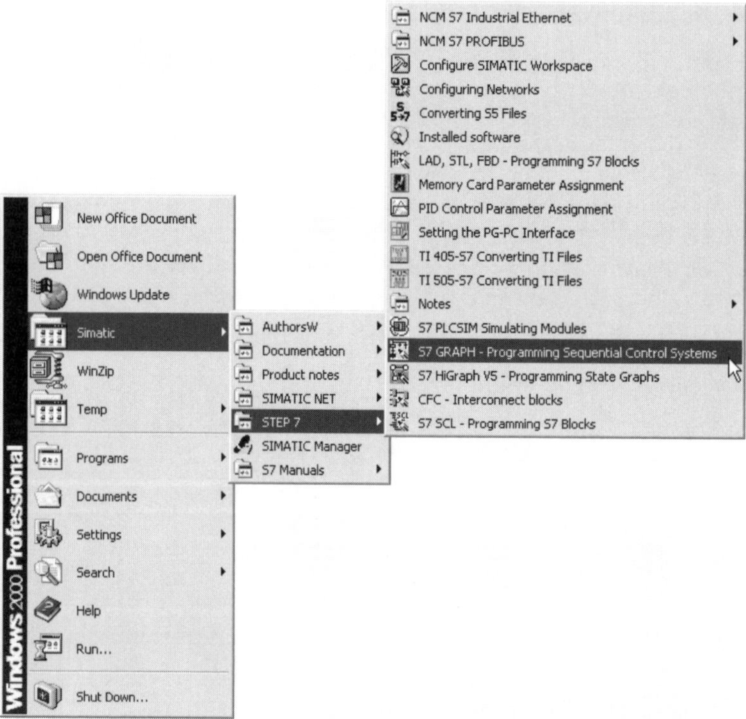

Figure 1-3. SIMATIC Start Menu with Standard and Optional Tools

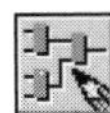

S7-CFC

S7-CFC, or *Continuous Function Chart*, is a graphical programming tool, suited for writing the code for operations that are continuous in nature. CFC programs, also referred to as charts, are characterized by the interconnecting of canned operations having defined inputs and outputs. New functions may be created, however most commonly required functions including arithmetic and logic, timers and counters, comparisons and conversions, trigonometric functions and many others are already available. CFC allows complex programming of process operations without focusing on the details of the programming method.

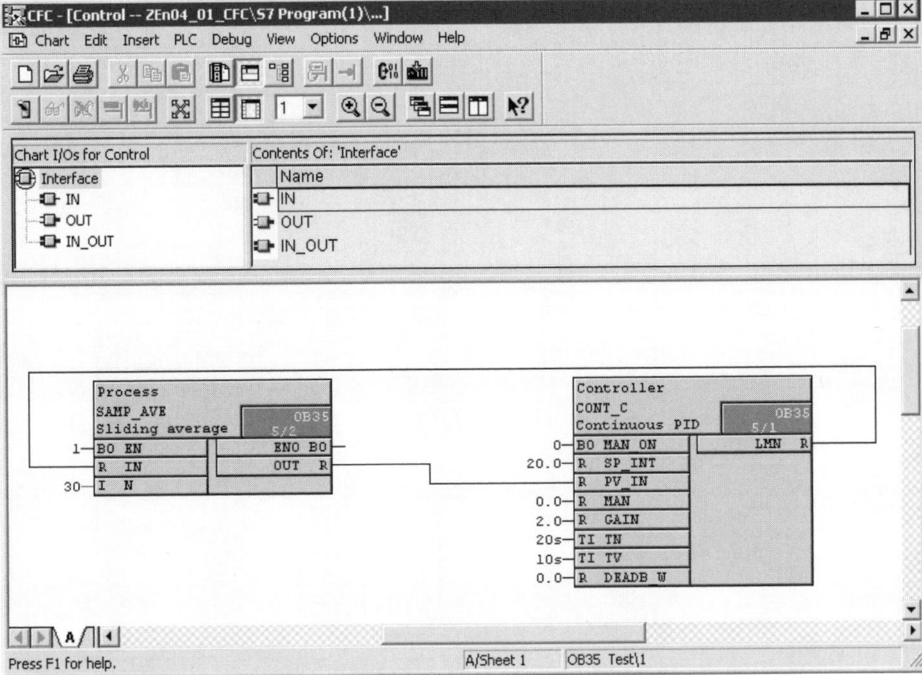

Figure 1-4. Window with Continuous Function Chart (CFC) programming.

S7-Graph

S7 Graph is a flowchart-like language involving sequences that describe the operations of a machine or process as individual steps and the transitions between each step. It is especially suited to sequential operations. A sequence may consist of both sequential and concurrent paths. The code describing the operations of a step and the transitions between steps may be programmed in the language of choice (e.g., LAD, FBD, STL, or SCL).

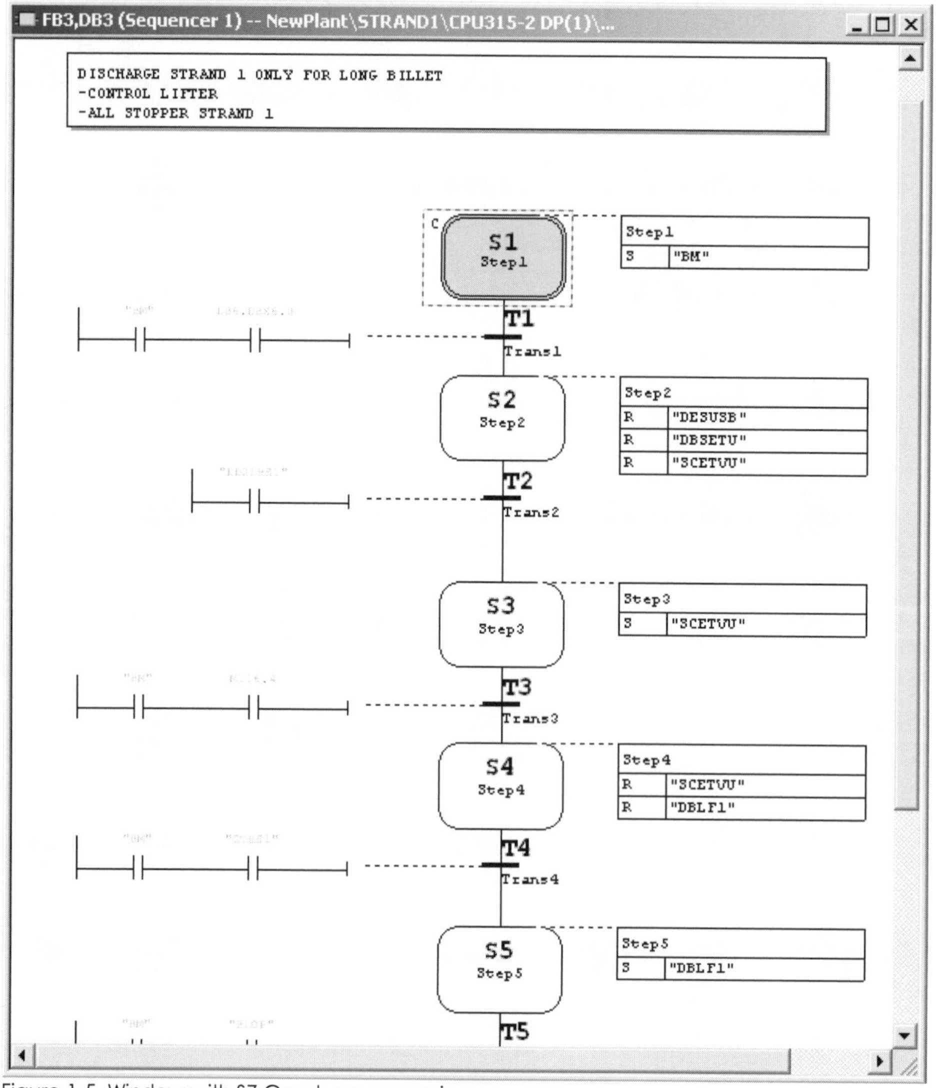

FB3,DB3 (Sequencer 1) -- NewPlant\STRAND1\CPU315-2 DP(1)\...

DISCHARGE STRAND 1 ONLY FOR LONG BILLET
-CONTROL LIFTER
-ALL STOPPER STRAND 1

S1 Step1 — Step1: S "BM"
T1 Trans1
S2 Step2 — Step2: R "DESUSB"; R "DBSETU"; R "SCETVU"
T2 Trans2
S3 Step3 — Step3: S "SCETVU"
T3 Trans3
S4 Step4 — Step4: R "SCETVU"; R "DBLF1"
T4 Trans4
S5 Step5 — Step5: S "DBLF1"
T5

Figure 1-5. Window with S7 Graph programming.

S7-SCL

Structured Control Language (SCL) is a textual high-level language which supports PASCAL-like operations like FOR-NEXT Loop, IF-THEN-ELSE, CASE, DO-WHILE statements, and variable declarations. S7-SCL simplifies the programming of loops and conditional branches, and is quite suited for formula calculations, complex optimization algorithms, or the management of large quantities of data. S7-SCL supports development of source files that are compiled into blocks that can be combined in an S7 program with other blocks.

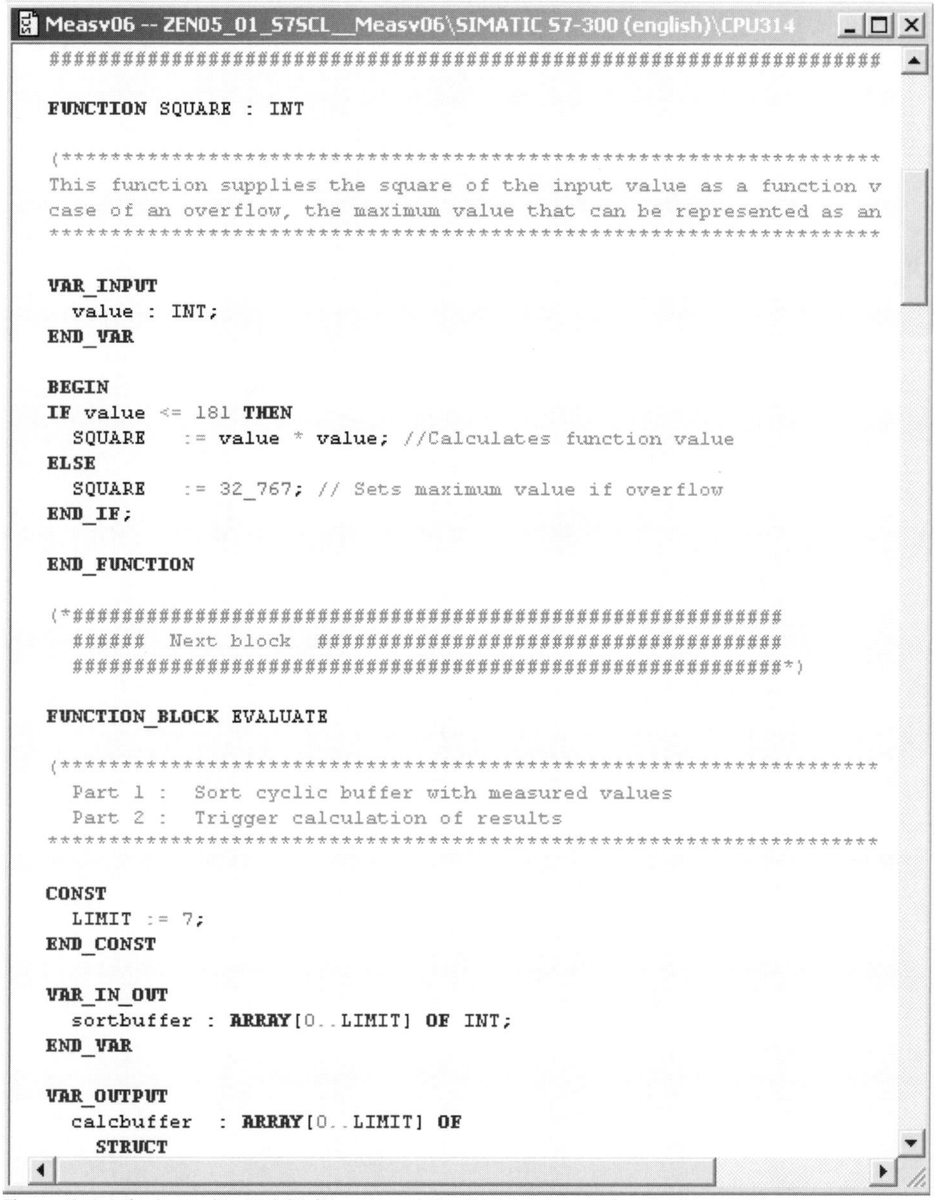

Measv06 -- ZEN05_01_S7SCL__Measv06\SIMATIC S7-300 (english)\CPU314

```
##################################################################

FUNCTION SQUARE : INT

(*****************************************************************
This function supplies the square of the input value as a function v
case of an overflow, the maximum value that can be represented as an
*****************************************************************

VAR_INPUT
  value : INT;
END_VAR

BEGIN
IF value <= 181 THEN
  SQUARE   := value * value; //Calculates function value
ELSE
  SQUARE   := 32_767; // Sets maximum value if overflow
END_IF;

END_FUNCTION

(*##########################################################
  ######  Next block  #####################################
  ########################################################*)

FUNCTION_BLOCK EVALUATE

(*****************************************************************
  Part 1 :  Sort cyclic buffer with measured values
  Part 2 :  Trigger calculation of results
*****************************************************************

CONST
  LIMIT := 7;
END_CONST

VAR_IN_OUT
  sortbuffer : ARRAY[0..LIMIT] OF INT;
END_VAR

VAR_OUTPUT
  calcbuffer  : ARRAY[0..LIMIT] OF
    STRUCT
```

Figure 1-6. Window with S7-SCL (Structured Control Language); a text-based language.

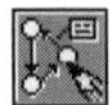

S7-HiGraph

S7-HiGraph is a graphical language, based upon what is known as the "state machine" or "state diagram" method of programming. HiGraph is especially suited for asynchronous operations. With this method, a system is broken down into autonomous functional units, each of which can accept various states. The actual behavior of each functional unit is described by what is called a state graph, and transitions are defined to control the switching among these state graphs.

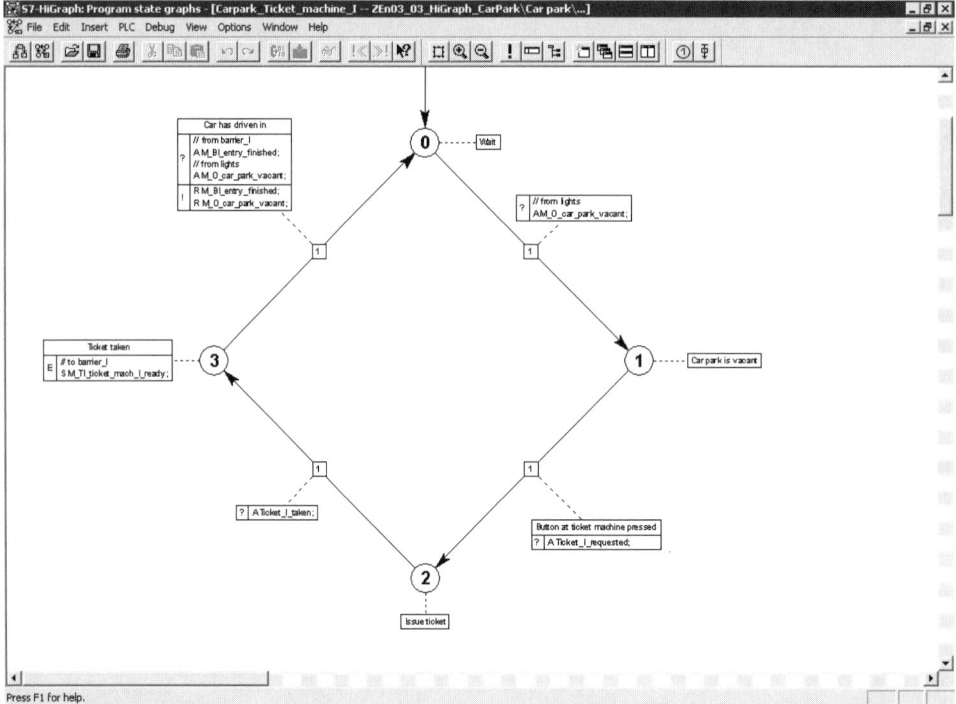

Figure 1-7. Window with S7 Hi-Graph; a state machine programming language.

Comments on Getting Started with STEP 7

In your initial working with STEP 7 certain tasks when performed up front, will help to ensure getting off to a good start. Whether your project is small or large, proper planning of your software installation, storage locations for projects, libraries, and archives is important for good organization in maintaining and protecting your work. Items you should consider performing before you get started are outlined below. Examples of each of these tasks have been presented, in a step-by-step manner, in the remainder of this chapter.

Once you have installed your software, become somewhat familiar with navigating the SIMATIC Manager software, and performed some of these important preliminary tasks, you'll be ready to proceed to the first step of Working with your STEP 7 Project.

Checklist: Getting Started with STEP 7

- *Ensure sufficient disk space is on the same drive of the PC programming system, for both STEP 7 Basic and optional tools.*
- *Install NCM for Industrial Ethernet and NCM for Profibus, during the installation if these networks will be installed.*
- *Install authorization key for the STEP 7 BASIS software.*
- *Install Authorization keys for NCM Industrial Ethernet and NCM Profibus, if these packages were installed.*
- *Configure the STEP 7 workstation environment to support single-user or multi-user project development.*
- *Create Windows folders for holding STEP 7 projects, libraries, and archives.*
- *From the SIMATIC Manager, define STEP 7 language, storage path, and archiving, and other operating preferences.*
- *Establish a physical connection between the CPU, of your initial work, and the programming system (PG/PC) using serial cable and PC Adapter (adapter only for PC).*
- *Set the PG/PC interface to PC Adapter, for STEP 7 online connections to the programming system.*

Installing the STEP 7 Software

Basic Concept

STEP 7 is pre-installed on SIMATIC PGs, but must be installed on PCs intended for use as a programming system. Suitable PCs include Pentium-based PCs, having a Windows operating system (95/98, NT/2000, XP, ME). The STEP 7 Basic package contains the *LAD/FBD/STL* programming editor, which supports Ladder, Function Block Diagram, and Statement List languages for programming S7 controllers. Optional languages and tools, which must be installed after STEP 7 is installed, must be considered since they should be installed on the same drive.

Essential Elements

In addition to the *Basic STEP 7* package, STEP 7 releases after 5.0 include: (1) *Acrobat Reader*, software for opening and viewing user manuals, and other related documentation; (2) *NCM for Industrial Ethernet*, the Ethernet configuration software; (3) *NCM for Profibus*, the Profibus configuration software; and (4) *AuthorsW*, the Windows program for installing and removing authorizations for standard and optional tools.

Application Tips

If Ethernet and Profibus network modules are not visible from the SIMATIC Manager project tree or available in the hardware configuration, then it may be that NCM Profibus and Ethernet packages have not been installed. Remember, although included on the standard CD, a separate license must be purchased for these NCM packages.

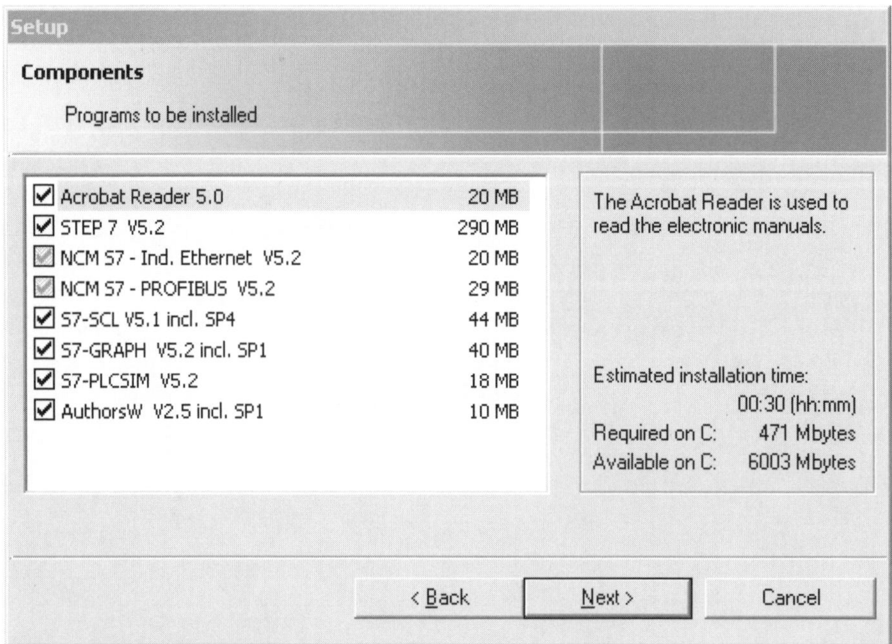

Figure 1-8. STEP 7 Setup dialog for selecting install components.

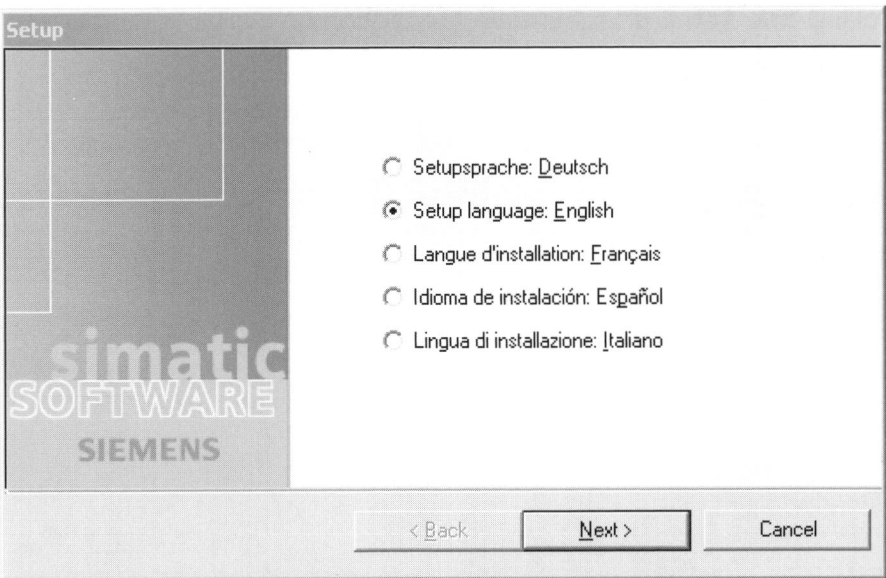

Figure 1-9. STEP 7 setup dialog for selecting native languages to install.

Quick Steps: Installing the STEP 7 Software

STEP	ACTION
1	Logon as Administrator or Power User, to install STEP 7 under Windows NT/2000.
2	If the installation is an upgrade to a STEP 7 version older than V3.2, then the previous version and all optional packages must be removed prior to the new installation.
3	Ensure that your PC hard drive has adequate space for the installation. Depending on your component selection from 200-380 MB will be required.
4	From the Windows Explorer navigate to the SETUP.exe installation program on the STEP 7 CD and double-click on the program to start the installation.
5	From the STEP 7 Setup dialog Setup Language, click and select the language of choice to use for the STEP 7 installation setup dialogs.
6	From the STEP 7 Setup dialog Components, place a check mark in each box to select what software components should be installed.
7	From the STEP 7 Setup dialog Install Language - Specific Files, place a check mark in the box besides each native language in which STEP 7 should be available.
8	From the STEP 7 Setup dialog Start-up Language, click and select the language of choice to start-up with when STEP 7 is launched.
9	Answer remaining questions as the setup program guides you through the installation.
10	Toward the end of the installation, you will be given the option to allow the setup program to install the STEP 7 authorization. Simply insert the diskette when prompted.

Installing and Removing STEP 7 Authorizations

Basic Concept

To run without interruption the STEP 7 software, and all optional packages, must be authorized by installing the appropriate authorization. An authorization, which acts like a software key, is required on each computer where STEP 7 is used. The program for installing and removing authorizations, "AuthorsW," is installed during the STEP 7 installation. To run a single license copy of STEP 7 on a different computer, you must uninstall or transfer the license back to the authorization disk and reinstall it on the new computer.

Essential Elements

The yellow STEP 7 Authorization disk, provided with STEP 7, supports management of multiple authorization keys as shown in the dialogue below. Each optional package requires authorization and is provided with its own authorization disk if purchased separately. Once an authorization is installed on the drive, when later removed it may be moved back to the STEP 7 multi-authorization disk for central management.

Application Tips

By default, authorizations are placed in a hidden folder, C:\AX NF ZZ. This folder must not be deleted, copied, or moved. Authorizations should be removed before performing disk operations such as ghost-imaging or de-fragmenting. If a drive is compressed, authorizations should be placed on the host drive. In the event the authorization for STEP 7 BASIS is lost, a 14-day Emergency key is provided on the yellow disk, to be used until the lost key is replaced.

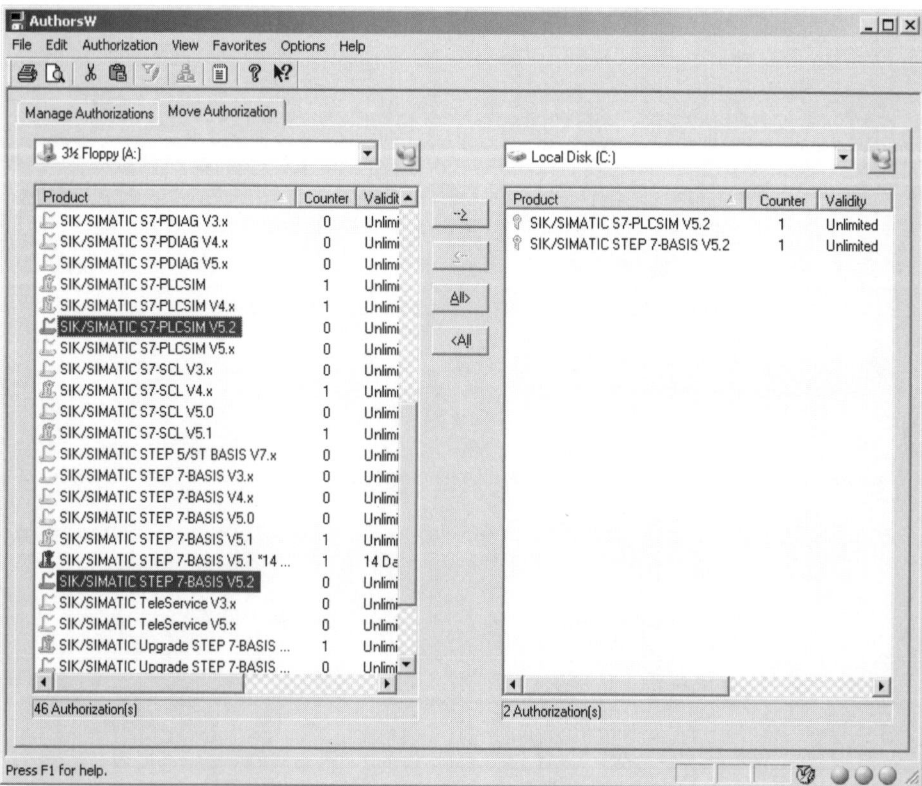

Figure 1-10. Authorization dialog for installing and removing software authorizations.

Quick Steps: Installing and Removing STEP 7 Authorizations

STEP	ACTION
	Install Authorization:
1	Place the STEP 7 authorization disk in the disk drive.
2	From the Windows **Start** (button), select **SIMATIC,** ➢ **AuthorsW** folder, ➢ **AuthorsW.**
3	Select the **Transfer** tab, to list the authorization keys found on the diskette, if it is not already displayed. The counter value for the key must be "1" on the diskette.
4	To install the authorization on the hard drive, find and select from the window displaying the authorizations, the correct version of STEP 7-BASIS for unlimited use.
5	With the key selected, press the install button ➔ to place the authorization on the hard drive.
6	Remove the authorization disk from the drive and store until required again.
	Remove Authorization:
1	Place the STEP 7 authorization disk in the disk drive.
2	From the Windows **Start** (button), select **SIMATIC** ➢**AuthorsW.**
3	Select the **Transfer** tab, to list the authorizations, if it is not already displayed.
4	To remove the authorization from the hard drive, select the STEP 7-BASIS key from the window displaying the authorizations found on the hard drive. The counter value for the key must be "0" on the diskette and "1" on the hard drive.
5	Press the remove button 🡸 to return the authorization to the authorization disk.
6	Remove the authorization disk from the drive and store until required again.

Opening SIMATIC Software Components

Basic Concept

Once STEP 7 is installed, a SIMATIC program group is placed in the Windows Start menu and is accessed from the Start button. The SIMATIC folder contains the main components of the STEP 7 BASIC package, which is also presented as a menu of items that may be selected. These items include component folders *AuthorsW*, *Documentation*, *Product Notes*, *SIMATIC NET*, *S7 Manuals*, and a shortcut for launching the *SIMATIC Manager*.

Essential Elements

From the SIMATIC menu, the AuthorsW folder contains the *AuthorsW* program for installing and removing licenses for standard and optional tools. The *Documentation* folder presents language-based folders that contain shortcuts to documents and manuals available in the installed languages. *Product Notes* presents language-based folders, which contain shortcuts to product information not included in user manuals. SIMATIC NET presents selections for PC configuration software tools, for *Industrial Ethernet* and for *Profibus*. The STEP 7 folder presents a menu of installed standard and optional tools. The *S7 Manuals* folder presents other related manuals that have been installed (e.g., optional software tools); and finally, selecting *SIMATIC Manager* launches the main project management tool.

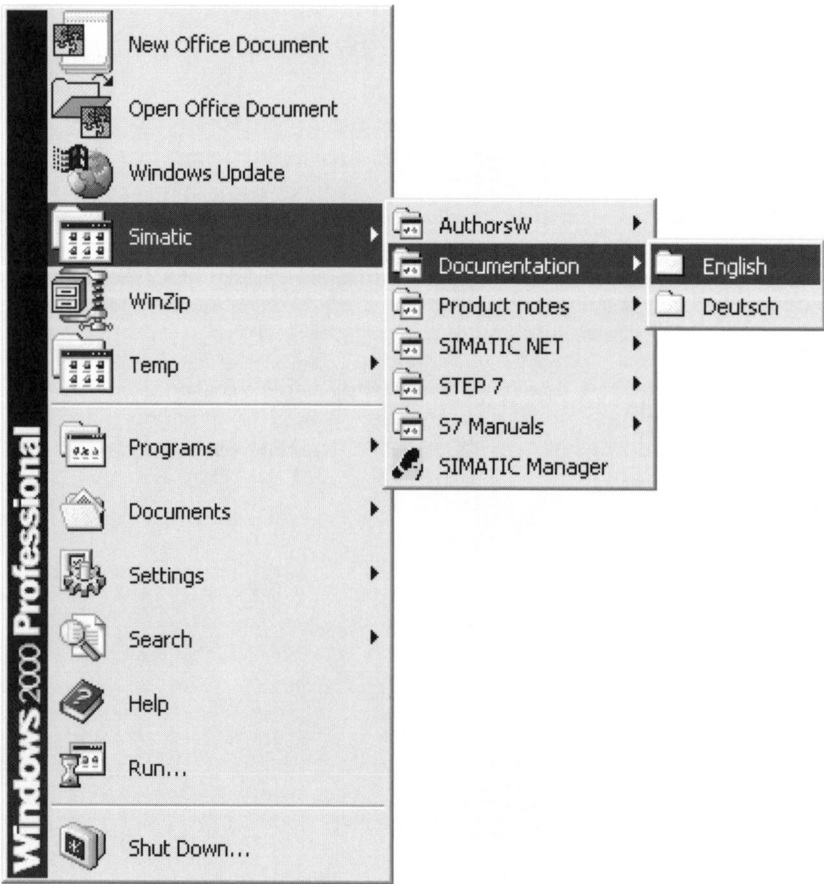

Figure 1-11. Opening SIMATIC Software Components from Windows Start Menu.

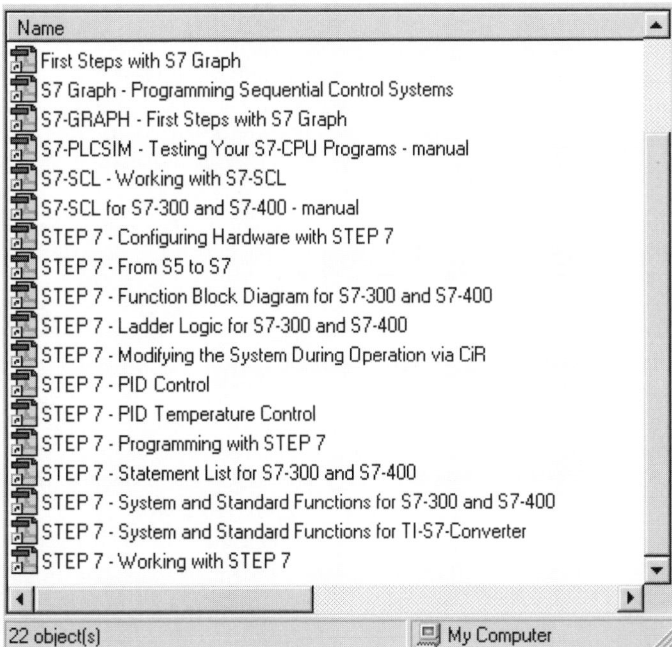

Figure 1-12. Shortcuts to typical installed S7/STEP 7 documentation.

Quick Steps: Opening SIMATIC Software Components

STEP	ACTION
1	Click **Start** ➢ **SIMATIC** ➢ **AuthorsW** ➢ **AuthorsW** to open the utility is used to install and remove software authorizations.
2	Click **Start** ➢ **SIMATIC** ➢ **Documentation** ➢ **English** (or **Deutsch, French, Italian, Spanish**) to locate STEP 7 documentation in one of the installed native languages.
3	Click **Start** ➢ **SIMATIC** ➢ **Product Notes** ➢ **English** (or **Deutsch, French, Italian, Spanish**) to locate important product notes not included in the user manual at release.
4	Click **Start** ➢ **SIMATIC NET**; to access configuration or diagnostic tools for Industrial Ethernet, Profibus, or other installed SIMATIC NET components.
5	Click **Start** ➢ **SIMATIC** ➢ **STEP 7** ➢ to access standard editors and utilities (e.g. **LAD/FBD/STL**) or optional editors (e.g., **CFC**) and utilities if they have been installed.
6	Click **Start** ➢ **SIMATIC** ➢ **SIMATIC Manager**; or from your desktop double-click on the SIMATIC Manager icon to open the SIMATIC Manager program.
7	Click **Start** ➢ **SIMATIC** ➢ **S7 Manuals** to locate other related user manuals that have been installed (e.g., optional packages).

Configuring a SIMATIC Workstation

Basic Concept

You may configure a STEP 7 workstation for use in a single- or multi-user environment. Single-user mode is the default configuration after the STEP 7 installation. Configuring workstations in a multi-user environment allows several users to access and work concurrently on a project.

Essential Elements

The multi-user configuration is possible for workstations on NT and Novell networks, after each workstation is configured using the *Configure SIMATIC Workspace* utility, to operate in a multi-user environment. Shared projects may be placed on a local drive, of one workstation and accessed by all other stations, or on a network server, where other workstations access the server. Shared projects may also be distributed among the local drives and one or more network servers, and all workstations have access to all projects.

Application Tips

The multi-user configuration is possible for workstations on NT and Novell networks, after each workstation is configured using the *Configure SIMATIC Workspace* utility, to operate in a multi-user environment. The shared project or projects may be placed on a local drive of one workstation and accessed by all others; or on a network server, and other workstations access the server; or distributed among the local drives and one or more network servers, and all workstations have access to all projects.

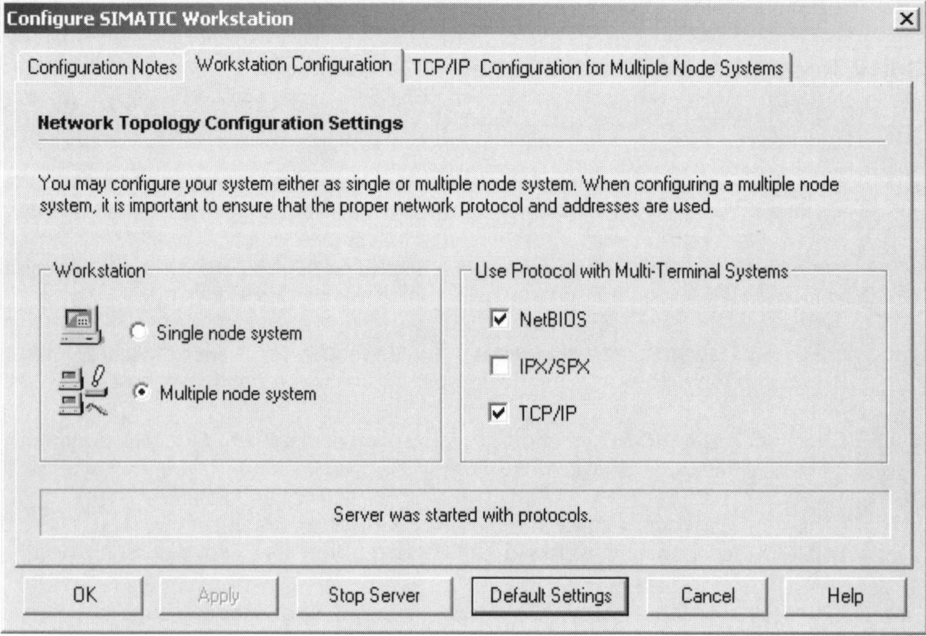

Figure 1-13. Configure workstation dialog: setting single/multiple

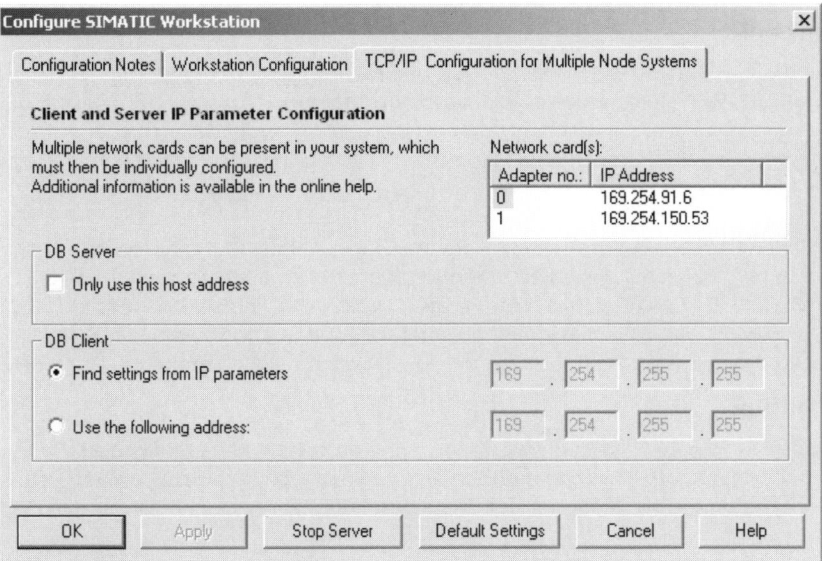

Figure 1-14. Configure workstation dialog: defining IP address parameters

Quick Steps: Configuring a SIMATIC Workstation

STEP	ACTION
1	From the Windows **Start** button, select **SIMATIC ➢ STEP 7 ➢ Configure SIMATIC Workspace** to open the workstation configuration dialog.
2	From the Configuration Workstation dialog, select the *Workstation Configuration* tab.
3	Select **Single Node System**, if the workstation will work alone on STEP 7 projects on the local drive, and not access projects on a central server or on other workstations.
4	Select **Multiple Node System**, if the workstation will operate as on of multiple workstations having access to projects on the local drive or on a central server.
5	Select the communication protocols that may be used for communication by this workstation (e.g., *TCP/IP*, *Net-BIOS*, or *IPX/SPX)*.
6	If TCP/IP is used, select the **TCP/IP Configuration for Multiple Node Systems** tab; select the network adapter if multiple cards are installed and set the workstation IP address
7	In the multi-user environment, a DB server facilitates communication between the STEP 7 client and other workstations. If more than one network adapter is available, you may specify the host address of the network in which the DB server should work exclusively. Under **DB Server**, place a check mark in the box **Only Use this Host Address**, and then specify the host address of the network. Otherwise, the default setting allows the DB server to use all of the installed network adapters.
8	Under the **DB Client** box, choose the setting **Find settings from IP Parameters**, to allow the workstation client to obtain its IP parameters dynamically; or use the field designated **Use Following Address**, to set the IP address for the client. This setting allows you to specify the broadcast address to be used for the corresponding subnet when more than one subnet exists. If several cards exist, adapter 0 is used to determine the broadcast address.

Defining Storage Path Preferences

Basic Concept

When working with STEP 7 you will create projects and perhaps libraries of programs and blocks. By default, STEP 7 stores all projects in a subfolder named *S7Proj*, and all libraries in a subfolder named *S7Libs*, both of which are located in the STEP 7 folder. To have your projects and libraries automatically stored in default locations of your choosing, you must first create then specify these locations.

Essential Elements

The Essential Elements of this task involve defining or, if necessary, creating the default directory paths for STEP 7 projects and libraries. Since by default STEP 7 stores projects in Siemens\STEP 7\S7_Proj, and libraries in Siemens\STEP 7\S7_Lib, you may want to ensure that your project work and libraries are placed in the right location in your computer's file system.

Application Tips

Each newly created project and library is created in the default paths currently set in STEP 7. To avoid loss of project data it is recommended that you place project folders on a drive other than the one where the Windows swap file is located.

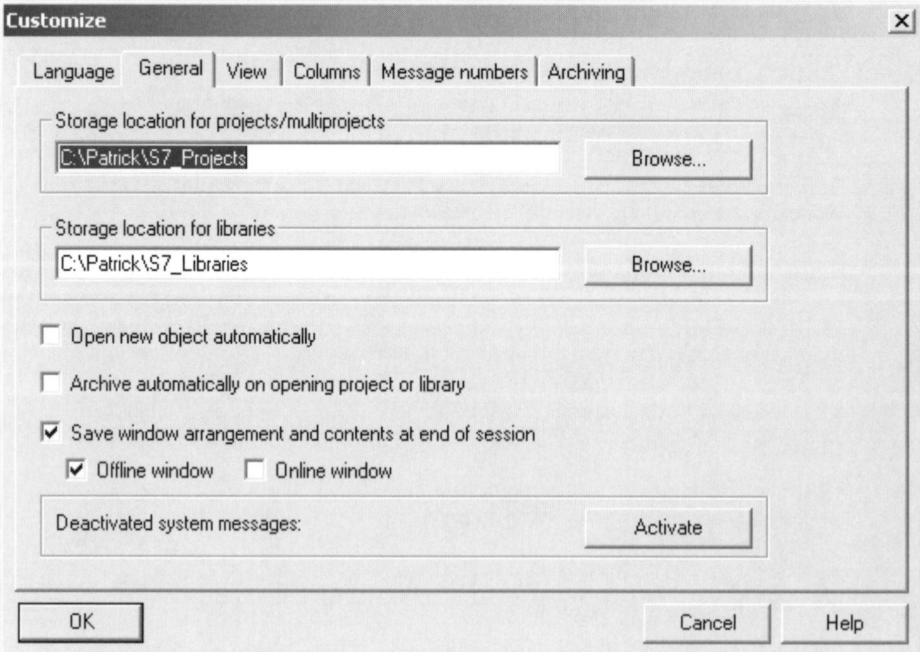

Figure 1-15. STEP 7 Customize dialog for defining Project/Library storage paths.

Quick Steps: Defining Storage Path Preferences

STEP	ACTION
1	Create a new folder for your STEP 7 projects (e.g., C:\S7_Projects). Each newly created project is generated in a subfolder, inside of the main project folder that you create.
2	Create a new folder for your STEP 7 libraries (e.g., C:\S7_Libraries). Each newly created library is generated in a subfolder, inside of the main library folder that you create.
3	Launch the SIMATIC Manager and from the menu select **Options** ➢ **Customize**.
4	Open the **General** tab to define your default storage path settings.
5	Type a folder path in the "Storage Location for Projects" field, or use the Browse button to find and select a folder to hold new projects (e.g., folder created in Step 1).
6	Type a folder path in the "Storage Location for Libraries" field, or use the Browse button to find and select a folder to hold new libraries (e.g., folder created in Step 2).
7	Activate the **Open New Objects Automatically** check box, if you wish to have each newly generated object (e.g., station, block, or subnet) open automatically in its associated editor or tool after it is created.
8	Place a check in the **Archive Automatically** check box, if you wish to have a backup copy made of the last revision of your project or library on each new opening of the project or library.
9	Close all STEP 7 tools and the SIMATIC Manager after any changes to the General tab, for changes to take affect; then reopen STEP 7.

Defining Language Preferences

Basic Concept

STEP 7 supports installation of multiple native language, all of which are distributed on a single CD. Any one, or a combination, of the languages may have been installed during the installation. The language setting will affect both the native language used on the program's dialogs, as well as the mnemonics that appear in the instructions and addresses of you user programs. At any time later, you may switch to any language of choice as required.

Essential Elements

This task involves setting the native language to use if multiple languages have been installed; and to ensure that address mnemonics to be used in the program are set to your choice. The mnemonics setting determines whether German or English instruction operands and address identifiers will be used as you create and display your program. If the mnemonics are set differently from what you expect and are using, errors will be reported as you enter your program.

Application Tips

By default, after initial installation, the language setting for the program mnemonics is German. English is the only other choice. If any language change is made, STEP 7 and all of its components must be closed and restarted for the change to take affect.

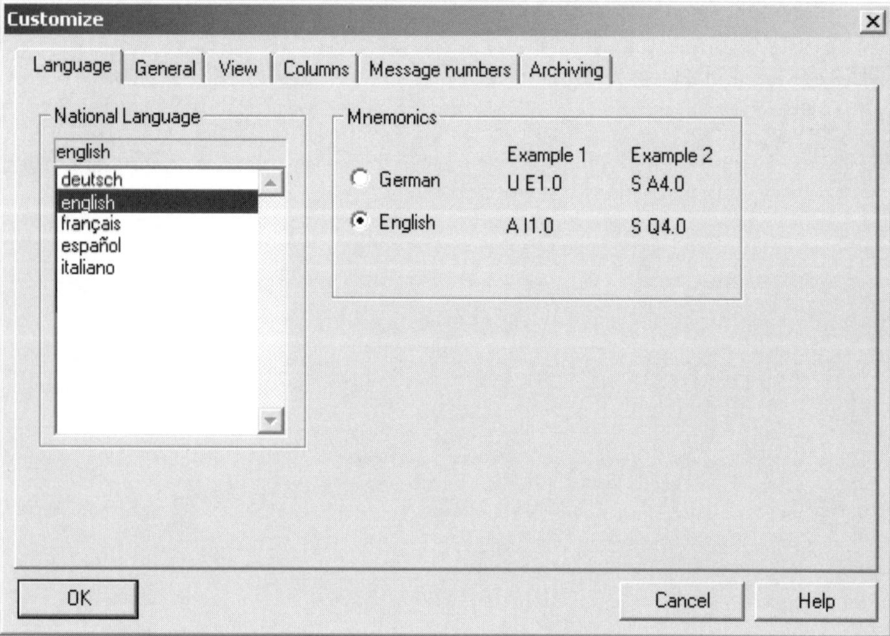

Figure 1-16. STEP 7 Customize dialog for defining Language preferences.

Quick Steps: Defining Language Preferences

STEP	ACTION
1	Launch the SIMATIC Manager and from the menu select **Options ➢ Customize**.
2	Open the **Language** tab to define your STEP 7 language preferences.
3	Under the **National Language** list box, if multiple languages were installed you will see the current setting in the field just above the list box.
4	Select the native language of choice (e.g. English) to use within the STEP 7 application.
5	From the group box labeled **Mnemonics**, select English or German, to determine what instruction mnemonics are to be used in your programs (e.g., set for English I = Inputs, Q = Outputs; set for German, E = Inputs, A = Outputs).
6	Click the **OK** button to close the dialog.
7	Close all STEP 7 tools and the SIMATIC Manager after any changes to the Language tab, for changes to take affect; then reopen STEP 7.

Defining Archiving Preferences

Basic Concept

STEP 7 allows the manual or automatic backup of your projects or libraries by a process known as archiving. To archive is to use a compression utility such as WinZip or PKZip to compress the files in order that they require less storage space. In the course of your STEP 7 development you may wish, at various stages, to make a backup of your work to date. Setting the archiving preferences will define certain parameters that that will determine how STEP 7 handles this operation.

Essential Elements

Setting preferences for archiving involves defining and, if necessary, creating the default archive directory path for storing backups of STEP 7 projects and libraries; and defining the default directory path in which projects should be placed upon retrieving and archived copy. In addition the program (e.g., WinZip) to use for archiving projects or libraries must be specified.

Application Tips

For changes to take affect, after options settings have been modified, STEP 7 and all of its components must be completely closed and restarted.

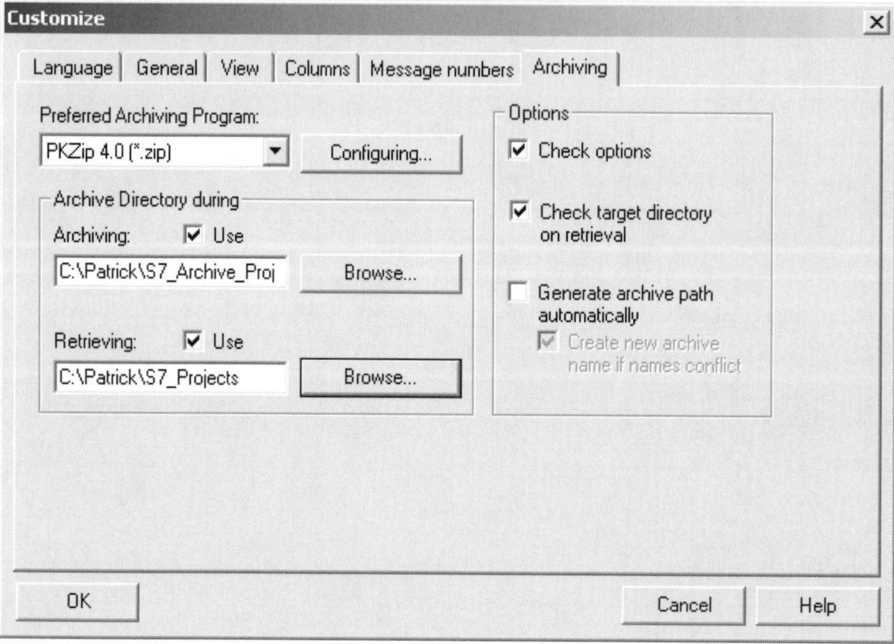

Figure 1-17. Customize dialog for defining Project/Library archiving preferences.

Quick Steps: Defining Archiving Preferences

STEP	ACTION
1	If you intend to have STEP 7 automatically archive your projects, you may wish to create a new folder for this purpose if you have not already done so (e.g., S7_Archive_Proj).
2	Launch the SIMATIC Manager and from the menu select **Options** ➢ **Customize**.
3	Open the **Archiving** tab to define your STEP 7 archiving preferences.
4	From the "Preferred Archive Program" drop list, select the program to use in creating archives (e.g., WinZip). Use the **Configuring** button to find and point to its location.
5	In the **Archiving** field, define a target directory and activate the **Use** checkbox if this directory should be targeted first for storing archives (e.g., folder created in Step 1). If the "Use" check box is not activated, then archive projects will be targeted for the location last used for storing archives.
6	In the **Retrieving** field, define a target directory and activate the **Use** checkbox if this directory should be searched first when retrieving an archive project; If the "Use" check box is not activated, the search for archives will start in the directory used last.
7	Activate the **Check Options** check box, to cause the archiving dialog to be presented each time you archive files. This setting is ignored if LHARC, PKZIP 2.50, or WinZip is set as the "Preferred archive program".
8	Activate the check box for **Check Target Directory,** to cause a directory selection dialog presented on each archive retrieval. This dialog allows you to specify a directory in which to restore a retrieved archive project. Otherwise, archived projects and libraries are restored to their respective default STEP 7 project or library directory.
9	Activate the **Generate Archive Path Automatically** check box, to have the name of the archive file automatically derived from the name of the project to be archived. The directory selection dialog is not presented.

Establishing a Direct CPU Programming Connection

Basic Concept

If you are using a PC as the STEP 7 programming system, the PC adapter is required for direct connection to the CPU if you are establishing a connection for the first time. This connection, which supports direct CPU connection for downloading your offline program, will also be required to download your initial hardware configuration to the CPU. The connection is between the serial port of your PC and the DP/MPI port on the S7-300 or S7-400 CPU. All S7 CPUs have an MPI port.

Essential Elements

With the PC adapter (PN 6ES7-972-00A22-0X), and the standard serial cable (PN 6ES7-901-1BF00-0X) you have all that is required for the online connection to a CPU via its DP/MPI port. An off-the-shelf serial cable with 9-pin connectors and the appropriate gender may be substituted for the previously mentioned serial cable. The PC Adapter performs the conversion to allow the standard RS232 serial port on your PC to support online connection to the MPI/Profibus-DP port of an S7-300 or S7-400 CPU.

Application Tips

The interface parameter set (i.e., protocol driver) for the PC Adapter is automatically installed during the STEP 7 installation, as one of the interfaces available for selection as an S7 online interface connection. All that is required, if not already done, is that you choose the PC Adapter as the interface to use for the S7ONLINE connection. This setting is made using the "Setting the PG/PC Interface" utility. STEP 7 PGs come equipped with an MPI card which supports either a direct or MPI network connection to a CPU, and does not require use of the PC adapter.

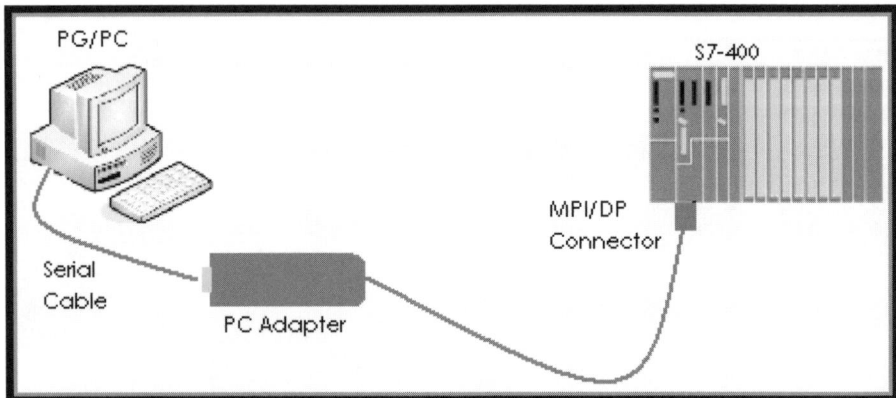

Figure 1-18. PG/PC-to-CPU programming connection using the PC-Adapter.

Quick Steps: Establishing a Direct CPU Programming Connection

STEP	ACTION
1	Set the slide switch on the PC adapter to either 19.2 or 38.4, according to your required transmission speed. This setting must match the transmission set in the **MPI Parameters**.
2	Connect one end of the serial cable to the RS-232 connector on the PC Adapter.
3	Connect the opposite end of the serial cable to the RS-232 serial port of your PC.
4	Connect the opposite or cable end of the PC Adapter, marked MPI/DP, to the MPI/DP port of the CPU to which the online connection is being established.
5	With the cable in place the STEP 7 online interface must be set to use the PC Adapter.
6	From the SIMATIC Manager menu, select **Options ➢ Set PG/PC Interface** to launch the utility for setting the online Interface for the PC or PG programming system.
7	Ensure that the field labeled "Access Point of the Application" is set as follows: **S7ONLINE (STEP 7) ➔ PC Adapter**.

Setting the PG/PC Interface for PC Adapter

Basic Concept

The standard direct connection to an S7 CPU involves use of the PC Adapter and a serial cable from your PC to the DP/MPI port on the CPU. With the cable in place and the appropriate switch setting on the adapter, all that is required is to ensure that the appropriate interface settings have been made using the *"Setting the PG/PC Interface"* utility of STEP 7. This utility allows you to select any one of the available interfaces by which online connection to a CPU can be established. It also allows you to define specific protocol settings for these interfaces.

Essential Elements

Setting the PG/PC Interface is to assign an "Interface Parameter" set to the S7ONLINE access point. The *"Access Point of the Application,"* is the unique name, known to an application (in this case STEP 7), that gives it the connection information by which it makes an online connection. The Access Point is linked to an *Interface Parameter Assignment*, which defines a specific Interface (i.e., protocol driver and/or network interface) and its set properties. The interface parameter in this case is "PC Adapter." The assignment then would be S7ONLINE (STEP7), bound to the PC Adapter (MPI); the selection is shown as **S7ONLINE (STEP7) → PC Adapter (MPI)**.

Application Tips

A direct connection may also be required if network connections that allowed remote online connection to a CPU have been lost. Direct connection is not only possible with the PC adapter, but also in the case where the PG/PC has an installed MPI card that allows connection to a CPU's MPI programming port.

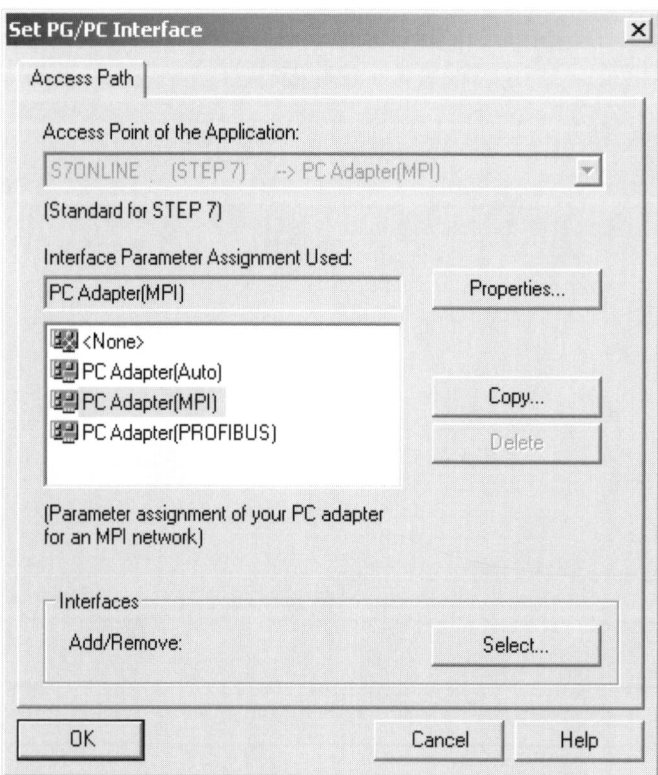

Figure 1-19. Dialog for defining PG/PC Interface connection.

Quick Steps: Setting the PG/PC Interface for PC Adapter

STEP	ACTION
1	Launch the "Setting the PG-PC Interface" utility from the SIMATIC Manager menu, select **Options ➢ Set PG/PC Interface**.
2	Once the *Setting the PG/PC Interface* dialog is presented, ensure that the **S7ONLINE (STEP 7)** access point is displayed in the field labeled "Access Point of the Application."
3	The current setting for the *Interface Parameter Assignment* is displayed to the right of the **S7ONLINE (STEP 7)** access point, after the symbol (**➔**). If no interface has been selected, what should be displayed is '**S7ONLINE (STEP 7**) **➔ None**'.
4	From the **Interface Parameter Assignment Used**" list box, click on and select **PC Adapter** (**MPI**) as the interface parameter to set as the S7ONLINE interface.
5	Once the selection is made the Access Point of the Application will reflect the newly selected interface. The setting should show **S7ONLINE (STEP 7**) **➔ PC Adapter (MPI)**.
6	Select the **Properties** button to set operating parameters of the PC Adapter interface (e.g., Transmission rate). This setting must match the switch setting on the adapter. If you set the adapter to 38.4 K Baud, then the dialog setting must be set to match.
7	Press **OK to** save properties and set the interface for the new S7ONLINE connection.

Step

2

Working with STEP 7 Projects and Libraries

Objectives

- Introduce STEP 7 Project Concept
- Introduce the SIMATIC Manager Tool
- Introduce Objects of a STEP 7 Project
- Manage Projects with SIMATIC Manager
- Find STEP 7 Projects and Libraries
- Archive and Retrieve Projects and Libraries
- Create New Projects and Libraries
- Use Copy and Paste as Productivity Enhancement Tools

The SIMATIC Manager

The SIMATIC Manager is the top-level program of STEP 7, and the primary tool used in all aspects of development and management of S7-300 and S7-400 projects. As the main tool for STEP 7, you will use the SIMATIC Manager for much of your work with STEP 7. From this main tool, you will have access to other standard and optional tools and utilities (e.g., LAD/FBD/STL programming editor and the *NetPro Network Configuration* tool), which you may launch from either the SIMATIC Manager menu or toolbar.

A Project and Library Manager

The SIMATIC Manager is similar to the Windows Explorer in that it supports the management and storing of files, and the launching of various software tools. The files in this case, are projects. Instead of working directly with files, however, your work with the SIMATIC Manager is reduced to the handling of logical objects that correspond to the various components that makes up your real-world system. In STEP 7, the main objects of your work will involve either a *Project* or *Library* object. Outside of creating storage paths for projects and libraries, in the Windows Explorer, your work should remain completely inside of STEP 7.

The STEP 7 Project

As the primary object of your STEP 7 work the *Project* will contain (1) hardware configuration data for PLCs, (2) module parameters and settings, (2) control programs and data (including documentation), and (4) the configuration data of network operating parameters and communications connections. As you work with the SIMATIC Manager, each project is opened in its own *project window*. The objects of the project may be opened and viewed, and edited by opening the object in its associated editor.

In the SIMATIC Manager, multiple projects may be opened simultaneously. Two STEP 7 project windows are shown opened below. As seen in the two windows, a project may vary in size from a single machine involving one PLC, to an entire plant involving many PLCs and networks. Also illustrated are some of the objects that you will work with under the SIMATIC Manager (e.g., SIMATIC PLCs, PG/PC programming systems, and networks.

Figure 2-1. (a) Project with Single PLC. (b) Project with multiple PLCs and networks.

Project Objects

Under the SIMATIC Manager, the objects of each project are presented in a split window. Objects containing other objects are connected in the tree structure of the left pane of the project window, while its contents are displayed in the right-pane. The hierarchy of the objects in the tree is based on the physical relationship they have within a real project. The **Project**, which contains all other objects, is at the top. Next is the **SIMATIC Station**, which represents a PLC. A project may contain several stations. Each station contains a **CPU**; and each CPU contains an **S7 Program**. Each program contains a **Source Files** and a **Blocks** folder. The Source folder contains user-created source text files; the Blocks folder contains *blocks* of the user program.

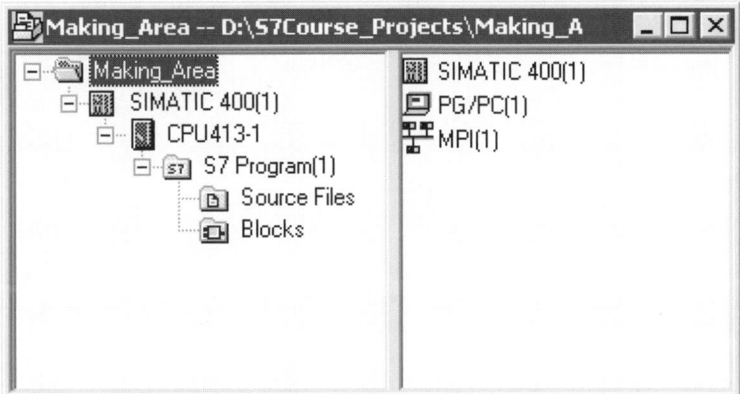

Figure 2-2. Hierarchy of STEP 7 project objects.

The STEP 7 Library

A **Library** object is used to store and mange reusable STEP 7 programs. In essence, the library is like a project stripped of everything except the S7 Program folder. The S7 Program object of a Library is the same as a program within a Project, except for the fact that the library program is not associated with a CPU.

Programs in a library are simply a method of maintaining code that you wish to keep and perhaps use at another time. Like a project, a library is developed in the SIMATIC Manager, where you may insert as many program folders as needed. Program folders may also be copied from a project to a library.

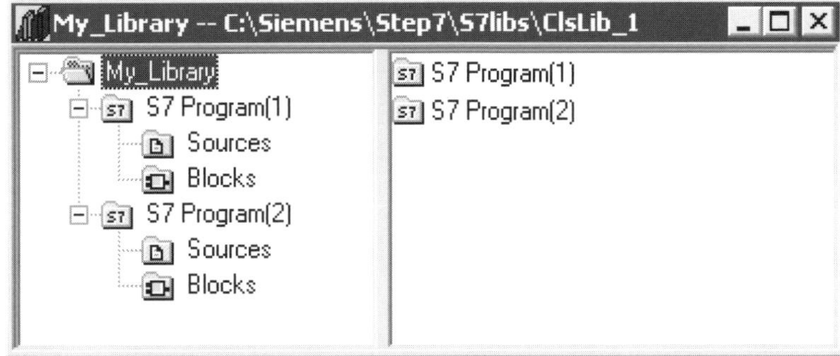

Figure 2-3. STEP 7 Library containing two S7 programs.

In addition to libraries that you might create, there are also libraries of canned functions available for purchase to S7 users. There is also the so-called *Standard Library*, which is included in the STEP 7 Basic package. The Standard Library is shown opened in the window below. The library has seven program folders, which have been renamed according to the type of blocks that each contains.

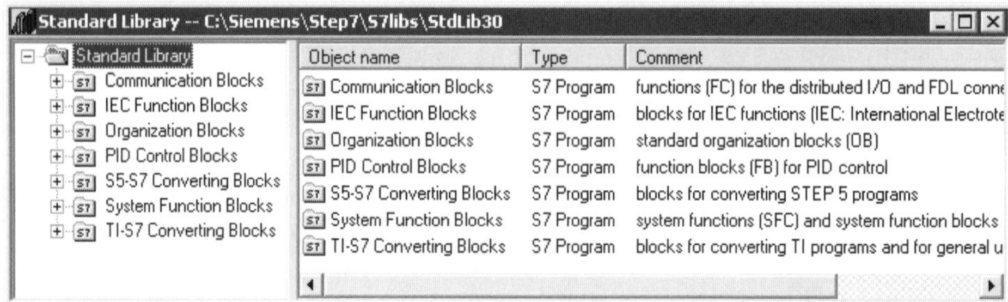

Figure 2-4. STEP 7 Standard Library, delivered with the STEP 7 Basic package.

Blocks that are integrated as part of S7 CPUs, for example organization blocks and system blocks, are included in the Standard library so that they may be used during offline development of your control program. Since these blocks are actually incorporated in the S7 CPU operating system, the library versions of these blocks are not actually code. The library contains the calling interface in the case of System Functions (SFCs) and System Function Blocks (SFBs), and the declaration table in the case of organization blocks.

Table 2-1. Standard STEP 7 Libraries.

Library Program	Brief Description
Communications Blocks	Loadable functions for controlling communications modules.
IEC Function Blocks	Loadable functions for editing variables with complex data types (STRING, DATE_AND_TIME).
Organization Blocks	Templates for the standard organization blocks; essentially the required 20-byte variable declaration table for each OB's start information. The actual code of an OB is in the S7 operating system.
PID Control Blocks	Loadable functions for implementing closed-loop control.
S5-S7 Converting Blocks	Loadable functions for S5 -to- S7 conversion. The blocks also replace S5 standard function blocks as part of the conversion.
System Blocks	Contains the calling interfaces for S7 System Functions (SFCs) and System Function Blocks (SFBs), to allow incorporation into the user program during offline development.
TI-S7 Converting Blocks	Loadable functions for the TI 505 -to- S7 conversion. The blocks replace TI standard function blocks as part of the conversion.

SIMATIC Manager Menu and Toolbar

The SIMATIC Manager has functionality similar to most Windows applications, and as such, its menu and toolbar looks and feels much the same. Standard *File* operations include New, Open, Close, Save As, and Delete. The file, of course is always the STEP 7 project or library. The New Project Wizard, also included in the File Menu, allows a new project to be created with assistance from STEP 7. The *Edit* menu allows standard Cut, Copy, and Paste operations, on the currently selected STEP 7 object. If the Project object is selected, then Stations, which are contained in projects may be inserted, copied, cut, paste or deleted.

Step 7 Toolbar buttons, as briefly described below, represent the most frequently used operations or tools of the SIMATIC Manager. Toolbar buttons are available for performing File, PLC, Edit, and window View operations. There are also buttons to launch some of the important software tools like the *NetPro Network Configuration* and *Find Active MPI Nodes*.

Table 2-2. SIMATIC Manager Toolbar Icons.

Icon	Toolbar Function
	Create New Project or Library
	Open Existing Project or Library
	Find Active MPI Nodes
	Open Memory Card Window
	Cut Selected Object
	Copy Object to Clipboard
	Paste Object from Clipboard
	Download to CPU
	Open Project Window Offline

Icon	Toolbar Function
	Open Project Window Online
	Display Objects as Large Icons
	Display Objects as Small Icons
	Display Project Objects as List
	Display Objects with Detail Info
	Filter Project Objects
	Open Network Configuration Tool
	Toggle Simulation Tool ON/OFF
	STEP 7 Help

Project Windows

The SIMATIC Manager allows the STEP 7 project and its objects to be opened and displayed. When navigating downward through the STEP 7 object tree, when an object containing other objects is a selected, the object opens as a folder with its contents displayed in the right-pane. Object at the lowest level of a container object do not contain other objects, and are always listed in the right-pane. Double-clicking on an object at the lowest level of a container causes the object to open in its associated editor. Block objects, which are contained in the Blocks folder for example, are opened in the LAD/FBD/STL editor, unless the block was compiled from a source editor, in which case the associated editor opens.

Offline Window

The *offline window* of a project, displays components of the project that are contained in offline files stored on your PG/PC. All of the documentation associated with the programs and configuration data of a project are stored in the offline files of the project. Therefore, when you go online with a CPU program without the use of the project, offline comments and data are not generally available. Although in some later CPU, comments and symbol address information may be stored in the CPU.

Making_Area -- D:\S7_Projects\Making_A

Making_Area
SIMATIC 400(1)
CPU413-1
S7 Program(1)
Sources
Blocks

Object name	Symbolic name	Created in ...	Type	Last modified	Comment
System Data	---	---	SDB	18.03.2004 11:35:53	---
OB1		LAD	Organization ...	13.12.2003 12:24:50	
FB100		LAD	Function Block	18.11.2003 11:14:32	
FC1		LAD	Function	16.03.2004 21:51:39	Simple Ladder Lo
FC2		LAD	Function	08.11.2001 14:38:03	Simple Ladder Lo
FC3		LAD	Function	08.11.2001 14:40:05	Simple Ladder Lo
FC10	Box_Counters	LAD	Function	08.03.2004 15:45:42	Using Counter Bo
FC11	Coil_Counters	LAD	Function	04.01.2001 11:30:25	Examples: Counte
FC12		FBD	Function	08.03.2004 15:24:07	Using Timer Box I
FC100		STL	Function	31.12.2003 16:39:24	Simple Ladder Lo
DB10		DB	Data Block	13.12.2003 12:20:44	

Figure 2-5. Offline Project Window with detail view.

Online Window

The *online window* of a project is a separate window that displays the project components that are contained in the currently selected PLC station. An online connection is required in order to view the contents of a CPU, or to work with the project online. Some offline objects are not in the CPU and are not seen in the online window (e.g., Source files). Comments and documentation associated with a block or a hardware object and viewed online, are actually stored in the offline files of the project. When you go online with a CPU without the use of the project, offline comments and data is not available. In some later CPUs, however, certain offline comments and documentation may be stored on the CPU.

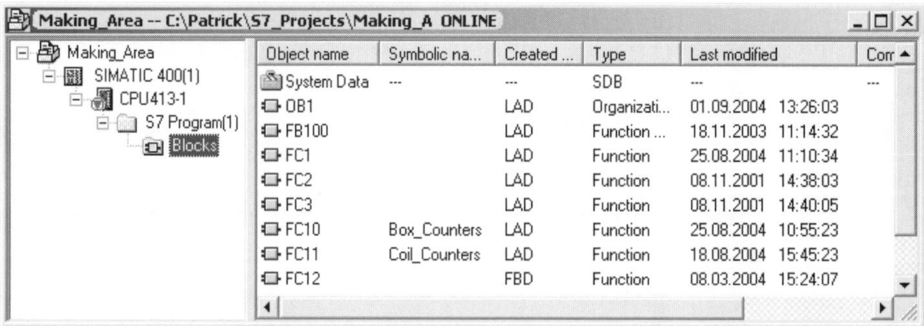

Making_Area -- C:\Patrick\S7_Projects\Making_A ONLINE

Making_Area
SIMATIC 400(1)
CPU413-1
S7 Program(1)
Blocks

Object name	Symbolic na...	Created ...	Type	Last modified	Com
System Data	---	---	SDB	---	---
OB1		LAD	Organizati...	01.09.2004 13:26:03	
FB100		LAD	Function ...	18.11.2003 11:14:32	
FC1		LAD	Function	25.08.2004 11:10:34	
FC2		LAD	Function	08.11.2001 14:38:03	
FC3		LAD	Function	08.11.2001 14:40:05	
FC10	Box_Counters	LAD	Function	25.08.2004 10:55:23	
FC11	Coil_Counters	LAD	Function	18.08.2004 15:45:23	
FC12		FBD	Function	08.03.2004 15:24:07	

Figure 2-6. Online Project Window with detail view.

Comments on Working with STEP 7 Projects and Libraries

In your project work with S7 equipment, your initial task will be to create a software Project. The project, a container object for all of the S7-related components of your automation system, is the basis of all the work you will perform. Whether your project contains one or several SIMATIC S7 PLCs, each is represented as a Station object that contains a CPU object, and each CPU contains an S7 Program object.

In building the project, you are essentially creating a framework of the basic container objects whose contents will be filled as you perform later tasks such as configuring the hardware, developing the program, and defining a network. These steps, which are outlined below, are described in the examples provided in this chapter. Once you have created the project and the basic components that it will contain, you will be ready to proceed to Working with the Hardware Configuration, the next step of developing the S7 project.

Checklist: Working with STEP 7 Projects and Libraries

- *If you will be working primarily with projects that already exist, then you may use the Open command from the SIMATIC Manager menu or toolbar to browse your directory structure and open STEP 7 projects.*
- *If you are developing STEP 7 projects, you might start by creating folders on your programming system (PG/PC) where S7 Projects and Libraries, will be stored. This is an optional task since you may also use STEP 7 default folders "S7Proj" and "S7Libs," which are located under the Siemens\STEP 7 folder.*
- *You may also wish to create a separate folder where archive projects and libraries may be stored.*
- *If you are starting by developing your own project, the first step is to create or generate the new project container, from the SIMATIC Manager.*
- *From the SIMATIC Manager, use the Options/Customize menu to ensure that you have defined the target storage paths for your S7 Projects and S7 Libraries.*
- *Create a Project using the New Project Wizard or Using the New Command.*
- *If you create the project using the New command, you will need to insert an S7-300 or S7-400 station from the SIMATIC Manager menu; if the Project Wizard is used, you will select the first station from one of the wizard dialogs.*
- *Open Hardware Configuration Tool and, as a minimum to get started, insert the required CPU.*
- *The S7 Program is generated when the CPU is inserted.*
- *The basic framework of the project is essentially complete when there is at least one **Station**, a **CPU**, and an **S7 Program**; you may then begin working with configuring your hardware.*

Navigating the STEP 7 Project Structure

Basic Concept

The STEP 7 objects, all of which may be edited, are presented in a split window of the SIMATIC Manager where navigation is much like that of the Windows Explorer. Since much of your STEP 7 work will be done from the SIMATIC Manager, fully understanding the hierarchy of the objects of a project will simplify and expedite your work.

Essential Elements

The **Project** object, at the top of the tree, contains the project's hardware stations, and networks; the **Station,** which represents a PLC, contains the configuration data for the station's hardware components. The **CPU** is contained within the station, along with other programmable modules (e.g., CPs and FMs). Each CPU contains an **S7 Program**; each S7 Program contains a **Sources** folder, a **Blocks** folder, and a *Symbols* object. The Sources folder will hold any text source files that *you create*; the Blocks folder will holds individual blocks (code blocks and data blocks) created for the user program. The Symbols object opens to a table of symbolic addresses assigned to absolute addresses used in the program.

Application Tips

Double-clicking on an object at the lowest level of a container will open that object in its associated editor. Block objects as shown in the window below, are at the lowest level of the *Blocks* folder of a program. Blocks, which will be described in the next chapter, contains STEP 7 code or data. A block is opened in the editor it was created in (e.g., LAD/FBD/STL).

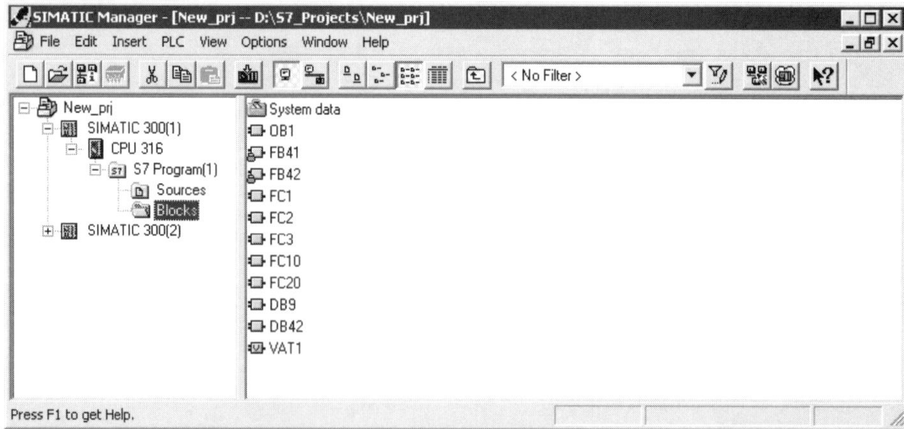

Figure 2-7. Offline Project Window, two Stations; Blocks folder of SIMATIC 300(1) selected.

Quick Steps: Navigating the STEP 7 Project Structure

STEP	ACTION
1	Clicking on the (+) sign of the **Project** displays S7-300 or S7-400 **Station** objects already inserted in the project; Stations also contain objects.
2	Clicking on the (+) sign of a **Station** object, reveals the **CPU** and other programmable modules already inserted in the station; programmable modules include CPUs, CPs, and function modules (FMs).
3	Clicking the (+) sign of a **CPU** object reveals an **S7 Program** object. Each CPU and other programmable modules contain a program. You will work primarily with CPU programs.
4	Clicking the (+) sign of the **S7 Program** object displays a folder for *Sources* and *Blocks*. The **Sources** folder will hold un-compiled source files (e.g., created using the STL or SCL text editors); the **Blocks** folder, will hold program code and data blocks.
5	Selecting the **Blocks** folder of a Program displays, in the right pane, any blocks that have been generated; OB1 is generated when the program is first generated; you will generate other blocks as you build your STEP 7 program incrementally.
6	Selecting the **Sources** folder of a Program displays, in the right pane, any source files that have been developed; this folder is initially empty and remains empty unless you create text source files using a standard or optional text editor.
7	Selecting the **S7 Program** displays, in the right pane, its *Sources* folder, *Blocks* folder, and *Symbols* object. Double-clicking on the Symbols object opens the symbols editor, in which symbolic names for absolute address are created and edited.
8	Selecting a **CPU** object in the tree structure displays its *S7 Program* folder and *Connections* object in the right pane. Double-clicking on the *Connections* object opens the network configuration tool, where communications connections are created or edited for the selected CPU.
9	Selecting any **Station** object will display the station's *Hardware* configuration object, and any programmable modules (e.g., CPUs, CPs, and FMs), already inserted in the station. Double clicking the *Hardware* object opens the station's hardware configuration, allowing you to create or edit the stations configuration.
10	Selecting the **Project** display its contents in the right pane. SIMATIC 300/400 stations, H-stations; S5 stations, PG/PCs, Other stations, and subnets may also be displayed.
11	The name of any object can be changed by right clicking on the object, and selecting **Rename**. Simply type over the name when it is highlighted
12	To edit the general information and comments on any object, right-click on the object and select **Object Properties**.
13	Finally, selecting **View ➢ Expand All,** from the menu, opens the folders of all container objects in the project tree; selecting **View ➢ Collapse All** closes all container object folders.

Navigating SIMATIC Manager Menus and Toolbar

Basic Concept

Navigating the menus and toolbar of the Hardware Configuration tool is much like navigating within the Windows' Explorer — it has the same look and feel. Instead of simply managing files, however, the SIMATIC Manager is an object-oriented tool allowing you to easily work with the objects of the STEP 7 world — objects that represent PLCs, CPUs, modules, networks, and so on. The NetPro network configuration tool and the hardware configuration tools are opened automatically when you double click on an object that was created using that tool.

Essential Elements

The complete of operations of the SIMATIC Manager may be accessed from its menus, including File, Edit, Insert, PLC, View, Options, Window, and Help. The toolbar, a set of buttons position directly beneath the menus, provide quick access more frequently used operations. As you move through the project tree, you will notice that the various command and options of the menus and toolbars change to reflect the current selection.

Application Tips

Multiple projects may be opened simultaneously, each being placed in its own window. Simplify using the Copy and Paste commands by opening multiple windows displayed horizontally or vertically. To copy from one project window you may drag and drop objects or use Cut/Paste. Make sure that before you paste, that you select the correct object to which you wish to paste. For example to copy a Station select the Project object in both windows, since stations are contained within the project and may only be copied to a project. Programmable modules (e.g., CPU, CP, FM), seen in the project tree, may only be copied and pasted while inside the Hardware Configuration Tool.

Move your mouse tip across the Icon and View buttons; a tool tip message defines each. Show objects as: Large icons; as Small icons; as a List; or with Detail information.

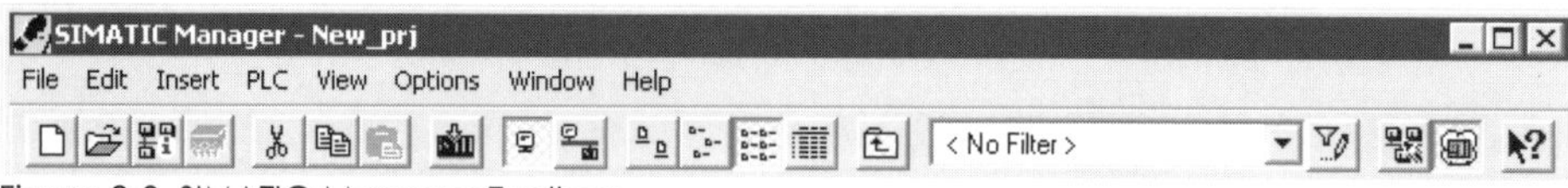

Figure 2-8. SIMATIC Manager Toolbar.

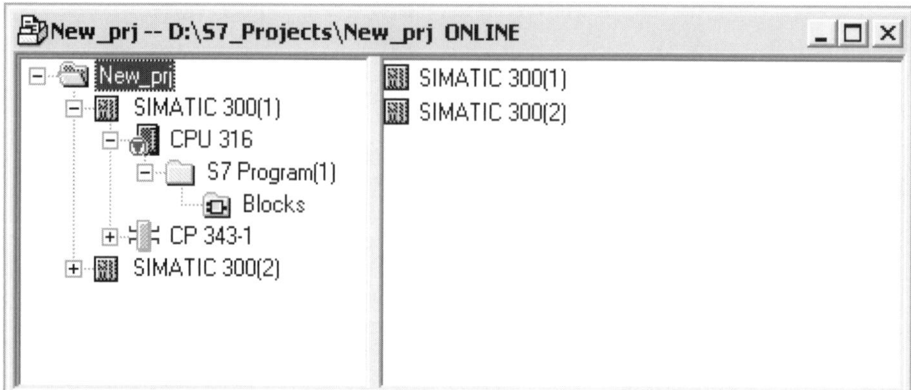

Figure 2-9. Online Project Window with detail view.

Quick Steps: Navigating SIMATIC Manager Menus and Toolbar

STEP	ACTION
	The **File** menu and toolbar options allow you to work with existing projects and libraries and to create and manage new ones.
1	By selecting **File ➢ Open,** you may open a project from the default project directory or browse for a project or library in another location on your computer.
2	**File ➢ New**, allows you to create a project or library in the default or in a new path.
	Edit menu and toolbar options allows properties of selected objects such as stations, networks, and blocks to be edited, as well as copied and pasted.
3	To copy a Station, you would select the station then select **Copy;** click inside the project window of the same project or in another project and then from the Edit menu, or using the right-click, select **Paste**.
4	Select any object, right-click and select **Object Properties** to view its properties.
	View and **Window** menus and toolbar options allow projects to be viewed online and offline, adjustments to icon size and window arrangement, and details given on a selected object.
5	With the NewProj window selected, select **View ➢ Online** to open the online window.
6	Selecting **Window ➢ Arrange ➢ Vertical,** places open project windows side-by-side; *Horizontal* is one above another, and *Cascade* is shifted so the title bar of each is visible.
	The **Insert** menu allows the insertion of new objects into the selected container object. Only objects that are correct for the selected container object may be inserted.
7	Selecting the Project folder, allows new **Stations**, **Subnets**, or **Programs** to be created.
8	Selecting the Blocks object allows new program blocks (e.g., FB, FC) to be created.
	PLC menu and toolbar buttons supports online CPU/PLC operations.
9	With the offline Blocks folder of an S7 Program selected, selecting **PLC ➢ Download** transfers offline blocks to the associated CPU.
	The **Options** menu allows you to set SIMATIC Manager operating preferences, and to launch standard and optional tools.
10	Selecting **Options ➢ Configure Network**, for example, opens the graphical network configuration tool.
11	Selecting **Options ➢ Customize** opens the dialogue for setting various SIMATIC Manager operating parameters (e.g., storage paths, and language preferences).

Finding STEP 7 Projects and Libraries

Basic Concept

When a project or library is opened for the first time, it is added to a list, by the SIMATIC Manager, and is thereafter displayed in the list whenever the **File ➤Open** command is used. S7 Projects and libraries may reside on a PC but not displayed if they have never been opened; they also may not appear in the project or library list if they have been hidden using the **File ➤Manage** option. You may search an entire drive or selected folders for STEP 7 projects and libraries that may exist, but do not appear in the project or library list.

Application Tips

If the search and find option is used from a programming editor (e.g., LAD/FBD/STL), then only user projects may be searched (and not sample projects or libraries). You may search for User projects, sample projects, and libraries from the SIMATIC Manager. User projects are projects that you create yourself; sample projects are the projects that were installed during the initial installation of the STEP 7. If you choose to, you may adapt and expand any of these sample projects, and re-save as user projects.

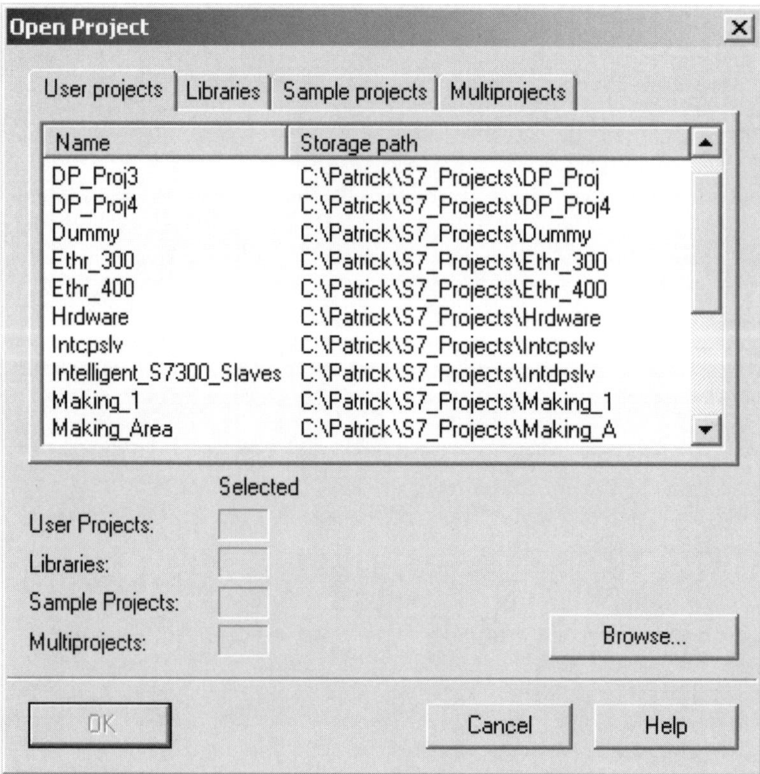

Figure 2-10. Dialog for opening STEP 7 projects and libraries.

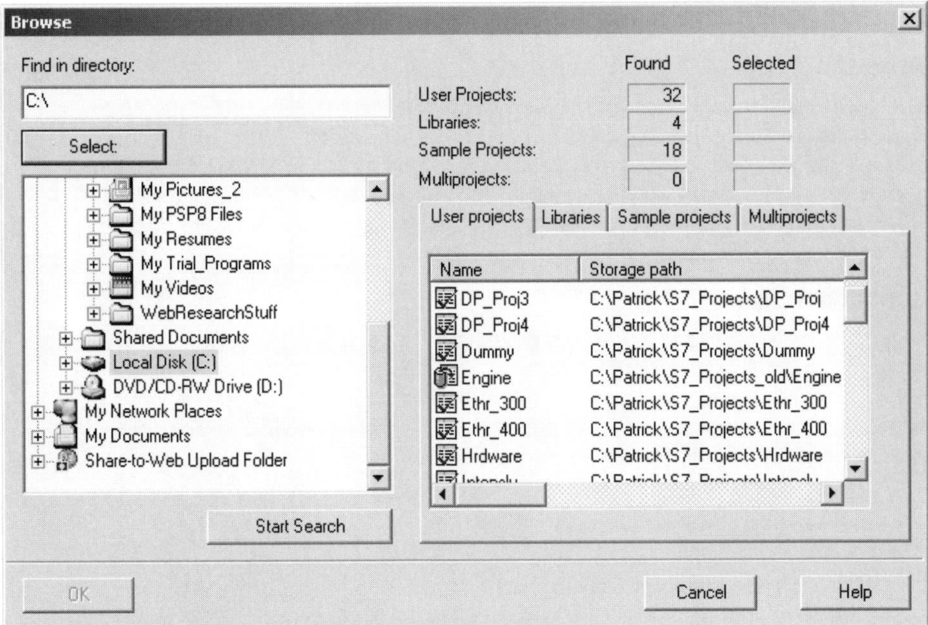

Figure 2-11. Dialog for searching STEP 7 projects and libraries.

Quick Steps: Finding STEP 7 Projects and Libraries

STEP	ACTION
1	Launch the SIMATIC Manager and select **File** ➢ **Open**. The User Projects tabs lists user projects currently in the project list and are known to the SIMATIC Manager.
2	Select the corresponding tab to display all known **S7 Libraries** or **Sample Projects**.
3	Click **Browse** button to search for unlisted *User Projects*, *Libraries*, and *Sample Projects*.
4	From the Browse dialog, select a specific drive or folder to search through.
5	Press the **Start Search** button.
6	When the search is completed, the number of *User Projects*, *Libraries*, or *Sample Projects* is listed under the **Found** column.
7	Select the appropriate tab to display found **User Projects**, **Libraries,** or **Sample Projects**.
8	The check mark symbol, shown to the left of the Project/library, indicates that the found entry is already in the project or library list.
9	The symbol shown to the left of the Project/library indicates that the found entry was marked as "Hidden," but can be added to the project or library list.
10	Select one or more entries to add to the project or library list; confirm the selection and close the dialog with **OK**.

Creating a Project using the New Project Wizard

Basic Concept

Before working with your hardware configuration, you must create a new STEP 7 project. Projects created using the project wizard will create a new project that will contain an S7-300 or S7-400 station based on your selection. The new station will contain the CPU of your choice, a STEP 7 program with the main organization block (OB1), and any other organization blocks you select. You will also be able to set the default language (i.e., LAD, FBD, or STL) for creating program blocks.

Application Tips

New projects are placed in the default project path for STEP 7 or in the path currently set as the default path. You may want to create a new STEP 7 project folder or set a new default project location prior to creating your project. Before completing the last dialogue of the wizard, you may backup and modify any of your selections. Modifications are also possible later, inside of STEP 7, after the project is generated.

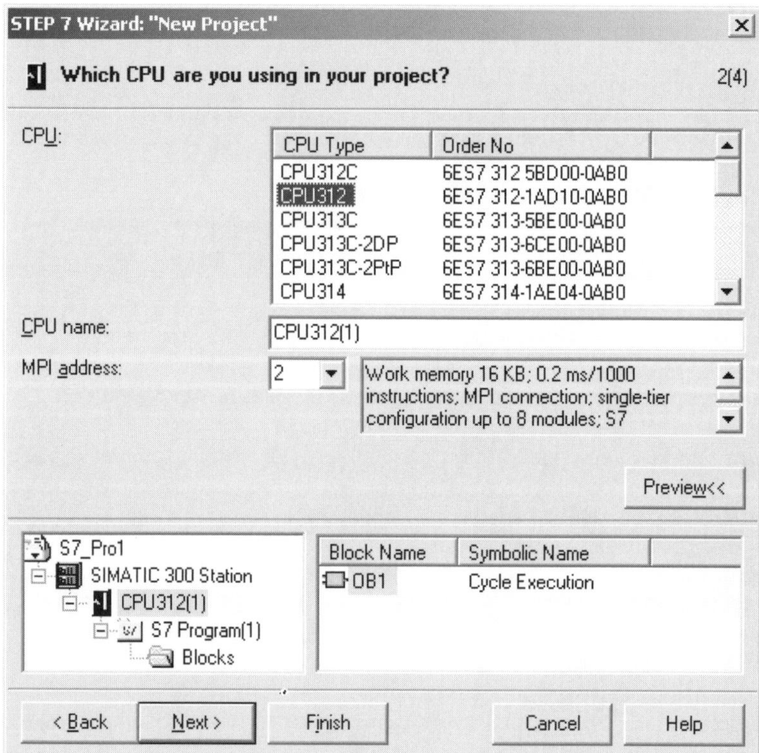

Figure 2-12. Project Wizard, Step 2; defining CPU, CPU name, and MPI address.

Quick Steps: Creating a Project using the New Project Wizard

STEP	ACTION
1	Launch the SIMATIC Manager and close any open projects.
2	If the New Project Wizard does not start automatically, from the menu, select **File** ➢ **New Project Wizard**.
3	Remove the check mark on the dialog, to disable the Wizard from opening each time you start the SIMATIC Manager, and click **Preview** button to open preview window.
4	Click the **Next** button to continue the New Project procedure.
5	Select from the **CPU Type** list, the S7-300 or S7-400 **CPU** to use in this project. You may change the default name of the CPU, displayed in the **CPU Name** field.
6	Select an **MPI Address** for your multipoint interface connection of this CPU, or accept the default address of 2, temporarily, and change the address later.
7	Click the **Next** button to continue the New Project procedure.
8	Select, from the **Blocks** window, organization blocks you wish generated for your first program. Leave OB1 (the main cyclical block of a STEP 7 program) checked.
9	Select the language representation (LAD, Ladder), (STL, Statement List), or (FBD, Function Block Diagrams) you wish to use for creating the OBs you selected.
10	Click the **Next** button to continue the New Project procedure.
11	In the final dialog, type a name for your project. Names longer than eight characters are used by the STEP 7 software, but will be shortened to eight characters in Windows.
12	Click the **Finish** button to generate the new project, otherwise you may press the **Back** button to back up and make changes or **Cancel** to abort the new project.
13	Use Windows Explorer to find the Project folder you created, to verify the newly created project folder.

Creating a Project Using the 'New' Command

Basic Concept

A STEP 7 project is generated automatically using the New Project Wizard, or manually using the *New* command. When created manually, you must generate the required objects of the project. The *New* command creates an empty project folder in which you must manually insert the basic objects of the project. These objects, which you will generate, are actually containers that will eventually hold the various data, such as configurations and programs of the project. Upon completion, you will have created a project, station, and program containers. The program container holds a source files container and a blocks container.

Essential Elements

A project generated with the New command is empty except for a default MPI subnet object. You must insert at least one PLC station, using the Insert ➢ Station command. Since an inserted S7-300 or S7-400 station does not generate a CPU object, you will have to open the hardware configuration tool and configure the station, at least partially, to include the appropriate rack and CPU (i.e., central rack). The Program object, which contains source file and offline block folder, is generated automatically when the CPU is inserted into the configuration. The essential objects of the project are thereby completed.

Application Tips

New projects are placed in a directory, currently set as the default project path. You may want to create a new STEP 7 project directory or set a new location as the default prior to creating your project. Remember, although STEP 7 supports a long name internally, the Windows Explorer will use the first eight characters of your long name as the project file name. To avoid possible problems, you should make all name changes inside of STEP 7.

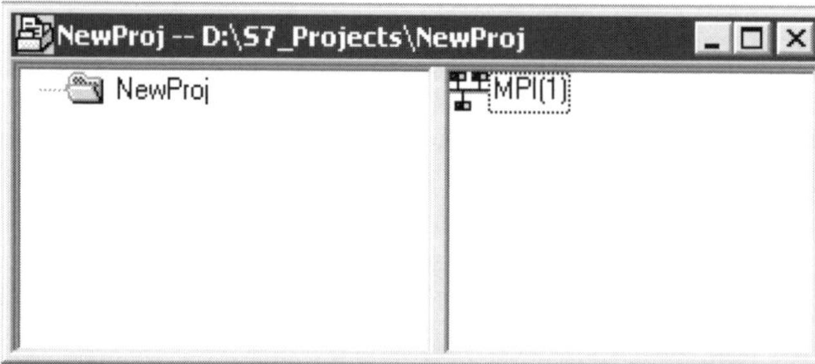

Figure 2-13. Empty Project object after using New command.

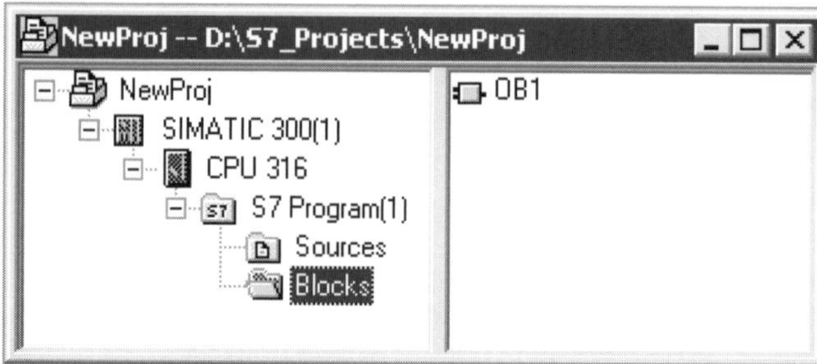

Figure 2-14. Complete Project structure, after inserting Station and CPU.

Quick Steps: Creating a Project Using the 'New' Command

STEP	ACTION
1	Launch the SIMATIC Manager and close any open projects.
2	From the menu select **File ➢ New,** or press the New Project icon on the toolbar.
3	Type in a name for the new project (e.g., Warehouse). Remember, STEP 7 supports long names, but Windows Explorer will only use 8-characters for the actual project file name.
4	From the menu, select **Insert ➢ Station ➢ SIMATIC 300 Station** (SIMATIC 400 for S7-400).
5	Click on the (+) sign of the project folder and then click on the S7-300 Station. The hardware object of the selected station is displayed in the right pane of the project window.
6	Double-click on the *Hardware* object, to open the Hardware Configuration tool.
7	If the Hardware Catalog is not in view on the screen, select **View ➢ Catalog**.
8	Click on the (+) sign of the **SIMATIC 300** (SIMATIC 400 for S7-400) catalog object.
9	Click on the (+) sign of the **Rack-300** (Rack-400 for S7-400) folder to open.
10	Select the **Rail** (CR or UR rack for S7-400) object, drag, and drop into the upper pane of the Configuration Window.
11	Click on the (+) sign of the **CPU-300** (CPU-400 for S7-400) catalog object to display the CPU parts folders.
12	Find the CPU folder that matches your CPU hardware and click on the (+) sign to open the folder. Select the CPU part number that matches the actual part number for your hardware, and drag and drop in slot-2 of the rack in the configuration window.
13	From the menu, select **Station ➢ Save** to save the configuration.
14	Return to the SIMATIC Manager to start working with your project.

Adding a New Station to a Project

Basic Concept

Once you have generated a project, manually or using the project wizard, each additional new station within the project must be added using the **Insert Station** command of the SIMATIC Manager. A station must be inserted before you can create its hardware configuration using the hardware configuration utility.

Essential Elements

Station objects that may be added to an S7 project include, *SIMATIC-300* or *SIMATIC-400 station*; SIMATIC *H-Station* for redundant configurations; *SIMATIC PC-Station*, for representing a PC or Siemens PG programming system; *S5 Station*, to represent each S5 PLC; or *Other Station,* to represent any non-Siemens system, as part of the STEP 7 project.

Application Tips

If several stations of the same type are required, use of the Cut and Paste features of the SIMATIC Manager will reduce your work. Simply select the station, copy and then paste in the right-pane of the project window. If the stations will be very similar or identical, however, you may wish to complete as much as possible of the configuration of the one station prior to copying it to new stations.

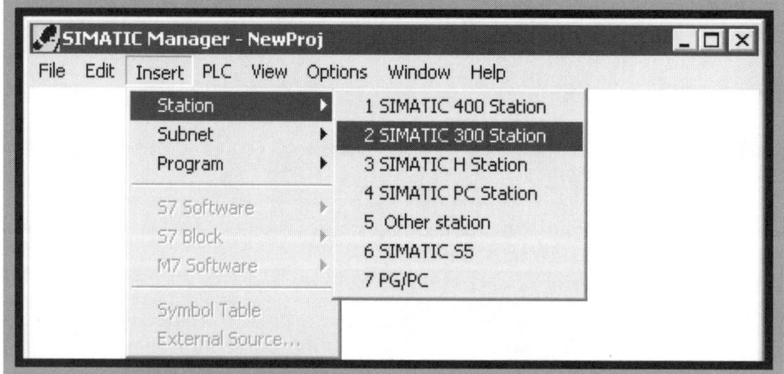

Figure 2-15. Adding new Stations to a Project from the SIMATIC Manager menu.

NewProj -- D:\S7_Projects\NewProj

NewProj

Object name	Symbolic name	Type	Size
MPI(1)	---	MPI	2816
SIMATIC 300(1)	---	SIMATIC 300 Station	---
SIMATIC 300(2)	---	SIMATIC 300 Station	---
SIMATIC 300(3)	---	SIMATIC 300 Station	---
SIMATIC 300(4)	---	SIMATIC 300 Station	---

Figure 2-16. Project window, after inserting four new Stations from the SIMATIC Manager.

Quick Steps: Adding a New Station to a Project

STEP	ACTION
1	From the SIMATIC Manager, open the required project and select the project folder.
2	From the menu select **Insert ➤ Station ➤ SIMATIC 300 Station** (or SIMATIC 400 Station); or right click on the project folder, select **Insert ➤ New Object,** and select a station type.
3	Right-click on the station object, then select **Object ➤ Properties** to modify the station name or other properties (e.g., enter comments on station's function in this project.)
4	Click in the Authors field and enter your name.
5	Click in the Comments field and enter a description of the station's role in this project.
6	Confirm dialog by pressing the **OK** button.
7	With the station installed, you are now ready to install the required racks and the modules that will be installed in each. Configuration is done using the hardware configuration tool.
8	Repeat Step 2 and Step 3 for as many stations as required, or consider creating the entire station module contents first, and then using the cut and paste features if the stations will be similar or identical.

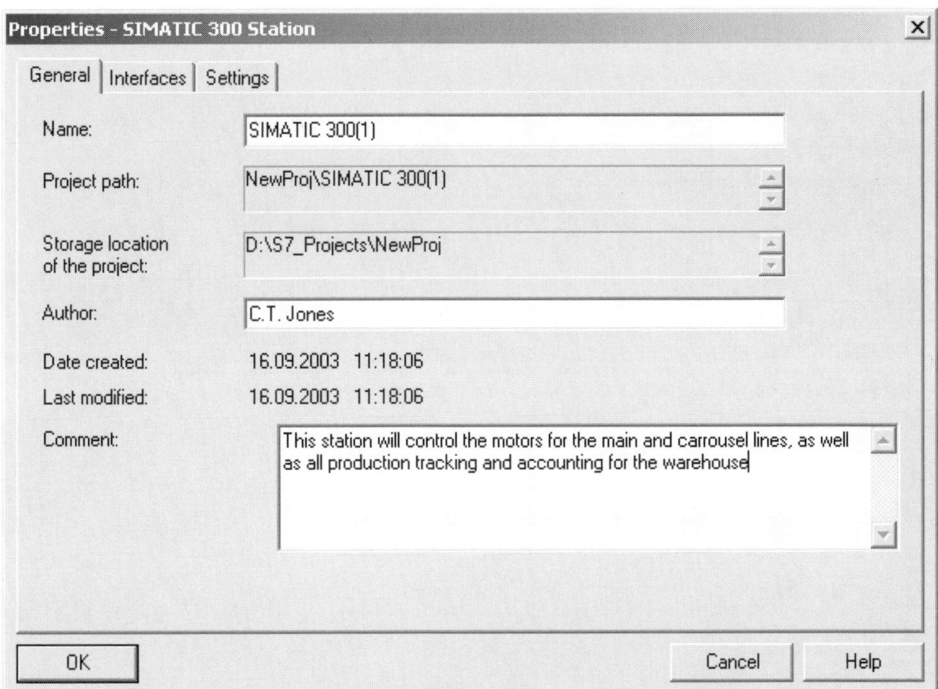

Figure 2-17. General Properties dialog for documenting a SIMATIC 300 station.

Archiving a Project or Library

Basic Concept

To archive a project or library is to generate a compressed backup of your project or library. Archiving provides a simple way of preserving your work as revisions, at various stages of development or operations. Creating archives also provide an extra measure of precaution. The allow you to make changes to a project, test the changes, and yet have a way of quickly reverting to a previously operating copy of the project.

Application Tips

For good organization and maintenance of your projects and backups, use of a separate folder for your archive projects is recommended. An archived project can be retrieved and restored as a fully expanded project using the *Retrieve* option from the STEP 7 File Menu.

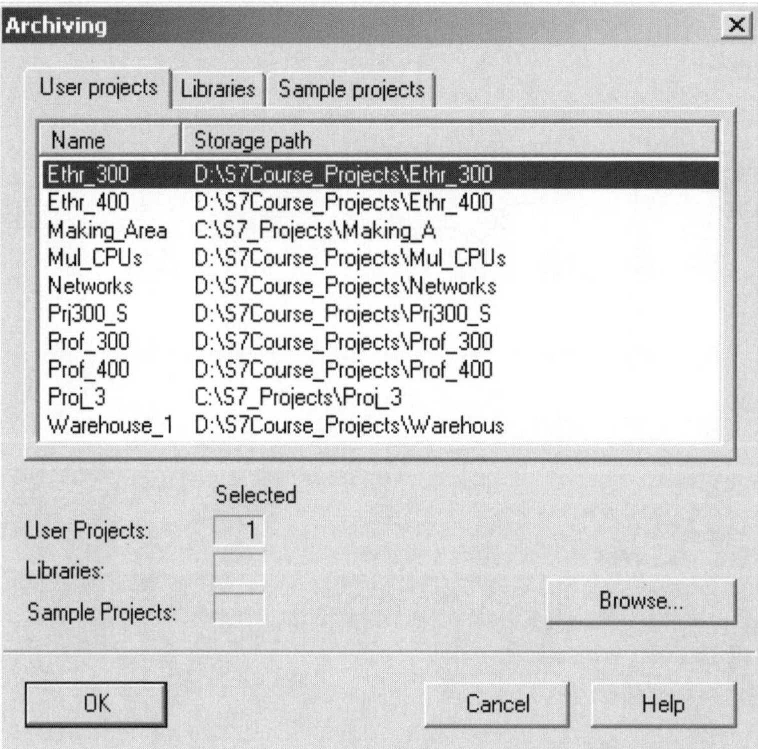

Figure 2-18. Archiving dialog for archiving STEP 7 Projects or Libraries.

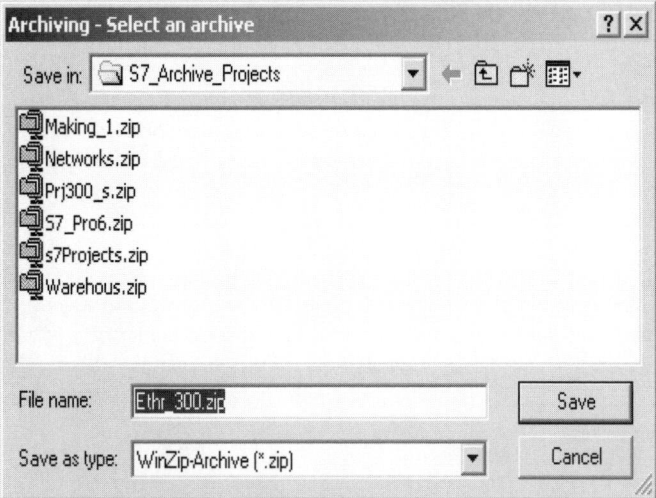

Figure 2-19. Dialog for setting archive name and storage location.

Quick Steps: Archiving a Project or Library

STEP	ACTION
1	Before using the SIMATIC Manager "Archive" menu command to store backup copies of projects, you may want to first review STEP 7 preference settings for Archiving.
2	Open the SIMATIC Manager and if the project you wish to archive opens automatically, because it was the last project open, then you must first close the file before it can be archived.
3	From the SIMATIC Manager menu, select **Archive...**
4	Select the appropriate tab and the desired file from "User Projects," "Libraries," or "Sample Projects", when The STEP 7 **Archiving** dialog is presented.
5	Press **OK** to continue with the archiving operation on the selected STEP 7 project or library. The Windows dialog "Archiving - Select an Archive" is presented for naming the archive file and the folder in which the archive should be stored.
6	Either accept the default directory and file name, or change as required. The storage location presented, is either the folder specified for "Archiving" or the last location where an archive was stored.
7	Confirm the archive operation with the **Save** button.

Retrieving an Archived Project or Library

Basic Concept

To retrieve an archived project is to uncompress the archive file and restore it as a fully expanded project that you can open in STEP 7. You should be aware as to whether or not the target directory already has a copy of the project. An existing copy of the project will be overwritten by the uncompressed archive.

Application Tips

It is important when retrieving an archived project that, when prompted to do so, you ensure to select the project folder where you wish to have the retrieved project stored. When your default storage path is displayed, you must be careful to select the project path and not the project file.

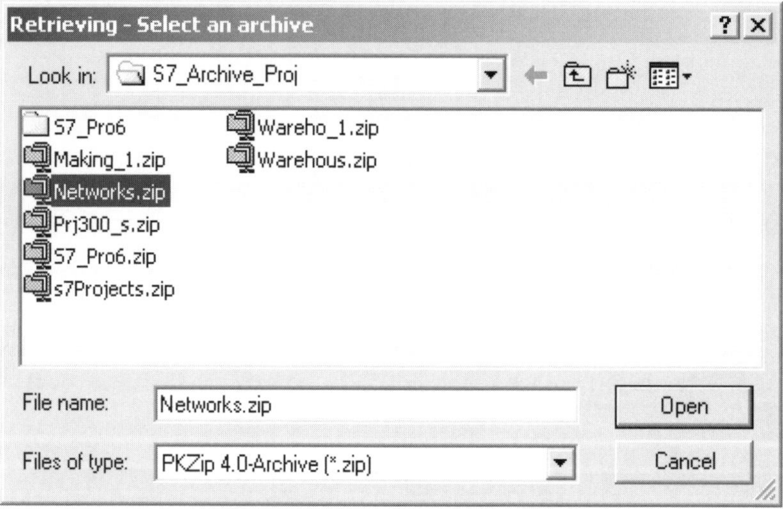

Figure 2-20. Dialog for retrieving an archived project or library.

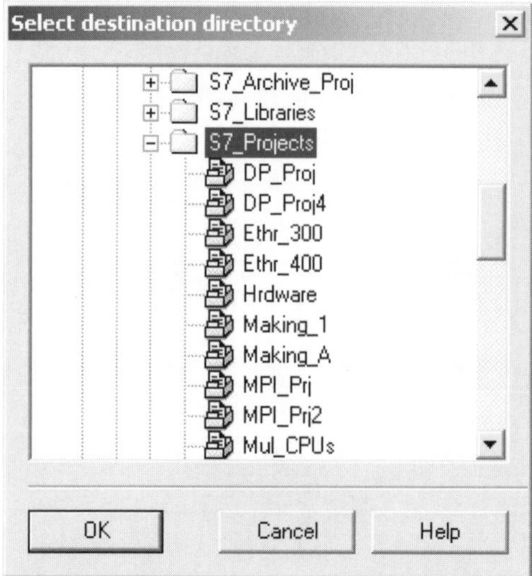

Figure 2-21. Selecting destination for restored project or library.

Quick Steps: Retrieving an Archived Project or Library

STEP	ACTION
1	Before using the SIMATIC Manager "Retrieve" menu command to retrieve a backup project or library, you may want to review STEP 7 preference settings for archiving.
2	Open the SIMATIC Manager and if the project you wish to retrieve opens automatically because of being the last project opened, you must close the project before it can be restored.
3	From the SIMATIC Manager menu, select **Retrieve...** the Windows dialog "*Retrieving - Select an Archive*" is presented for you to select the archive project file to restore. The folder presented is the location specified in Archive preferences or in which an archive was last stored.
4	From the listed files, select the archive project or library file to retrieve.
5	Press **Open** to continue retrieving the selected archive project or library. The target project/library storage path is presented in the "*Select Destination Directory*" dialog, along with stored projects.
6	Select the folder in which you wish to have the project or library restored; **do not** select the project file icon of the project you are restoring.
7	The uncompressed archive will be placed in the directory specified under the Archiving tab under Customize options, unless you specify otherwise.

Step 3

Working with Hardware Configurations

Objectives

- Introduce S7-300/S7-400 Components
- Introduce Hardware Configuration Tool
- Configure S7-300 and S7-400 Central Racks
- Configure S7-300/S7-400 I/O Expansion Racks
- Configure S7-300/S7-400 Profibus DP Masters
- Configure S7-300/S7-400 Profibus DP Slaves
- Configure Digital/Analog I/O Addressing
- Configure Standard CPU Properties
- Configure Multi-Computing CPU Properties
- Configure Digital/Analog Module Properties
- Upload and Download Station Configurations

S7-300/S7-400 Hardware Components Overview

The S7-300 and S7-400 controllers are both modular in nature, each built-up from the same basic types of components. These components include racks, power supplies, central processing units, interface modules, communications processors, signal modules, and function modules. Each of these component types, briefly described in the following discussion, represent the parts from which you will configure S7-300 and S7-400 systems.

Racks

Each S7-300 or S7-400 is built up using one or more racks that provides the mounting mechanism for modules. As a system expands, additional racks are interconnected using bus cables and interface modules (IMs). S7 rack systems uses two busses: a high-speed I/O signal bus (P-bus), and a communications bus (C-bus, or K-bus in some documents), which supports data transfer to and from communications modules and that connects the programming interface and CPU with installed programmable modules (e.g., CPs and FMs). Racks having a P-bus only (no C-bus) will only support the installation of signal modules (SMs).

Table 3-1. S7-400 Controller (CR), Universal (UR), and Expansion Racks (ER).

Rack Short Name	Brief Description
Rail (UR)	S7-300 Universal rack; a DIN standard rail for mounting all modules.
CR2	S7-400 Controller rack; 18-slot with two-segments (10-slot/8-slot).
CR3	S7-400 Controller rack; 4-slot, not suitable for redundant P.S.
ER1	S7-400 Expansion rack; 18-slot, no C-bus.
ER2	S7-400 Expansion rack; 9-slot, no C-bus.
UR1	S7-400 Universal rack; 18-slot, support for all modules.
UR2	S7-400 Universal rack; 9-slot, support for all modules.
UR2-H	S7-400 Universal rack; Two 9-slot segments (18-slots). Supports compact configuration with standard or redundant PLC systems.

Note: No C-Bus implies that only signal modules are supported in the rack. CR2, ER1, ER2, UR1, and UR2 are all available with and without support for redundant power supply.

S7-300 Rack

The S7-300 rack is a standard DIN Rail. The rail is available in different lengths and is used to configure both central and expansion racks. Before a module is installed, you must insert the molded U-shaped bus connector, provided with each module, into module's right-edge. Each module is installed by hooking it onto the top edge of the rail and firmly pressing downward until snapping into place. With each module in place, the right-edge of the connector, which extends out from the module will insert into the left-edge of the next module, thereby forming the S7-300 backplane bus. The bus is daisy-chained from module-to-module starting with the CPU. In the S7-300, a central rack and three expansion racks may be connected, using the appropriate interface modules.

S7-400 Central, Universal, and Expansion Racks

The S7-400 offers three types of racks from which to select. *Controller racks,* designated (CR), allow CPU and all module types to be installed. *Expansion rack (ER)*, having a P-bus only, only support installation of signal modules. *Universal racks (UR)*, as the name implies, support all module types and may serve either as a controller or as an expansion rack.

Controller racks are designed with a backplane that is divided in two segments — each segment having nine slots. The controller rack (CR2), which supports both the P-bus and C-bus, will allow a CPU and its associated I/O residing in each segment to operate in parallel, but independently. Universal racks include the UR1, UR2, and the UR2-H. The UR2-H must be used in S7-400H systems, but may also be used for standard S7-400 systems.

Interface Modules (IM)

Interface modules (IMs) allow the S7-300/S7-400 systems to expand beyond the central rack, forming local or remote configurations. The selection of IMs requires a *publisher/receiver pair*, as shown in the table below. *Publisher interface modules* are always placed in the CPU rack while *receiver interface modules* are placed in expansion racks. Each interface module pair is selected based on the distance that racks must be located and whether or not the racks must support installation of communications processors (communications bus is required).

Table 3-2. S7-300/S7-400 Interface Modules (IMs) for local and remote I/O expansion.

Interface Module Pair		Brief Description	Distance
CPU IM	Exp. Rack IM		
IM 365	IM 365	S7-300 Local Expansion (1-Tier Max. Expansion)	1 m
IM 360	IM 361	S7-300 Local Expansion (3-Tier Max. Expansion)	10 m
IM 460-0	IM 461-0	S7-400 Local Expansion; No 5V transfer; C-bus.	5 m
IM 460-1	IM 461-1	S7-400 Local Expansion with 5V transfer; No C-bus	1.5 m
IM 460-3	IM 461-3	S7-400 Remote Expansion; No 5V transfer; C-bus	100 m
IM 460-4	IM 461-4	S7-400 Remote Expansion; No 5V trans.; No C-bus	600 m
IM 463-2	IM 314	S7-400 Remote Expansion using S5 Racks & I/O	600 m

Note: No C-Bus implies that only signal modules will be supported in the expansion rack.

S7-300 IMs

The S7-300 may be expanded up to a maximum of three additional racks for a total of four tiers. Expansion to a maximum of one additional tier (total of 2) is accomplished using the IM 365 — two interface modules permanently attached by a fixed cable length of 1 meter. One interface module is placed in the central rack (CR) and the other in the expansion rack (ER). A maximum of eight modules may be placed in the expansion rack. Expansion up to three additional tiers (total of 4) is accomplished using the IM 360/IM 361 combination. A single IM 360 is placed in the central rack (CR) and a single IM 361 in each additional expansion rack (ER). A maximum of three expansion racks may be installed and a maximum of eight modules may be placed in each expansion rack.

S7-400 IMs

The S7-400 supports centralized I/O expansion from 1.5 meters to 3 meters and remote I/O expansion up to 100 meters, using interface modules (IMs). In Table 3-2, both the IM 460-0/IM 461-0 and the IM 460-1/IM 461-1 publisher/receiver interface module pairs support local expansion. The IM 460-3/IM 461-3 pair and the IM 460-4/IM 461-4 pair both support remote expansion. S7-400 remote I/O expansion up to 600 meters may also be configured using existing or newly installed S5 I/O racks that are connected using the IM 463-2. Altogether, a maximum of six interface modules of any combination may be placed in the central rack. The IM 460-3 may be installed in the central rack with up to a total of six other interface modules to form any combination of local and remote I/O expansion.

Power Supply (PS)

The power supply provides the internal operating voltages required by a rack and its complement of installed modules. The short name of S7-300 power supplies is PS 307; the S7-400 has both PS 405 and PS 407. Like with other S7 short names, each of these short names represent several part numbers. The PS 307 power supplies all require an input supply voltage of 120/230 VAC and outputs 24 VDC to the CPU and other installed modules. The unique features of each power supply can be determined when the part number is selected in the hardware catalog. If an S7-400 power supply is selected for its redundant-use capability, it must be installed in a rack that supports this feature. In the hardware catalog these modules for the S7-300/S7-400 are in folders PS-300 and PS-400 respectively.

Table 3-3. S7-300/S7-400 Power Supply (PS) Modules.

Module	Family	Input	Outputs
PS 307	S7-300	120/230 VAC	24 VDC/2 A
PS 307	S7-300	120/230 VAC	24 VDC/5 A
PS 307	S7-300	120/230 VAC	24 VDC/10 A
PS 407	S7-400	120/230 VAC	24 VDC/.5 A; 5 VDC/4 A
PS 407	S7-400	120/230 VAC	24 VDC/1A; 5 VDC/10 A
PS 407	S7-400	120/230 VAC	24 VDC/1 A; 5 VDC/20 A
PS 405	S7-400	24 VDC	24 VDC/.5 A; 5 VDC/4 A
PS 405	S7-400	24 VDC	24 VDC/1 A; 5 VDC/10 A
PS 405	S7-400	24 VDC	24 VDC/1 A; 5 VDC/20 A

Central Processing Unit (CPU)

Central processing unit (CPU) modules are responsible for storing the control program of S7-300/S7-400 controllers in its memory, and thereby managing the control of the associated machine or process. In the hardware catalog these modules for the S7-300/S7-400 are in folders CPU-300 and CPU-400 respectively. S7-300/S7-400 CPUs are further grouped into subfolders based on feature categories as briefly described in the table below.

Table 3-4. S7-300/S7-400 Central Processing Unit Modules (CPUs).

Short Name	Brief Description
CPU 31x-x	S7-300 Standard CPUs
CPU 31x-x IFM	CPUs with integrated I/O Functions (e.g., digital/analog, HS counters)
CPU 31x-x DP	S7-300; CPUs with integrated Profibus DP (DP master/DP slave port)
CPU 31xC	S7-300; Compact CPUs with integrated I/O Functions
CPU 31xF	S7-300; Fault Tolerant CPUs with integrated I/O Functions
CPU 41x-x	S7-400 Standard CPUs
CPU 41x-x DP	S7-400 Profibus DP CPUs; CPUs with integrated Profibus DP master.

Signal Modules (SM)

Signal modules include the various digital and analog input and output circuit cards used to interface standard current and voltage signals to the S7-300/S7-400. As a whole signal modules adapt incoming signals or outgoing signals to the proper levels. Both the S7-300 and S7-400 provide digital input (DI-300/DI-400) modules, digital output modules (DO-300/DO-400), analog input modules (AI-300/AI-400), and analog output modules (AO-300/AO-400). The S7-300 also offers digital and analog modules that contain both inputs and output circuits on the same module. In the hardware catalog these modules are in folders DI/DO-300 and AI/AO-300.

Function Modules (FM)

Function modules are intelligent I/O modules, designed to perform complex or time-critical I/O tasks independent of the CPU. Typical function modules included closed-loop PID control for standard and temperature loops; stepper- and servo positioning control, and high-speed counter functions. Programming of function modules is simplified though the use of supplemental software used to configure the application parameters of the module, and the calling of standard Function Blocks (FBs) from the user program. In the hardware catalog these modules for the S7-300/S7-400 are in folders FM-300 and FM-400 respectively.

Communications Processors (CP)

Generally, two types of communications processors for establishing links between the PLC and other intelligent devices — including other PLCs. The first class of communication processors supports *Local Area Network* (LAN) connections, for communication among multiple S7 PLCs and other devices on the same medium (e.g., *Industrial Ethernet, PROFIBUS,* and *AS-I*). The second class of communications processor is used to form an end-to-end serial link between an S7-300 or S7-400 and one other device. These are called *point-to-point CPs,* linking the S7 PLC with devices such as message displays, operator workstations, PCs, bar code readers, printers, or even another PLC. In the hardware catalog these modules for the S7-300/S7-400 are in folders CP-300 and CP-400 respectively.

Multipoint Programming Interface (MPI)

The multi-point programming interface (MPI) serves two functions: first, it is the primary programming interface to all S7 CPUs, communications processors (CPs), and function modules (FMs). Each S7 CPU, CP, FM and operator panel (OP), has an integrated MPI port which serves in the direct connection for programming these programmable modules. In addition to being a programming interface, the MPI interface may serve as a low cost network, supporting small amounts of data exchange among connected MPI nodes, without the need for network modules. As a network MPI supports up to 32 nodes, uses the same RS-485 cable components, media, and connection methods as Profibus. By placing a programming system (PG/PC) on the MPI subnet, all nodes with the MPI interface may be accessed for programming.

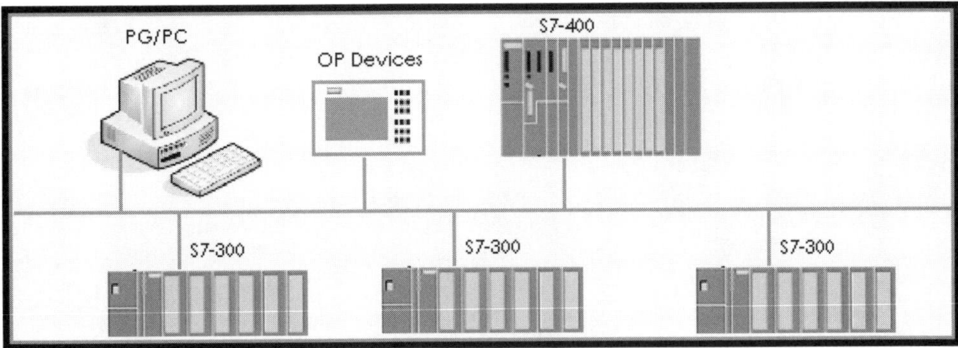

Figure 3-1 Multi-Point Interface programming interface and network.

The Hardware Configuration Tool

In STEP 7, creating a configuration of the hardware is to create a software model of the actual hardware components of the S7 system. The configuration includes the arranging of local and remote racks, along with their interconnecting interface modules (IMs); the installing of modules in each rack, including CPUs, CPs, SMs, FMs; and Profibus distributed I/O drops. Configuration also involves module addressing and the setting of module parameters.

Configuring the Hardware

Creation of the hardware configuration requires the *Hardware Configuration Tool*, which is started from the SIMATIC Manager. After a station is created, the configuration tool is launched by opening the Station folder and then double-clicking on the hardware object. With the hardware configuration tool, an object model of your actual hardware arrangement is developed, with each component of your hardware installation having a matching object in the hardware configuration. Once developed, a configuration may be copied to other STEP 7 projects, modifying it as required.

The completed configuration is checked for errors, compiled and then saved. The configuration is saved to the *System Data* object, which is placed in the offline *Blocks* folder. The System Data object may then be downloaded to the CPU, thereby providing the CPU with complete information of the hardware configuration. The CPU in turn transfers configuration parameters to the installed modules. Once created and downloaded, the hardware configuration is a powerful diagnostic tool. The status of all configured components may then be viewed online, showing the correct or malfunctioning operation.

Menus and Toolbar

Menu headings of the Hardware Configuration tool include *Station*, *Edit*, *Insert*, *PLC*, *View*, *Options*, *Window*, and *Help*. Station operations allow you to create, open, save, compile and check the configuration for errors. Standard Cut, Copy, and Paste operations from the Edit menu or from the right click allow racks, modules and other station objects to be edited as required. Standard online operations such as configuration upload and download, monitoring and diagnostic tools are supported by PLC operations. View and Window operations allow components of the configuration window to be displayed, hidden, or arranged to your convenience. The toolbar buttons, listed below, represent some of the most frequently used menu operations, and buttons that launch other STEP 7 software utilities.

Table 3-5. Hardware Configuration Toolbar Buttons.

Icon	Toolbar Function	Icon	Toolbar Function
	Create New Station in Project		Paste Object
	Open Station Offline Window		Download Station Configuration
	Open Station Online Window		Upload Station Configuration
	Save Configuration		View Address Overview of Station
	Save and Compile Configuration		View Hardware Catalog
	Print Station Configuration		Open Network Configuration Tool
	Copy Selected Object		STEP 7 Help

Hardware Catalog Window

The hardware catalog window contains the various component objects used to create a configuration of your automation solution. When the configuration tool is opened, the catalog, which can be hidden from view, can be displayed by selecting **View ➢ Catalog** from the menu. You Dock and undock the catalog by double-clicking above the word "Profile" if the window is docked or in the title bar if un-docked. When undocked, you can be resize the window and move it around to suit your convenience. You may dock the catalog on either the left side or the right side of the configuration window by dragging and dropping it on the left or right edge of the window.

The major component containers of the catalog, *Profibus DP, Profibus PA, SIMATIC 300, SIMATIC 400, SIMATIC PC Based Control,* and *SIMATIC PC Station,* are presented in a tree. Profibus-DP and Profibus-PA containers include part for DP or PA slave devices. The SIMATIC 300 and SIMATIC 400 containers each contain folders of the various part numbers for racks and modules used to build an S7-300 or S7-400 configuration. The subfolders of the S7-300 and S7-400 have similar names and include individual folders that contain communications processors (CPs), central processing units (CPUs), function modules (FM), interface modules (IMs), power supplies (PS), module racks (Rack), and signal modules (SMs).

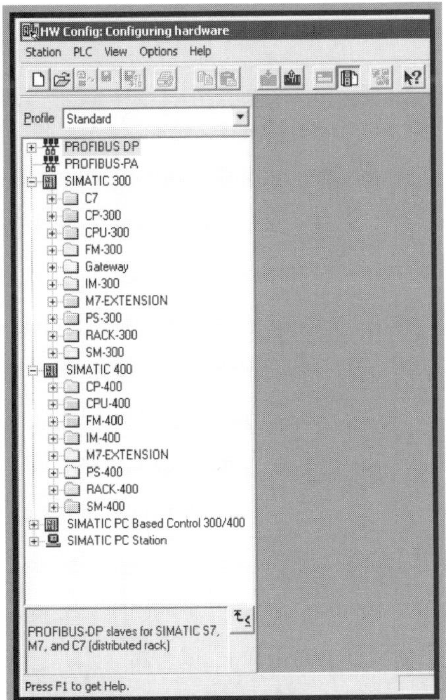

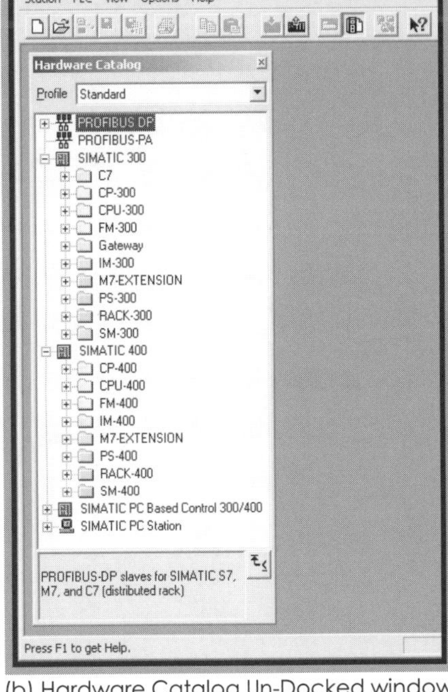

Figure 3-2 (a) Hardware Catalog Left-Docking. (b) Hardware Catalog Un-Docked window.

Station Window - Rack Arrangement Pane

As you develop your configuration, the upper pane of the station window will contain the central and expansion racks of a station as well as any Profibus-DP distributed drops. During station configuration, racks (including DP/PA slave drops) are selected from the catalog and dragged and dropped into the station window. Each rack is represented as a 2-column table showing the empty slots where modules will be installed. Profibus-DP distributed drops are shown as symbol objects, based on the type of drop (e.g., modular or integrated).

As the configuration is developed and completed, the arrangement pane will show the interconnection between interface modules in the central rack and in each expansion rack; P drops will be shown connected to a DP master. As the configuration progresses, each new station is opened in a separate station window. Multiple stations may be opened simultaneously and components may be copied and pasted between stations.

Station Window - Configuration Tables Pane

When a central rack, an expansion rack, or a DP station is selected in the upper pane of the station window, a detailed view of its contents is displayed in a configuration table in the lower pane of the station window. Each slot and the module it contains are listed in a row of the table. The columns include information such as *slot*, *module*, *order number*, *address*, and *comment*. STEP 7 automatically assigns module addresses, which may be modified if the station can be addressed freely.

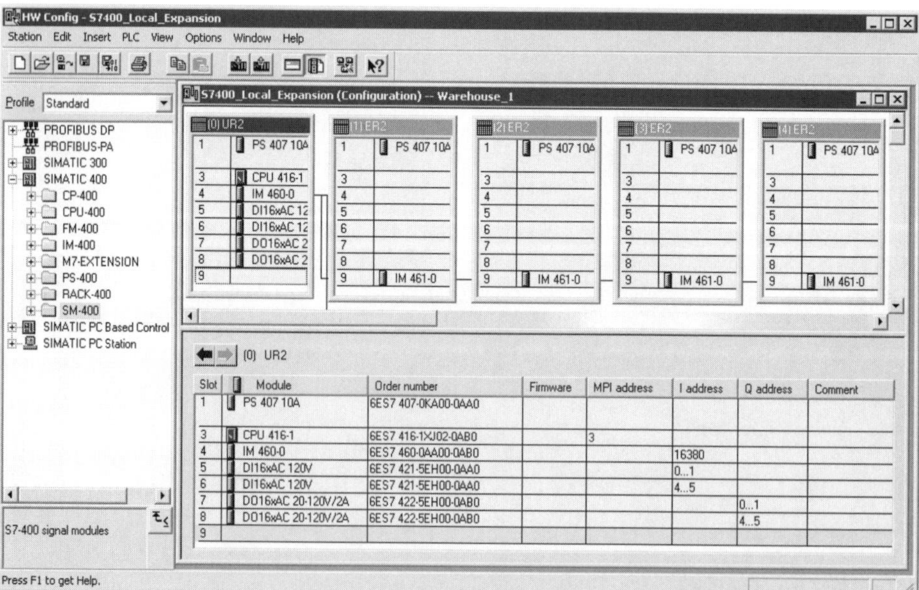

Figure 3-3. Hardware Configuration tool with docked catalog window; project window with rack arrangement in top pane and configuration table of selected rack in bottom pane.

Comments on Working with Hardware Configurations

A well-planned and documented hardware configuration will help to ensure a smooth installation and start-up of your hardware. It will also enhance the maintenance task later. Tasks associated with configuring your hardware are truly straightforward and simple. If done with a keen eye for detail, the Hardware Configuration will pay huge dividends later in terms of start-up and its ability to provide fast pinpointing of process or system-related faults.

Regardless of your actual station hardware or arrangement – you can accomplish the configuration task using the following checklist. Several configuration examples are presented in a step-by-step manner, in the remainder of the chapter. As you use this chapter keep in mind that, while there are examples of practically every type of configuration task, you only have to be concerned with your configuration. Once the configuration is completed or at least completed to the point needed to establish online communications with a CPU (e.g., central rack configuration), you'll be able to download your configuration and then proceed to the next step of Working with the STEP 7 Program.

Checklist: Working with Hardware Configurations

- *Add stations as required from the SIMATIC Manager (e.g., Insert ➢ Station ➢ SIMATIC 300) to match the actual installation.*
- *Open the Hardware Configuration Tool and do the following task.*
- *Create Central Rack Configuration to match actual installation.*
- *Configure local I/O racks to match the actual installation.*
- *Configure remote I/O racks to match the actual installation.*
- *Configure Profibus DP slave drops to match actual installation.*
- *Insert and arrange modules to match actual installation (e.g., PS, CPU, IM, CP, SM, and FM modules).*
- *Modify default I/O address assignments if required.*
- *Modify default module parameters if required (i.e., CPU, CP, SM, and FM).*
- *Use Symbols editor from the Hardware Configuration tool for fast assignment of symbolic addresses to Input/Output addresses.*
- *Perform consistency check for errors in hardware configuration.*
- *Save and compile the hardware configuration of each PLC.*
- *Download the hardware configuration to each PLC.*

Navigating the Hardware Catalog

Basic Concept

The hardware catalog contains the component objects used to configure your S7-300/S7-400 systems. As you initially build your configuration to match the hardware arrangement, or as you edit the configuration to reflect added components, you will select these parts from the hardware catalog. Each catalog object is defined by a unique part number that is the same for the actual part. As you learn to navigate the catalog you will become increasingly familiar with the application, capability, and nomenclature of S7-300/S7-400 components.

Essential Elements

Like project components, displayed in the SIMATIC Manager, components of the hardware catalog are presented in a tree structure comprised of object containers. The major containers include **Profibus DP**, **Profibus PA**, **SIMATIC 300**, **SIMATIC 400**, **SIMATIC PC Based Control**, and **SIMATIC PC Station**. When a container is opened, its parts categories are displayed as folders. The SIMATIC 300 container, for example, contains **CP-300**, **CPU-300**, **FM-300**, **IM-300**, **PS-300**, **Rack-300**, and **SM-300** folders. Having basically the same categories, the SIMATIC-400 folders will have similar names ending in 400 (e.g., **CPU-400**). As you navigate toward a specific part, you may actually traverse two or more levels of folders that further categorize the parts. Whenever a catalog folder or part is selected, a brief description appears in the bottom pane of the catalog window.

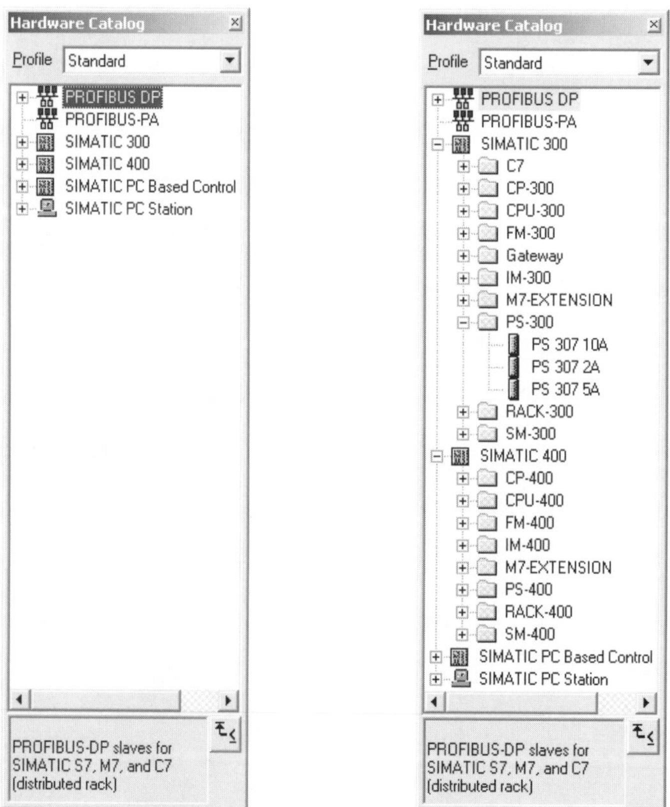

Figure 3-4. Hardware Catalog, first displaying main object containers, then displaying opened SIMATIC 300 and SIMATIC 400 object containers.

Quick Steps: Navigating the Hardware Catalog

STEP	ACTION
1	With the Hardware Configuration tool open, selecting **View** ➢ **Catalog** displays the catalog if it is not already onscreen.
2	The catalog may be docked or undocked and positioned for your working convenience; Double-click on the title bar of the catalog window for un-docking.
3	To dock catalog on the left side, drag and drop onto the main window's left edge.
4	To dock catalog on the right, drag and drop onto the main window's right edge.
5	From the hardware catalog, click on the plus sign (+) of the **SIMATIC 300** or the **SIMATIC 400** object to access its component folders and parts.
6	Some folders, for example the **CP-300** and **CP-400,** will have more than one level of subfolders to further categorize the components. CP folders include subfolders for *Profibus*, *Industrial Ethernet*, and *Point-to-Point* communications processors.
7	Selecting any folder or subfolder displays a brief description of its contents in the bottom pane of the catalog window.
8	Select a specific component in order to view its part number and a brief description of its features in the bottom pane of the catalog window.
9	Open the **Rack-300** or **Rack-400** folder, to view and select module racks. The S7-300 has a single rack type. S7-400 rack types include: CR, for central racks, ER, for expansion racks; and UR, for universal racks (usable as central or expansion rack).
10	View and select power supply modules in the **PS-300** or **PS-400** folder. The S7-400 has subfolders for *Standard* and *Redundant* power supplies; its DC power supplies are prefixed with PS-405 and AC power supplies are prefixed with PS-407.
11	Open the **CPU-300** or **CPU-400** folder and subfolders, to view and select CPUs. Subfolders contain standard and Profibus DP (**DP**) CPU modules; the S7-300 also contains point-to-point (**PtP**) and integrated function (**IFM**) CPUs.
12	View and select interface modules for local and remote rack expansion for the S7-300 in the **IM-300** folder or for the S7-400 in the **IM-400** folder.
13	Open the **FM-300** or **FM-400** folder, to view and select function modules. Subfolders contain modules for Cam, PID, High Speed Counter, and Positioning control.
14	View and select signal modules in the **SM-300** or **SM-400** folder. Both folders include subfolders for digital input (**DI**), digital output (**DO**), analog input (**AI**), and analog output (**AO**). The S7-300 also contains folders for digital input/output (**DI/DO**) and analog input/output (**AI/AO**) modules.

Navigating the Hardware Configuration Menu and Toolbar

Basic Concept

The Hardware Configuration tool is an object-based tool that supports easy configuring of hardware stations. During use, one or more stations may be opened simultaneously for viewing or for editing. In addition to facilitating offline station configuration, established configurations may also be opened online for monitoring and diagnostic purposes. Information and data on programmable modules including CPUs, CPs, and FMs may also be viewed and modified while the station is opened online.

Essential Elements

The complete operations of the Hardware Configuration tool are accessed from its main menus including *Station*, *Edit*, *Insert*, *PLC*, *View*, *Options*, *Window*, and *Help*. Toolbar buttons allow quick access to more commonly used commands. Edit operations allow you to copy and paste racks and modules, and to edit other object properties. The PLC menu supports online functions with the CPU and other programmable modules. View and Window menus allow manipulation of the main and station windows and provide access to each station's I/O address usage. Finally, the Options menu supports customizing of the configuration tool and access to optional tools (e.g., Symbols Editor, and Network Configuration).

Application Tips

Display multiple windows horizontally or vertically, to simplify use of drag and drop or the use of Copy and Paste. For example to copy a rack select the rack and paste in the same station window or drag and drop in another station window. Modules (e.g., CPU, CP, FM, SM, and PS) may also be copied and pasted to the same rack or to another rack (i.e., table).

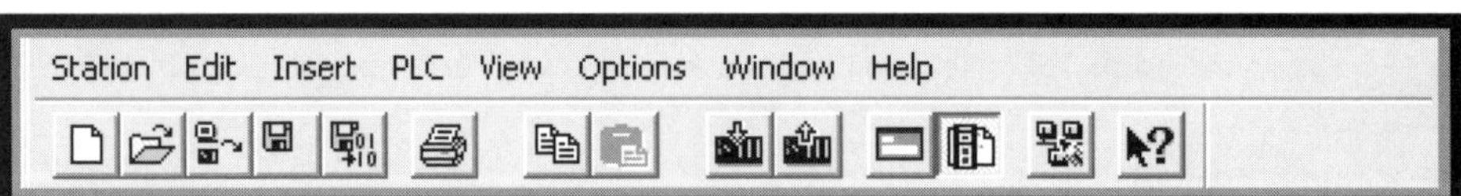

Figure 3-5. Hardware Configuration Menu and Toolbar.

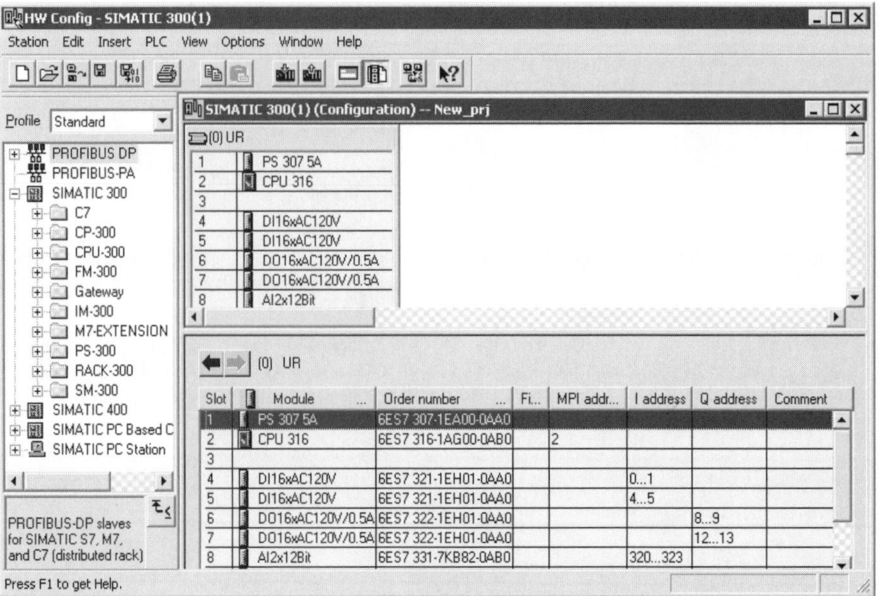

Figure 3-6. Hardware Configuration Tool: Catalog and Station Configuration windows.

Navigating the Hardware Configuration Menu and Toolbar

STEP	ACTION
	The **Station** menu or toolbar options allow you to work with existing Stations online or offline; to create new stations and to save and compile completed configurations.
1	Selecting **Station** ➢ **Open** allows you to open a station configuration from the current project, or browse and open a station configuration from another project.
2	**Station** ➢ **New**, allows you to create a station in the current or in another project.
3	Press the **Online** ⋞⋟ **Offline** button to switch the station between online and offline.
	The **Edit** menu and toolbar options allow copying and pasting of racks and modules, as well as viewing and editing of a selected object's properties.
4	To copy one or more modules select the modules, then select **Copy**; select a slot in the same rack or in another rack, then select **Paste** from the menu or the right-click.
5	To copy a whole rack you would click on its title bar and select **Copy**; click anywhere inside the configuration pane of the same station window or inside another station window, then select **Paste** using the menu, or the right-click.
6	Select any object then right click and **Object Properties** to view its properties.
	The **PLC** menu and toolbar allow configurations to be uploaded or downloaded.
7	Once a station is configured and compiled without errors, selecting **PLC** ➢ **Download** allows the station to be downloaded over an established connection.
8	Select **PLC** ➢ **Upload** to upload the PLC configuration to the programming system.
	The **View** and **Window** menus allow viewing of a station's assigned addresses, and the arranging of station windows and racks to suit your working convenience.
9	If multiple station windows are open, they may be arranged accordingly by selecting **Window** ➢ **Arrange** (*Cascade*, *Vertically*, or *Horizontally*), from the menu.
10	Selecting **View**, allows you to show or hide the hardware configuration tool's Catalog, *Toolbar*, or the *Status Bar*.
	Options allow Hardware Configuration tool preferences to be customized and optional tools like the Symbol Editor and Network Configuration to be launched.
11	Selecting **Options** ➢ **Customize**, for example, opens the dialogue for setting various display and operating parameters of the Hardware Configuration tool.
12	**Options** ➢ **Configure Network** opens the tool for configuring networks; **Options** ➢ **Symbol Table** opens the symbolic address editor.

Viewing Station Configuration Details

Basic Concept

A station's configuration involves arrangement of central and expansion racks and modules, rack and module addressing, and the assigning of parameters to each module where the default parameters will not be used. The completed configuration should reflect the actual arrangement of racks and modules of the real hardware. The details of a project's hardware configuration can be viewed station-by-station, rack-by-rack, module-by-module.

Essential Elements

The Hardware Configuration tool displays each open station in a separate window. Each central rack, local and remote expansion rack, or distributed I/O drop (DP slaves) of a station is depicted in the top pane of the configuration window. Each rack is shown as a as a two-column table, whose title bar is labeled with the rack-type and number. Each DP slave is shown as an object labeled with its slave type, its assigned Profibus address, and that is either attached or not attached to a DP Master. When a rack or DP slave is selected in the top pane, its details are shown as a table in the bottom pane of the station window. The rows of each table lists the modules contained in the rack, according to the actual slot.

Application Tips

The default or assigned properties of any object may be viewed by selecting that object, then right-clicking and selecting **Object Properties**. From the Hardware Configuration, an overview of addresses assigned to a CPU is obtained by selecting **View ➢ Address Overview**.

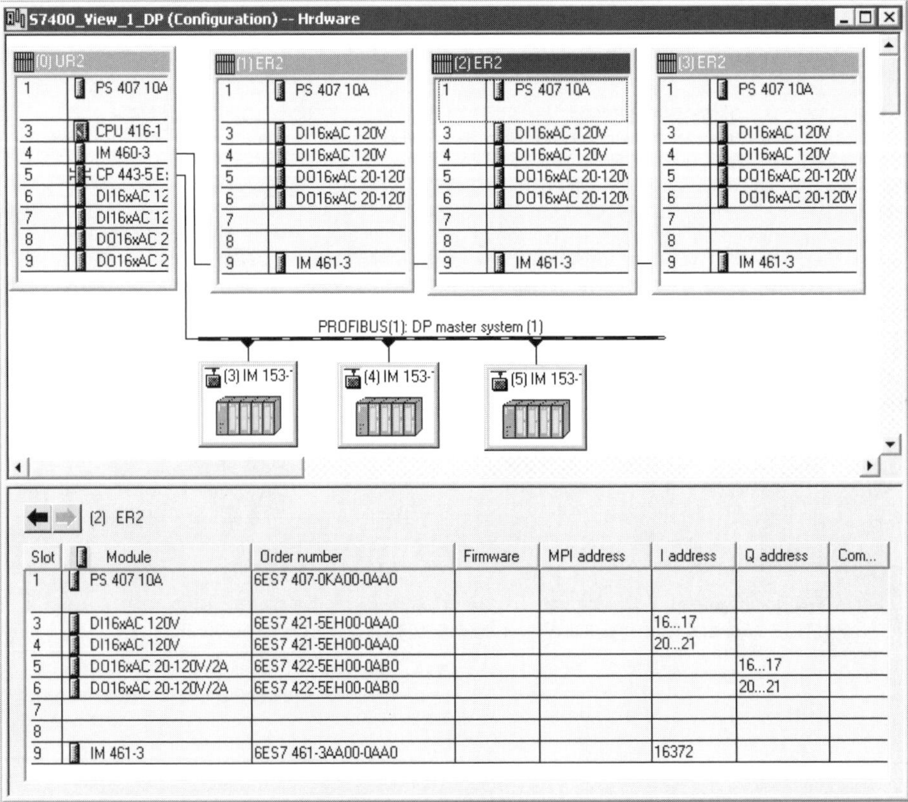

Slot	Module	Order number	Firmware	MPI address	I address	Q address	Com...
1	PS 407 10A	6ES7 407-0KA00-0AA0					
3	DI16xAC 120V	6ES7 421-5EH00-0AA0			16...17		
4	DI16xAC 120V	6ES7 421-5EH00-0AA0			20...21		
5	DO16xAC 20-120V/2A	6ES7 422-5EH00-0AB0				16...17	
6	DO16xAC 20-120V/2A	6ES7 422-5EH00-0AB0				20...21	
7							
8							
9	IM 461-3	6ES7 461-3AA00-0AA0			16372		

Figure 3-7. S7 Configuration; rack-2 selected in top pane, rack-2 modules in bottom pane.

Quick Steps: Viewing Station Configuration Details

STEP	ACTION
1	With the desired project open, double click on the *Hardware* object of the station you wish to view.
2	With the station window in view, the central rack is usually placed first in the top pane. The rack number, always Rack-0, is enclosed in parentheses and followed by the rack type. The address is (0) UR for a universal rack, (0) CR for a central rack type.
3	If expansion racks are configured, they too are displayed as tables in the top pane. Each rack is assigned according to the rack type and the next available number. The designation UR is for the universal rack type and ER for the expansion rack type.
4	If expansion racks have been connected to an interface module (IM) a line is shown linking each rack. In the S7-400, the IM channel connections may be viewed by selecting the IM in the central rack, right-clicking, selecting **Object Properties**, then selecting the *Connection* tab. The racks in this configuration are all connected on channel-1 (C1). Selecting **Cancel** or **OK** closes the view.
5	If Profibus-DP slaves are configured, each slave object is shown as being either attached to a DP Master system or not attached. When attached, the assigned Profibus Address is shown for each DP slave.
6	When any rack, is selected in the top pane, its details are shown in the table in the bottom pane. Module details include *Slot*, *Module* description, *Order Number*, *Firmware version*, *MPI address* if applicable, *Input (I) address*, and *Output (Q) address*.
7	Details of each selected slave may also be viewed by selecting the DP-object. If the drop, for instance, is a modular drop, its module details are displayed in the bottom pane as with a normal rack.
8	You can view the assigned or default properties of any configured object (e.g., Rack, Module, or DP master system) by selecting the object, a right-click, then selecting **Object Properties**.
9	View assigned addresses of a CPU by selecting **View** ➢ **Address Overview**, from the menu of the Hardware Configuration tool.
10	Other unopened stations of the same project may be opened for viewing by selecting **Station** ➢ **Open** from the menu, then selecting the station from the dialog.
11	If multiple station windows are open, they may be arranged accordingly by selecting **Window** ➢ **Arrange** (*Cascade*, *Vertically*, or *Horizontally*), from the menu.

Building a Station Configuration

Basic Concept

A station configuration is created for each S7-300/S7-400 station in a project. The configuration data, which is eventually developed and downloaded to the PLC, will contain an exact image of each rack, its arrangement of modules, as well as the parameters set for each module. Once loaded to the CPU, the configuration is available for diagnostic purposes. On each start-up in the S7-400 and in some S7-300s (e.g., CPU 318-2), the configuration can be compared to the actual arrangement for any discrepancies.

Essential Elements

Building the configuration for each PLC involves (1) selecting each component from the Catalog, and (2) dragging and dropping the component into the Station window. Typically a configuration involves a central rack and perhaps some arrangement of expansion racks, each of which is configured with some combination of modules. Modules may include a power supply (PS), digital/analog signal modules (SM), communications processors (CP), and function modules (FM), all of which must reflect the actual physical arrangement. Each module is dragged from the catalog to its matching slot number in the configuration table.

Application Tips

When a hardware configuration is completed, and saved using the save and compile command, the compiled configuration is saved in system data blocks (SDBs). The configuration data, eventually downloaded to the PLC, is represented by the *System Data* object located in the offline blocks folder of the S7 program. After the download, an online copy of the *System Data* object is also contained in the online blocks folder.

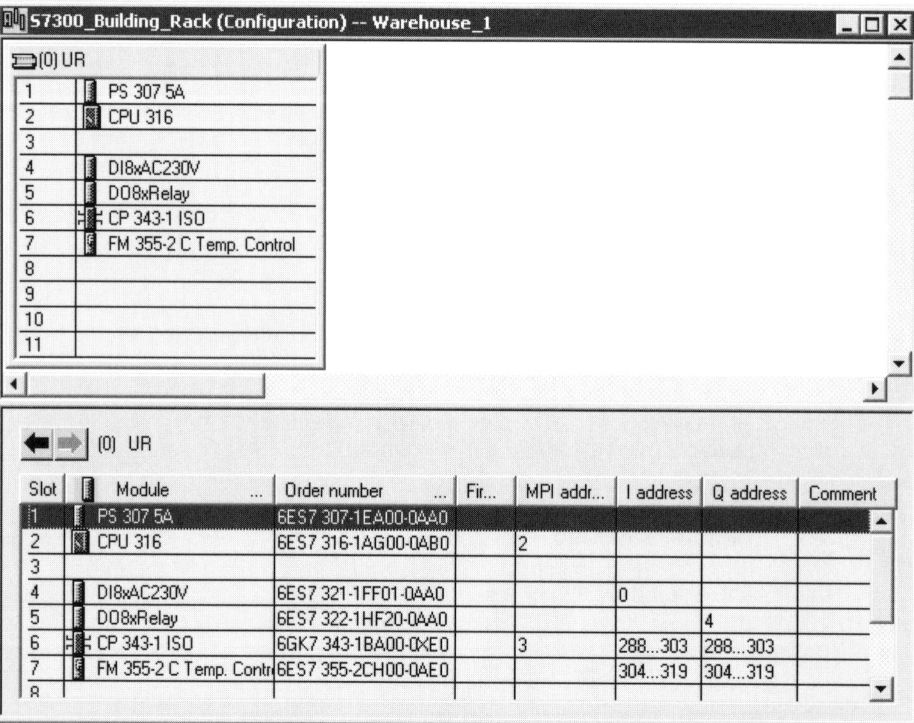

Figure 3-8. A rack selected in top pane, shows modules displayed in the bottom pane.

Quick Steps: Building a Station Configuration

STEP	ACTION
1	From the SIMATIC Manager, open the required project and either select an existing station or create a new station for which a hardware configuration is to be created.
2	With the station selected in the left pane of the project window, double-click on the station's *Hardware* object in the right pane, to open the configuration tool.
3	Maximize the Hardware Configuration window, and if the Hardware Catalog is not already in view select **View ➤ Catalog**, from the menu to open the catalog.
4	From the catalog, click on the plus sign (+) to open the **SIMATIC 300** or **SIMATIC 400** object to access the individual component folders for the required PLC.
5	Open the **Rack-300** or **Rack-400** folder and drag a central rack to the upper configuration pane (the S7-400 has CR and UR rack options; S7-300 uses the Rail).
6	Open the **PS-300** or **PS-400** folder, select the desired power supply and drag the part to slot-1 of the central rack.
7	Open the **CPU-300** or **CPU-400** folder, then the appropriate CPU subfolder to select the desired CPU. Drag the selected CPU to an acceptable slot in the central rack.
8	In a similar fashion, signal modules (**SM-300/SM-400**), communications processors (**CP-300/CP-400**), and function Modules (**FM-300/FM-400**) that match the actual part numbers of your S7-300/S7-400 are dragged and dropped into the configuration table.
9	After completing a station, from the menu, select **Station ➤ Consistency Check** to check for errors. The configuration will not compile for downloading unless found free or errors.
10	From the menu bar, select **Station ➤ Save** to save the configuration; or if done, use **Save and Compile** to generate the System Data for downloading to the CPU.

Downloading a Station Configuration

Basic Concept

After a station's configuration has been generated, found free of errors, and compiled — then and only then may it be download to the PLC. The configuration will reflect the station's local, remote and distributed I/O rack arrangement, as well as the parameters of all configurable modules (e.g., CPU, CPs, FMs, SMs, etc.). Configuration parameters are subsequently transferred to the appropriate modules during the CPU startup.

Essential Elements

For a station configuration to compile for download, certain perquisites must be met. Authorizations must be installed for any optional packages used; network modules in a subnet must have unique addresses; the station hardware and the network configuration must match the actual configuration, and be found consistent and free of errors. Once found free of errors, a System Data object (SDB), containing the configuration, is generated and placed in the *Offline Blocks* folder of the associated station (i.e., CPU program folder).

Application Tips

Whether downloading a configuration for the first time or otherwise, the complete PLC configuration is always downloaded to the CPU. A partial configuration download is not possible. If the actual station configuration is not yet completed or will be completed in stages then you should consider creating a temporary project, whose hardware configuration is developed in stages, reflecting only the portion of the hardware that is physically installed to date (e.g., central controller rack and modules only).

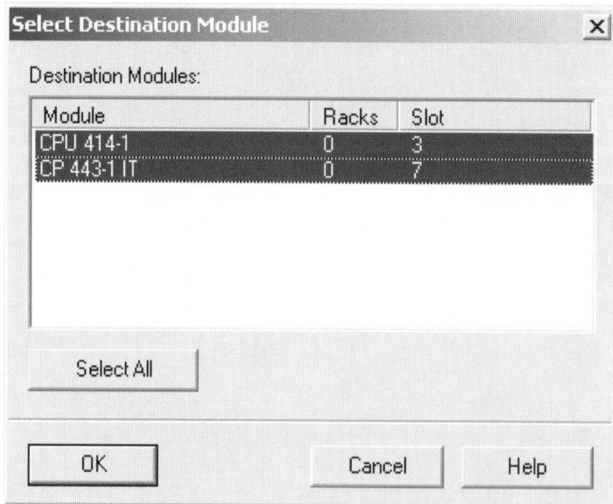

Figure 3-9. Module Download list with all modules targeted to receive configuration data by default.

Quick Steps: Downloading a Station Configuration

STEP	ACTION
	Download Configuration: From SIMATIC Manager:
1	First, ensure that an online connection is established from the PG/PC to the station to which you wish to download. A direct MPI connection is recommended.
2	If downloading a configuration for the first time, an overall reset of the CPU memory is recommended before downloading. The CPU should be in the STOP mode.
3	With the required project open and the appropriate S7 Program folder expanded, select the offline Blocks folder; then from the right-pane select the *System Data* object. The *System Data* is available only if the station configuration was compiled.
4	With the *System Data* object selected, from the menu, select **PLC** ➢ **Download**, or press the **Download** icon from the toolbar.
5	When prompted to select if the System Data in the CPU should be deleted, confirm with **Yes** to overwrite any existing configuration.
6	Check the CPU LED indicators; if the downloaded and actual configuration part numbers match, then no fault LEDs should be illuminated.
	Download Configuration: From Hardware Configuration Tool:
1	First, ensure that an online connection is established from the PG/PC to the station to which you wish to download. A direct MPI connection is recommended.
2	If downloading a configuration for the first time, an overall reset of the CPU memory is recommended before downloading. The CPU should be in the STOP mode.
3	After completing the configuration for a station, from the menu, select **Station** ➢ **Consistency Check** to check for errors.
4	When the consistency check reports back with no errors, from the menu, select **Station** ➢ **Save and Compile** to generate system data blocks (SDBs).
5	With the appropriate station open, from the menu, select **PLC** ➢ **Download**, or press the **Download Module** icon from the toolbar. A dialog is presented, listing modules for which download configuration data exists.
6	Press the **Select ALL** button to ensure that configuration data is downloaded to all modules for which configuration data exists.
7	Check the CPU LED indicators; if the downloaded and actual configuration part numbers match, then no fault LEDs should be illuminated.

Uploading a Station Configuration

Basic Concept

Each station configuration is generally built, from start to finish, using the Hardware Configuration tool. In some cases, however, you may wish to start the configuration by uploading a configuration from an existing PLC. Such a case may exist when the hardware is available and perhaps even installed before any configuration is started; or a new station you wish to create is similar to or identical to one that already exists. An uploaded station is only meant to serve as a quick start for building a new station configuration.

Essential Elements

A station may be uploaded to an existing project, however an empty project is recommended. When performing the upload a new station object, based on the uploaded station, is automatically generated in the open project — thereby eliminating the possibility of overwriting or corrupting any offline data. In the case of the S7-300, the central and all expansion rack configuration data is uploaded; in the case of the S7-400, the central rack and module configuration data is uploaded, but no expansion racks are included. No configuration data is uploaded for distributed I/O drops. Although the uploaded configuration uses default addressing for both the S7-300 and S7-400, the addressing may be modified in the S7-400 and in some S7-300 stations, depending on the CPU.

Application Tips

Stations that rely on a dependent relationship to one or more other stations (e.g., DP Master/Intelligent slave) should always be uploaded together. If dependent stations are not all uploaded, the station that you are interested in will remain inconsistent when you attempt to save and compile. If component part numbers appear incomplete after a station upload, it may be that the specific parts are not recognized by STEP 7 and that the components do not appear in the Catalog. You may enter the "incomplete" order numbers when you configure the hardware using the menu command **Options > Specify Module**.

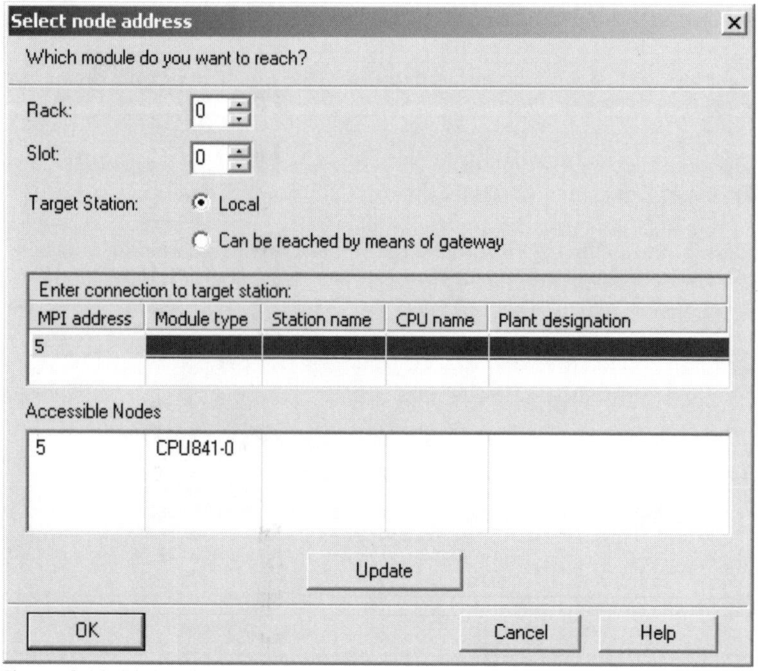

Select node address

Which module do you want to reach?

Rack: 0

Slot: 0

Target Station: Local

Can be reached by means of gateway

Enter connection to target station:

MPI address	Module type	Station name	CPU name	Plant designation
5				

Accessible Nodes

5	CPU841-0			

Update

OK Cancel Help

Figure 3-10. Dialog for accessing a station targeted for upload.

Quick Steps: Uploading a Station Configuration

STEP	ACTION
1	From the SIMATIC Manager, open the project to which the station configuration will be uploaded (a newly created or empty project is recommended).
2	Ensure that there is an established online connection from the PG/PC to the station you wish to upload (e.g., via a direct MPI connection, via Profibus, or via Ethernet).
3	From the SIMATIC Manager menu, select **PLC** ➢ **Upload Station**. A dialog box that lists accessible stations on the local subnet and that can be reached via a gateway is presented.
4	In the dialog box, enter the rack and slot number of the module via which the configuration should be read (generally the CPU). Define whether the station is on the local subnet or can be reached via a gateway; if the station can be reached by way of several modules, select the connection method. Confirm with "**OK**."
5	When the upload is complete the station configuration data is contained in a new station which has been given a default name. Click on the new station, right-click, and select **Object Properties** to assign a new station name and to add descriptive comments.
6	Click on the new station in the project tree, and from the right-pane double-click on the *Hardware* object to open the station in the hardware configuration tool.
7	Select the CPU module then right-click and select **Object Properties**; from the *General tab*, press the **Properties** button to assign a new MPI address.
8	Modify other MPI addresses, for modules in the station, if required (e.g., CPs and FMs), and network station addresses as required (e.g., Profibus, Industrial Ethernet).
9	Remember, since certain components of the configuration may not have been uploaded (e.g. expansion racks and modules,) you may have to complete the configuration of the new station, using the hardware configuration tool.
10	Where it appears that a component part number is incomplete, double-click on the component or select **Options** ➢ **Specify Module** from the menu — a dialog will appear from which you may select the actual order number; end the dialog with **OK** and a parameter dialog box is presented for entering module parameters.
11	When done, from the menu, select **Station** ➢ **Consistency Check** to check for errors.
12	If there are no errors, from the menu, select **Save and Compile** to save the configuration and to generate the new offline System Data object.

Assigning Symbolic Addresses to Input/Output Modules

Basic Concept

Symbolic addresses are alphanumeric names given to the various S7 memory locations. By assigning names to addresses, that are similar to the contents of the memory location, symbol addresses aid in the understanding and troubleshooting of the STEP 7 program. Symbolic addresses are generally entered directly into the *Symbol Table,* which is opened from the SIMATIC Manager. You may choose to enter the symbolic addresses for I/O modules, however, can be entered quickly from the Hardware Configuration tool.

Essential Elements

To create a symbol address is to assign a meaningful label of up to 24-characters, as a substitute address for an absolute address. As each digital input or digital output module and analog input or output module is inserted, a symbolic address may be assigned to each individual I/O address. The absolute memory area for digital inputs is identified with "**I**", with "**Q**" for digital outputs, with "**PIW**" for analog inputs, and with "**PQW**" for analog outputs. As the dialog is opened for each module, the individual input or output addresses for the module are listed in the **Address** column. You enter a symbolic address in the **Symbol** column and its description is entered in the **Comment** column.

Application Tips

You may also enter symbolic addresses for I/O modules from the Symbols editor, which you may start from the SIMATIC Manager, by double-clicking on the Symbol Table object. A symbol editor icon is also available on the toolbar in the LAD/FBD/STL editor, and in the Monitor/Modify tool. Symbolic addresses entered in the Symbols editor are global symbols; whenever you reference a global symbol, a pound sign must precede it (e.g., #TEMP_1).

Edit Symbols - DI16xAC120/230V

Address	Symbol	Data Type	Comment
I 0.0	START_PB_00	BOOL	
I 0.1	START_PB_01	BOOL	
I 0.2	START_PB_02	BOOL	
I 0.3	START_PB_03	BOOL	
I 0.4	START_PB_04	BOOL	
I 0.5	START_PB_05	BOOL	
I 0.6	START_PB_06	BOOL	
I 0.7	I0.7	BOOL	
I 1.0	STOP_PB_00	BOOL	
I 1.1	STOP_PB_01	BOOL	
I 1.2	I1.2	BOOL	
I 1.3	STOP_PB_03	BOOL	
I 1.4	STOP_PB_04	BOOL	
I 1.5	I1.5	BOOL	
I 1.6	STOP_PB_06	BOOL	

Add Symbols | Delete Symbol | Sort: Address Ascending

The symbol table is updated with 'OK' or 'Apply'

OK | Apply | Cancel | Help

Figure 3-11. Entering Symbolic Addresses in the Hardware Configuration Tool - Digital I/O.

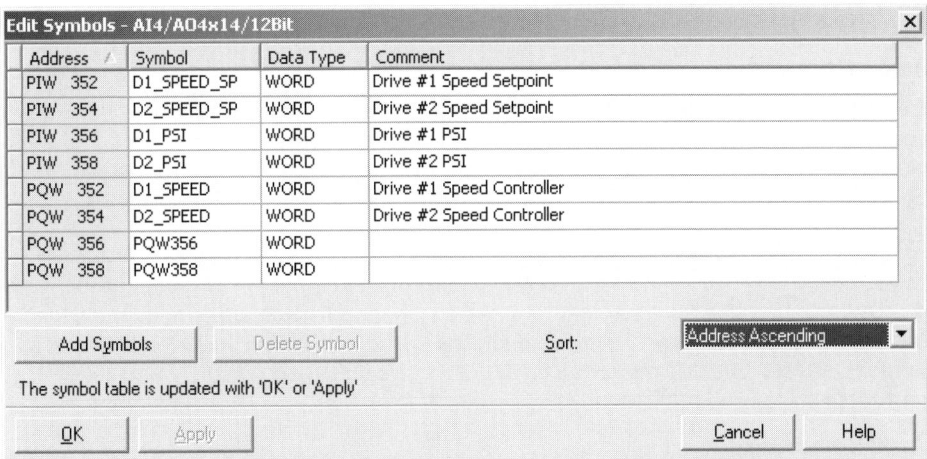

Figure 3-12. Entering Symbolic Addresses in the Hardware Configuration Tool - Analog I/O.

Quick Steps: Assigning Symbolic Addresses to Input/Output Modules

STEP	ACTION
1	Open the desired station in the Hardware Configuration tool, and then select the rack containing the modules to which symbolic addresses should be assigned.
2	To enter symbol addresses for a module, select the module, right-click and select **Edit Symbolic Addresses**. The symbol table is presented like a spreadsheet, having a column for the absolute **Address**, **Symbol** address, **Data Type** of the address, and a **Comment**. The absolute addresses assigned to the open module are already listed in the absolute Address column.
3	Click in the **Symbol** field and enter up to 24-alphanumeric characters for the symbolic address of an input address; enter a comment using up to a maximum of 80-characters. The data type of each absolute address is automatically inserted.
4	Press the **Add Symbols** button to use the absolute address in the **Symbol** field for any input or output address that you have not yet defined with a symbol address.
5	Click on the **Sort** drop arrow and select a method of sorting the symbol addresses (e.g., *Address Ascending, Address Descending,* or *Symbol Ascending*).
6	Press the **Apply** button to accept entries and leave dialog open to resume with entering symbol addresses; when done, confirm your entries with the **OK** button.

Configuring an S7-300 Central Rack

Basic Concept

The central rack is the main rack of an S7-300 station. This rack, sometimes called the controller rack, will always contain a CPU and power supply, and perhaps some arrangement of other modules. You will probably get started with this rack in order to connect with the CPU and start with your initial programming and configuration work.

Essential Elements

As you build the S7-300 central rack configuration, remember that the CPU and power supply modules you use in the configuration, must match the actual hardware arrangement. The S7-300 only has one rack type — it is called a rail. The rail is considered a universal rack since it supports all types of S7-300 modules [i.e., CPU, signal modules (SM), function modules (FM), and communications processors (CP)].

Application Tips

If you need to connect to the CPU to test portions of your code, the minimum configuration you will need to get started is shown in the figure below. You will also need to install backup batteries to the CPU and connect a source to the power supply. As you build your hardware configuration to match your actual hardware arrangement, remember that you must install modules for the S7-300 without leaving any empty slots. The exception is with slot-3, which is used only by interface modules (IMs). Slot-3 is left empty even if only one rack is installed.

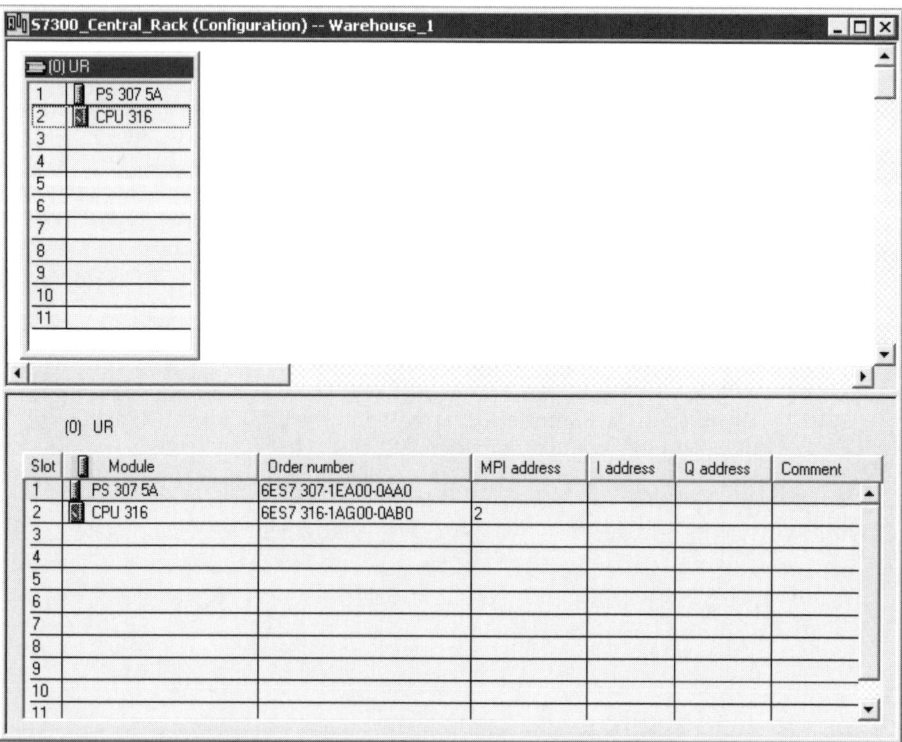

Figure 3-13. S7-300 Central Rack Configuration, with CPU and Power Supply.

Quick Steps: Configuring an S7-300 Central Rack

STEP	ACTION
1	From the SIMATIC Manager open the required project, and from the project tree select the S7-300 Station for which a central rack configuration is to be created.
2	From the right pane, double click on the Hardware object to open the station in the Hardware configuration tool.
3	Open the Hardware Catalog window, from the View menu, if it is not already opened.
4	Open the **SIMATIC 300** catalog object to view the S7-300 component folders.
5	Open the **Rack-300** folder, select and drag the rail to the top pane of the station window to install the S7-300 I/O rail. The central rack is labeled 0(UR), since it is rack-0 and the rail is always a universal rack type.
6	Open the **PS-300** folder, select the desired power supply and drag the part to slot-1 of the central rack.
7	Open the **CPU-300** folder and then the appropriate CPU subfolder to select the desired CPU. Drag the selected CPU to slot-2 of the central rack.
8	From the menu, select **Station ➢ Consistency Check**, to check for errors. The configuration cannot be compiled for downloading unless it is found free of errors.
9	From the menu bar, select **Station ➢ Save** to save the configuration; or if done, use **Save and Compile** to generate the System Data that are downloaded to the CPU.

Configuring S7-300 Local I/O Expansion

Basic Concept

The S7-300 supports local I/O expansion of an S7-300 central rack to allow additional I/O modules. You may connect up to three additional I/O rails to the S7-300 central rack, using components referred to as interface modules (IMs). Adding local I/O expansion involves placement of a sender interface module in the S7-300 central rack and its corresponding receiver interface in each expansion rack.

Essential Elements

The IM 360 is a sender interface module placed in the central rack of the S7-300. The corresponding receiver module of the IM 360 is the IM 361, which is placed in each additional expansion rack. Interface modules of the S7-300 are always installed in slot-3.

Application Tips

Up to three additional I/O racks are supported by the S7-300 and you may mount them either horizontally or vertically. A maximum distance of 30 meters is allowed from the central rack to the last expansion rack. The maximum distance between each rack is 10 meters. Although this expansion supports thirty-two module slots (slot-4 to slot-11 in each rack), the actual maximum I/O count remains CPU dependent.

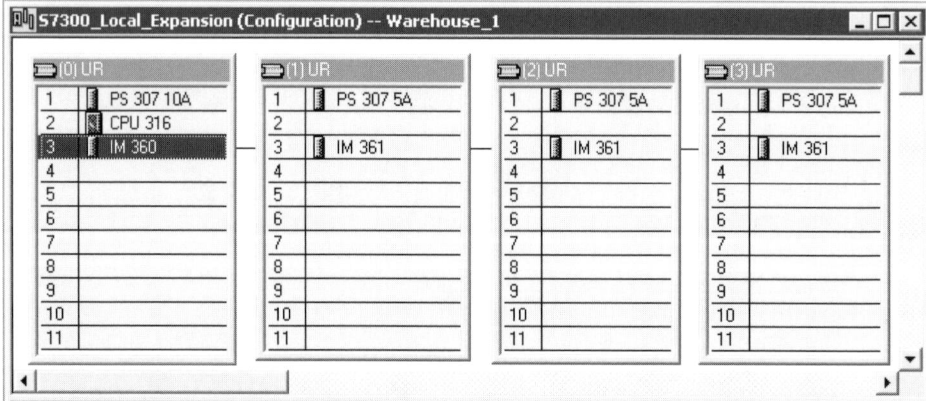

Figure 3-14. S7-300 Local I/O Expansion up to 4-racks, using IM 360/IM 361 pair.

Quick Steps: Configuring S7-300 Local I/O Expansion

STEP	ACTION
1	From the SIMATIC Manager open the required project, and from the project tree select the S7-300 Station for which local I/O expansion is required.
2	From the right pane, double click on the Hardware object to open the station in the Hardware configuration tool.
3	Open the Hardware Catalog window, from the View menu, if it is not already opened.
4	Open the **SIMATIC 300** catalog object to view the S7-300 component folders.
5	Open the **Rack-300** folder, select and drag the I/O rail to the top pane of the station window. If the central rack is already installed it is labeled 0(UR); otherwise it must be configured before adding expansion racks.
6	Install up to 3 additional expansion racks. The racks will be labeled from 1(UR) to 3(UR).
7	Open the **IM-300** folder, to show the S7-300 expansion interface modules. IMs designated for sending (S) are installed only in slot-3 of the central rack. IMs designated for receiving (R) are installed only in slot-3 of the expansion racks.
8	From the hardware catalog, select the correct IM-360 interface module and drag it to slot-3 of the CPU rack.
9	Select the correct IM-361 interface module and drag to slot-3 of each expansion rack.
10	Open the **PS-300** folder; select the required power supply and drag the part to slot-1 of each expansion rack.
11	In a similar fashion, I/O signal modules (**SM-300** folder), communications processors (**CP-300** folder), and function Modules (**FM-300** folder) are added to your rack by matching the actual part numbers with actual physical components.
12	From the menu, select **Station ➢ Consistency Check**, to check for errors. The configuration cannot be compiled for downloading unless it is found free of errors.
13	From the menu bar, select **Station ➢ Save** to save the configuration; or if done, use **Save and Compile** to generate the System Data that are downloaded to the CPU.

Configuring S7-300 Single-Tier I/O Expansion

Basic Concept

This configuration task involves adding a single rack expansion to an S7-300 central rack configuration. Upon completion, this will be a fixed configuration, involving the central rack and the one expansion rack. No further expansion beyond this is possible without changing the interface module (IM). The fixed configuration supports up to 16 I/O modules, with a maximum of eight modules in the central rack and in the expansion rack. The connection between the central and the expansion rack is a fixed 1-meter cable.

Essential Elements

The IM 365 is a sender/receiver interface module that is placed in the central rack and in one expansion rack, to support the fixed-tier configuration. Interface modules of the S7-300 are always installed in slot-3. No additional power supply is required in the expansion rack of this configuration since power is supplied from the central rack, through the connecting bus cable.

Application Tips

The two-tier expansion configuration is typically used on a small installation in which expansion beyond this fixed amount of I/O is not expected. Only digital and analog signal modules may be placed in this expansion rack. Later expansion would require removal of the IM-365 pair and use of the IM-360/IM-361 combination.

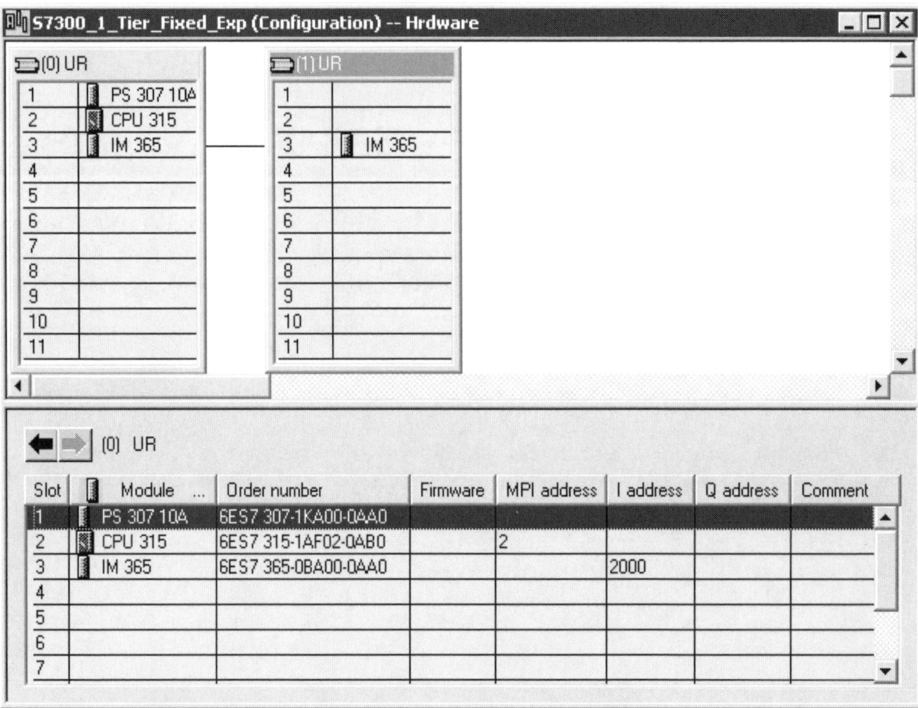

Figure 3-15. S7-300 Two-Tier Fixed I/O expansion, using IM 365 pair.

Quick Steps: Configuring S7-300 Two-Tier I/O Expansion

STEP	ACTION
1	From the SIMATIC Manager open the required project, and from the project tree select the S7-300 Station for which single tier I/O expansion must be added.
2	From the right pane, double click on the Hardware object to open the station in the Hardware configuration tool.
3	Open the Hardware Catalog window, from the View menu, if it is not already opened.
4	Open the **SIMATIC 300** catalog object to view the S7-300 component folders.
5	Open the **Rack-300** folder, select and drag the rail to the configuration window. If the central rack is already installed it is labeled 0(UR); otherwise, it must be configured first.
6	Drag and drop one additional rack into the configuration window. The single expansion rack will be labeled 1(UR) for the first expansion rack.
7	Open the **IM-300** folder, to show the S7-300 expansion interface modules. IMs designated for sending (S) are installed in slot-3 only of the central rack.
8	Select the IM-365 interface module from the catalog and drag to slot-3 of rack 0(UR), the central rack.
9	Select the IM-365 interface module from the catalog and drag it to slot-3 of the expansion rack 1(UR)
10	Open the **PS-300** folder, select the desired power supply. Click and drag the part to slot-1 of the expansion rack if required. A power supply in the expansion rack is only required if the main supply is not adequate to support the complete I/O combination.
11	From the menu, select **Station ➢ Consistency Check**, to check for errors. The configuration cannot be compiled for downloading unless it is found free of errors.
12	From the menu bar, select **Station ➢ Save** to save the configuration; or if done, use **Save and Compile** to generate the System Data that are downloaded to the CPU.

Configuring an S7-300 as DP Master

Basic Concept

The S7-300 supports distributed I/O, using the DP component of Profibus, which operates much like remote I/O. The DP configuration may involve one or more *DP Master Stations*, each of which manages the I/O of one or more assigned *DP Slave Stations*. Depending on the requirements of your application, as the DP master, you may either choose an S7-300 with DP master capability or a Profibus communications processor (CP) with DP capability.

Essential Elements

The DP master can be configured using S7-300 CPU, with an integrated Profibus-DP interface (e.g., CPU 315-2 DP or CPU 316-2 DP); or using a Profibus communications processor with DP master capability (e.g., CP342-5 or CP342-5 FO, for fiber optic links), in conjunction with a CPU that supports this feature. In any case, the DP master will perform the same function.

In configuring the DP Master, the Profibus interface of the DP master must be attached to the desired Profibus Subnet and assigned a unique Profibus Address (e.g., 3-125). If not previously set, then the subnet operating parameters (e.g., transmission rate, profile) will also need to be set. See configuration task, *Chapter 7, Adding and Configuring a Profibus Subnet*.

Application Tips

Typically, a DP subnet will involve a single DP master. This is referred to as a Mono-Master configuration. You may also have several mono-master systems within a project, by simply creating multiple PROFIBUS DP subnets and attaching a DP master to each. A Multi-Master configuration, involving two or more masters on a single subnet, though not typical, is also possible. You would do this by attaching the DP masters to the same DP subnet.

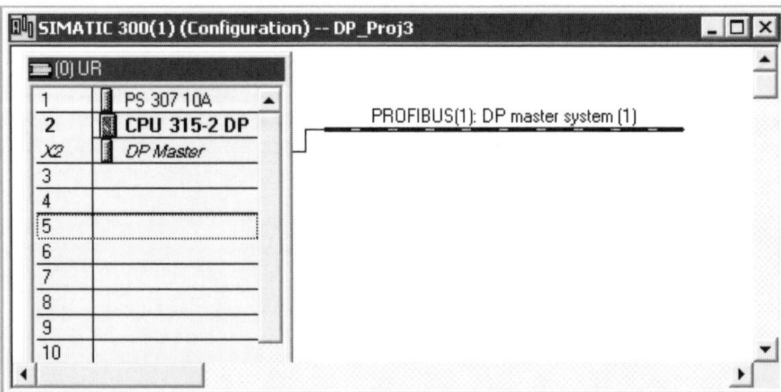

Figure 3-16. S7-300 station with the CPU 315-2 DP selected as DP master.

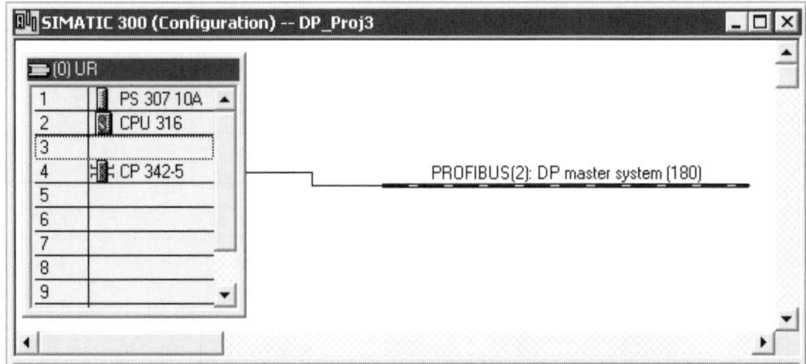

Figure 3-17. S7-300 station with the CP-342-5 selected as DP master.

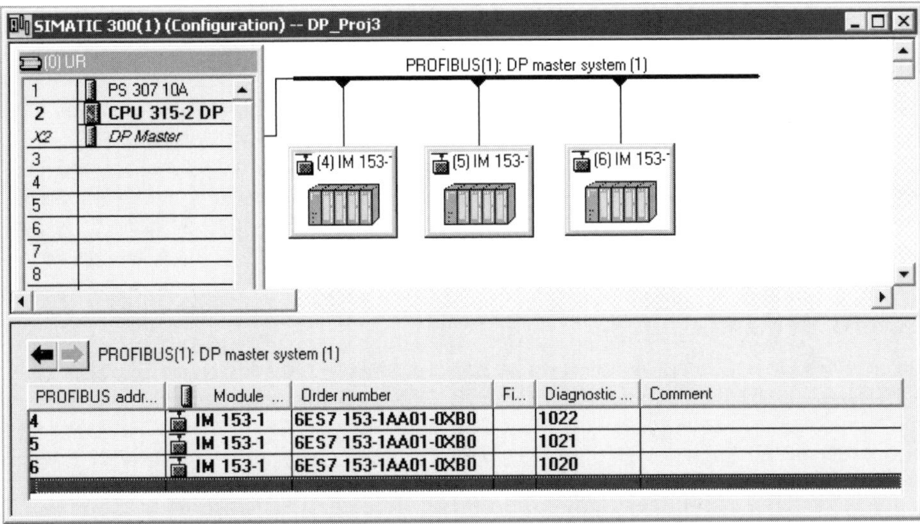

Figure 3-18. S7-300 CPU 315-2 DP master with attached ET-200 M modular I/O drops.

Quick Steps: Configuring an S7-300 DP Master

STEP	ACTION
1	From the SIMATIC Manager open the required project, and S7-300 station to serve as DP Master. The S7-300 must be inserted to the project if it not already installed.
2	With the S7-300 station opened in the hardware configuration window select the central rack. Insert the required rack and power supply if it is not already installed.
3	If a CPU will serve as the DP master then the correct Profibus-DP CPU must be selected (e.g., CPU 315-2 DP, or CPU 316-2 DP). From the catalog open the **CPU-300** folder, select and drag the required CPU to slot-2 of the central rack.
4	When the Profibus interface dialog is presented, select the Profibus subnet to which the DP master should be attached; set or accept the suggested Profibus address (e.g., 3-125), then click on the Properties button to set the subnet **Transmission Rate** and other properties if they have not yet been set.
5	After the DP properties dialog is saved, a DP Master system object should appear as a black and white dashed line extended from the DP master. If the object is not shown, right-click on the *DP master* object in row *2X2*, and select **Add Master System**.
6	You can save your work, but at least one DP slave must be attached to compile without errors. See the various tasks on *Configuring and Attaching Modular, Compact, and Intelligent DP Slaves*.

If a communications processor (CP) should serve as the DP master, then the appropriate selection must be made from the **CP-300** folder under the Profibus folder. From the **CP342-5** or **CP342-5 FO** subfolder, find the required communications processor and drag it to slot-4 or higher of the central rack. The Profibus interface dialog will be presented and the configuration may be continued with Steps 4, 5, and 6.

Configuring and Attaching Modular DP Slaves

Basic Concept

The S7-300 supports distributed I/O, using the DP component of Profibus, which operates much like remote I/O drops. The DP configuration involves one or more *DP Master Stations,* each of which manages the I/O of one or more assigned *DP slave Stations*. A *modular DP slave* station, is one in which individual I/O modules are installed in an I/O base. Examples of modular DP subsystems include ET-200M, ET-200L SC, ET-200S, and ET-200iS. Attaching configured modular DP slaves is simplified if the Profibus subnet and DP master are already configured.

Essential Elements

The DP master to which the DP slaves will be attached, will be provided by an S7-300 or S7-400 CPU, with a DP interface (e.g., CPU 315-2 DP, CPU 416-2 DP); or formed by a communications processor (CP) with DP master capability (e.g., CP342-5 or CP443-5 Ext.) and a compatible CPU. Each slave is attached to a DP master system using a modular DP slave interface module (IM) that is installed on the S7-300 mounting rail, along with an arrangement of I/O modules. Each IM is assigned a unique Profibus address (e.g., 3-125).

Several modular I/O styles are available. Your choice of IM will determine the number and type of usable modules. Modules supported by a given IM are contained in the hardware catalog, under the specific interface module (IM) folder. ET-200M, based on several choices of the IM-153, for example, uses S7-300 modules. The ET-200S family of *Smart Connect* modules connects using the IM-151. Each family of modular DP slaves is found in a corresponding folder of the hardware catalog.

Application Tips

As you attach each slave to the DP Master system, STEP 7 automatically assigns the next available address. You should allow STEP 7 to assign the address, unless otherwise necessary. At least one modular DP slave drop must be assigned to the DP master before the master will save and compile without errors. You must configure each modular slave with at least one module. If multiple DP master systems (i.e., DP subnets) are in the project, ensure that you attach the each slave to the correct subnet.

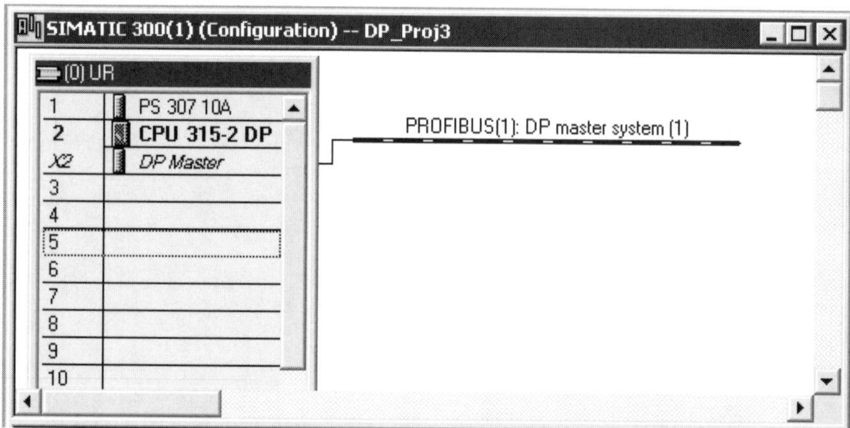

Figure 3-19. S7-300 DP master to which modular DP slaves will be attached.
The DP master system object is represented as a black and white dashed line.

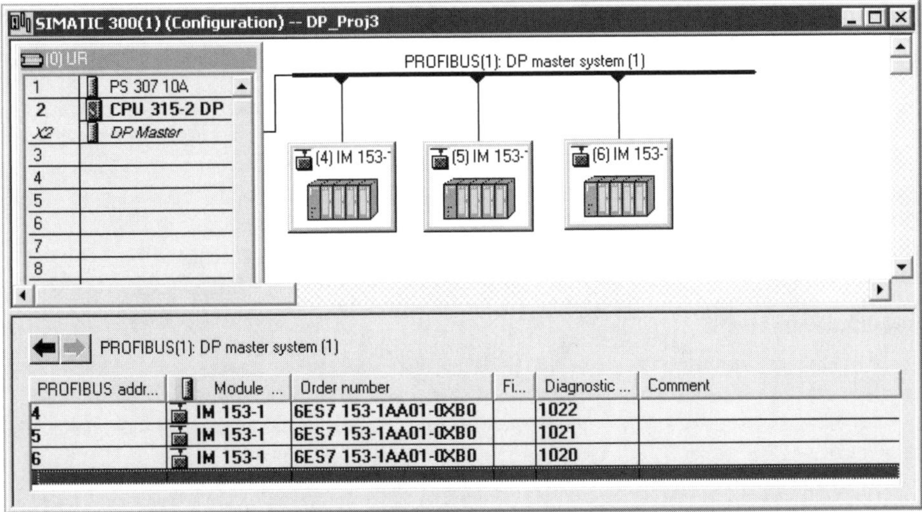

Figure 3-20. S7-300 CPU 315-2 DP master with ET-200M modular DP slaves.

Quick Steps: Configuring and Attaching Modular DP Slaves

STEP	ACTION
	Configuring the Modular Profibus DP Slaves with ET-200M
1	From the SIMATIC Manager open the required project, and from the project tree select the Station that will serve as DP master for the modular slaves you will attach.
2	With the DP master selected, open the Hardware configuration tool by double-clicking on the station Hardware object in the right pane of the project window.
3	With the DP master displayed, a DP master system object should be shown as a black and white dashed line. If the DP master system object is not shown, then right-click on the *DP master* object in row *2X2*, and select **Add Master System**.
4	With the configuration window open to the DP master, from the Profibus DP folder of the Catalog window, open the ET-200M folder and drag the required IM 153 interface module to the configuration window, and drop while touching the mouse tip to the DP master system object.
5	When the DP slave dialog is presented, select the Profibus subnet to which the DP master will be attached; set or accept the suggested Profibus address (e.g., 3-125).
6	Select the IM 153 bus interface module in the top pane, to display its configuration table in the bottom pane and to install required I/O modules (e.g., signal modules).
7	Drag and drop each additional IM 153 bus interface, as required, until done.
8	From the Station menu, perform a **Consistency Check**, then **Save and Compile** the configuration.

Configuring and Attaching Compact DP Slaves

Basic Concept

The S7-300 supports distributed I/O, using the DP component of Profibus, which operates much like remote I/O. The DP configuration involves one or more *DP Master Stations*, each of which manages the I/O of one or more assigned *DP Slave Stations*. A compact DP slave typically involves an encased design that combines the DP slave processor, a bus interface, a fixed number of I/O circuits and terminal blocks. Examples of compact DP subsystems include ET-200B, ET-200C, and ET-200L. Attaching configured compact DP slaves will be simplified if the Profibus subnet and DP master are already configured.

Essential Elements

The DP master to which the DP slaves will be attached, will be provided by an S7-300 or S7-400 CPU with an integrated DP interface (e.g., CPU 315-2 DP, CPU 416-2 DP); or formed by a communications processor (CP) with DP master capability (e.g., CP342-5 or CP443-5 Ext.) and a compatible CPU.

Compact DP slaves incorporate both the DP-interface and integrated I/O. Therefore, there is no interface module (IM), or mounting base as with modular slaves. The DP-interface of each compact slave must be attached to the DP master system, and assigned a unique Profibus address (e.g., 3-125). The combination of I/O circuits of a device is selected based upon signal requirements at the drop location. Compact slaves, and the integrated I/O each offers may be viewed in the Profibus-DP folder of the hardware catalog.

Application Tips

As you attach each slave to the DP Master System object, STEP 7 automatically assigns the next available address. You should allow STEP 7 to assign the address, unless otherwise necessary. You must configure and attach at least one slave drop to the DP master, in order for the Master to saved and compiled without errors. If multiple DP master systems (i.e., DP subnets) are in the project, ensure that you attach each slave to the correct subnet.

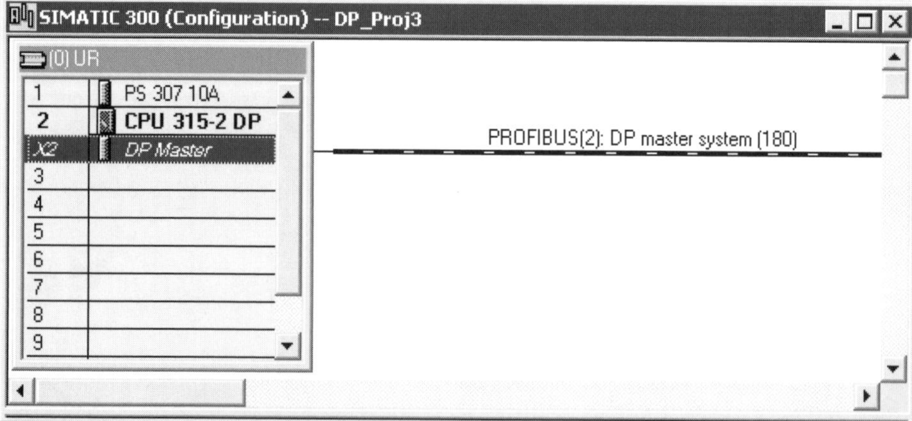

Figure 3-21. S7-300 DP master to which compact DP slaves will be attached. Note the DP master system object (for Profibus subnet), shown as a black and white dashed line.

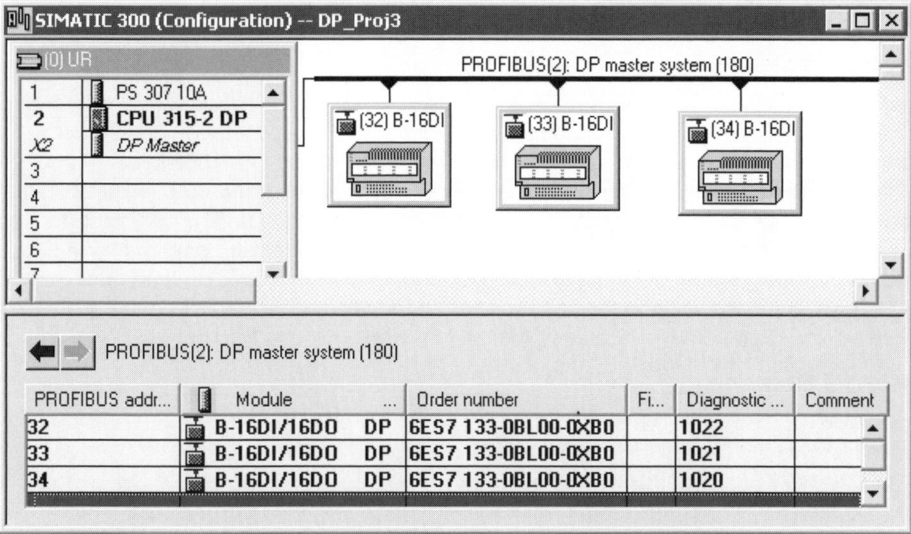

Figure 3-22. S7-300 CP-342-5 as DP master with compact DP slave drops.

Quick Steps: Configuring and Attaching Compact DP Slaves

STEP	ACTION
	Configuring Compact Profibus DP Slaves with ET-200B
1	From the SIMATIC Manager open the required project and from the project tree select the Station to serve as the DP master for the compact slaves you will attach.
2	With the DP master selected, open the Hardware configuration tool by double-clicking on the station Hardware object in the right pane of the project window.
3	With the DP master displayed, a DP master system object should be shown as a black and white dashed line. If the DP master system object is not shown, then right-click on the *DP Master* object in row *2X2*, and select **Add Master System**.
4	From the Profibus-DP folder of the catalog, open the ET-200B folder and select the correct part number for the compact slave; drag and drop the part onto the DP master System object.
5	When the DP slave dialog is presented, select the Profibus subnet to which the DP master will be attached; set or accept the suggested Profibus address (e.g., 3-125).
6	Select each compact I/O drop to display its configuration table in the bottom pane and to configure parameters of the drop if required (e.g., Profibus address).
7	Drag and drop each additional compact I/O block, as required, until done.
8	From the Station menu, perform a **Consistency Check**, then **Save and Compile** the configuration.

Configuring the S7-31x-2 DP as an Intelligent Slave

Basic Concept

An S7-300 can serve as an intelligent DP slave when a CPU with DP capability is configured to operate in DP slave mode. Operating as an intelligent slave, the S7-300 can perform its normal control duties as well as operate as a slave device reporting to a DP master. Such a configuration may be required in applications where certain field input/output signals of the slave may require some form of pre-processing locally at the slave, before the DP master accesses the data. Where intelligent slaves are involved, the DP master does not have direct access to the field signals of the slave. Instead, the DP master accesses an area of memory in the slave where the slave writes data that the master reads as field inputs, and the slave receives input data from the master, which it in turn maps onto the field outputs.

Essential Elements

If the slave is based on the CPU of the S7-300 then a central station must be created just as it would be normally, except that a CPU with DP capability must be installed (e.g., CPU 315-2 DP or CPU 316-2 DP). The DP interface of the CPU must be set to operate in *DP Slave mode*, attached to the DP master System, and assigned a unique Profibus address (e.g., 3-125). The slave must also be set to "Active", if programming functions (PG/PC) and standard communications via the DP interface should be allowed. After the S7-300 PLCs are configured as slaves, they will be listed as *configured slave controllers,* as shown in the dialog below. You may then logically connect the DP slaves to a configured DP master.

Application Tips

Working with intelligent slaves is simplified when the Profibus subnet and the DP master are already configured. If multiple DP master systems (i.e., DP subnets) are in the project, ensure that you attach each slave to the correct subnet. Once you have assigned the slaves to a DP master, the final step is to define the master-slave data exchange areas. This step is described in the task "*Configuring Master-Intelligent Slave Data Exchange Area*."

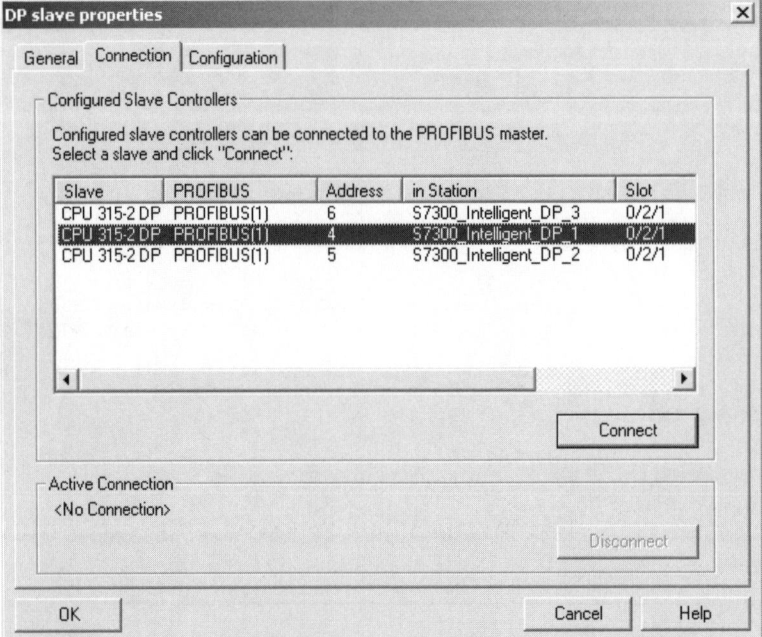

Figure 3-23. Connection tab of DP Slave properties shows configured slave controllers — allows intelligent slaves to be logically connected to a DP master.

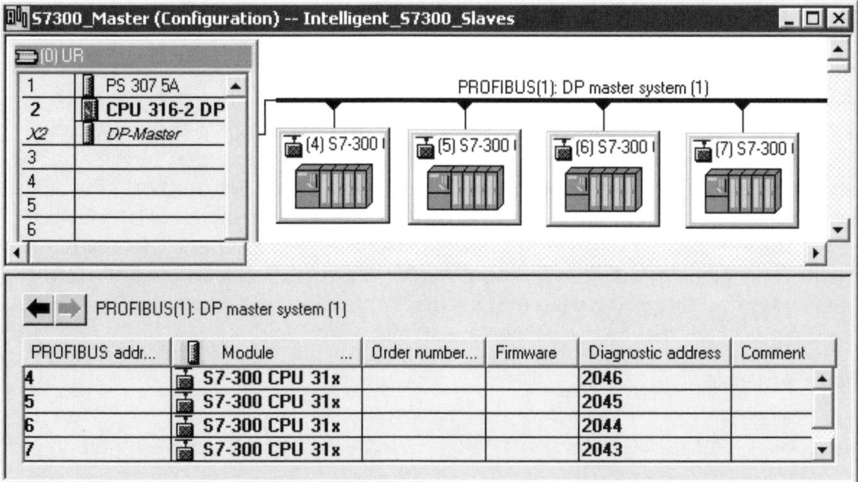

Figure 3-24. S7-300 CPU 316-2 DP as DP master with S7-300 PLCs as intelligent DP slaves.

Quick Steps: Configuring the CPU 31x-2 DP as an Intelligent DP Slave

STEP	ACTION
1	The S7-300 station that will serve as an intelligent slave must be created if not already in the project. If necessary insert a new S7-300 station, open the station in the hardware configuration tool, and insert the required power supply.
2	From the catalog, open the **CPU-300** folder; drag the required CPU 315-2 DP to slot-2 of the central rack.
3	The Profibus interface dialog will open when the CPU 315-2 DP is inserted. Select the subnet, to which the DP slave will be attached, then set or accept the suggested Profibus address (e.g., 3-125). Confirm entries by pressing the **OK** button.
4	Double-click on the DP object in row 2x2; when the dialog is displayed select the Operating Mode tab and activate the CPU for DP slave mode; confirm with **OK**.
5	From the SIMATIC manager select and open the DP Master station in the Hardware Configuration tool. A DP master system object should appear as a black and white dashed line extended from the *DP master* object, in row *2X2*. If the DP master system object is not shown, right-click on row *2X2* and select **Add Master System**.
6	From the Profibus folder of the hardware catalog, open the *Configured Stations* subfolder; drag and drop the **CPU 31x** object onto the DP master system object.
7	The DP slave properties dialog is opened to the *Connection* tab, which lists the configured intelligent slaves. Select each slave you wish to attach to the DP master; press the **Connect** button, then press the **OK** button to confirm connections.
8	You may save and compile the intelligent station, but the DP master will not compile until each slave has at least one area defined for Master-Slave I/O data exchange. See the task *Configuring Master-Intelligent Slave I/O Exchange Areas.*

Configuring the CP-342-5 as an Intelligent Slave

Basic Concept

A Profibus CP-342-5 communications processor can be used to allow the S7-300 to serve as an intelligent DP slave. Operating as an intelligent slave, the S7-300 can perform its normal control duties as well as operate as a slave device reporting to a DP master. Such a configuration may be required in applications where certain field input/output signals of the slave may require some form of pre-processing locally at the slave, before the DP master accesses the data. Where intelligent slaves are involved, the DP master does not have direct access to the field signals of the slave. Instead, the DP master accesses an area of memory in the slave where the slave writes data that the master reads as field inputs, and the slave receives input data from the master, which it in turn maps onto field outputs.

Essential Elements

If the slave is based on the CP-342-5, then the S7-300 central station in which the communications processor will be installed must be created just as it would be normally, except that the CP-342-5 must be installed. The DP interface of the CP-342-5 must be set to operate in the *DP Slave mode*, attached to the DP master System, and assigned a unique Profibus address (e.g., 3-125). The CP-342-5 must also be set to "Active," if programming functions (PG/PC) and standard communications via the DP interface should be allowed. After the CP-342-5 is configured as an intelligent slave, it will be listed as a *configured slave controller* as shown in the dialog below. You may then logically connect the DP slaves to an already configured DP master station.

Application Tips

Working with intelligent slaves is simplified when the Profibus subnet and the DP master are already configured. If multiple DP master systems (i.e., DP subnets) are in the project, ensure that you attach each slave to the correct subnet. Once you have assigned the slaves to a DP master, the final step is to define the master-slave data exchange areas. This step is described in the task "*Configuring Data Exchange Areas for CP-345-2 Intelligent Slaves.*"

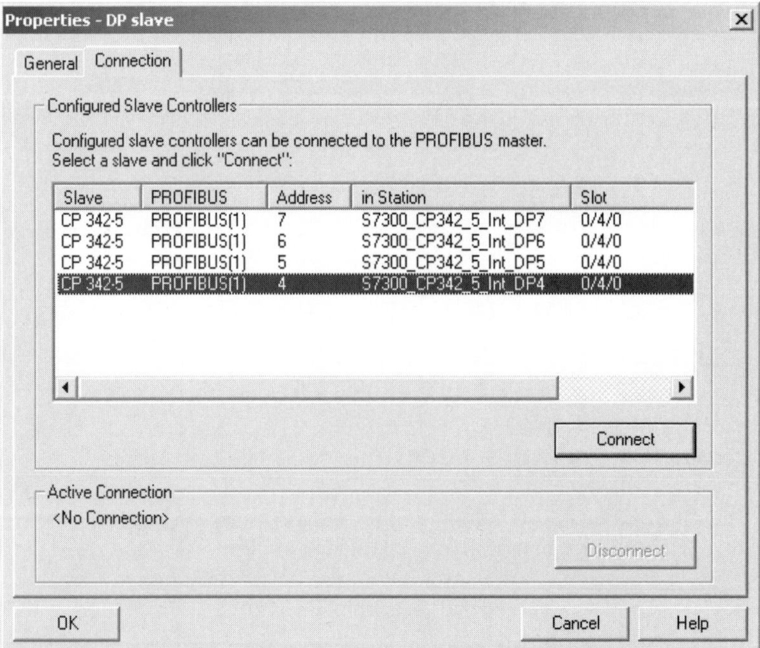

Figure 3-25. Connection tab of DP Slave properties shows configured slave controllers — allows intelligent slaves to be logically connected to a DP master.

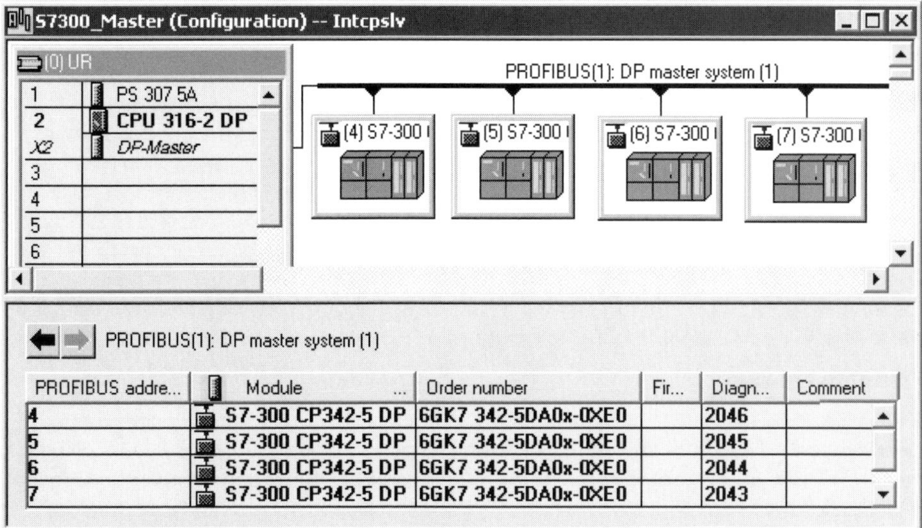

Figure 3-26. S7-300 CPU 316-2 DP as DP master with S7-300s/CP 342-5 as intelligent slaves.

Quick Steps: Configuring the CP-342-5 as an Intelligent DP Slave

STEP	ACTION
1	The S7-300 station that will serve as an intelligent slave must be created if not already in the project. If necessary, insert a new S7-300 station, open the station in the hardware configuration tool, and insert the required power supply.
2	From the hardware catalog, open the **CP-300** folder and the **Profibus** subfolder; then drag the required CP342-5 to slot-4 or higher of the central rack.
3	The Profibus interface dialog is opened when the CP342-5 is inserted. Select the subnet to which the DP slave will be attached; then set or accept the suggested Profibus address (e.g., 3-125), and confirm the selection by pressing the **OK** button.
4	Double-click on the *DP master* object in row 2x2; when the dialog is opens select the Operating Mode tab and activate the CP342-5 for DP slave; confirm with **OK**.
5	From the SIMATIC manager select and open the DP Master station in the Hardware Configuration tool. A DP master system object should appear as a black and white dashed line extended from the *DP master* object, in row *2X2*. If the DP master system object is not shown, right-click on row *2X2* and select **Add Master System**.
6	From the Profibus folder of the hardware catalog, open the *Configured Stations* subfolder, drag and drop the **CP 342-5** object onto the DP master system object.
7	The DP slave properties dialog is opened to the *Connection* tab with configured DP slaves listed. Select each intelligent slave you wish to attach to the DP master; press the **Connect** button, then press the **OK** button to confirm the connections.
8	You may save and compile the intelligent station, but the DP master will not compile until each slave has at least one area defined for Master-Slave I/O data exchange. See the task *Configuring Exchange Area for CP 342-5 Intelligent Slaves*.

Configuring the ET-200X (BM 147/CPU) as an Intelligent Slave

Basic Concept

The BM 147/CPU Basic Module is a Basic language processor and CPU combination. Operating as an intelligent slave, the BM 147/CPU can perform its normal control duties as well as operate as a slave device reporting to a DP master. Such a configuration may be required in applications where certain field input/output signals of the slave may require some form of pre-processing locally at the slave, before the DP master accesses the data. Where intelligent slaves are involved, the DP master does not have direct access to the field signals of the slave. Instead the DP master accesses an area of memory in the slave where the slave outputs data to be read by the master as field inputs, and receives input data from the master to be mapped onto the field outputs.

Essential Elements

The configuration of the BM 147/CPU Basic Module, as an intelligent slave, is started by creating a new S7-300 station. Instead of placing a rack in the empty window, you will drag the BM 147/CPU Basic Module to the window. The BM 147/CPU Basic Module is in the Profibus folder of hardware catalog, contained in the ET 200X subfolder. Like other Profibus nodes, the BM 147/CPU slave must be attached to the DP master system, and assigned a unique Profibus address (e.g., 3-125). Once configured, the BM 147/CPU will be listed as a configured slave controller in the connection dialog of the DP slave properties. You may then logically connect the DP slave to an already configured DP master station.

Application Tips

Working with intelligent slaves is simplified when the Profibus subnet and the DP master are already configured. If multiple DP master systems (i.e., DP subnets) are in the project, ensure that you attach each slave to the correct subnet. Once you assign the slaves to a DP master, the final step is to define the master-slave data exchange areas. This step is described in the task "*Configuring Master-Intelligent Slave Data Exchange Area.*"

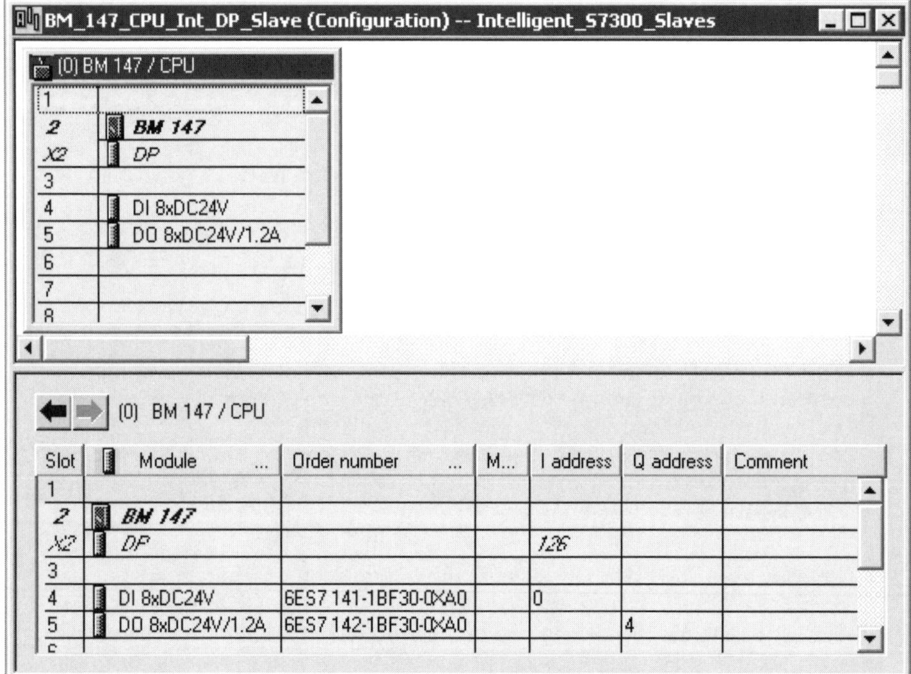

Figure 3-27. The BM 147/CPU configured as an intelligent DP slave. When a modular slave station is selected, its modules are listed in the bottom pane.

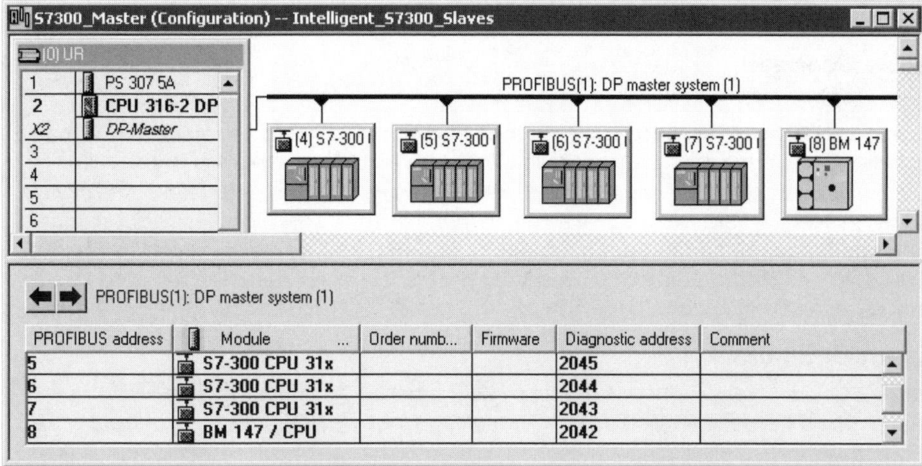

Figure 3-28. When the DP master system object is selected, the intelligent slaves are listed in the bottom pane.

Quick Steps: Configuring the ET-200X (BM 147/CPU) as an Intelligent Slave

STEP	ACTION
1	Open the required project and create a new S7-300 station to serve as the intelligent slave, and then open the station in the Hardware Configuration tool.
2	From the catalog, open the Profibus DP folder, the ET 200X subfolder, then drag and drop the BM 147/CPU object to the empty top pane of the station window.
3	The Profibus interface dialog is opened when the BM 147/CPU object is inserted; select the subnet to which the DP slave will be attached; set or accept the suggested Profibus address (e.g., 3-125), and confirm entry with the **OK** button.
4	From the catalog, expand the BM 147/CPU folder and from the signal module subfolders, drag the required expansion sub-modules to the appropriate rack slot.
5	Once the required expansion sub-modules are entered, save the configuration.
	From the SIMATIC manager, select and open the DP Master station in the Hardware Configuration tool. A DP Master system object should appear as a black and white dashed line extended from the *DP master* object, in row *2X2*. If the DP master system object is not shown, right-click on row *2X2* and select **Add Master System**.
6	From the Profibus folder of the catalog, open the *Configured Stations* subfolder; drag and drop the **X-BM 147/CPU** object onto the DP master system object.
7	The DP slave properties dialog is opened to the *Connection* tab with configured DP slaves listed. Select each slave (e.g., **X-BM 147/CPU**) you wish to attach to the DP master and press the **Connect** button; confirm connections with the **OK** button.
8	You may save and compile the intelligent station, but the DP master will not compile until each slave has at least one area defined for Master-Slave I/O data exchange. See the task *Configuring Master-Intelligent Slave I/O Exchange Area.*

Configuring Master-Intelligent Slave Data Exchange Area

Basic Concept

I/O data exchange between a DP master and an intelligent DP slave is different from the normal master-slave interaction, with modular and compact slaves. Normally, the DP master directly accesses the slave I/O — cyclically writing to outputs and reading inputs. With intelligent slaves, the I/O is not accessed directly. Instead, DP data areas are defined in the slave to allow I/O data exchange with the master. The DP master reads (*input data*) from areas defined as outputs in the slave and writes (*output data*) to areas defined as inputs in the slave. The user program in the slave is responsible for continuously reading data from the data area and writing data to the data area to handle I/O data exchange with the master.

Essential Elements

When defining a data exchange area, an address defined as **output** *in the slave,* will be read from an address defined as *input in the DP master*. An address defined as **input** *in the slave,* is written to from an address defined as *output in the DP master*. Depending on the application, a single contiguous input area and a single contiguous output area may be required, or several I/O areas of one or more bytes or words may be required. In either case, you must specify the data **unit** (i.e., bytes or words), **length**, and **consistency**. The length is the number of data units, and consistency defines how the data should be kept together when transferred. If the slave CPU, or the DP master supports **process image partitions** (e.g., S7-400), then you may specify the partition number to which the area should be assigned.

Application Tips

When specifying a data area for an S7-300 slave, you must not use input/output addresses assigned to installed modules. If the data area is for the BM 147/CPU, then the useable address range for the slave is from byte128 through 159 for both the input and output areas.

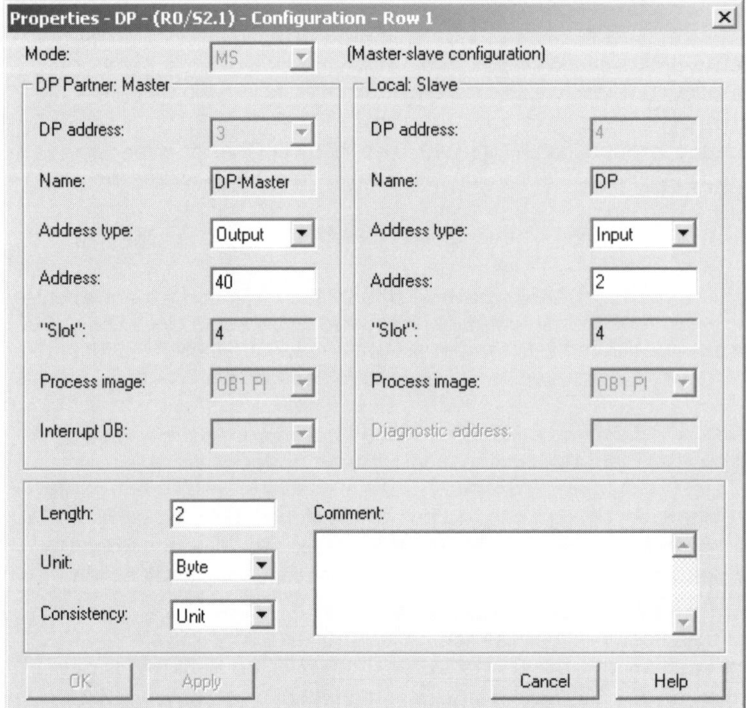

Figure 3-29. Dialog for configuring master-intelligent slave data exchange area.

Quick Steps: Configuring Master-Intelligent Slave Data Exchange Area

STEP	ACTION
1	With the DP Master station open in the Hardware Configuration tool, in the top pane, double click on the intelligent DP slave object for which you wish to define data exchange areas.
2	When the DP slave Properties dialog is presented, select the *Configuration* tab to open the dialog for defining master-slave exchange areas. Exchange areas that have already been defined will appear as rows in a table. Each row represents -- In the end, the number of rows will depend on the number of separate areas you need to define.
	Defining an Intelligent Slave **Input Area:**
3a	To define a slave input area, select *Output* as the **Address Type** in the DP master column, and in the **Address** field enter a start byte number for the logical area from which data will be sent to the DP slave. Select *Input* as the **Address Type** in the DP slave column, and in the **Address** field enter a start byte number in which data will be received from the DP master partner.
4a	In the lower part of the dialog, define the **length** of the data area; select the **unit** of the data as *bytes* or *words*; and define the **consistency** of the data according to the *unit* or *total length*. The consistency defines how the data should be kept together and maintained during transfer (e.g., byte, word, total length).
5a	Click in the **Comment** field to enter information associated with the data area.
6a	Confirm the entry by pressing the **OK** button to enter the defined logical area. The defined area will be entered as a *Configuration Row* on *the Configuration* tab.
	Defining an Intelligent Slave **Output Area:**
3b	To define a slave output area, select *Input* as the **Address Type** in the DP master column, and in the **Address** field enter a start byte number for the logical area in which data will be received from the DP slave. Select *Output* as the **Address Type** in the DP slave column, and in the **Address** field enter a start byte number from which data will be sent to the DP master partner.
4b	In the lower part of the dialog, define the **length** of the data area; select the **unit** of the data as *bytes* or *words*; and define the **consistency** of the data according to the *unit* or *total length*. The consistency defines how the data should be kept together and maintained during transfer (e.g., byte, word, total length).
5b	Click in the **Comment** field to enter information associated with the data area.
6b	Confirm the entry by pressing the **OK** button to enter the defined logical area. The defined area will be entered as a *Configuration Row*. The process may then be repeated to define another logical data exchange area.

Configuring Data Exchange for CP 342-5 Intelligent Slave

Basic Concept

I/O data exchange between a DP master and an intelligent slave is different from the normal master-slave interaction, with modular and compact slaves. Normally, the DP master accesses the slave I/O directly — cyclically writing to outputs and reading inputs. When there is an intelligent slave, the I/O is not accessed. Instead, DP data areas are defined in the slave to allow I/O data exchange with the master. The DP master reads (input data) from areas defined as outputs in the slave and writes (output data) to areas defined as inputs in the slave. The user program in the slave is responsible for continuously reading data from the data area and writing data to the data area to handle I/O data exchange with the master.

Essential Elements

When defining the data exchange area, an address defined as **output** *from the slave* will be read to an address defined as *input to the DP master*. An address defined as **input** *to the slave* is written to from an address defined as *output from the DP master*. Depending on the I/O application, a single contiguous input area and a single contiguous output area may be required, or several areas of one or more byte or word units may be required. In either case, the data **unit** (i.e., bytes or words), and **consistency** must be specified. Consistency defines how the data should be kept together when transferred; by the unit (e.g., **byte**, or **word**), or by the total length. If the slave CPU or DP master supports using **process image partitions** (e.g., in S7-400), then a partition number to which area should belong may be specified.

Application Tips

When defining the input/output address areas, you must not use address bytes that have been assigned to installed modules.

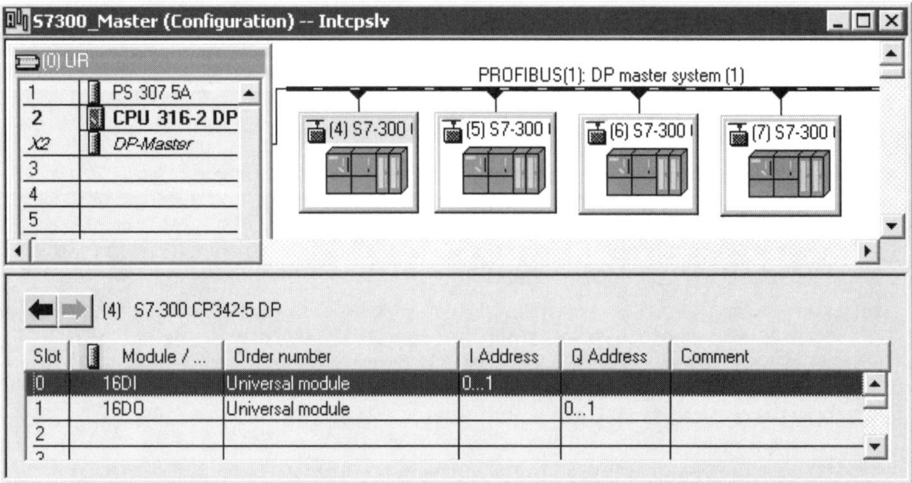

Figure 3-30. S7-300 Intelligent slave with Universal modules for configuring data exchange.

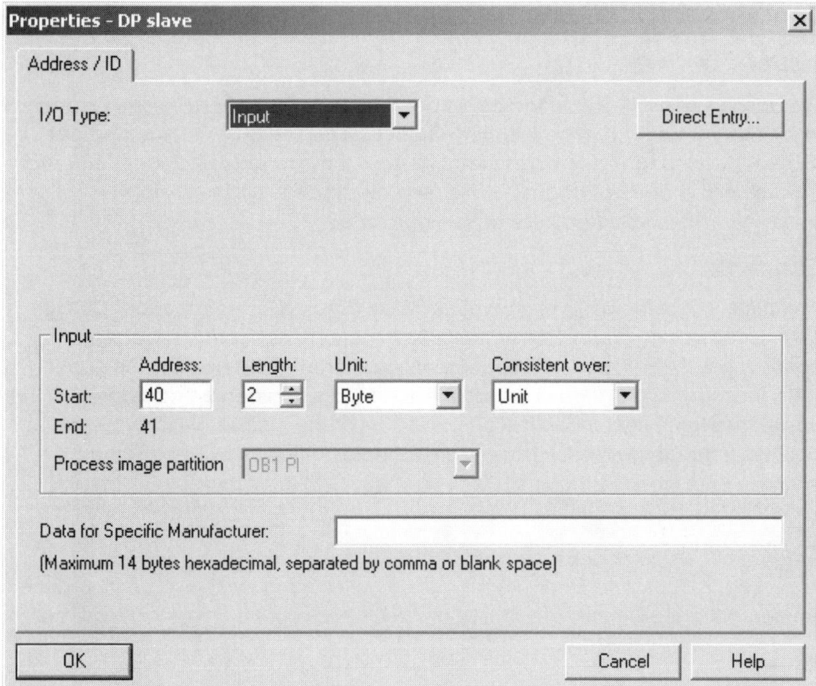

Figure 3-31. Configuration dialog for I/O data exchange with CP-342-5 slave.

Quick Steps: Configuring Data Exchange for CP 342-5 Intelligent-Slave

STEP	ACTION
1	With the DP Master station open in the Hardware Configuration tool, in the top pane, and select the intelligent DP slave object for which you wish to define data exchange areas.
2	From the hardware catalog and under the Configured Stations folder, open the **S7-300 CP-342-5 DP** folder; then find the Universal Module and drag one or more modules to the configuration table in the bottom pane. The number of modules installed will depend on the number of separate areas you need to define.
3	Double-click on a module to open the configuration dialog for defining the data exchange area.
4	Select the **I/O Type** as *Input*, if the module receives input data from the master; as *Output*, to allow the module to send data to the master. Select *Output-Input*, to allow the module to accept input from the master and to send output data to the master. Select *Empty Slot* to leave an unused area within the defined area.
5	Next, define the **length** of the data area; select the **unit** of the data as *bytes* or *words*; and define the **consistency** of the data according to the *unit* or *total length*. The consistency defines how the data should be kept together and maintained during transfer (e.g., byte, word, total length).
6	Confirm the entry by pressing the **OK** button.

Configuring S7-300 CPU General Properties

Basic Concept

The General properties of an S7-300 CPU module are viewed in the properties dialog of the CPU with a right-click on the CPU, then selecting **Object Properties**. Basic information about the CPU, which may vary depending on the relative age of the module, includes items such as type and location, a short list of characteristic features, and for modules with communications interfaces — related information is provided.

Essential Elements

The ***Short Description***, the same as given in the hardware catalog, gives important CPU features such as work memory size, cycle time/1000 instructions, digital I/O capacity, supported connections, DP/MPI ports, multi-computing and routing capability. The ***Order Number*** reflects the module assigned in the configuration tool and can be compared to the installed module. The ***Name*** field shows the default name of the module, which you can modify as required. Changes to the CPU name are reflected in the SIMATIC Manager.

The ***Interface*** parameter group, generally on modules with communications interfaces, shows basic information such as node address and whether or not the node is attached. The Properties button provides quick access to parameters of the interface; but interface properties (e.g., MPI, Profibus, and Ethernet) are generally set in the hardware configuration tool. The **Comment** field allows more details about the CPU application to be documented.

Application Tips

The user-defined ***Plant Designation*** identifier, available with some CPUs, is a user-defined identifier that can be evaluated in start-up OBs the user program. Examples of evaluating the "plant designation" are provided in the STEP 7 Sample programs.

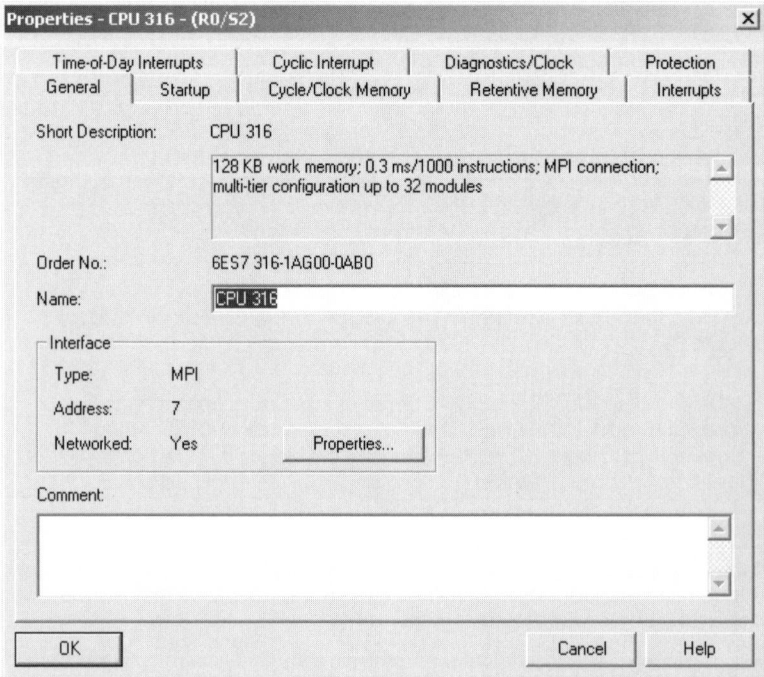

Figure 3-32. CPU Properties: General parameters dialog (e.g., CPU 316).

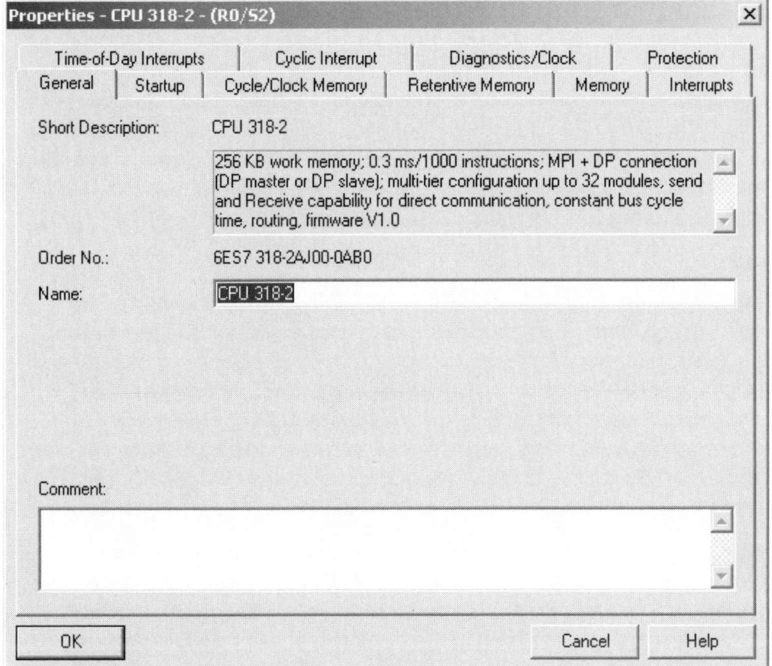

Figure 3-33. CPU Properties: General parameters dialog (e.g., CPU 318-2).

Quick Steps: Configuring S7-300 CPU General Properties

STEP	ACTION
1	With a station open in the Hardware Configuration tool select the CPU, right-click and select **Object Properties,** then select the *General* tab.
2	The **Short description** of the module provides a brief summary of the unique features of the installed CPU module; it cannot be modified.
3	The **Order Number** of the module is the specific number used to identify the exact CPU that should be installed in the rack. The order number determines the unique features of the CPU, which are summarized under the Short Description.
4	Click in the **Name** field to modify the default name assigned to the module. The default name is based on the short name of the selected CPU type (e.g., CPU 318).
5	The **Interface** static group, shows the module's set MPI address (default = 2), and whether or not the module is attached to the network; the Properties button allows the module's Interface parameters to be accessed for viewing or editing.
6	Click in the **Comment** field to document the module purpose or application.
7	Confirm any changes by pressing the **OK** button.

Configuring S7-300 CPU Start-Up Properties

Basic Concept

S7-300 CPU modules may be started without any changes to the default start-up parameters. There are a few adjustable parameters, however, that allow you to determine CPU start-up characteristics or whether or not the CPU should even be allowed to start. These parameters are checked by the CPU at the various startups (e.g., power ON, or the transition from STOP-to-RUN), before beginning to process the main cyclical program (OB1).

Essential Elements

The parameter ***Startup if Setpoint and Actual Configuration Differ***, determines if the CPU should be allowed to start-up if the setpoint and actual configurations differ — the default behavior in the S7-300, is to start even if the setpoint and actual configurations differ. The parameters ***Reset Outputs at Hot Restart*** and ***Disable Hot Restart by Operator*** are both grayed since the "hot restart" does not apply to the S7-300. The ***Startup After Power-On*** parameter is generally grayed out and fixed to "Warm Restart," since it is generally the only start-up mode possible in S7-300 CPUs — in newer S7-300 CPUs like the CPU 318, it is possible to select whether a start-up after Power- On will initiate a "Cold Restart," or a "Warm Restart."

The Monitoring Time parameter, ***Finished Message***, specifies the maximum CPU wait-time, at power-up, to receive ready-signals for all configured modules. The parameter ***Transfer of Parameters to Modules*** specifies the maximum time for modules, including DP-Interface modules, to acknowledge receipt of their configuration parameters (the wait-time starts after receipt of the "Finished Message"). If either of these two monitoring parameters are not acknowledged before the set time expires, the preset and actual configurations will be considered different. The CPU reaction is then determined by the parameter "Startup if Preset and Actual Configuration Differ." Finally, the ***Hot Restart*** parameter applies only in the S7-400.

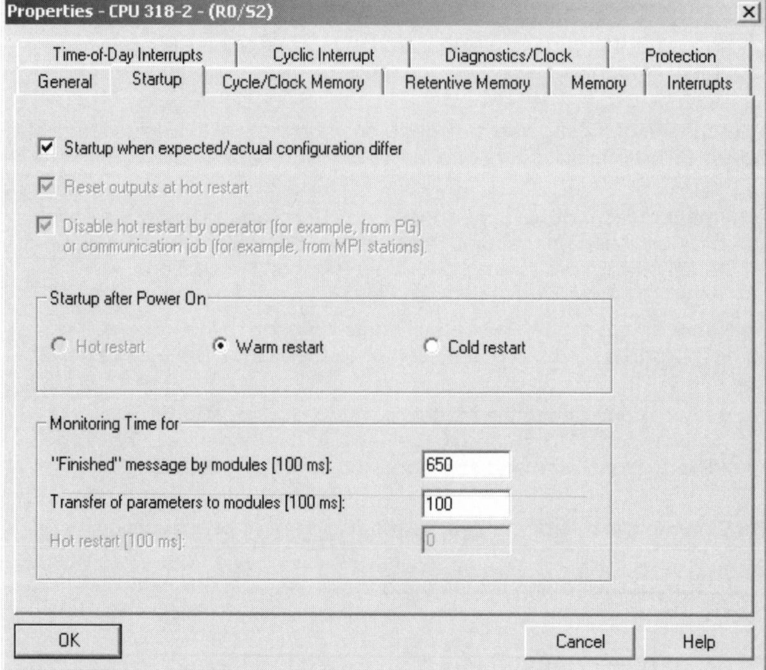

Figure 3-34. S7-300 CPU Properties: Start-up parameters dialog.

Quick Steps: Configuring S7-300 CPU Start-up Properties

STEP	ACTION
1	With the Hardware Configuration tool open to the desired project and station; select the CPU module, right-click and select **Object Properties,** then select the *Startup* tab. Parameters that are grayed are either not possible in the S7-300 or in the open CPU.
2	Activate the check box **Startup if Preset and Actual Configuration Differ,** to allow the CPU to start even if the preset and setpoint configurations differ. With this setting, central rack or distributed I/O modules are not checked; PROFIBUS-DP interface modules are checked and must be inserted for the CPU to start. This parameter is activated by default in older S7-300 CPUs and cannot be changed.
3	De-activate the check box to disable **Startup if Preset and Actual Configuration Differ**. If at least one configured slot differs, then the CPU switches to STOP. If module-slots other than those originally configured have inserted modules, they are not compared. This setting is generally not possible with older S7-300 CPUs.
4	The parameters "**Reset Outputs at Hot Restart**" and "**Disable Hot Restart by Operator**" are grayed since they are only possible in the S7-400.
5	Where allowed, select "Cold Restart" or "Warm Restart" as the mode of **Startup After Power On**. The option is fixed to "Warm Restart" in older S7-300 CPUs.
6	Set the parameter **Finished Message**, in milliseconds, to specify the maximum CPU wait-time for configured modules to signal ready for operation after power-up.
7	Set the parameter **Transfer of Parameters to Modules**, in milliseconds, to specify the maximum time for all modules, including DP slaves, to acknowledge receipt of their configuration parameters.
8	The **Hot Restart** parameter is grayed, since Hot Restart is not supported in the S7-300.
9	Confirm the configuration parameters with **OK**.
10	Use the toolbar **Save and Compile** button to generate the system data blocks.

Configuring S7-300 CPU Cycle and Clock Memory

Basic Concept

The Cycle and Clock Memory parameters allow for the cyclical processing time of the CPU to be influenced and for a single byte to be defined where the CPU may output periodic clock pulses. Modifying default parameters of this dialog is not essential for S7-300 operation; however they do allow adjustments that optimize use of CPU processing time.

Essential Elements

The ***Update Process Image Cyclically*** parameter determines if the process image of inputs and outputs should be updated cyclically. This parameter is fixed to always update the PII and PIQ in the S7-300, but may be altered in the CPU 318. The ***Scan Cycle Monitoring Time*** is the timer value that if exceeded by the CPU cycle, the CPU is stopped. The ***Minimum Scan Cycle***, not applying to the S7-300, is grayed. The ***Scan Cycle Load from Communications*** allows a percentage of the total processing time to be allocated to communications.

Adjusting the ***Size of the Process Image Inputs*** *and* ***Process Image Outputs***, perhaps to reflect actual usage, is only relevant in the CPU 318. The parameter ***OB 85 Call Up at I/O Access Error***, is generally grayed and by default is fixed such that I/O access errors do not result in a call of OB85, nor is an entry made in the diagnostic buffer. In later CPU revisions, other options for calling OB 85 include "Only for Incoming and Outgoing Errors", or "On Each Individual Access." Finally, the ***Clock Memory*** parameter allows a byte from Bit memory (M) to be specified as the location where the CPU will output eight individual clock-pulses.

Application Tips

Clock pulses, are accessed in the user program, via the eight bits of the specified byte. From the most-significant to the least-significant bit, the output frequency is 0.5, 0.625, 1.0, 1.25, 2.0, 2.5, 5.0, and 10 Hz. These pulses are used instead of timers, where periodic signals are needed. Typical uses include flasher-circuits, alarm synchronization, or where events must be continuously triggered at periodic intervals. Clock pulses have an on/off ratio of 1:1.

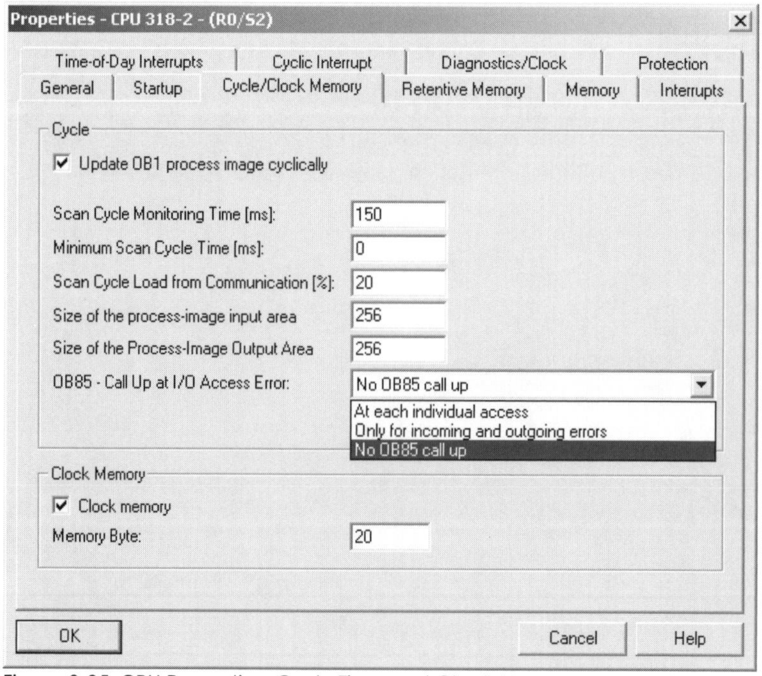

Figure 3-35. CPU Properties: Cycle Time and Clock Memory parameters.

Quick Steps: Configuring S7-300 CPU Cycle and Clock Memory

STEP	ACTION
1	With the Hardware Configuration tool open to the desired project and station; select the CPU module, right-click and select **Object Properties,** then select the *Cycle/Clock Memory* tab. Parameters that are grayed are either not possible in the S7-300 or in the open CPU.
2	Activate the checkbox **Update OB 1 Process Image Cyclically** for cyclic updates of the process image tables (PII/PIQ). This default setting is generally fixed in most S7-300 CPUs and cannot be changed. De-activate the checkbox to disable normal cyclic updates of the process image; I/O update must be handled by calling blocks.
3	In the **Scan Cycle Monitoring Time** field, specify in milliseconds, the cycle time that the CPU is not to exceed. If the actual cycle time exceeds this watchdog time, the CPU enters the STOP mode, unless OB 80 is loaded in CPU memory.
4	In the **Minimum Scan Cycle Time** field specify, in milliseconds, the minimum time that should be used to process the program. If the actual cycle time is less than the specified minimum, the CPU waits until the minimum cycle time expires, or if OB 90 is in memory, this remaining time is used by the CPU for background processing.
5	In the **Scan Cycle Load from Communications** field, enter a value from 10 to 50, as a percent of the set cycle time to allocate for communications processing.
6	Click in the **Size of the Process Image Inputs** field and enter an end byte-address for the PII that reflects the actual number of input-bytes (PII) installed. The PII update time is thereby minimized. The parameter is only relevant in some S7-300 CPUs.
7	Click in the **Size of the Process Image Outputs** field and enter an end byte-address for the PIQ that reflects the actual number of output-bytes (PIQ) installed. The PIQ update time is thereby minimized. The parameter is only relevant in some S7-300 CPUs.
8	Where the **OB 85 Call at I/O Access Error** parameter is allowed, select the option "No Call of OB 85" if OB85 should not be called when an I/O access error occurs during update of the process images (PII/PIQ); select the option "Only for Incoming and Outgoing Errors," to minimize the impact on the CPU cycle. The option "On each individual access" can result in an increase in CPU cycle time.
9	To use clock memory, activate the **Clock Memory** check box and then enter a byte (e.g., MB 97) from Bit Memory, to use for CPU output of eight individual clock pulses.
10	Confirm the configuration parameters with the **OK** button.
11	Use the toolbar **Save and Compile** button to generate the system data blocks.

Configuring S7-300 CPU Retentive Memory

Basic Concept

The CPU Retentive Memory parameter settings allow areas of memory to be defined as retentive. Retentive memory retains its stored contents when the CPU operating mode is switched from RUN to STOP and even upon power loss. Modifying the default parameters of this dialog is not essential for operation of S7-300 CPUs, unless you require that certain memory bytes, timers, counters, and data areas retain their contents on loss of power.

Essential Elements

In the S7-300 parts of bit memory (M), timer (T), and counter (C) areas may be defined as retentive. The first three parameters require an entry of the number of bytes, timers, or counters to reserve starting from byte-0, timer-0, or counter-0 respectively. An entry of 4 in the bit memory field reserves byte-0, 1, 2, and 3 as retentive. In addition to these retentive memory areas, the S7-300 also allows Data Block areas to be reserved as retentive. The defined areas will retain its data given power loss and even the loss of battery backup. Finally, the maximum size that can be reserved as retentive is dependent on the CPU.

Application Tips

Recall that under normal conditions involving loss of power, the contents of data blocks are always retained by battery backup. The maintenance free S7-300 feature, described here, is to protect against loss of the areas defined as retentive not only given a loss of power, but particularly a loss of battery backup.

Properties - CPU 316 - (R0/S2)

Time-of-Day Interrupts | Cyclic Interrupt | Diagnostics/Clock | Protection
General | Startup | Cycle/Clock Memory | Retentive Memory | Interrupts

Retentivity

Number of Memory Bytes Starting with MB0: 16
Number of S7 Timers Starting with T0: 0
Number of S7 Counters Starting with C0: 8

Areas

	DB No.	Byte Address	Number of Bytes
Retentive Area 1:	1	0	0
Retentive Area 2:	1	0	0
Retentive Area 3:	1	0	0
Retentive Area 4:	1	0	0
Retentive Area 5:	1	0	0
Retentive Area 6:	1	0	0
Retentive Area 7:	1	0	0
Retentive Area 8:	1	0	0

OK | Cancel | Help

Figure 3-36. CPU Properties: Retentive Memory parameters.

Quick Steps: Configuring S7-300 CPU Retentive Memory

STEP	ACTION
1	With the Hardware Configuration tool open to the desired project and station; select the CPU module, right-click and select **Object Properties,** then select the *Retentive Memory* tab to define required retentive memory areas.
2	Click in the **Memory Bytes** field, and enter the total number of bytes to define as retentive. The defined area starts at MB 0 and will include all bytes up to MB n-1.
3	Click in the **S7 Timers** field and enter the number of timers to define as retentive. The defined retentive area starts at T0, and will include all timers up to Tn-1.
4	Click in the **S7 Counters** field and enter the number of counters to define as retentive. The defined retentive area starts at C0, and will include all counters up to Cn-1.
5	To define retentive data block areas as required, start with Retentive Area 1and enter the data block number in the **DB No.** field; enter the starting **Byte Address** in the second field; and enter the total **Number of Bytes** in the last field. Using an even number for both the start address and the total number of bytes will help to avoid overwriting data areas.
6	Define as many retentive data block areas as required and confirm the parameters with the **OK** button.
7	Use the toolbar **Save and Compile** button to generate the system data blocks.

Configuring S7-300 CPU Local Memory

Basic Concept

Local memory is the S7 memory area used when temporary (TEMP) variables, which were declared in a code block, are accessed during program execution. This memory, also called the local stack or L-stack, is only available to the current block while the block is being processed. After the block is done the local stack is available to the next called block, which overwrites the previous data. Each CPU has a fixed local memory size that in most S7-300 CPUs cannot be altered, but may be re-allocated in the CPU 318 and some newer CPUs.

Essential Elements

Each CPU has a fixed amount of local memory that is equally divided among the priorities classes (1-29). As will be discussed in more detail later, all code blocks (e.g., FB, FC, SFB, and SFC) are called directly or indirectly from an organization block (OB) for processing. Furthermore, all organization blocks are called by the operating system based on a pre-assigned priority class. Since local memory is allocated to the priority of OBs, the local memory available to each code block is based on the OB from which the block is called.

In CPUs where local memory may be re-allocated, the memory assigned to a priority can be re-distributed to where it is needed. In some cases for instance, a priority class is not in use or all of its local memory is not needed. Re-allocation of the local memory is useful since priority classes do not all need the same local stack size. High priority classes, for example, do not normally involve large data requirements or more than two levels of block calls.

Application Tips

When allocating local memory, all temporary variables of the OB and of the blocks called from that OB must be considered. If too many nesting levels (blocks called in a horizontal direction) are used, the L-stack may overflow. The local stack requirements of each program should be tested. The local data requirements of synchronous error OBs must always be taken into consideration. All organization blocks require a minimum of 20-bytes of local memory to store the 20-bytes of start information entered by the CPU.

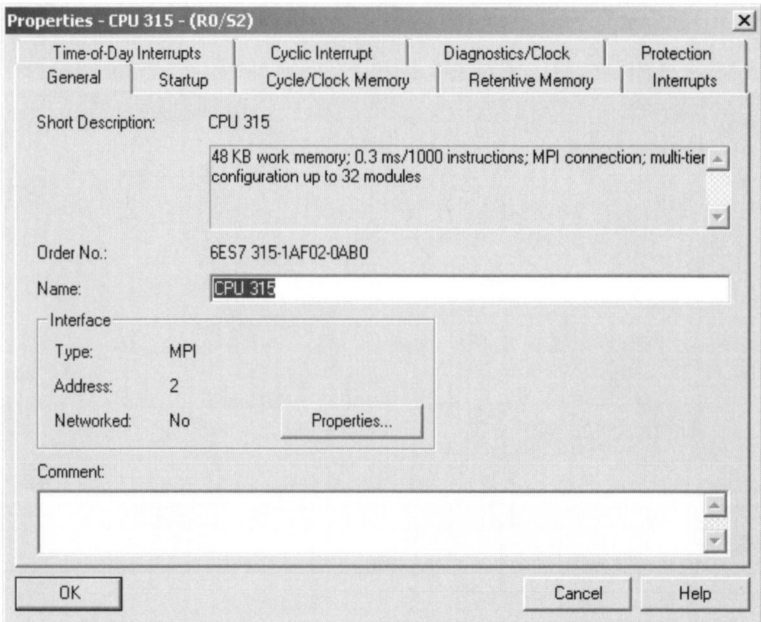

Figure 3-37. CPU 315 - No Memory tab. All priorities allocated 256-bytes.

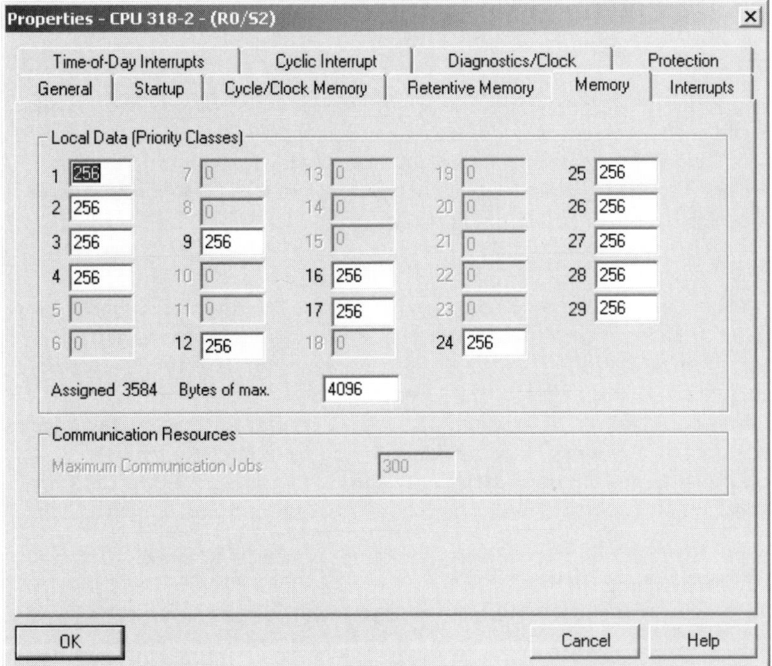

Figure 3-38. CPU 318 Local memory parameters. Memory re-allocation allowed.

Quick Steps: Configuring S7-300 CPU Local Memory

STEP	ACTION
1	With the Hardware Configuration tool open to the desired project and station; select the CPU module, right-click and select **Object Properties,** then select the *Memory* tab to make required changes to the local memory areas.
2	Click in a **Priority** field and enter the total number of bytes to define for each priority class. A minimum of 20 bytes must be entered to satisfy the minimum requirement of every organization block for storing 20 bytes of start information used by the CPU. The value must be divisible by 2.
3	Modify as many priority classes as required and confirm settings with **OK**.
4	Use the toolbar **Save and Compile** button to generate the system data blocks.

Configuring S7-300 CPU Diagnostics and Clock Properties

Basic Concept

Each S7-300 is equipped with a real-time clock that is required for using *runtime meters* and for triggering *Time-of-Day interrupts*. Each CPU also incorporates a *Diagnostic Buffer* — an S7 memory area where Diagnostic events are stored in the order they occur. Modifying the default Diagnostics and Clock properties is not essential; however these parameters allow you to influence operating characteristics of the diagnostics buffer and of the CPU clock.

Essential Elements

Three parameters influence the diagnostic buffer. Activating the **Extended Functional Scope** causes the CPU to enter events other than standard errors into the diagnostic buffer. For example, each start of an organization block is considered a diagnostic event. The increase in diagnostic events will cause the buffer to overflow sooner, and make it more likely that important messages will be overwritten. In some CPUs the parameter **Number of Messages in Buffer**, allows the diagnostic buffer size to be adjusted. The buffer normally holds 100 messages. The parameter **Report Cause of STOP**, if activated, causes each CPU STOP to be entered in the buffer and reported to a designated PG/PC or operator panel.

The clock can be synchronized periodically based on a user-defined **synchronization mode**, **time interval** and **correction factor**. If the CPU is one of several clock-equipped modules in the local rack or in a network, the synchronization mode allows the clock to be designated as a master and responsible for setting other clocks; or as a slave and set by another clock. Furthermore, your selection will determine if the clock is synchronized internally from the PLC bus, or externally over the MPI port. Whether or not a clock can be synchronized externally or act as either master or slave is CPU-dependent.

Application Tips

The extended scope cannot be activated in STEP 7 after V3.1. If the parameter was previously active in a project now being converted from V2.x to V3.1 or higher, then the setting can now be downloaded to the CPU or de-activated.

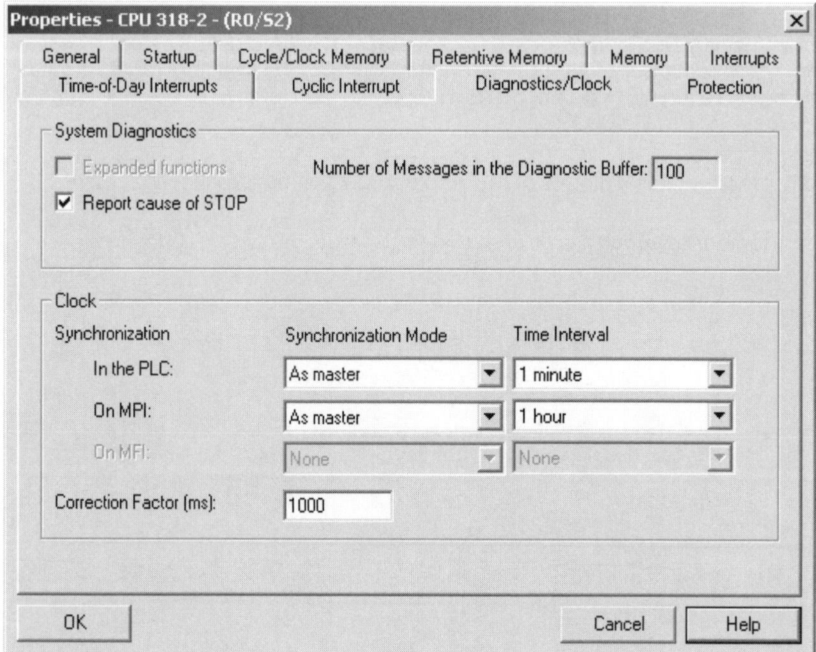

Figure 3-39. CPU Properties: CPU Diagnostics and Clock parameters.

Quick Steps: Configuring S7-300 CPU Diagnostic and Clock Properties

STEP	ACTION
1	With the Hardware Configuration tool open to the desired project and station; select the CPU module, right-click and select **Object Properties,** then select the *Diagnostics/Clock* tab to make the required changes.
2	Activate the **Report Cause of STOP** check box, to have all causes of stops to be entered to the diagnostic buffer and to be sent to a designated PG/PC or operator panel. Otherwise deactivate the checkbox to minimize the number of events entering the buffer.
3	In the PLC field under **Synchronization Mode**, select *As Master*, if the clock will synchronize other clocks; select *As Slave*, if the clock will be set by another local clock; or select *None* if the local clock is not to be synchronized.
4	In the PLC field under **Time Interval,** select the synchronization interval, if the clock was set to synchronize other clocks. In most S7-300 CPUs, the synchronization mode can only be set for Master.
5	In the MPI field under **Synchronization Mode**, select *As Master*, if the clock will synchronize other clocks; select *As Slave*, if the clock will be set by another local clock; or set *None* if the local clock is not to be synchronized. In most S7-300 CPUs, the external synchronization mode can only be set for Slave.
6	In the MPI field under **Time Interval,** select the synchronization interval, if the clock was set to synchronize other clocks.
7	Confirm the configuration parameters with the **OK** button.

Configuring S7-300 CPU Access Protection

Basic Concept

Some S7-300 CPUs support the ability to use password protection to limit access to the CPU's control program and other operations. This protection is in addition to the protection provided by the CPU key-switch. With password protection enabled, the control program and its data are protected from unauthorized changes (i.e., write access protection). It is even possible to prevent read access to program blocks that are considered proprietary (i.e., read access protection). Use of online functions such as upload/download blocks may also be prohibited to non-password holders.

Essential Elements

Using CPU access protection involves setting the protection level and if required, defining a password. Once configured, the protection level and password is downloaded to the module with the configuration data. Complete read and write access is possible for all password holders, regardless of the key-switch position. Otherwise, three levels of protection may apply to non-password holders. **Level 1** is the default setting in which standard key-switch access is in operation. **Level 2** provides *write-protection*; read access is possible, but write access is denied without the password. **Level 3** provides *read/write protection*, which means that both read and write access are denied, independent of the key-switch position.

Application Tips

On CPUs of series S7-31xC, the operating mode switch is not a key-switch. In this case, the switch has only two positions, RUN and STOP. Both of these positions have no restrictions. For this reason, setting a password for protection level 1 is not possible.

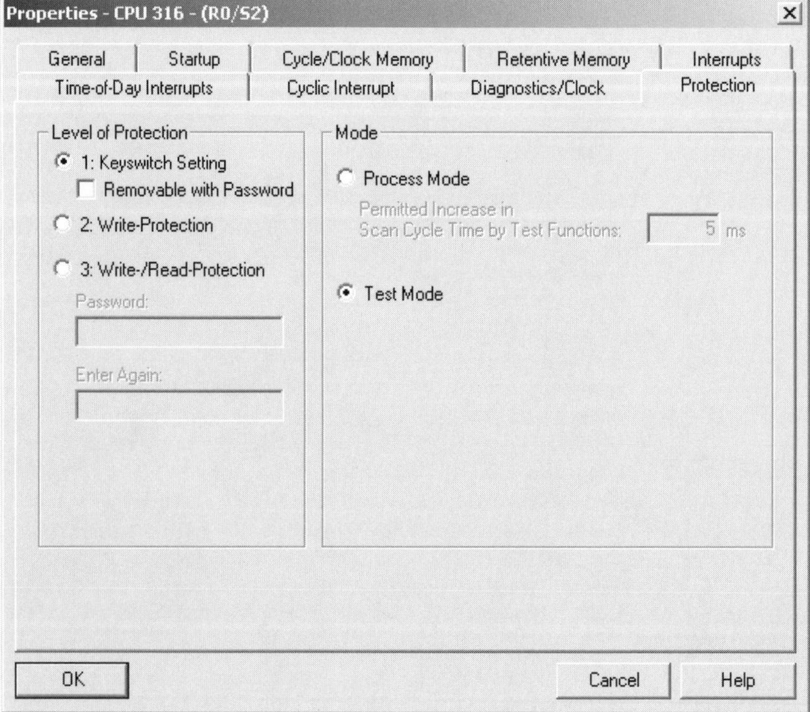

Figure 3-40. CPU Properties: CPU Access Protection tab.

Quick Steps: Configuring S7-300 CPU Access Protection

STEP	ACTION
1	With the Hardware Configuration tool open to the desired project and station; select the CPU module, right-click and select **Object Properties,** then select the *Protection* tab.
2	By default, *Level 1* or *Key-switch* protection is activated; at this level, password protection is not enabled. There is no read or write access restriction in either the RUN-P or STOP position; no write access is permitted in the RUN position. The key-switch must be moved to STOP or RUN-P before any download or data write (e.g., modify variables) operation is allowed.
3	Activate the **Removable with Password** option to allow read and write access to password holders, independent of the key-switch position. A password must be entered in the **Password** field and in the **Enter Again field** as confirmation.
4	Select *Level 2 Protection* by first activating the **Write Protection** radio button; then enter and confirm the desired password in the appropriate fields. In this mode, data and programs may be monitored, but write operations such as download or modify variables is only permitted to password holders.
5	Select *Level 3 Protection* by first activating the **Write-Read Protection** radio button; then enter and confirm the desired password in the appropriate fields. In this mode, read and write operations are permitted to password holders only.
6	Activate the **Process Mode** of operation in order to restrict PG/PC test functions such as program status or monitor/modify variables; the intent is that the set permitted scan cycle time increase is not exceeded. In this mode, testing with breakpoints and single-step program execution are not possible.
7	Activate the **Test Mode** of operation if there should be no time restrictions placed on test functions via the PG/PC. Realize that these operations may result in considerable increase in the CPU cycle time.
8	Confirm the configuration parameters with the **OK** button.
9	Use the toolbar **Save and Compile** button to generate the system data blocks.

Configuring S7-300 CPU Interrupt Properties

Basic Concept

The S7-300 CPU Interrupts tab lists four categories of interrupts. *Hardware Interrupts* include those generated by machine or process inputs using an interrupt input module or those triggered by modules (e.g., CP, FM, or analog SM) that have onboard events that can generate an interrupt (e.g., over limit, or setpoint reached). *Time-Delay Interrupts* are used when the main cyclical program (OB1) must be interrupted after a precise time delay, generated in the use program, has expired. *Asynchronous Error Interrupts* are generated by system-related faults (e.g., battery failure, power failure, wire break on analog input). DPV1 interrupts can be triggered by DPV1 slave devices, to ensure that the CPU in the DP master processes the event in the slave.

Essential Elements

Each S7 interrupt has an associated organization block that is called by the operating system for processing whenever the interrupt occurs. Which of the listed OBs are actually available is CPU-dependent. OBs that are not supported in a CPU are grayed out. In most S7-300 CPUs, for example, OB 40 is the only hardware interrupt; OB 20, the only time-delay interrupt. The CPU 318, and some newer CPUs, support additional interrupts. The priority is generally fixed in the S7-300, but where changes are allowed only the allowed alternates should be used. Hardware and time-delay interrupts alternates include 2-to-24; asynchronous interrupts allow 24-to-26; and DPV1 interrupts allow 2, 3, 4, 9, 12, 16, and 17. A priority of zero essentially deactivates an interrupt OB. To avoid loss of interrupts, you should not assign a priority twice.

Application Tips

When an interrupt occurs, the main cyclical program (OB1) is interrupted. Responding to an interrupt requires writing the STEP 7 code for the OB, and downloading the block to the CPU. The OB is then ready to act as the service routine — called when the interrupt is triggered. The code written in the OB should be kept as short as possible (e.g., activate alarm, send message, halt machine). Remember, time-delay interrupts are triggered from the program. You will need to call the system function (SFC 32, SRT_DINT) for starting a time delay interrupt.

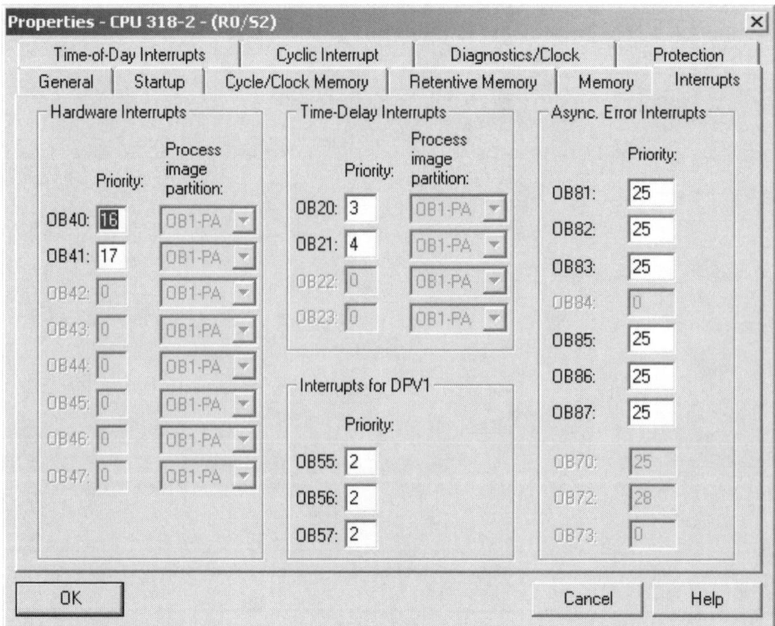

Figure 3-41. S7-300 CPU Properties: Interrupts Parameters dialog.

Quick Steps: Configuring S7-300 CPU Interrupt Properties

STEP	ACTION
1	With the Hardware Configuration tool open to the desired project and station; select the CPU module, right-click and select **Object Properties,** then select the *Interrupts* tab to make your desired priority-class changes (optional).
2	OB40 is the only Hardware Interrupt OB in most S7-300 CPUs and it has a fixed priority (priority-16). If the open CPU supports modifying the default priority then you may enter a new value of 0, or 2-24. A value of zero may disables an OB.
3	OB20 is the only Time-Delay Interrupt OB in most S7-300 CPUs and it has a fixed priority (priority-3). If the open CPU supports modifying (e.g., CPU 318) the default priority then you may enter a value of 0, or 2-24. A value of zero disables an OB.
4	OB81 is the only Asynchronous Interrupt OB in most S7-300 CPUs and it has a fixed priority (priority-25). If the open CPU supports modifying the default priority then you may enter a value of 0, 2-4, 9, 12, 16, 17, or 24-26. Using the default value or a value in the range of 24-26 ensures that asynchronous error OBs are not be interrupted by other interrupt events.
5	Where the priority of DPV1 Interrupt OBs is alterable (e.g., CPU 318), you may enter a new value of 0, 2-4, 9, 12, 16, 17, or 24. A value of 0 is in effect disabling the OB.

Configuring S7-300 CPU Time-of-Day Interrupts

Basic Concept

Configuring a *time-of-day (TOD) interrupt* is only required to have a part of your STEP 7 program execute at a specific date and time. When the interrupt is triggered, the main cyclical program in OB 1 will be interrupted at the configured time and date and the associated service routine (organization block) of the interrupt will be called for processing. Time-of-day interrupt can be configured so that upon reaching the target date and time, the associated OB is executed once and only once, or every minute, hourly, daily, weekly, monthly, every end of month, or yearly.

Essential Elements

The properties tab for time-of-day Interrupts lists Organization Blocks OB10 through OB17. These are the eight time-of-day interrupts supported by S7, however OB 10 is the only time-of-day interrupt supported by most S7-300 CPUs. The CPU 318, and perhaps some newer CPUs, will support additional interrupts. The CPU does not support OBs that are grayed out. Configuring a time-of-day interrupt requires activating the OB, selecting the execution interval, and setting the start date and time. The priority is generally fixed in the S7-300, but in CPUs that support changes alternate priorities 3-24 must be used.

Application Tips

Programming a time-of-day interrupt involves setting the start parameters, activating the OB, writing the code for the OB, and downloading the OB to the CPU. The OB is then ready to act as the service routine — called when the interrupt is generated. The parameters of time-of-day interrupts may be set here in the hardware configuration tool, or from within the STEP 7 program, using standard system functions. TOD interrupt application might include generating operator messages or production reports, or scheduled process readings.

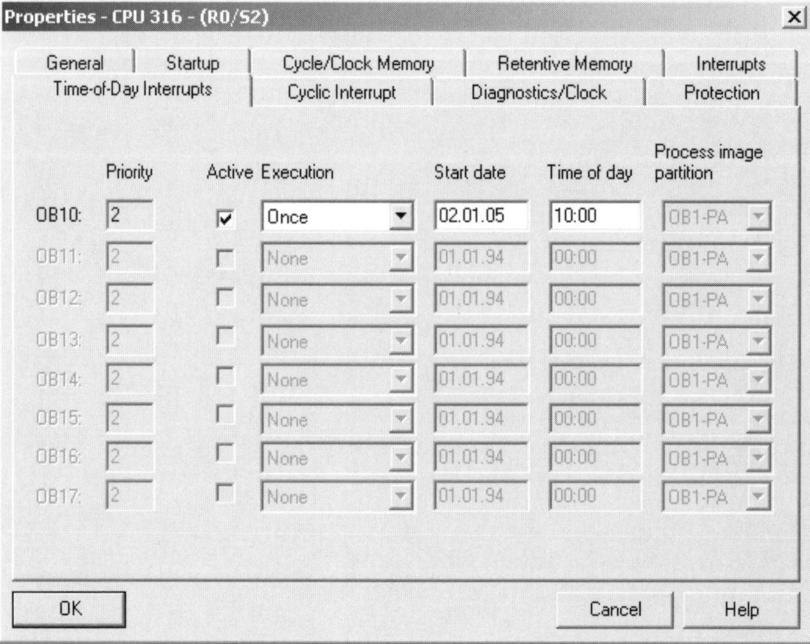

Figure 3-42. S7-300 CPU Properties: Time-of-Day Interrupt tab — CPU 316/earlier CPUs.

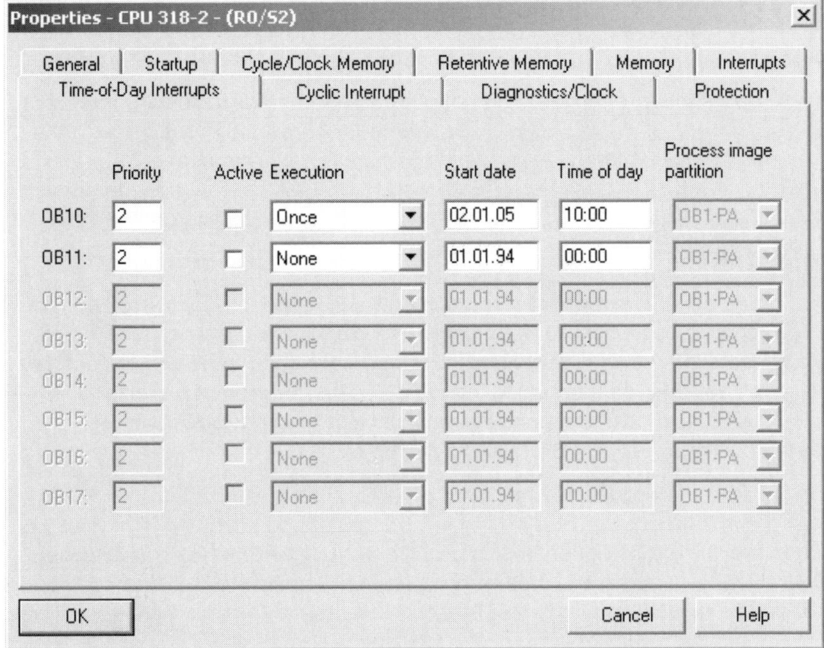

Figure 3-43. S7-300 CPU Properties: Time-of-Day Interrupt dialog CPU 318.

Quick Steps: Configuring S7-300 CPU Time-of-Day Interrupts

STEP	ACTION
1	With the Hardware Configuration tool open to the desired project and station; select the CPU module, right-click and select **Object Properties,** then select the *Time-of-Day Interrupts* tab.
2	Select the time-of-day interrupt OB whose parameters you wish to alter: to modify the **Priority** class click in the field and enter a valid alternate priority of 3-to-24.
3	Click inside the **Active** check box to activate the OB. Time-of-day interrupts may also be activated from the STEP 7 program using system function SFC 30 (ACT_TINT).
4	Click the drop arrow in the **Execution** field, and select whether the OB should be called once, every minute, every hour, every day, every week, every month, every month-end, or every year. Time-of-day interrupts may also be set from the STEP 7 program, using system function SFC 28 (SET_TINT).
5	Next, click in the **Start Date** field and enter the start date for the time-of-day interrupt. If the start date is not entered on the properties dialog, it can be set from the STEP 7 program using the system function SFC 28 (SET_TINT).
6	Click in the **Time of Day** field and enter the start time for the time-of-day interrupt. The start time can be entered here on the properties dialog, or from the STEP 7 program, using the system function SFC 28 (SET_TINT).
7	Confirm the configuration parameters with the **OK** button and use the toolbar **Save and Compile** button to generate the system data blocks.

Configuring S7-300 CPU Cyclic Interrupts

Basic Concept

A *cyclic* or "*watchdog*" *interrupt*, is generated at a fixed periodic interval, and triggers execution of a specific cyclic interrupt OB. Cyclic interrupts are used to interrupt the CPU's normal processing at precise intervals (e.g., every 100 ms, every 500 ms, or every 5 sec) to execute code that must be processed on a regular basis. Typical uses of cyclic interrupts include communications, control loops, temperature control, and sampling analog inputs.

Essential Elements

The properties tab for cyclic interrupts lists Organization Blocks OB30 through OB38. These are the nine cyclic interrupt OBs supported by S7. Each OB is assigned a priority class (7-15), as well as a default cyclic interval at which each is processed. Which of these OBs are actually available, however, depends on the CPU. In most S7-300 CPUs, for example, only OB 35 is available. The CPU 318, and perhaps some newer CPUs, will support additional cyclic interrupts. The CPU does not support OBs that are grayed out.

Configuring changes to default parameters is not necessary unless required. The priority is generally fixed in the S7-300, but in CPUs that allow changes only the allowed alternates (i.e., 3-24) should be used. It is also important that a priority is not assigned twice, so as to avoid loss of an interrupt. It is also possible to modify the period of execution and to enter a Phase offset value. If several cyclic interrupts are active, a phase offset will ensure that interrupts are distributed throughout the CPU cycle and do not trigger simultaneously.

Application Tips

Starting a cyclic interrupt simply involves determining which OB will be used, modifying the default parameters if required, creating, and downloading the organization block code as part of your program. The OB will be processed according to the specified interval, which is timed from when the mode selector is moved from STOP to RUN. Finally, it is essential that the code programmed in a cyclic OB is executed in significantly less time than the call interval.

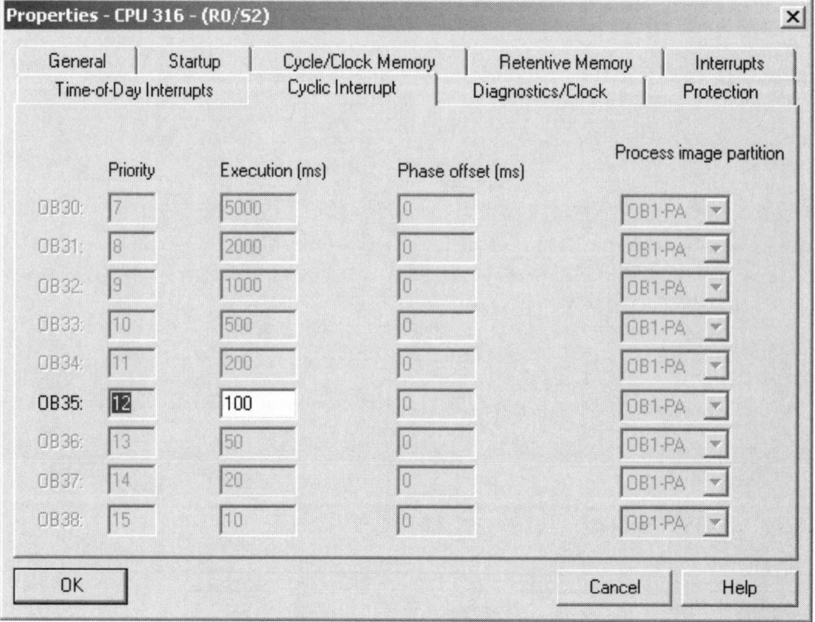

Figure 3-44. S7-300 CPU Properties: Cyclic Interrupt dialog CPU 316 or earlier CPUs.

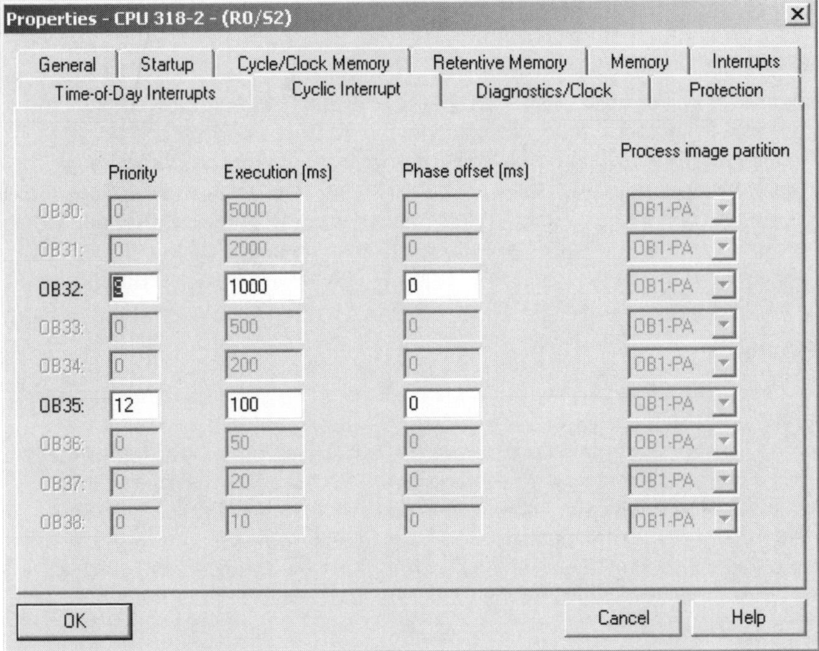

Figure 3-45. S7-300 CPU Properties: Cyclic Interrupt dialog of CPU 318.

Quick Steps: Configuring S7-300 CPU Cyclic Interrupts

STEP	ACTION
1	With the Hardware Configuration tool open to the desired project and station; select the CPU module, right-click and select **Object Properties,** then select the *Cyclic Interrupts*. Grayed parameters are not available in the opened CPU.
2	Select the cyclic OB whose parameters you wish to alter, and enter a new **Priority** class if you wish to modify the default priority; click in the field and enter a valid alternate (3-24). Assigning a priority twice should be avoided to prevent loss of interrupts due to multiple (>12) interrupts of the same priority occurring simultaneously.
3	Enter a new **Execution** value if you wish to modify the default processing interval for the OB; click in the field, and enter a valid value in the range of 10 milliseconds to 6 seconds (i.e., 10-to-60,000).
4	If multiple cyclic interrupt OBs will be used, enter a **Phase Offset** value in milliseconds, for each. The phase offset will ensure that processing of the OBs is distributed somewhat evenly throughout the total CPU cycle. A valid range for this field is (0 -to- 60000). The default value of 0 represents no phase offset.
5	Confirm the configuration parameters with the **OK** button and use the toolbar **Save and Compile** button to generate the *System Data* object.

Configuring S7-300 Digital Input Module Properties

Basic Concept

The general properties of standard digital input modules in the S7-300 are viewed or edited from the Hardware Configuration. When inserted into the hardware configuration digital input modules may be started without any modifications to the default properties, unless the module is an interrupt input type (Also see *Configuring S7-300 Interrupt Input Properties*). Basic information about the module is provided on the *General* tab. The addresses assigned to the module are provided on the *Addresses* tab (Also see *Configuring S7-300 Digital I/O Module Addresses*). While addresses are generally fixed in the S7-300, some CPUs do support address changes (e.g., CPU 316-2 DP).

Essential Elements

The ***Short Description***, the same as given in the hardware catalog, gives important features of the digital input module such as number of input circuits, supply voltage, and number of circuits per group (i.e., per common). For example a 120/230 VAC input module may have 16-inputs with four groups of four circuits (grouping of 4). The ***Order Number*** reflects the currently open module, which is assigned to a specific slot in the configuration — it should match the part number of the physically installed module. The ***Name*** field shows the default name of the module. Although the name can be changed, it is not recommended. The **Comment** field allows more details about the module or its application to be documented.

Application Tips

The writing area on a module connector limits what descriptive information that can be written for each digital input. Use the comment field to provide detailed information on the use of the digital inputs of a module.

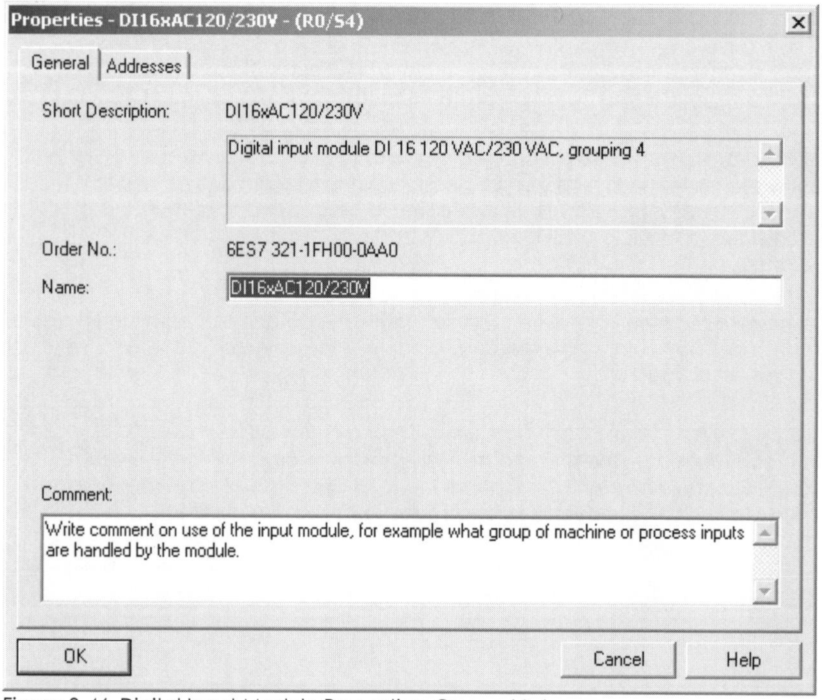

Properties - DI16xAC120/230V - (R0/S4)
General | Addresses
Short Description: DI16xAC120/230V
Digital input module DI 16 120 VAC/230 VAC, grouping 4
Order No.: 6ES7 321-1FH00-0AA0
Name: DI16xAC120/230V
Comment:
Write comment on use of the input module, for example what group of machine or process inputs are handled by the module.
OK | Cancel | Help

Figure 3-46. Digital Input Module Properties: General tab.

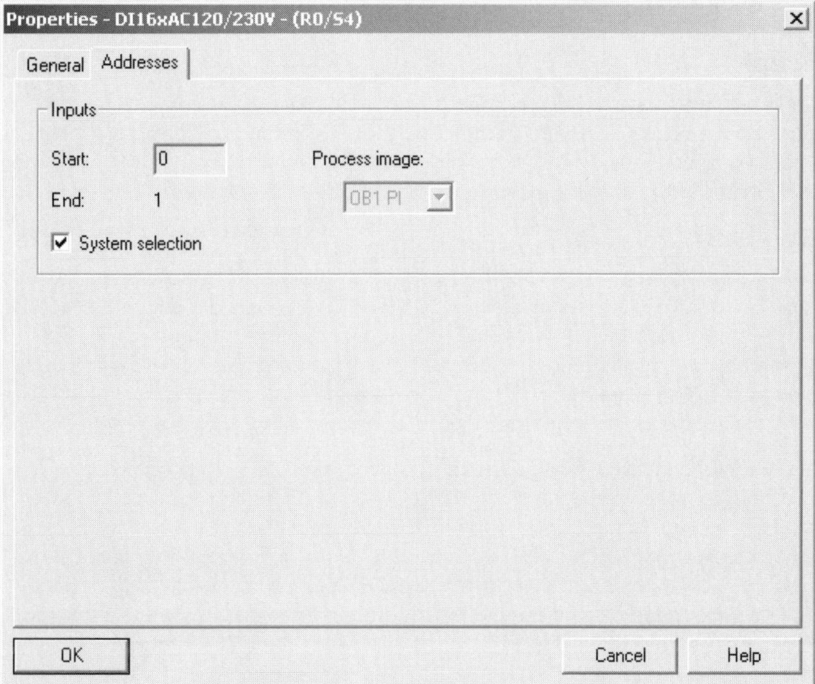

Figure 3-47. Digital Input Module Properties: Address tab.

Quick Steps: Configuring S7-300 Digital Input Module Properties

STEP	ACTION
1	With a station open in the hardware configuration, select a rack to view its installed modules; double click on a digital input module whose properties you wish to view or edit. Not all properties may be edited (e.g., *Order Number*, *Short Description*).
2	On the *General* tab, the **Short description** of the module provides a brief summary of the module features (e.g., number of inputs, supply voltage, circuit grouping). The **Order Number** is the part number of the open module in the configuration and should reflect the digital input module installed in the same slot in the physical rack.
3	Click in the **Name** field to enter a new name for the module; the default name is based on the number of inputs and the supply voltage, prefixed with **DI** for digital input.
4	Use the **Comment** field to enter a description of the module inputs or application.
5	Select the *Address* tab to view or modify the start byte address for the module (e.g., CPU 318). Most S7-300 CPUs do not allow the start byte address to be changed so the **System Selection** check box will be activated and grayed. See *Configuring S7-300 Digital I/O Module Addresses*.
6	Confirm the configuration parameters with the **OK** button and use the toolbar **Save and Compile** button to generate the *System Data* object.

Configuring S7-300 Interrupt Input Properties

Basic Concept

The general and address properties of the interrupt input module are the same as for the standard digital input module, however, a third dialog tab is provided for setting the interrupt module's interrupt properties. By default, hardware interrupts are disabled so if this module is used then it will be necessary to enable interrupt capability from the properties dialog.

Essential Elements

Activating the **hardware interrupt** parameter enables all of the interrupt inputs for use. An **input delay** can be specified for the entire module, based on the type of signals connected. The input delay should closely match the delay characteristics of the input device in order to protect against false input signals. Finally, in groups of two, the interrupts may be set to trigger on the **rising edge** or **falling edge** of the signal, or on both the rising and falling edge. If neither parameter is activated, then the associated channels will not trigger an interrupt.

Activating the **diagnostic interrupt** enables the module to generate a diagnostic interrupt. The diagnostic function varies depending on the module, but on the interrupt input module monitors the supply voltage of the connected sensors. The input channels are divided into groups of eight inputs each (i.e., inputs 0-7, and inputs 8-15). By activating the group check box, each group can be monitored for **no sensor supply**. If monitoring is activated and the supply voltage fails, the group diagnostic event is entered into the diagnostic data storage area of the module. The data can be read from the module, in the STEP 7 program, using the system function to "read system status list" i.e., SFC 51 RDSYSST), or evaluated in OB 82.

Application Tips

Response to hardware interrupts in the user program is handled using organization block 40 in most S7-300 CPUs. In the CPU 318, OB 40 and OB 41 are available. The interrupt OB is assigned to the module on the *Addresses* tab. The interrupt OB must be programmed and downloaded to the CPU. When a hardware interrupt is triggered, the operating system interrupts the main program block (OB1) and calls the designated interrupt OB program.

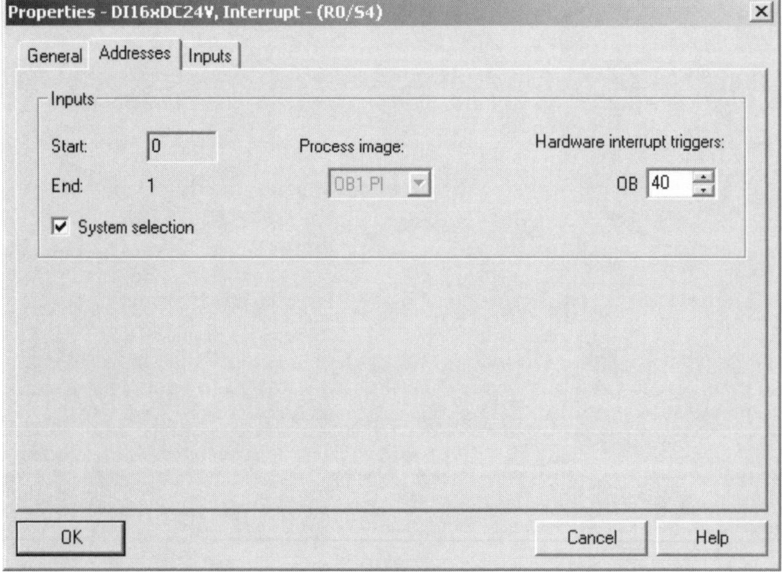

Figure 3-48. S7-300 Interrupt Input Module Properties: Addresses tab.

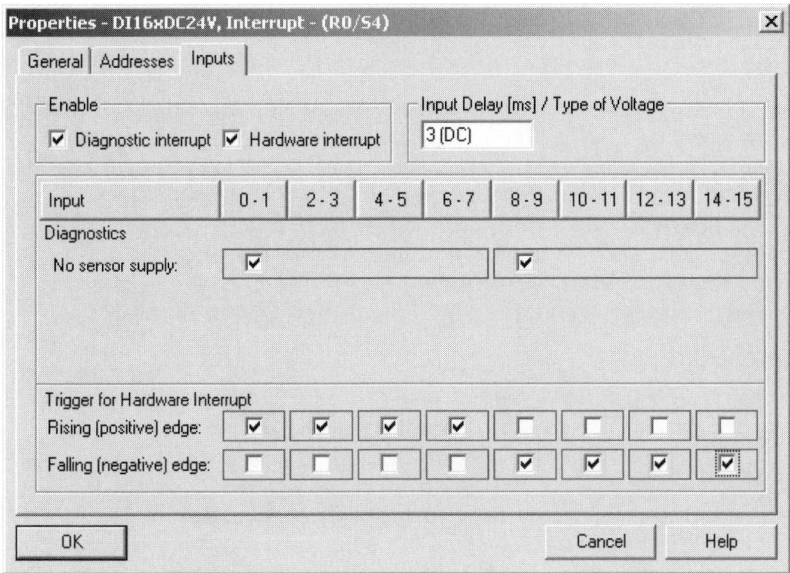

Figure 3-49. S7-300 Interrupt Input Module Properties; Input Parameters tab.

Quick Steps: Configuring S7-300 Interrupt Input Properties

STEP	ACTION
1	With the required station open in the hardware configuration, select the rack where the module is installed; double click on the interrupt input module to open the properties dialog, then select the *Inputs* tab.
2	Activate the **Diagnostic Interrupt** box to enable the module to monitor for loss of the supply voltage for the connected sensor inputs (*No Sensor Supply*).
3	Activate the **Hardware Interrupt** check box to enable use of the sixteen interrupt inputs. When any one of the inputs are triggered, the main program block (OB 1) is interrupted and the assigned interrupt OB (OB 40 - OB 41) is called for processing. The interrupt OB is assigned to the module on the module *Addresses* tab.
4	Click in the **Input Delay** field and select the typical signal delay time (in milliseconds) of the connected input signals. This parameter affects all inputs.
5	If the diagnostics are enabled, then activating the related **No Sensor Supply** check box causes the module to monitor for loss of the supply voltage in channel group 1 (inputs 0-7), or channel group 2 (inputs 8-15). A suitable program in OB82, can be used to evaluate which error has occurred and on which input the error occurred.
6	Activate **Rising Edge** check box, if the associated inputs, above the box, should trigger on a 0-to-1 transition; or Activate **Falling Edge** check box, if the inputs should trigger on a 1-to-0 transition. Each check box affects two inputs (e.g., input 0, 1).
7	Confirm the configuration parameters with the **OK** button and use the toolbar **Save and Compile** button to generate the *System Data* object.

Configuring S7-300 Digital Output Module Properties

Basic Concept

The properties of standard digital S7-300 digital output modules are viewed or edited from the Hardware Configuration. As well, when inserted into the configuration these modules may be started without any changes to the default properties, unless you wish to modify the start address or the module has output signal parameters you wish to influence. Like the digital input module, basic information about the module is provided on the *General* tab, addresses assigned to the module are provided on the *Addresses* tab (Also see *Configuring S7-300 Digital I/O Module Addresses*), and on some digital output modules an *Output* tab, contains parameters that allow the behavior of the outputs on the module to be influenced.

Essential Elements

The ***Short Description*** gives important features of the digital output module such as number of output circuits, supply voltage, and number of circuits per group (i.e., per common). For example a 24 VDC/.5A output module may have 8-outputs with one group of eight circuits (grouping of 8). The ***Order Number*** reflects the module assigned to a specific slot in the configuration — it should match the part number of the physically installed module. The ***Name*** field shows the default name of the module. Although the name can be changed, it is not recommended. The **Comment** field allows details about the module or its application to be documented.

A digital output module may include an *Output* tab if the module has **diagnostic interrupt** features that monitors and reports certain diagnostic events that have been enabled on the module. Diagnostic functions may include monitoring for *wire break, no load voltage, short circuit to ground,* and *short circuit to supply.* This digital output also has the ability to set a **substitute value** that determines the state that each output should be set to when the CPU goes to STOP.

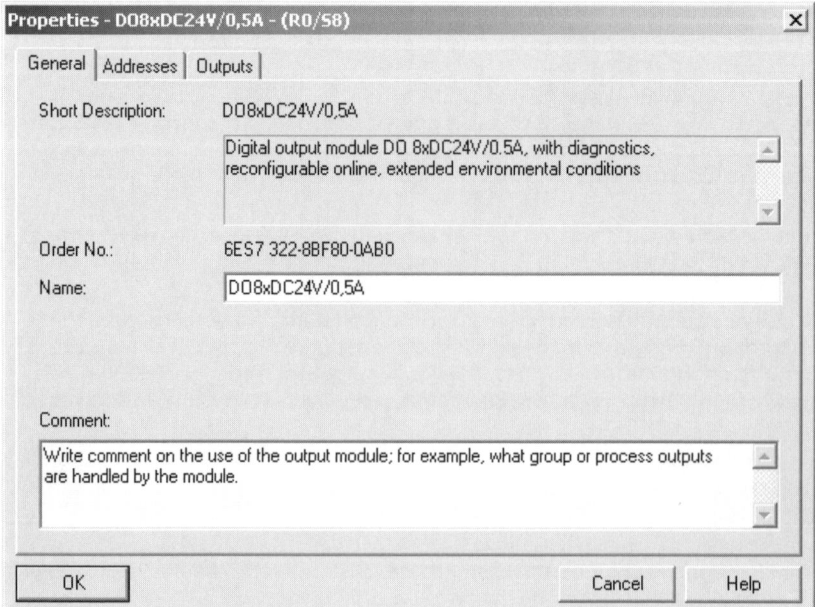

Figure 3-50. S7-300 Digital Output General Properties dialog.

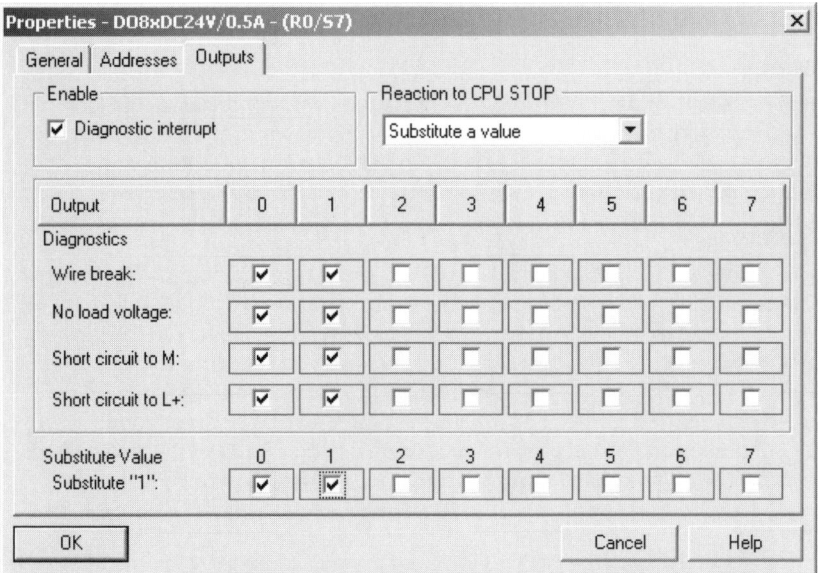

Figure 3-51. S7-300 Digital Output Properties dialog: Output parameter settings.

Quick Steps: Configuring S7-300 Digital Output Module Properties

STEP	ACTION
1	With a station open in the hardware configuration, select a rack to view its installed modules; double click on a digital output module whose properties you wish to view or edit. Not all properties may be edited (e.g., *Order Number*, *Short Description*).
2	On the *General* tab, the **Short description** provides a brief summary of the module features (e.g., number of outputs, supply voltage, circuit grouping). The **Order Number** is the part number of the open module in the configuration and should reflect the digital output module installed in the same slot of the physical rack.
3	The default **Name**, which is modifiable, is based on the number of outputs, the supply voltage, and the output current, all prefixed with **DO** for digital output.
4	Use the **Comment** field to enter a description of the module outputs or application.
5	Select the *Output* tab to set operating parameters for the outputs of the module.
	In the **Reaction to CPU STOP** field, select Keep Last Valid State (logic-0 or logic-1); or select Substitute a Value. If you select "substitute a value," then for each output you wish to be set to logic 1, activate the corresponding check box for Substitute "1". Boxes not activated will be set to logic 0.
7	Activate the **Diagnostic Interrupt** box to enable the module's ability to generate a diagnostic interrupt when any one of its diagnostic monitoring events occurs.
8	To activate diagnostic monitoring (e.g., Wire Break, No Load Voltage, Short Circuit to Ground (M), Short Circuit to Supply (L+), or Blown Fuse) for an output, activate the corresponding check box beneath each output circuit you wish to monitor.

Configuring S7-300 Digital I/O Addresses

Basic Concept

As each digital module is inserted in an S7-300 configuration, the addresses are assigned automatically to the module, based the slot in which the module is inserted. Each slot, as shown in the figure below, reserves four-bytes starting with byte-0 in slot 0. Reserving four bytes, allows each slot to support modules with a maximum of 32 input or outputs.

Essential Elements

The figure below shows an S7- 300 with maximum expansion. Each slot, as shown in the figure, reserves four-bytes starting with byte-0 in slot 0. Slot-4 reserves bytes 0-3; slot-5 reserves bytes 4-7; and so on. If an input module is installed in a slot, the reserved bytes are in the process image of inputs (PII); if an output module is installed, the reserved bytes are in the process image of outputs (PIQ). In the S7-300 the default starting byte-address assigned to each module (*See the configuration table in Figure 3-53*) is generally fixed and cannot be changed. CPUs that support free assignment of addresses (e.g., CPU 315 V 1.1 and higher), also allow input and output modules to use the same starting byte addresses.

Application Tips

Digital inputs are accessed from the PII using the input memory identifier '**I**' as a prefix to the address; digital outputs are accessed in the PIQ, using the output memory identifier '**Q**' as a prefix to the address. Both inputs and outputs are referenced as "**byte.bit**." A 32-point output module in slot-4 would use addresses **Q 0.0** through **Q 3.7**, where Q0.0 corresponds to the first output circuit; an input module would use **I 0.0** through **I 3.7**, where **I 0.0** corresponds to the first input circuit. Inputs and outputs may also be accessed as bytes, words, or double words.

		I/Q	I/Q	I/Q	I/Q	I/Q	I/Q	I/Q	I/Q
Rack-3	**Slot-3**	**Slot-4**	**Slot-5**	**Slot-6**	**Slot-7**	**Slot-8**	**Slot-9**	**Slot-10**	**Slot-11**
Power Supply	IM 361	96.0 to 99.7	100.0 to 103.7	104.0 to 107.7	108.0 to 111.7	112.0 to 115.7	116.0 to 119.7	120.0 to 123.7	124.0 to 127.7

		I/Q	I/Q	I/Q	I/Q	I/Q	I/Q	I/Q	I/Q
Rack-2	**Slot-3**	**Slot-4**	**Slot-5**	**Slot-6**	**Slot-7**	**Slot-8**	**Slot-9**	**Slot-10**	**Slot-11**
Power Supply	IM 361	64.0 to 67.7	68.0 to 71.7	72.0 to 75.7	76.0 to 79.7	80.0 to 83.7	84.0 to 87.7	88.0 to 91.7	92.0 to 95.7

		I/Q	I/Q	I/Q	I/Q	I/Q	I/Q	I/Q	I/Q
Rack-1	**Slot-3**	**Slot-4**	**Slot-5**	**Slot-6**	**Slot-7**	**Slot-8**	**Slot-9**	**Slot-10**	**Slot-11**
Power Supply	IM 361	32.0 to 35.7	36.0 to 39.7	40.0 to 43.7	44.0 to 47.7	48.0 to 51.7	52.0 to 55.7	56.0 to 59.7	60.0 to 63.7

		I/Q	I/Q	I/Q	I/Q	I/Q	I/Q	I/Q	I/Q
Rail 0	**Slot-3**	**Slot-4**	**Slot-5**	**Slot-6**	**Slot-7**	**Slot-8**	**Slot-9**	**Slot-10**	**Slot-11**
CPU and Power Supply	IM 360	0.0 to 3.7	4.0 to 7.7	8.0 to 11.7	12.0 to 15.7	16.0 to 19.7	20.0 to 23.0	24.0 to 27.7	28.0 to 31.7

Figure 3-52. Default Digital Addressing of the S7-300, starting at byte-0. Each slot always reserves 4-bytes, allowing for up to 32-inputs or outputs per module.

Quick Steps: Configuring S7-300 Digital I/O Addresses

STEP	ACTION
1	From the SIMATIC Manager, open the required project and select the S7-300 Station for which addressing is required.
2	With the station selected, in the right pane of the station window double click on the Hardware object to open the station in the Hardware configuration tool.
3	Select a rack whose modules you wish to view the installed digital I/O modules or insert new digital I/O modules.
4	View the starting byte address of each input module in the **I-Address** column or each output module in the **Q-Address** column.
5	To modify the starting byte address of a module, select the module, right-click and select **Object Properties**, then select the Address tab.
6	If the module addresses of this CPU can be modified, the **System Selection** checkbox will have a default check mark which can be removed; otherwise, the box will be activated and grayed out or not available at all. To modify a start byte address, first de-activate the **System Selection** check box to disable system address determination.
7	Enter a new start byte-address and confirm by pressing **OK**. In the S7-300, digital I/O start byte addresses, after byte-0, must be a multiple of 4 (e.g., 4, 8, 12, 16, 24, etc.). In CPUs where you may assign addresses freely (e.g., CPU 315 V1.1 and higher), input and output modules may use the same starting byte address.
8	The **Process Image** field shows the process image responsible for updating the module outputs. By default digital outputs are cyclically updated by the CPU.
9	Confirm the configuration parameters with the **OK** button and use the toolbar **Save and Compile** button to generate the *System Data* object.

(0) UR

Slot	Module	Ord...	Fi...	MPI...	I address	Q address	Comment
1	PS 307 10A	6ES7 30					
2	CPU 316	6ES7 31		2			
3	IM 360	6ES7 36			2000		
4	DI16xDC24V	6ES7 32			0...1		16-point Input Card: 2-bytes used, 4-bytes reserved in PII.
5	DI16xDC24V	6ES7 32			4...5		16-point Input Card: 2-bytes used, 4-bytes reserved in PII.
6	DI32xDC24V	6ES7 32			8...11		32-point Input Card: 4-bytes used, 4-bytes reserved in PII.
7	DI32xDC24V	6ES7 32			12...15		32-point Input Card: 4-bytes used, 4-bytes reserved in PII.
8	DO16xDC24V/0.5A	6ES7 32				16...17	16-point Output Card: 2-bytes used, 4-bytes reserved in PIQ.
9	DO16xDC24V/0.5A	6ES7 32				20...21	16-point Output Card: 2-bytes used, 4-bytes reserved in PIQ.
10	DO32xDC24V/0.5A	6ES7 32				24...27	32-point Output Card: 4-bytes used, 4-bytes reserved in PIQ.
11	DO32xDC24V/0.5A	6ES7 32				28...31	32-point Output Card: 4-bytes used, 4-bytes reserved in PIQ.

Figure 3-53. Default digital addresses in S7-300 configuration. Note that although 4-bytes (32-bits) are always reserved, the actual byte usage of the module is displayed in the I-address and Q-address columns.

Configuring S7-300 Analog Input Module Properties

Basic Concept

The general properties of analog input modules in the S7-300 are viewed or edited from the Hardware Configuration. Basic information about the module, like part number and a brief description, is provided on the *General* tab. The addresses assigned to the module are provided on the *Addresses* tab (See *Configuring S7-300 Analog I/O Module Addresses*). While addresses are generally fixed in the S7-300, some CPUs do support address changes (e.g., CPU 316-2 DP). Before analog input modules are started it may be necessary to set certain module operating parameters (See *Configuring S7-300 Analog Input Signal Parameters).*

Essential Elements

The ***Short Description***, the same as given in the hardware catalog, gives important features of the analog input module such as acceptable input signals, bit resolution, and number of input channels. For example an eight channel analog input module may have 14-bit resolution and accept 0-20 mA/4-20 mA current signals and several voltage ranges. The ***Order Number*** reflects the currently open module, inserted in a specific slot in the configuration — it should match the part number of the actual installed module. The ***Name*** field shows the default name of the module. Although the name can be changed, it is not recommended. The **Comment** field allows details of the module's application to be entered.

Application Tips

The writing area on the module connector limits what descriptive information can be written for each input channel. Use the comment field to provide detailed information on the use of the analog inputs of a module.

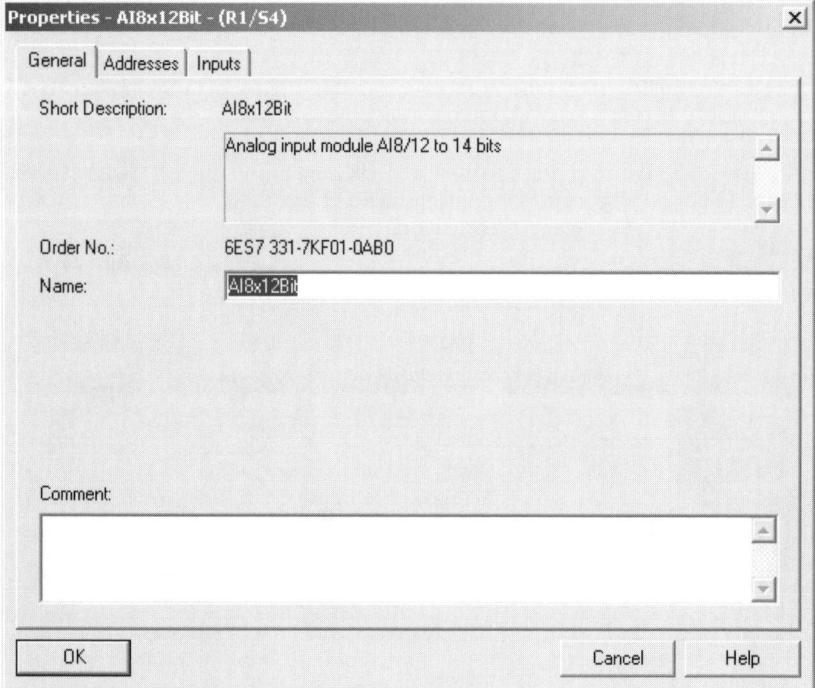

Figure 3-54. S7-300 Analog Input General Properties dialog.

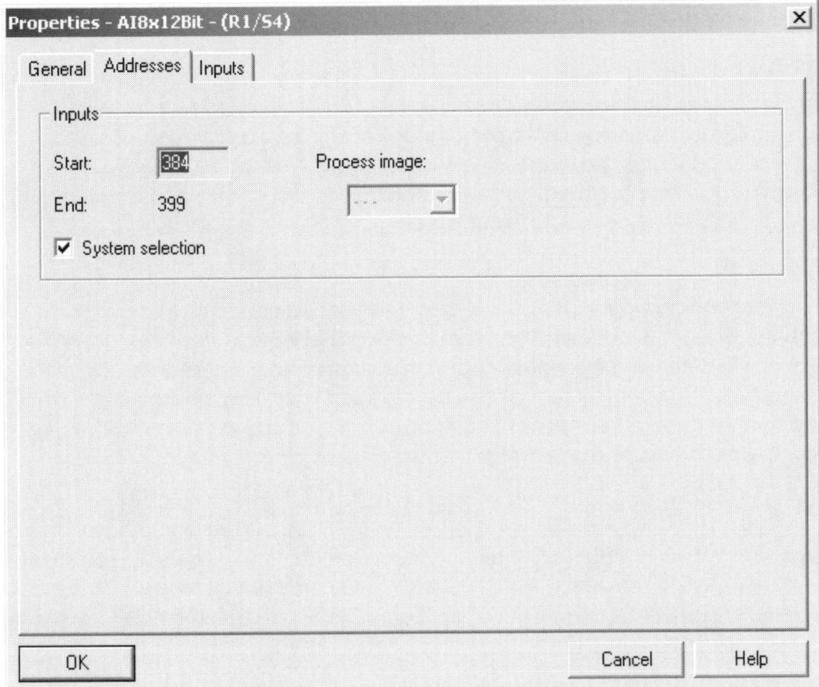

Figure 3-55. S7-300 Analog Input Module Properties: Addresses tab.

Quick Steps: Configuring S7-300 Analog Input Module Properties

STEP	ACTION
1	With a station open in the hardware configuration, select a rack to view its installed modules; double click on an analog input module whose properties you wish to view or edit. Not all properties may be edited (e.g., *Order Number, Short Description*).
2	On the *General* tab, the **Short description** provides a brief summary of the module features (e.g., module type, number of inputs, bit resolution). The **Order Number** is the part number of the open module in the configuration and should reflect the analog input module installed in the corresponding slot of the actual rack.
3	To alter the module name, click in the **Name** field to enter a new name; the default name is the number of analog input channels and the bit resolution prefixed with **AI** for analog input.
4	Use the **Comment** field to enter a description of the module inputs or application.
5	Select the *Addresses* tab to view or modify the start byte address for the module. Most S7-300 CPUs do not allow the start byte address to be changed so the **System Selection** check box will be activated and grayed out. See *Configuring S7-300 Analog I/O Module Addresses*.
6	Confirm the configuration parameters with the **OK** button and use the toolbar **Save and Compile** button to generate the *System Data* object.

Configuring S7-300 Analog Input Signal Parameters

Basic Concept

In addition to the *General* and *Addresses* properties tabs used to configure analog input modules, most analog input modules will also have one or two Input properties tabs. The *Input-Part 1* or *Input-Part 2* tabs allow you to view or configure the module's Hardware Interrupt or Diagnostic Interrupt features as well as determine input channel parameters that affect the type of signals may be used at each input.

Essential Elements

The configuration of analog input signal parameters involves two concerns. In the Enable Section you can activate the module's hardware or diagnostic interrupt features; and in the Input Section, you will determine the input signal characteristics for each channel. On most analog input modules a hardware interrupt can be triggered when user-defined signal limits are exceeded and when all of the input channels have been read and converted (i.e., whenever new measured values are at all input channels).

Depending on the module, diagnostic events may be activated for individual channels or channel groups and include *wire break, configuration parameter assignment error*, or *common mode error*. When a diagnostic interrupt occurs, on a channel or within a channel group, the module reports the event to the CPU (diagnostic buffer entry). Module-related diagnostic information as well as channel-specific diagnostic events can be read from the module using system functions (SFCs).

Application Tips

On modules that use a plug-in range card to allow the various input signal ranges, ensure that the range card is inserted for the required channels. Also ensure that the encoding position on the module matches the letter (e.g., A, B, C) displayed on the input dialog (below each channel) in the fields labeled *Position of Measuring Range Selection Module*.

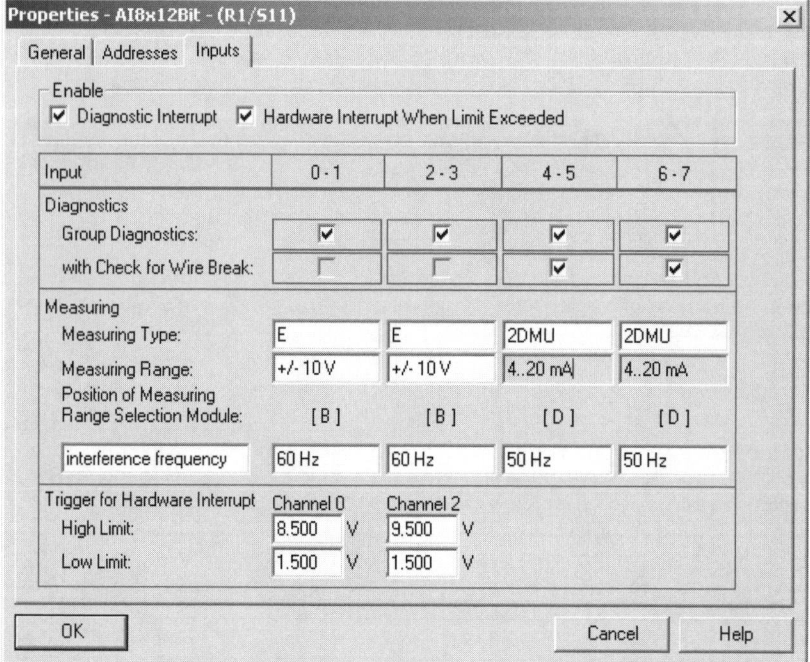

Figure 3-56. Analog Input Module Properties: Input parameters tab.

Quick Steps: Configuring S7-300 Analog Input Signal Parameters

STEP	ACTION
1	With the required station open in the hardware configuration, select the rack where the module is installed; double click on the analog input module to open the properties dialog; then select the *Inputs* tab to view or edit input signal parameters.
2	Activate the check box **Diagnostic Interrupt** in order to use the module's ability to generate an interrupt on certain module-specific diagnostic events.
3	Activate the check box **Hardware Interrupt when Limit Exceeded**, to allow inputs to generate an interrupt when the input signal exceeds (goes above or falls below) the defined limits.
4	Activate the check box **Hardware Interrupt at End of Scan Cycle** (if feature available) to allow the module to signal the CPU with a hardware interrupt when all of the input channels on the module have acquired new measured values.
5	If the "Diagnostic Interrupt" is enabled then for each channel or channel group, you may activate the check box for **Group Diagnostics** and for other channel diagnostics you wish to enable (e.g., **Wire Break**). If a diagnostic event occurs, a diagnostic interrupt is triggered and module-dependent diagnostic information is stored in the module's data area.
6	Click in the **Measurement Type** field directly below each channel or channel group, and from the list of available measurement types (e.g., voltage (E), current, T/C, RTD) select one for the associated channel (s). If a channel or channel group is not connected, select "Deactivated."
7	Click in the **Measurement Range** field directly below each channel or channel group, and from the list of available measurement ranges (e.g. ±10V, ±5V, and ±1V) select a range for the associated channel (s).
8	If the hardware interrupt is enabled for input signal limits exceeded, then for each channel that supports this feature (only certain channels can monitor the input value limits), click in the corresponding fields, and enter the **Upper limit** value and **Lower Limit** value. The module will trigger an interrupt when the input value exceeds the set "Upper limit," or when the input signal falls below the set "Lower Limit."
9	Confirm the configuration parameters with the **OK** button and use the toolbar **Save and Compile** button to generate the *System Data* object.

Configuring S7-300 Analog Output Module Properties

Basic Concept

The general properties of analog output modules in the S7-300 are viewed or edited from the Hardware Configuration. Basic information about the module, like part number and a brief description, is provided on the *General* tab. The addresses assigned to the module are provided on the *Addresses* tab (Also see *Configuring S7-300 Analog I/O Module Addresses*). While addresses are generally fixed in the S7-300, some CPUs do support address changes. Finally, there is the *Outputs* parameters tab, for viewing or setting the operating parameters of the module and individual output channels.

Essential Elements

The ***Short Description***, same as given in the hardware catalog, is a gives a shortlist of features of the analog output module such as output signal ranges, bit resolution, and number of output channels. The ***Order Number*** reflects the module assigned in the configuration tool and can be verified with part number of the physically installed module. The ***Name*** field shows the default name of the module. Changes to the module name are reflected in the configuration table. The **Comment** field allows more detail information about the module (e.g., how it is used) to be documented.

You will be able to configure, the *Outputs* parameters tab, for viewing or setting the signal type and range used by each output channel. The choices for the **Type of Output** are voltage and current;

Application Tips

The writing area on an analog output module connector limits what descriptive information that can be written for each output. Use the comment field to provide detailed information on the use of the analog output channels of a module.

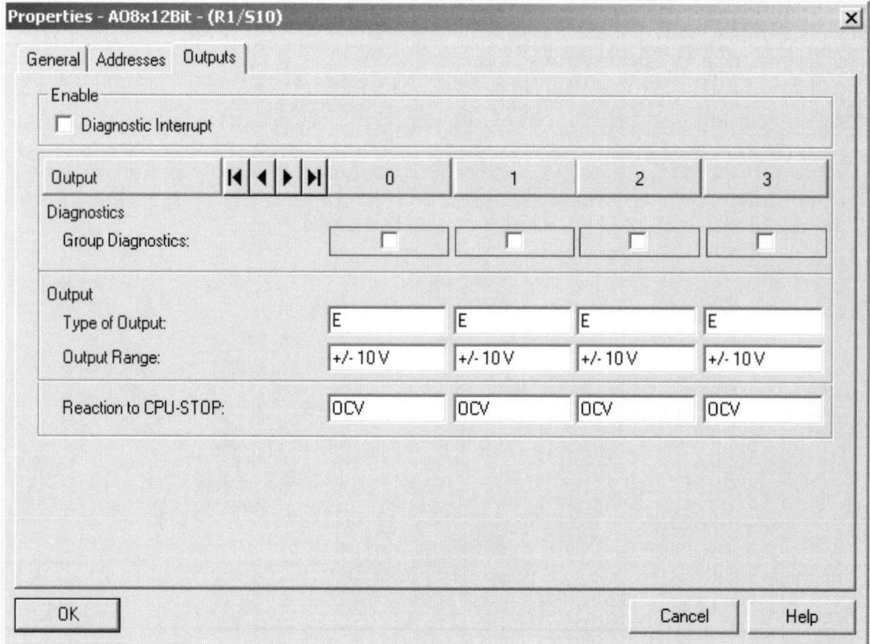

Figure 3-57. S7-300 Analog Output Module Properties: Output parameters tab.

Quick Steps: Configuring S7-300 Analog Output Module Properties

STEP	ACTION
1	With the required station open in the hardware configuration tool, select the central or expansion rack, or DP slave to display its installed I/O modules; select the analog output module and double click to open the properties dialog.
2	On the *General* tab, the **Short description** provides a brief summary of the module features (e.g., number of outputs, digital resolution). The **Order Number** is the part number of the open module in the configuration and should reflect the analog output module installed in the corresponding slot of the physical rack.
3	The default **Name**, which is modifiable, is based on the number of analog output channels and the bit resolution prefixed with **AO** for analog output.
4	Use the **Comment** field to enter a description of the module outputs or application.
5	Select the *Addresses* tab to view or modify the start byte address for the module. Since most S7-300 CPUs do not allow the start byte addresses to be changed, the **System Selection** check box will be activated and grayed out. Otherwise, you may de-activate the check box. See *Configuring S7-300 Analog I/O Module Addresses*.
6	Select the *Outputs* tab to view or modify the operating parameters for the module and for each analog output channel.
7	Activate the check box **Diagnostic Interrupt** in order to use the module's ability to generate an interrupt on module-specific diagnostic events.
8	If the "Diagnostic Interrupt" is enabled then you may activate the check box for **Group Diagnostics** for each channel you wish to have monitored (e.g., Wire break for current outputs, and Short circuit for voltage outputs). If a diagnostic event occurs, a diagnostic interrupt is triggered and module-dependent diagnostic information is stored in the module's data area.
9	Click in the **Measurement Type** field directly below each channel, and from the list of available measurement types [e.g., voltage (E), current (I)] select one for the associated output channel. If a channel is not connected, select "Deactivated."
10	Click in the **Measurement Range** field directly below each channel, and from the list of available measurement ranges (e.g. ±10V, 1-5V, and 0-10V) select a range for the associated output channel.
11	In the **Reaction to CPU STOP** field, select Keep Last Value (KLV); or select Zero Current/Voltage (0CV). If you select "0CV," then for each current or voltage output channel will output a value of zero.
12	Confirm the configuration parameters with the **OK** button and use the toolbar **Save and Compile** button to generate the *System Data* object.

Configuring S7-300 Analog I/O Addresses

Basic Concept

Default analog I/O addressing in the S7-300, like with digital I/O, is predetermined by the physical slot in which the module is inserted. Each slot, as shown in the figure below, reserves sixteen bytes of the peripheral memory (P) area, starting with byte 256 in slot 0. Since each analog input or output channel uses two bytes to handle its digital data, an allocation of sixteen bytes per slot supports modules of 2, 4, or 8 analog I/O channels.

Essential Elements

The figure below shows an S7- 300 with maximum expansion. Each slot reserves sixteen bytes of analog peripheral memory, starting with byte-256 in slot 0. Slot-4 reserves bytes 256 -to- 271; slot-5 reserves bytes 272 -to- 287; and so on. If an input module is inserted, each input address is prefixed with **PIW** for peripheral input word; if an output module is inserted, each output channel address is prefixed with **PQW** for peripheral output word. The first analog input is **PIW 256** (bytes 256-257); the second input is **PIW 258**. The first analog output is **PQW 256** (bytes 256-257); the second analog output is **PQW 258**. Default starting byte-addresses assigned in the S7-300 cannot be changed, except with the CPU 315 version 1.1 and higher.

Application Tips

Unlike digital I/O, that is updated cyclically and placed in the image tables (PII/PIQ), analog I/O must be handled in the program. This is true since with peripheral memory, modules are directly accessed over the I/O bus. In LAD/FBD an analog input is read by specifying the input address as the source in a move instruction; an analog output is written by specifying the output address as the destination of the move operation. In STL, a load operation reads an analog input by specifying the input address (e.g., L PW 258). A value is sent to an analog output with a transfer operation that specifies the output (e.g., T PW 260).

		PIW/PQW	PIW/PQW	PIW/PQW	PIW/PQW	PIW/PQW	PIW/PQW	PIW/PQW	PIW/PQW
Rack-3	Slot-3	Slot-4	Slot-5	Slot-6	Slot-7	Slot-8	Slot-9	Slot-10	Slot-11
Power Supply	IM 361	640 to 655	656 to 671	672 to 687	688 to 703	704 to 719	720 to 735	736 to 751	752 to 767
Rack-2	Slot-3	Slot-4	Slot-5	Slot-6	Slot-7	Slot-8	Slot-9	Slot-10	Slot-11
Power Supply	IM 361	512 to 527	528 to 543	544 to 559	560 to 575	576 to 591	592 to 607	608 to 623	624 to 639
Rack-1	Slot-3	Slot-4	Slot-5	Slot-6	Slot-7	Slot-8	Slot-9	Slot-10	Slot-11
Power Supply	IM 361	384 to 399	400 to 415	416 to 431	432 to 447	448 to 463	464 to 479	480 to 495	496 to 511
Rail 0	Slot-3	Slot-4	Slot-5	Slot-6	Slot-7	Slot-8	Slot-9	Slot-10	Slot-11
CPU and Power Supply	IM 360	256 to 271	272 to 287	288 to 303	304 to 319	320 to 335	336 to 351	352 to 367	368 to 383

Figure 3-58. Default S7-300 analog addressing, starting at byte-256 of peripheral memory.

Quick Steps: Configuring S7-300 Analog I/O Addresses

STEP	ACTION
1	From the SIMATIC Manager, open the required project and select the S7-300 Station for which addressing is required.
2	With the station selected, in the right pane of the station window double click on the Hardware object to open the station in the Hardware configuration tool.
3	Select a rack whose modules you wish to view the installed analog I/O modules or insert new analog I/O modules.
4	View the starting byte address of each input module in the **I-Address** column or each output module in the **Q-Address** column.
5	To modify the starting byte address of a module, select the module, right-click and select **Object Properties**, then select the Address tab.
6	If the module addresses of this CPU can be modified, the **System Selection** checkbox will have a default check mark which can be removed; otherwise, the box will be grayed out or not available at all. To modify a start byte address, first de-activate the **System Selection** check box to disable system address determination.
7	Enter a new start byte-address and confirm by pressing **OK**. In the S7-300, analog start byte addresses after byte-256 must be a multiple of 16. In CPUs that support free assignment of addresses (e.g., CPU 315 and higher), input and output modules may use the same starting byte addresses (e.g., PIW 256, PQW 256).
8	The **Process Image** field has no meaning with analog modules and is grayed.
9	Confirm the configuration parameters with the **OK** button and use the toolbar **Save and Compile** button to generate the *System Data* object.

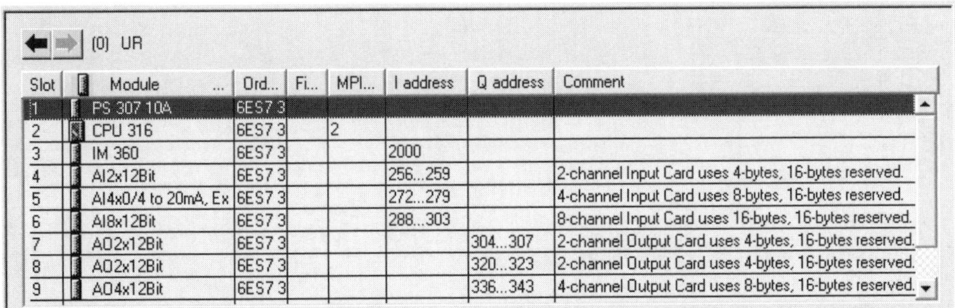

(0) UR

Slot	Module ...	Ord...	Fi...	MPI...	I address	Q address	Comment
1	PS 307 10A	6ES7 3					
2	CPU 316	6ES7 3		2			
3	IM 360	6ES7 3			2000		
4	AI2x12Bit	6ES7 3			256...259		2-channel Input Card uses 4-bytes, 16-bytes reserved.
5	AI4x0/4 to 20mA, Ex	6ES7 3			272...279		4-channel Input Card uses 8-bytes, 16-bytes reserved.
6	AI8x12Bit	6ES7 3			288...303		8-channel Input Card uses 16-bytes, 16-bytes reserved.
7	AO2x12Bit	6ES7 3				304...307	2-channel Output Card uses 4-bytes, 16-bytes reserved.
8	AO2x12Bit	6ES7 3				320...323	2-channel Output Card uses 4-bytes, 16-bytes reserved.
9	AO4x12Bit	6ES7 3				336...343	4-channel Output Card uses 8-bytes, 16-bytes reserved.

Figure 3-59. Default analog addresses in S7-300 setpoint configuration. Note, that although 16-bytes are always reserved, the actual byte usage of the module is displayed under the I-address and Q-address columns.

Configuring an S7-400 Central Rack

Basic Concept

The central rack is the main rack of an S7-400 station. This rack, sometimes called the controller rack, will always contain a CPU and power supply, and perhaps some arrangement of other modules. You will probably get started with this rack in order to connect with the CPU and start with your initial programming and configuration work.

Essential Elements

If a CPU has not been selected, you will need to select one that meets the known requirements of the application. Options for the main rack type will include the choice of a central rack (CR) or a universal rack (UR), both of which support signal (SM), communications (CP), and function modules (FM). Rack options will include number of slots, and depending on the particular part, support for redundant power supply. The power supply choice is based upon the power requirements of the combination of modules installed in the rack.

Application Tips

If you need to connect to the CPU to test portions of your code, the minimum configuration you will need to get started is shown in the figure below. In the initial setup of this station you will also need to install optional backup batteries, required memory cards and provide a power source to the installed power supply. It is important to remember that your hardware configuration must match your actual hardware arrangement.

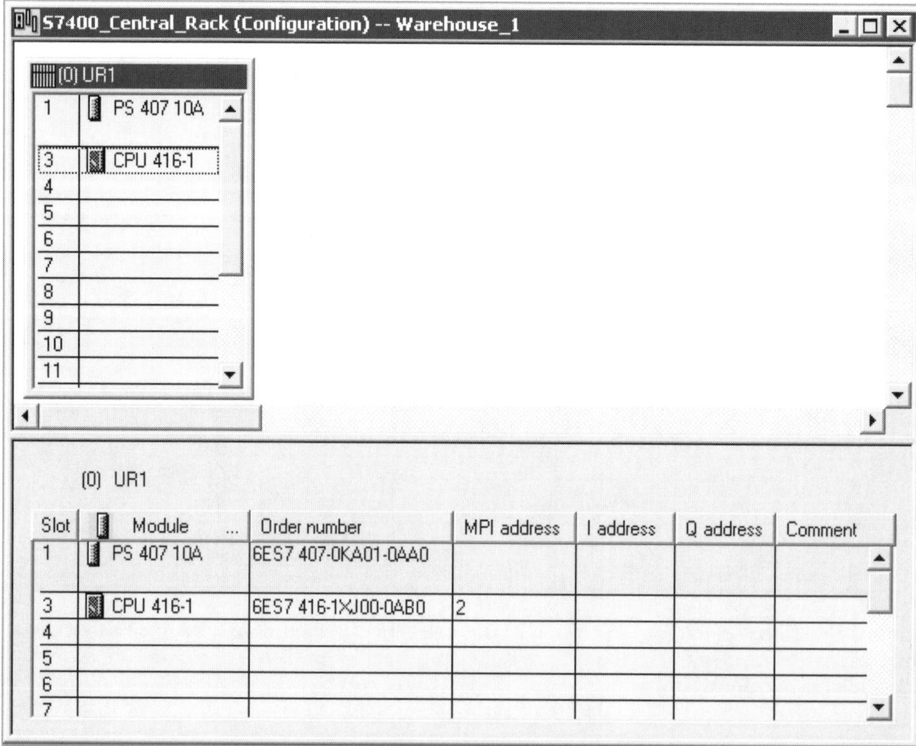

Figure 3-60. S7-400 Central Rack Configuration, with CPU and Power Supply.

Quick Steps: Configuring an S7-400 Central Rack

STEP	ACTION
1	From the SIMATIC Manager, open the required project and from the project tree select the S7-400 Station for which a central rack configuration is to be created.
2	From the right pane, double click on the Hardware object to open the station in the Hardware configuration tool.
3	Open the Hardware Catalog window, from the View menu, if it is not already opened.
4	Open the **SIMATIC 400** catalog object to view the S7-400 component folders.
5	Open the **Rack-400** folder to list the S7-400 racks.
6	Select and drag the appropriate central rack (CR) or universal rack (UR) from the folder to the top pane of the station window. The rack will be labeled 0(CR) or 0(UR).
7	Open the **PS-400** folder, select the desired power supply and drag the part to slot-1 of the central rack.
8	Open the **CPU-400** folder and then the appropriate CPU subfolder to select the required CPU. Drag the selected CPU to slot-3 of the central rack.
9	From the menu, select **Station** ➢ **Consistency Check**, to check for errors.
10	From the menu bar, select **Station** ➢ **Save** to save the configuration; or if done, use **Save and Compile** to generate the System Data that are downloaded to the CPU.

Configuring an S7-400 Multi-Computing Central Rack

Basic Concept

Multi-computing in the S7-400 can involve use of up to four CPUs in a central rack, operating in an independent but synchronized manner. Synchronized operation means that the participating CPUs switch operational modes (e.g., starting, stopping) together. The CPUs can exchange data via their MPI backplane connection (i.e., C-bus).

Dividing the tasks of very large machines or processes into smaller independent control tasks can result in better response times, increased reliability, and uncomplicated programs. Smaller programs are also easier to manage and to troubleshoot. An example case might be to separate all control from data processing and communications functions. All process control loops, for example, may be assigned to a CPU.

Essential Elements

The multi-computing configuration requires the use of CPUs and racks that support this capability. This arrangement is implied when two or more multi-computing CPUs are placed in a rack supporting this operation. A maximum of four CPUs may be installed in the central rack (e.g., UR1). Your specific choice of central rack will depend on the number of slots required and whether or not redundant power supplies are used.

Application Tips

All modules associated with this station, including expansion I/O, must be assigned to a specific CPU in the multi-computing configuration. A module is assigned to a CPU (e.g., CPU1, CPU2, CPU3, or CPU4) under the module's object properties.

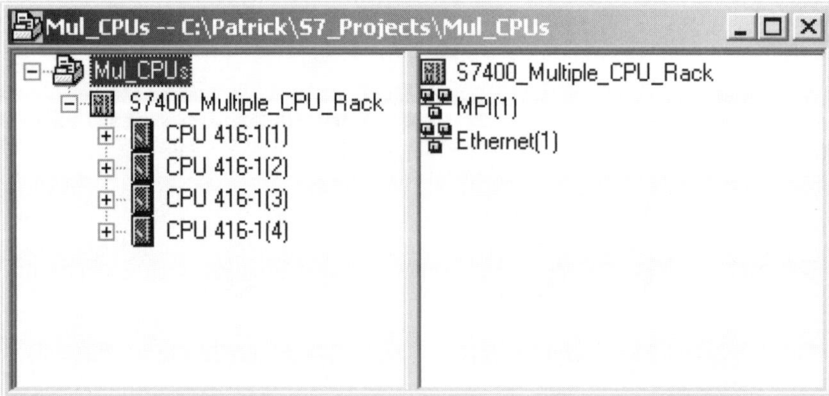

Figure 3-61. Project folder containing multi-computing station with four CPUs.

Quick Steps: Configuring an S7-400 Multi-Computing Central Rack

STEP	ACTION
1	From the SIMATIC Manager, open to the required project and from the project tree select the S7-400 Station to which multi-computing will be added, or first add a new S7-400 station if it does not already exist.
2	With the station selected, from the right pane of the project window double click on the Hardware object to open the station in the Hardware configuration tool.
3	Open the Hardware Catalog window, from the View menu, if it is not already opened.
4	Open the **SIMATIC 400** catalog object to view the S7-400 component folders.
5	Open the **Rack-400** folder to list the S7-400 racks.
6	Select the appropriate universal rack (UR) from the folder, that supports multi-computing and fits the total slots and power supply requirements then drag it to the top pane of the station. The rack will be labeled 0(UR).
7	Open the **PS-400** folder, select your power supply and drag the part to slot-1 of the central rack.
8	Open the **CPU-400** folder and then the appropriate CPU subfolder to select your multi-computing CPU. Drag the selected CPU to slot-3 of the central rack. Add up to four CPU modules as required.
9	Configure parameters for each CPU.
10	From the menu, select **Station ➢ Consistency Check** to check for errors.
11	From the menu bar, select **Station ➢ Save** to save the configuration; or if done, use **Save and Compile** to generate the System Data that are downloaded to the CPU.
12	Download the complete station configuration to all CPU modules.

Configuring the S7-400 as a DP Master

Basic Concept

The S7-400 supports distributed I/O, using the DP component of Profibus, which operates much like remote I/O. The DP configuration may involve one or more *DP Master Stations,* each of which manages the I/O of one or more assigned *DP Slave Stations*. Depending on the requirements of your application, as the DP master, you may either choose an S7-400 with DP master capability or a Profibus communications processor (CP) with DP capability.

Essential Elements

The DP master can be configured using an S7-400 CPU, with an integrated Profibus-DP interface (e.g., CPU 414-2 DP, CPU 416-2 DP); or using a Profibus communications processor with DP master capability (e.g., CP443-5 Ext.), in conjunction with a CPU that supports this feature. In any case, the DP master will control one or more assigned DP slaves.

In configuring the DP Master, the Profibus interface of the DP master must be attached to a DP Master System, and assigned a unique Profibus Address (e.g., 3-125). If not previously set, then the subnet operating parameters (e.g., transmission rate, profile) will also need to be set. See configuration task *Adding and Configuring a Profibus Subnet in Chapter 7.*

Application Tips

Typically, a DP subnet will involve a single DP master. This is referred to as a Mono-Master configuration. You may also have several mono-master systems within a project, by simply creating multiple PROFIBUS DP subnets and attaching a DP master to each. A Multi-Master configuration, involving two or more masters on a single subnet, though not typical, is also possible. You would do this by attaching the DP masters to the same DP subnet.

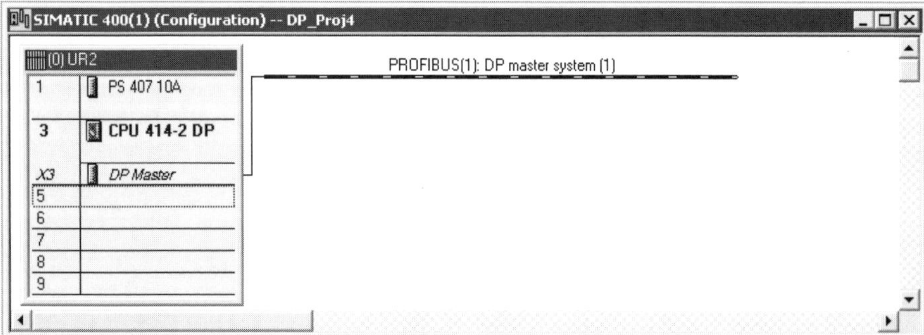

Figure 3-62. S7-400 CPU 414-2 DP selected and inserted as DP master. The black and white dashed line is the Profibus DP Master system object to which DP slaves will be attached.

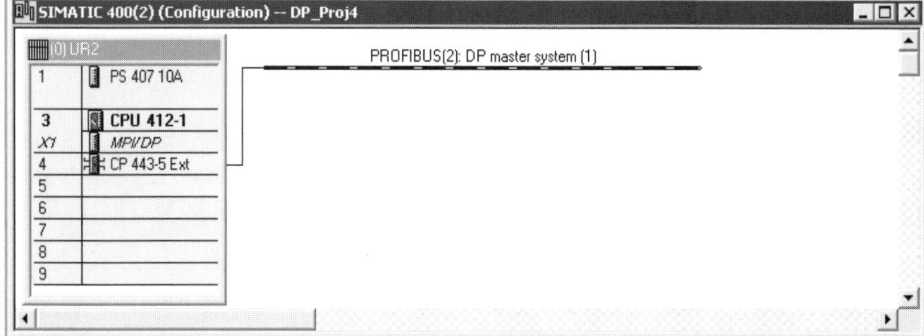

Figure 3-63. S7-400 CP 443-5 Ext. selected and inserted as DP master. The black and white dashed line is the Profibus DP Master system object to which DP slaves will be attached.

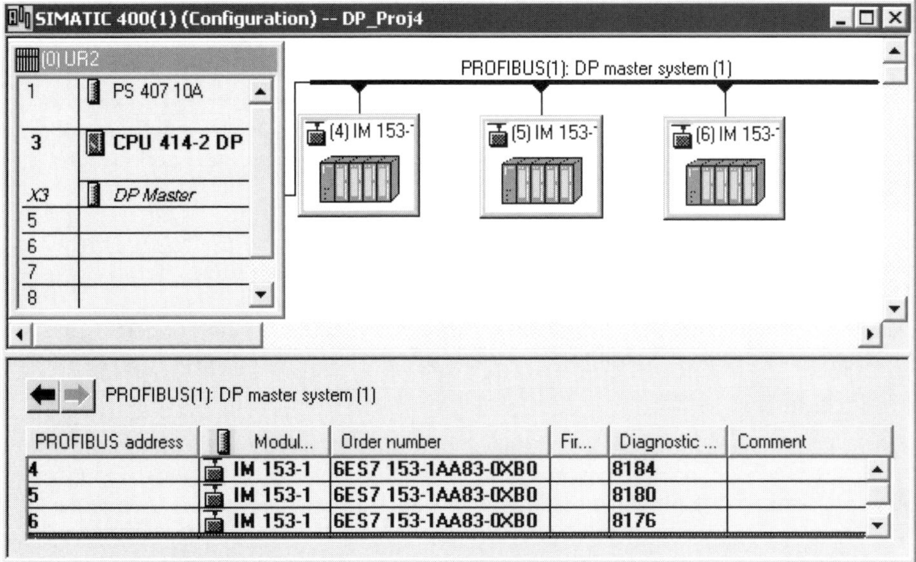

Figure 3-64. S7-400 CPU 414-2 DP -master with ET-200 M modular I/O drops.

Quick Steps: Configuring the S7-400 as a DP Master

STEP	ACTION
1	From the SIMATIC Manager open the required project, and S7-400 station to serve as DP Master. The S7-400 must be inserted to the project if it not already installed.
2	With the S7-400 station opened in the hardware configuration window select the central rack. Insert the required rack and power supply if it is not already installed.
3	If a CPU will serve as the DP master then the correct Profibus-DP CPU must be selected (e.g., CPU 412-2 DP or CPU 416-2 DP). From the catalog open the **CPU-400** folder, select and drag the required CPU to slot-2 of the central rack.
4	When the Profibus interface dialog is presented, select the Profibus subnet to which the DP master should be attached; set or accept the suggested Profibus address (e.g., 3-125), then click on the Properties button to set the subnet **Transmission Rate** and other properties if they have not yet been set.
5	After the DP properties dialog is saved, a DP Master system object should appear extended from the DP master as a black and white dashed line. If the object is not shown, right-click on the *DP* object in row *2X2,* and select **Add Master System**.
6	You can save your work, but at least one DP slave must be attached to compile without errors. See the earlier tasks on *Configuring and Attaching Modular, Compact, and Intelligent DP Slaves.*

If a communications processor (CP) should serve as the DP master, then the appropriate selection must be made from the **CP-400** folder under the Profibus folder. Find the required CP443-5 Extended communications processor and drag it to slot-3 or higher of the central rack. The Profibus interface dialog will be presented and the configuration may be continued with Steps 4, 5, and 6.

Configuring S7-400 Local I/O Expansion

Basic Concept

The S7-400 supports local I/O expansion from 1.5 meters to 3 meters using the appropriate interface modules (IMs). Local I/O expansion involves placing a publisher interface module in an S7-400 central rack and a corresponding receiver interface in each expansion rack.

Essential Elements

The IM460-0 and IM 460-1are both publisher interface modules placed in the central rack to for local expansion. Whichever of these publisher interface modules is used, its matching receiver module is placed in each expansion rack. The IM 461-0 is the receiver for the IM 460-0; the IM 461-1 is the receiver for the IM 460-1. These modules are selected according to distance requirements and whether or not the expansion racks will support communications.

Application Tips

Each IM 460-0 module has two channels, each of which supports up to 4-racks. The IM 460-1 module has two channels, each of which are limited to connection of only one expansion rack; only two IM 460-1interface modules may be placed in the central rack. Altogether, a maximum of six interface modules of any combination may be placed in the central rack.

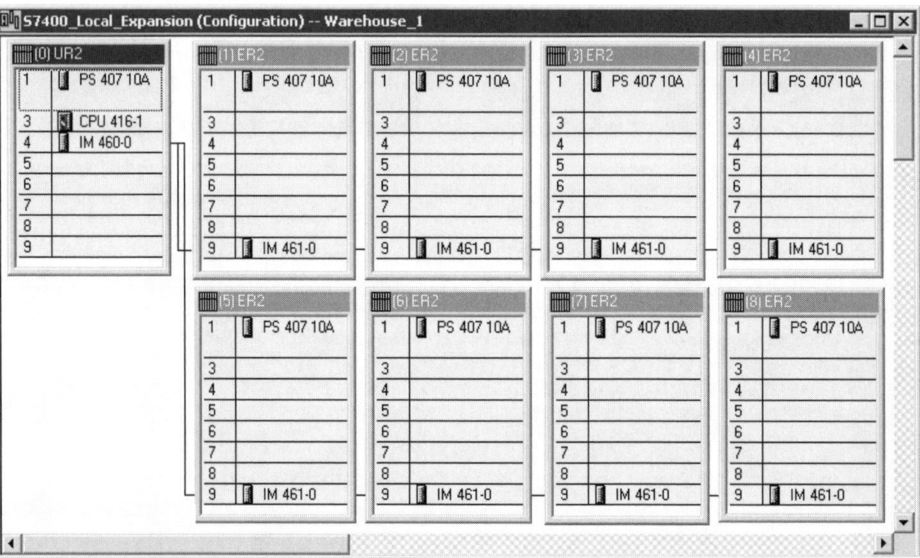

Figure 3-65. S7-400 Local I/O Expansion, using the IM 460-0/IM 461-0 pair.

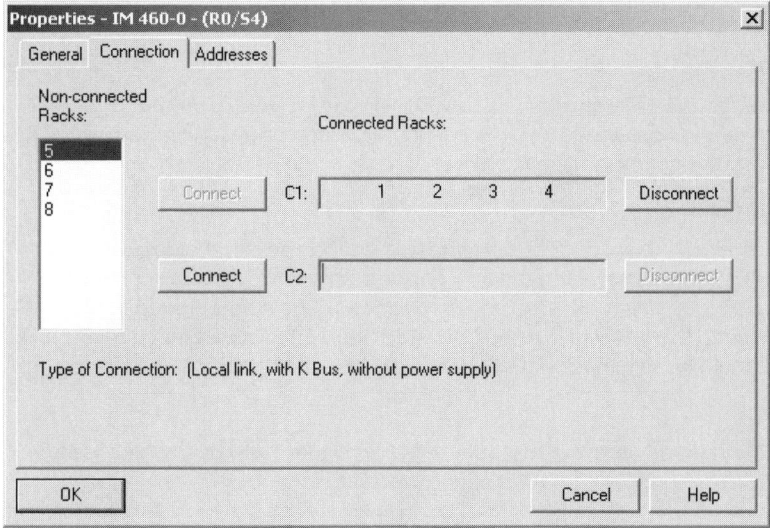

Figure 3-66. S7-400 Local I/O Expansion, using the IM 460-0/IM 461-0 pair.

Quick Steps: Configuring S7-400 Local I/O Expansion

STEP	ACTION
1	From the SIMATIC Manager open the required project, and from the project tree select the S7-400 Station for which local I/O expansion must be configured.
2	With the station selected, from the right pane of the project window double click on the Hardware object to open the station in the Hardware configuration tool.
3	Open the **SIMATIC 400** catalog object to view the S7-400 component folders.
4	Open the **Rack-400** folder, select and drag the appropriate expansion racks (ER) or universal racks (UR) to the top pane of the station window. If the central rack is already installed it is labeled 0(UR) or 0(CR); otherwise it must be configured first.
5	Open the **PS-400** folder, select the desired power supply. Click and drag the part to slot-1 of each expansion rack as required.
6	Open the **IM-400** folder by clicking the plus sign. The S7-400 expansion interface modules are shown. Publisher IMs are installed only in the CPU rack (slot-3 and higher). Receiver IMs are installed in I/O expansion racks (slot 9- only, or slot-18 only).
7	Select an IM-460-0 interface module from the catalog and drag to slot-3 or slot-4 of the CPU rack 0(UR). Select an IM-461-0 interface from the catalog and drag to slot-9 or slot-18 of each installed expansion rack.
8	Select the IM-460-0 in the central rack and right click and select Object Properties to connect the expansion racks to the desired channel. From the *Connection* tab, assign each **Non-Connected** rack to either channel 1 or channel 2, by pressing the corresponding **Connect** button. Use **OK** button to save.
9	From the menu, select **Station ➢ Consistency Check** to check for errors, and then use **Save and Compile** to save the configuration before downloading to the CPU.

Configuring S7-400 Remote I/O Expansion

Basic Concept

The S7-400 supports remote I/O expansion up to 100 meters using expansion interface modules (IMs). Remote I/O expansion involves placing a publisher interface module in an S7-400 central rack and the corresponding receiver interface in one or more expansion racks.

Essential Elements

The 460-3 is a publisher interface module, placed in the central rack. Its matching receiver module, the IM 461-3, is placed in each expansion rack. Each IM 460-3 has two channels, each of which supports up to 4-racks. Each installed rack must be connected to either channel 1 or channel 2. The IM 460-3 may be installed in the central rack with up to a total of six other interface modules to form any combination of local and remote I/O expansion.

Application Tips

S7-400 remote I/O expansion up to 600 meters may be configured using existing or newly installed S5 I/O racks that are connected using the IM 463-2.

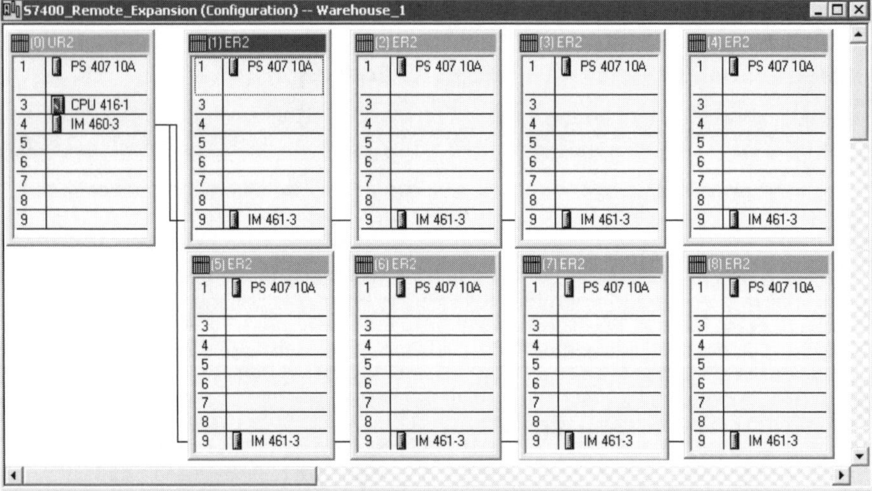

Figure 3-67. S7-400 Remote I/O Expansion, using IM 460-3/IM 461-3 pair.

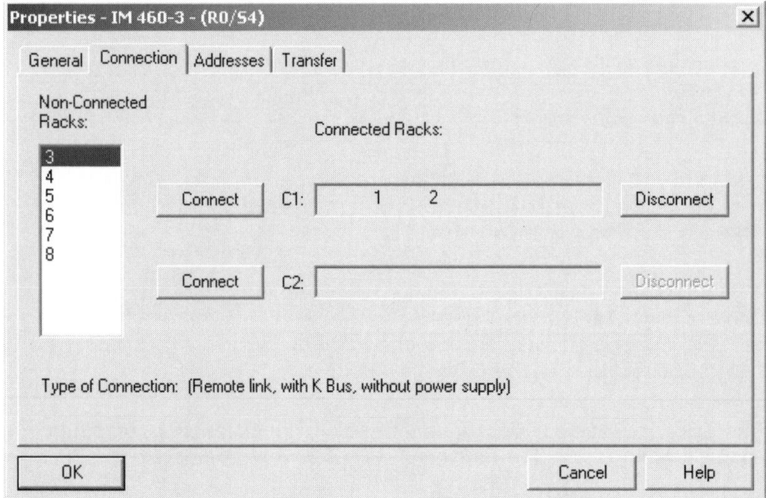

Figure 3-68. IM 460-3 Properties dialog; configuring rack connections.

Quick Steps: Configuring S7-400 Remote I/O Expansion

STEP	ACTION
1	From the SIMATIC Manager open the required project, and from the project tree select the S7-400 Station for which remote I/O expansion must be configured.
2	With the station selected, from the right pane of the project window double click on the Hardware object to open the station in the Hardware configuration tool.
3	Open the **SIMATIC 400** catalog object to view the S7-400 component folders.
4	Open the **Rack-400** folder, select and drag the appropriate expansion racks (ER) or universal racks (UR) to the top pane of the station window. If the central rack is already installed it is labeled 0(UR) or 0(CR); otherwise it must be configured first.
5	Open the **PS-400** folder, select the desired power supply. Click and drag the part to slot-1 of each expansion rack as required.
6	Open the **IM-400** folder by clicking the plus sign. The S7-400 expansion interface modules are shown. Publisher IMs are installed only in the CPU rack (slot-3 and higher). Receiver IMs are installed in I/O expansion racks (slot 9- only, or slot-18 only).
7	Select an IM-460-3 interface module from the catalog and drag to slot-3 or slot-4 of the CPU rack 0(UR); select an IM-461-3 interface module from the catalog and drag it to slot-9 or slot-18 of each of your installed expansion racks.
8	Select the IM-460-3 in the central rack and right click and select Object Properties to connect the expansion racks to the desired channel. From the *Connection* tab, assign each **Non-Connected** rack to either channel 1 or channel 2, by pressing the corresponding **Connect** button. Use the **OK** button to save.
9	From the menu, select **Station ➢ Consistency Check** to check for errors, and then use **Save and Compile** to save the configuration before downloading to the CPU.

Configuring S7-400 CPU General Properties

Basic Concept

The General properties of an S7-400 CPU module are viewed in the properties dialog of the CPU with a right-click on the CPU, then selecting **Object Properties**. Basic information about the CPU, which may vary depending on the relative age of the module, includes items such as type and location, a short list of characteristic features, and for modules with communications interfaces — related information is provided.

Essential Elements

The ***Short Description***, the same as given in the hardware catalog, gives important CPU features such as work memory size, cycle time/1000 instructions, digital I/O capacity, supported connections, DP/MPI ports, multi-computing and routing capability. The ***Order Number*** reflects the module assigned in the configuration tool and can be compared to the installed module. The ***Name*** field shows the default name of the module, which you can modify as required. Changes to the CPU name are reflected in the SIMATIC Manager.

The ***Interface*** parameter group, generally on modules with communications interfaces, shows basic information such as node address and if the node is attached to a subnet. The Properties button provides quick access to view interface parameters, but the interface parameters (e.g., MPI, Profibus, and Ethernet) are set in the hardware configuration tool. The **Comment** field allows information about the module (e.g., application) to be documented.

Application Tips

The user-defined ***Plant Designation*** identifier, available with some CPUs, is a user-defined identifier that can be evaluated in start-up OBs the user program. Examples of evaluating the "plant designation" are provided in the STEP 7 Sample programs.

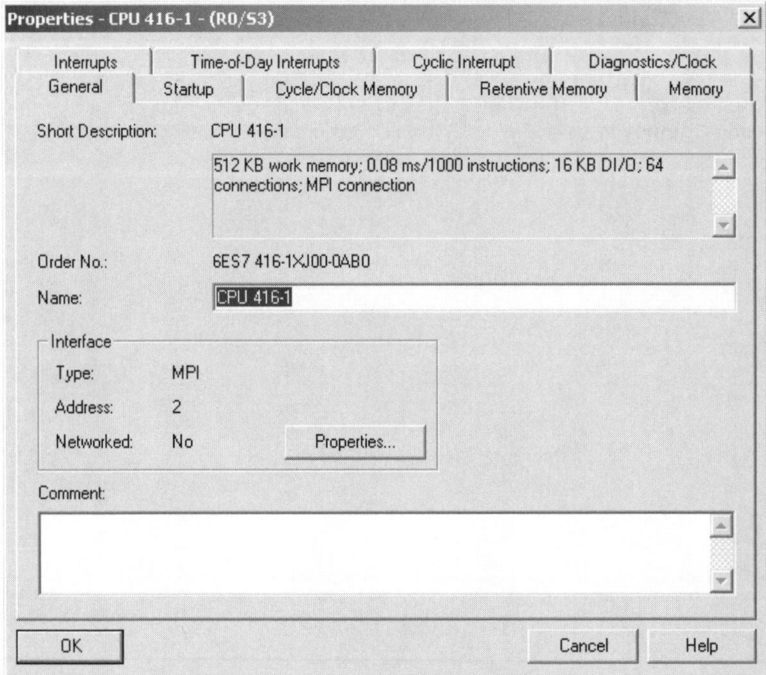

Figure 3-69. CPU Properties: General parameters dialog.

Quick Steps: Configuring S7-400 CPU General Properties

STEP	ACTION
1	With a station open in the Hardware Configuration tool select the CPU, right-click and select **Object Properties,** then select the *General* tab.
2	The **Short description** of the module provides a brief summary of the unique features of the installed CPU module; it cannot be modified.
3	The **Order Number** of the module is the specific number used to identify the exact CPU that should be installed in the physical rack. The order number determines the unique features of the CPU, which are summarized under the Short Description.
4	Click in the **Name** field to modify the default name assigned to the module. The default name is based on the short name of the particular CPU type [e.g., CPU 416-1(1)] and a concatenated numeric value in parentheses, that represents the number of the installed CPU. A number is assigned to the CPU since the S7-400 supports up to four CPUs in a rack.
5	The **Interface** static grouping, shows the module's set MPI address (default = 2), and whether or not the module is attached to the network; the Properties button allows the module's Interface parameters to be accessed for viewing or editing.
6	Click in the **Comment** field to enter a description of the module's application — especially since the S7-400 supports up to four CPU's in one rack.
7	Confirm the configuration parameters with the **OK** button.

Configuring S7-400 CPU Start-Up Properties

Basic Concept

S7-400 CPU modules may be started without any changes to the default start-up parameters. There are a few adjustable parameters, however, that allow you to determine CPU start-up characteristics or whether or not the CPU should even be allowed to start. These parameters are checked by the CPU at the various startups (e.g., power ON, or the transition from STOP-to-RUN), before beginning to process the main cyclical program (OB1).

Essential Elements

The parameter ***Startup if Setpoint and Actual Configuration Differ***, determines if the CPU should be allowed to start-up if the setpoint and actual configurations differ. The default behavior is to start even if the setpoint and actual configurations differ. The parameter Reset Outputs at Hot Restart allows the CPU to reset the outputs, and the process image of the outputs (PIQ) at each hot restart. This parameter is essential since outputs are not normally reset on a Hot Restart. The parameter ***Disable Hot Restart by Operator*** prevents a hot restart from the PG/PC or otherwise. The parameter ***Startup After Power-On*** determines whether a start-up after a power-up will initiate a "Warm Restart," a Hot Restart," or a "Cold Restart."

The Monitoring Time parameter, ***Finished Message***, specifies the maximum CPU wait-time, at power-up, to receive ready-signals for all configured modules. The parameter ***Transfer of Parameters to Modules*** specifies the maximum time for modules, including DP-Interface modules, to acknowledge receipt of their configuration parameters (the wait-time starts after receipt of the "Finished Message"). If either of these two parameters are not acknowledged before the set time expires, the preset and actual configurations will be considered different. The CPU reaction is then determined by the parameter "Startup if Preset and Actual Configuration Differ." Finally, the ***Hot Restart*** parameter prevents the CPU from starting if the transition time from power off to power on, or from STOP to RUN is longer than the time specified for this parameter. A "0" value shuts monitoring off.

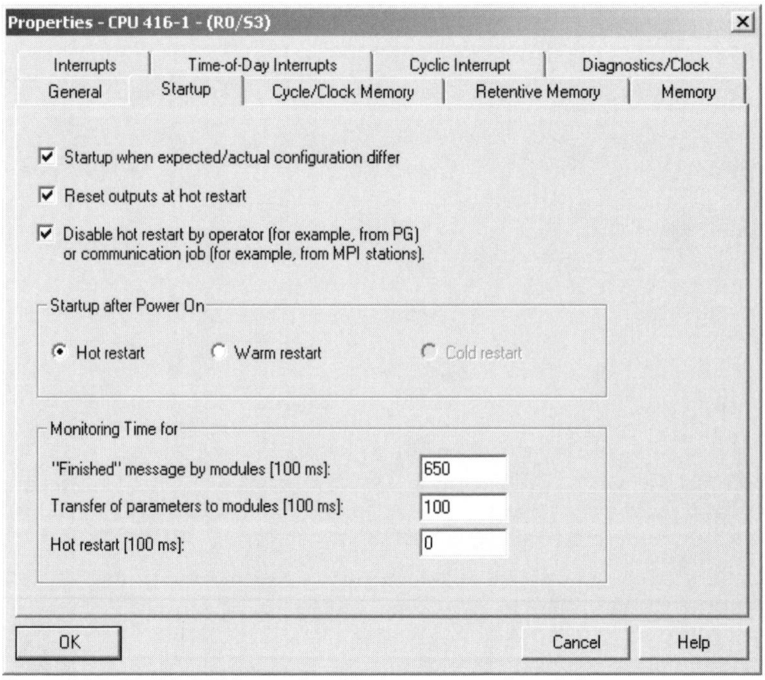

Figure 3-70. S7-400 CPU Properties: Start-up parameters dialog.

Quick Steps: Configuring S7-400 CPU Start-Up Properties

STEP	ACTION
1	With the Hardware Configuration tool open to the desired project and station, from the central rack select the CPU module, right-click and select **Object Properties**, then select the *Startup* tab. Parameters that are grayed are not possible in the opened CPU.
2	Activate the check box **Startup if Preset and Actual Configuration Differ,** to allow the CPU to start even if the preset and setpoint configurations differ. With this setting, central rack or distributed I/O modules are not checked; PROFIBUS-DP interface modules are checked and must be inserted for the CPU to start. This parameter is activated by default.
	De-activate the check box to disable **Startup if Preset and Actual Configuration Differ**. If at least one configured slot differs, then the CPU switches to STOP. If module-slots other than those originally configured have modules inserted, they are not compared.
3	Activate the "**Reset Outputs at Hot Restart**" to have all outputs and the PIQ to be cleared on each occurrence of a Hot Restart. De-activate the check box in order to have all outputs and the PIQ to maintain their last state on each Hot Restart.
4	Select "Warm Restart," "Hot Restart," or "Cold Restart," as the mode of **Startup After Power On**. The Cold Restart option is not available on older CPUs.
5	Set the parameter **Finished Message**, in milliseconds, to specify the maximum CPU wait-time for configured modules to signal ready for operation after power-up.
6	Set the parameter **Transfer of Parameters to Modules**, in milliseconds, to specify the maximum time for all modules, including DP slaves, to acknowledge receipt of their configuration parameters.
7	Set a value for the **Hot Restart** parameter in milliseconds. If the time that expires between a transition from power-off to power-on, or from STOP to RUN is longer than the time specified for this parameter, then CPU start-up will be disabled. The value "0" means no monitoring occurs.
8	Confirm the configuration parameters with the **OK** button.
9	From the menu, select **Save and Compile** to generate the System Data object before downloading to the CPU.

Configuring S7-400 CPU Cycle and Clock Memory

Basic Concept

The Cycle and Clock Memory parameters allow for the cyclical processing time of the CPU to be influenced and for a single byte to be defined where the CPU may output periodic clock pulses. Modifying default parameters of this dialog is not essential for S7-400 operation; they do, however, allow adjustments that optimize use of CPU processing time.

Essential Elements

The ***Update Process Image Cyclically*** parameter determines if the process image of inputs and outputs should be updated cyclically. By default this parameter is set to update the PII and PIQ cyclically. The ***Scan Cycle Monitoring Time*** is the timer value that if exceeded by the CPU cycle, the CPU is stopped. The ***Minimum Scan Cycle*** defines a minimum cycle time that CPU should always use - even if it must wait. The ***Scan Cycle Load from Communications*** sets a percentage of the total cycle time for servicing communications.

Adjusting the ***Size of the Process Image (PII/PIQ)***, perhaps to reflect actual usage of input/output bytes, is only relevant in certain S7-400 CPUs. The parameter ***OB 85 Call Up at I/O Access Error*** determines how OB 85 should be handled in the event of I/O access errors. The parameter is generally grayed and is fixed so that I/O access errors do not result in a call of OB85, nor are entries made in the diagnostic buffer. In later CPU revisions, other options for calling OB 85 include "Only for Incoming and Outgoing Errors", or "On Each Individual Access." The ***Clock Memory*** parameter allows a byte from Bit memory (M) to be specified as the location where the CPU will output eight individual clock-pulses.

Application Tips

Clock pulses, are accessed in the user program, via the eight bits of the specified byte. From the most-significant to the least-significant bit, the output frequency is 0.5, 0.625, 1.0, 1.25, 2.0, 2.5, 5.0, and 10 Hz. You may use these pulses instead of timers, where periodic signals are needed. Typical uses include flasher-circuits, alarm synchronization, or where events must be continuously triggered at periodic intervals. Clock pulses have an on-to-off ratio of 1:1.

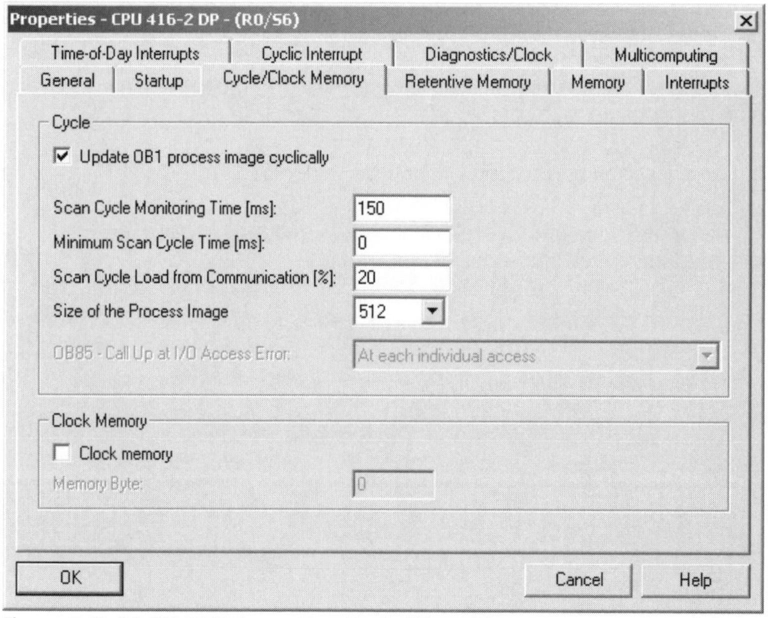

Figure 3-71. S7-400 CPU Properties: Cycle/Clock Memory parameters dialog.

Quick Steps: Configuring S7-400 CPU Cycle and Clock Memory

STEP	ACTION
1	With the Hardware Configuration tool open to the desired project and station, from the central rack select the CPU module, right-click and select **Object Properties**, then select the *Cycle/Clock Memory* tab. Parameters that are grayed are not possible in the opened CPU.
2	Activate the checkbox **Update OB 1 Process Image Cyclically** for cyclic updates of the process image tables (PII/PIQ). This is the default setting but can be changed.
	De-activate the checkbox **Update OB 1 Process Image Cyclically** to disable normal cyclic updates of the process image (PII/PIQ). If this parameter is set I/O updates must be handled by calling blocks.
3	In the **Scan Cycle Monitoring Time** field, specify in milliseconds, the cycle time that the CPU is not to exceed. If the actual cycle time exceeds this watchdog time, the CPU enters the STOP mode, unless OB 80 is loaded in CPU memory.
4	In the **Minimum Scan Cycle Time** field specify, in milliseconds, the minimum time that should be used to process the program. If the actual cycle time is less than the specified minimum, the CPU waits until the minimum cycle time expires, or if OB 90 is in memory, this remaining time is used by the CPU for background processing.
5	In the **Scan Cycle Load from Communications** field, enter a value from 10 to 50, as a percent of the set cycle time to allocate for communications processing.
6	In the **Size of the Process Image (PII/PIQ)** field, only relevant in some S7-400 CPUs and the CPU 318. Enter an end byte-address for the PII/PIQ to reflect the actual usage of digital input-bytes (PII) and digital output-bytes (PIQ). The PII/PIQ update time is thereby minimized.
7	Where the **OB 85 Call at I/O Access Error** parameter is allowed, select the option "No Call of OB 85" if OB85 should not be called when an I/O access error occurs during update of the process images (PII/PIQ); select the option "Only for Incoming and Outgoing Errors," to minimize the impact on the CPU cycle. The option "On each individual access" can result in an increase in CPU cycle time.
8	To use clock memory, activate the **Clock Memory** check box and then enter a byte (e.g., MB 97) from Bit Memory, to use for CPU output of eight individual clock pulses.
9	Confirm the configuration parameters with the **OK** button.
10	From the menu, select **Save and Compile** to generate the System Data object before downloading to the CPU.

Configuring S7-400 CPU Retentive Memory

Basic Concept

The CPU Retentive Memory parameter settings allow areas of memory to be defined as retentive. Memory is retentive if it retains its stored contents when the CPU operating mode is switched from RUN to STOP and even upon power loss. Modifying the default parameters of this dialog is not essential for operation of S7-400 CPUs, unless you require that certain memory bytes, timers, and counters retain their contents on loss of power.

Essential Elements

In the S7-400 parts of bit memory (M), timer (T), and counter (C) areas may be defined as retentive. The first three parameters require an entry of the number of bytes, timers, or counters to reserve starting from byte-0, timer-0, or counter-0 respectively. An entry of 4 in the bit memory field reserves byte-0, 1, 2, and 3 as retentive. Whereas in the S7-300 Data Block areas may also be reserved as retentive, this is not possible in the S7-400, therefore this part of the dialog is grayed out. Finally, the maximum size that can be reserved as retentive is dependent on the CPU.

Application Tips

Recall, that under normal conditions involving loss of power with battery backup, the contents of data blocks are already retained. The S7-300 feature is to protect against loss of these areas given a loss of power, but particularly a loss of battery backup. This maintenance free feature is not incorporated in the S7-400. Battery-backup must be in place in order to retain areas defined as retentive.

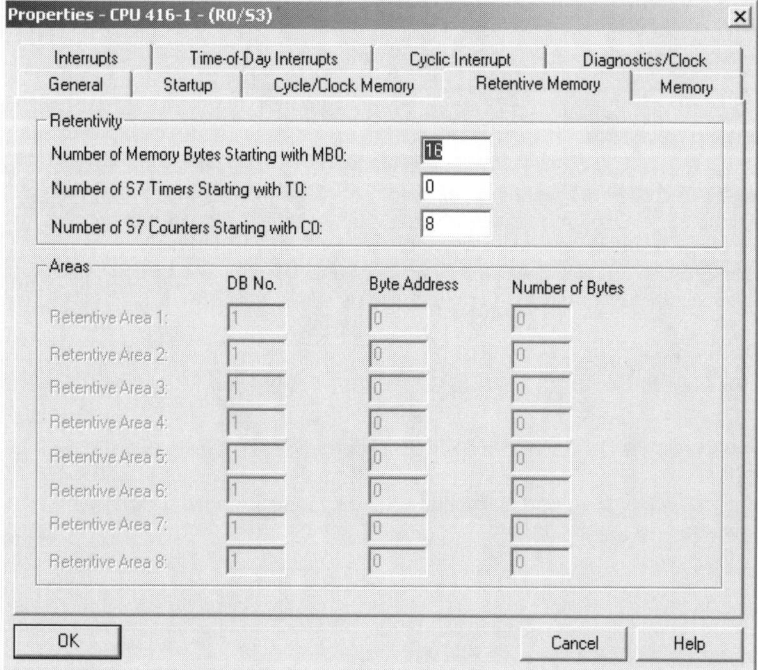

Figure 3-72. S7-400 CPU Properties: Retentive Memory parameters dialog.

Quick Steps: Configuring S7-400 CPU Retentive Memory

STEP	ACTION
1	With the Hardware Configuration tool open to the desired project and station, from the central rack select the CPU module, right-click and select **Object Properties**, then select the *Retentive Memory* tab to define retentive memory areas.
2	Click in the **Memory Bytes** field, and enter the total number of bytes to define as retentive. The defined area starts at MB 0 and will include all bytes up to MB n-1.
3	Click in the **S7 Timers** field and enter the number of timers to define as retentive. The defined retentive area starts at T0, and will include all timers up to Tn-1.
4	Click in the **S7 Counters** field and enter the number of counters to define as retentive. The defined retentive area starts at C0, and will include all counters up to Cn-1.
5	Confirm the configuration parameters with **OK**.
6	Use the toolbar **Save and Compile** button to generate the System Data object before downloading to the CPU.

Configuring S7-400 CPU Local Memory

Basic Concept

Local memory is the S7 memory used by the temporary (TEMP) variables of a code block, when accessed during program execution. The local memory, also called the local stack or L-stack, is only available to the current block while the block is being processed. After the block is done the local stack is available to the next called block, which overwrites the previous data. Each S7-400 CPU has a fixed amount of local memory that can be re-allocated, as required, from the Memory tab of the CPU properties dialog.

Essential Elements

Each CPU has a fixed amount of local memory that, by default, is equally divided among the priorities classes. In this way, every priority class is guaranteed its own local stack. As will be discussed in more detail later, all blocks (e.g., FB, FC, SFB, and SFC) are called directly or indirectly from an organization block for processing. Furthermore, all organization blocks are called by the operating system based on a pre-assigned priority class. Since local memory is allocated to the priority class of organization blocks, the local memory available to each code block is based on the organization block from which the block is called.

In S7-400 the total local memory, of the open CPU, may be re-allocated. That is to say that part of the local memory assigned to a priority can be re-distributed to where it is needed. This may be the case where a priority class is not in use or where all of its local memory is not needed. Re-allocation of the local memory is useful since priority classes do not all need the same local memory size. High priority classes (e.g., 25-28), for example, do not normally involve large data requirements or more than two levels of block calls.

Application Tips

When allocating local memory, all temporary variables of the OB and of the blocks called from that OB must be considered. If too many nesting levels (blocks called in a horizontal direction) are used, the L-stack may overflow. The local stack requirements of your program should be tested. The local data requirements of synchronous error OBs must always be taken into consideration. All organization blocks require a minimum of 20-bytes of local memory to store the 20-bytes of start information entered by the CPU.

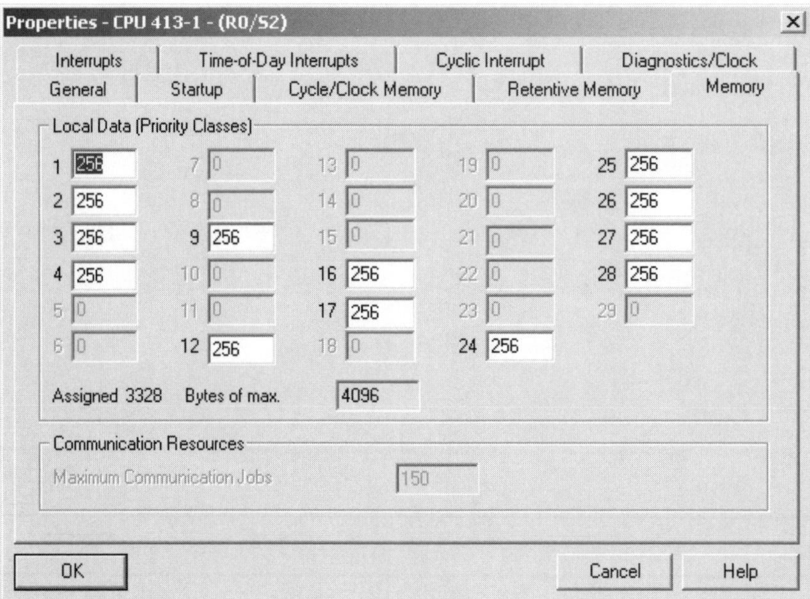

Figure 3-73. CPU 413-1 Local memory dialog; memory re-allocation allowed.

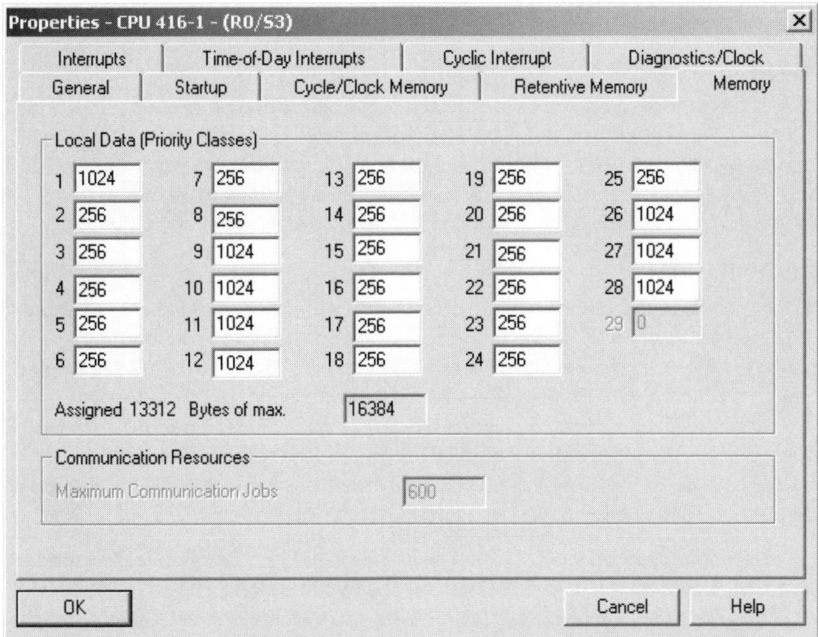

Figure 3-74. CPU 416-1 Local memory dialog; memory re-allocation allowed.

Quick Steps: Configuring S7-400 CPU Local Memory

STEP	ACTION
1	With the Hardware Configuration tool open to the desired project and station, from the central rack select the CPU module, right-click and select **Object Properties**, then select the *Retentive Memory* tab to define retentive memory areas.
2	Click in a **Priority** field and enter the total number of bytes to define for the priority class. Note: the number of bytes must be divisible by 2 and values 2-to-18 are not valid. A minimum of 20-bytes satisfies the minimum requirement that every organization block also requires 20 bytes of local data for its start information.
3	Modify as many priority classes as required and confirm settings with **OK**.
4	Use the toolbar **Save and Compile** button to generate the System Data object before downloading to the CPU.

Configuring S7-400 CPU Diagnostics and Clock Properties

Basic Concept

Each S7-400 is equipped with a real-time clock that is required for using *runtime meters* and for triggering *Time-of-Day interrupts*. Each CPU also incorporates a *Diagnostic Buffer* — an S7 memory area where Diagnostic events are stored in the order they occur. Modifying the default Diagnostics and Clock properties is not essential; however these parameters allow you to influence operating characteristics of the diagnostics buffer and of the CPU clock.

Essential Elements

Three parameters influence the diagnostic buffer. Activating the **Extended Functional Scope** causes the CPU to enter events other than standard errors into the diagnostic buffer. For example, each start of an organization block is considered a diagnostic event. The increase in diagnostic events will cause the buffer to overflow sooner, and make it more likely that important messages will be overwritten. In some CPUs the parameter **Number of Messages in Buffer**, allows the diagnostic buffer size to be adjusted. The buffer normally holds 100 messages. The parameter **Report Cause of STOP**, if activated, causes each CPU STOP to be entered in the buffer and reported to a designated PG/PC or operator panel.

The clock can be synchronized periodically based on a user-defined **synchronization mode**, **time interval** and **correction factor**. If the CPU is one of several clock-equipped modules in the local rack or in a network, the synchronization mode allows the clock to be designated as a master and responsible for setting other clocks; or as a slave and set by another clock. Furthermore, your selection will determine if the clock is synchronized internally from the PLC bus, or externally over the MPI port. Whether or not a clock can be synchronized externally or act as either master or slave is CPU-dependent.

Application Tips

The extended scope cannot be activated in STEP 7 after V3.1. If the parameter was previously active in a project now being converted from V2.x to V3.1 or higher, then the setting can now be downloaded to the CPU or de-activated.

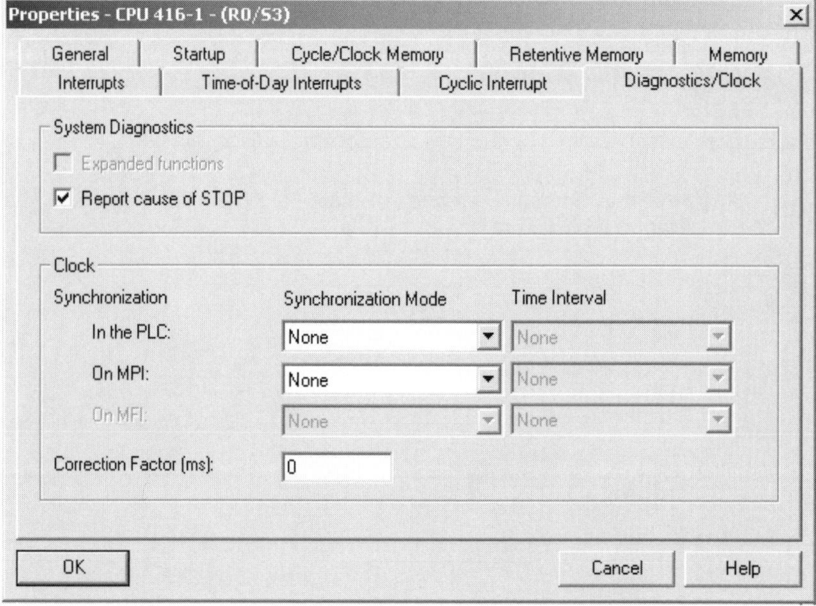

Figure 3-75. S7-400 CPU Properties: Diagnostics/Clock parameters dialog.

Quick Steps: Configuring S7-400 CPU Diagnostics and Clock Properties

STEP	ACTION
1	With the Hardware Configuration tool open to the desired project and station, from the central rack select the CPU module, right-click and select **Object Properties**, then select the *Diagnostics/Clock* tab to make the required changes.
2	Activate the **Report Cause of STOP** check box, to have all causes of stops to be entered to the diagnostic buffer and to be sent to a designated PG/PC or operator panel. Otherwise, deactivate the checkbox to minimize the number of events entering the buffer.
3	In the PLC field under **Synchronization Mode**, select *As Master*, if the clock will synchronize other clocks; select *As Slave*, if the clock will be set by another local clock; or select *None* if the local clock is not to be synchronized.
4	In the PLC field under **Time Interval,** select the synchronization interval, if the clock was set to synchronize other clocks.
5	In the MPI field under **Synchronization Mode**, select *As Master*, if the clock will synchronize other clocks; select *As Slave*, if the clock will be set by another local clock; or set *None* if the local clock is not to be synchronized.
6	In the MPI field under **Time Interval,** select the synchronization interval, if the clock was set to synchronize other clocks.
7	Confirm the configuration parameters with **OK**.
8	From the menu, select **Save and Compile** to generate the System Data object before downloading to the CPU.

Configuring S7-400 CPU Access Protection

Basic Concept

Some S7-400 CPUs support the ability to use password protection to limit access to the CPU's control program and other operations. This protection is in addition to the protection provided by the CPU key-switch. With password protection enabled, the control program and its data are protected from unauthorized changes (i.e., write access protection). It is even possible to prevent read access to program blocks that are considered proprietary (i.e., read access protection). Use of online functions such as upload/download blocks may also be prohibited to non-password holders.

Essential Elements

Using CPU access protection involves setting the protection level and if required, defining a password. Once configured, the protection level and password is downloaded to the module with the configuration data. Complete read and write access is possible for all password holders, regardless of the key-switch position. Otherwise, three levels of protection may apply to non-password holders. **Level 1** is the default setting in which standard key-switch access is in operation. **Level 2** provides *write-protection*; read access is possible, but write access is denied without the password. **Level 3** provides *read/write protection*, which means that both read and write access are denied, independent of the key-switch position.

Application Tips

Since online functions, like Monitor/Modify Variables or Upload/Download blocks, are affected by protection level 2, users are prompted to enter a password whenever such an operation is invoked. If the correct password is entered, access rights are permitted to modules for which a particular protection level was set during parameter assignment. The online connection is established to the protected module and the online functions of the protection level may be executed. An alternate method, one that will eliminate the need to continually re-enter a password, you may enter the password from the SIMATIC Manager.

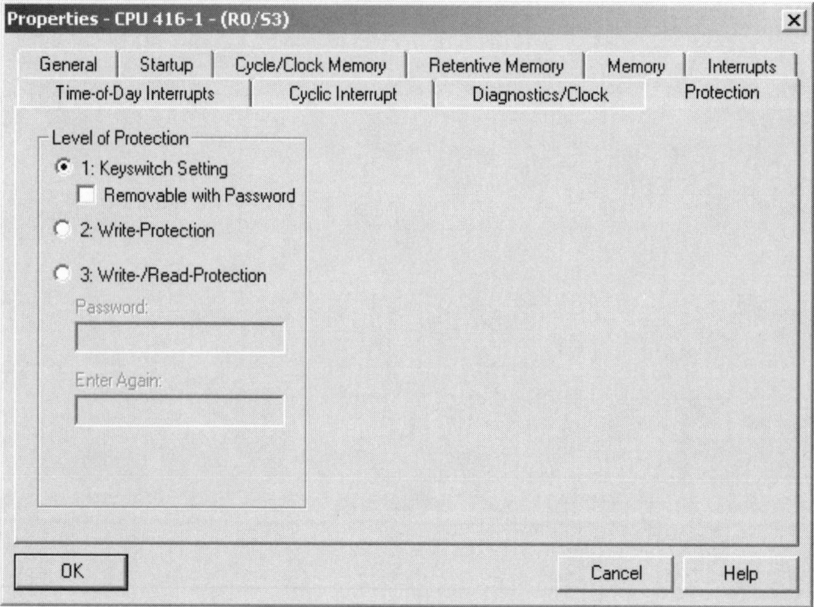

Figure 3-76. S7-400 CPU Properties: Protection parameters dialog.

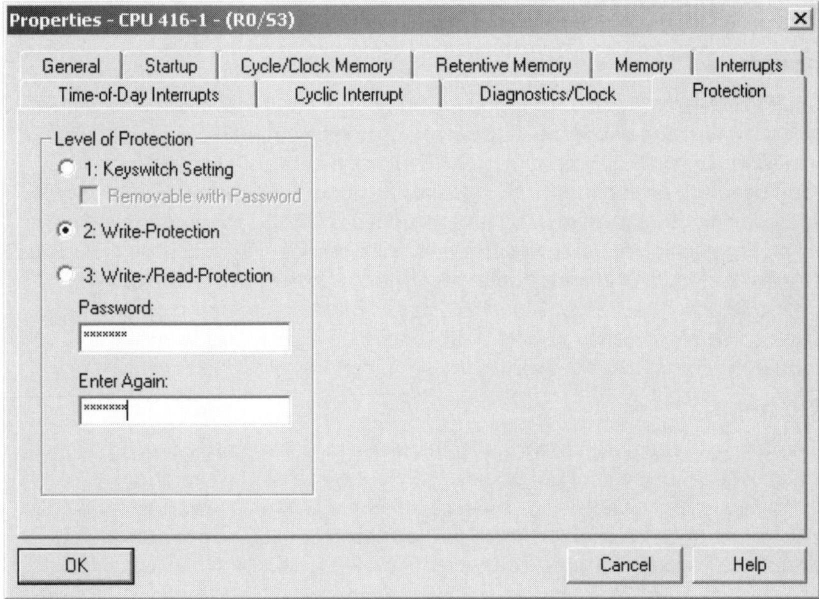

Figure 3-77. S7-400 CPU Properties: Protection parameters dialog.

Quick Steps: Configuring S7-400 CPU Access Protection

STEP	ACTION
1	With the Hardware Configuration tool open to the desired project and station, from the central rack, select the CPU module, right-click and select **Object Properties**, then select the *Protection* tab.
2	By default, *Level 1* or *Key-switch* protection is activated; at this level, password protection is not enabled. There is no read /write access restriction in either the RUN-P or STOP position, however, read access is permitted in the RUN position. The key-switch must be moved to STOP or RUN-P before any download or data write (e.g., modify variables) operation is allowed.
3	Select *Level 1* or *Key-switch* protection with the option **Removable with Password** enabled to allow the key-switch restrictions to be removed with password entry. In this mode, Read and write access is possible for password holders, independent of the key-switch position.
4	Select *Level 2 Protection* by first activating the **Write Protection** radio button; then enter and confirm the desired password in the appropriate fields. In this mode, data and programs may be monitored, but write operations such as download or modify variables is only permitted to password holders.
5	Select *Level 3 Protection* by first activating the **Write-Read Protection** radio button; then enter and confirm the desired password in the appropriate fields. In this mode, read and write operations are permitted to password holders only.
6	From the menu, select **Save and Compile** to generate the System Data object before downloading to the CPU.

Configuring S7-400 CPU Interrupt Properties

Basic Concept

The S7-400 CPU Interrupts tab lists four categories of interrupts. *Hardware Interrupts* include those generated by machine or process inputs using an interrupt input module or those triggered by modules (e.g., CP, FM, or analog SM) that have onboard diagnostic or status events that can generate an interrupt (e.g., over limit, or under limit). *Time-Delay Interrupts* are generated from the user program using system functions (SFCs) and are used when the main cyclical program (OB1) must be interrupted to call a service after a precise time delay has expired. *Asynchronous Error Interrupts*, are generated by system-related faults (e.g., battery failure, power failure, wire break on analog input), that are non-program related. DPV1 interrupts can be triggered by so-called DPV1 slave devices, to ensure that the controlling master CPU processes the event in the slave that triggered the interrupt.

Essential Elements

Each S7 interrupt has an associated organization block that is called for processing by the operating system whenever the interrupt occurs. Which of the listed OBs are actually available is CPU-dependent. OBs that are not supported in a CPU are grayed out. The only modifiable parameter of an interrupt OB is the default priority. If you need to change the priority, only use the allowed alternates. Hardware, time-delay, and DPV1interrupt priority alternates include 0, and 2-to-24; asynchronous interrupts allow 24-to-26. A priority of zero deactivates an interrupt OB. To avoid loss of interrupts, you should not assign a priority twice.

Application Tips

When an interrupt occurs, the main cyclical program (OB1) is interrupted. Responding to an interrupt requires writing the STEP 7 code for the OB, and downloading the block to the CPU. The OB is then ready to act as the service routine — called when the interrupt is triggered. The code written in the OB should be kept as short as possible (e.g., activate alarm, send message, halt machine). Remember, time-delay interrupts are triggered from the program. You will need to call the system function for starting the time delay (SFC 32, SRT_DINT).

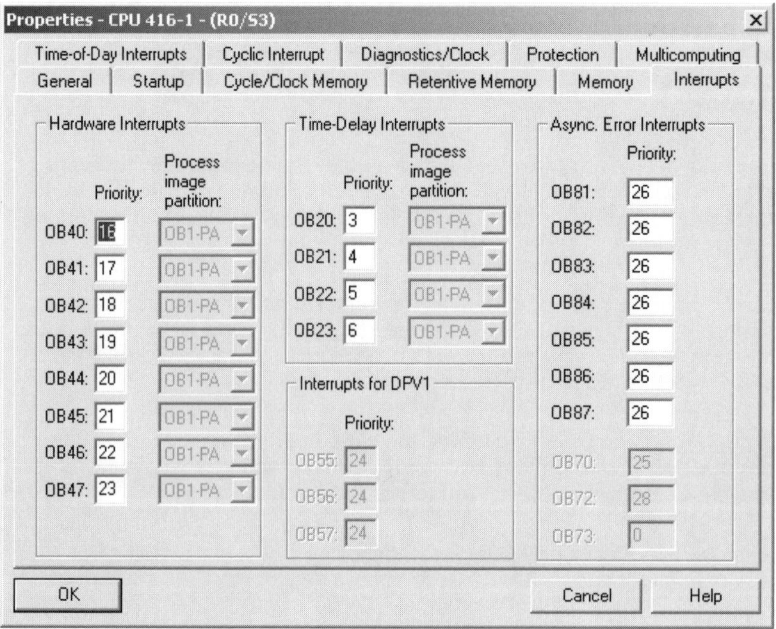

Figure 3-78. CPU Properties: Interrupt parameters dialog.

Quick Steps: Configuring S7-400 CPU Interrupt Properties

STEP	ACTION
1	With the Hardware Configuration tool open to the desired project and station; select the CPU module, right-click and select **Object Properties,** then select the *Interrupts* tab to make your desired priority-class changes (optional).
2	Up to eight Hardware Interrupts are supported in the S7-400. The associated organization blocks are OB40 to OB47. The corresponding default priorities are 16 to 23. To enter a new priority, click in the field and enter a new value (0, or 2-24.).
3	Up to four Time-Delay Interrupts are supported in the S7-400. The associated organization blocks are OB20 to OB23. The corresponding default priorities are 3 to 6. To enter a new priority, click in the field and enter a new value (0, or 2-24).
4	OB81 is the only Asynchronous Interrupt OB in most S7-300 CPUs and it has a fixed priority (priority-25). If the open CPU supports modifying the default priority then you may enter a value of 0, 2-4, 9, 12, 16, 17, or 24-26. Using the default value or a value in the range of 24-26 ensures that asynchronous error OBs are not be interrupted by other interrupt events.
5	Where the priority of DPV1 Interrupt OBs is alterable, you may enter a new value of 0, and 2-24. A value of 0 is in effect disabling the OB.
6	From the menu, select **Save and Compile** to generate the System Data object before downloading to the CPU.

Configuring S7-400 CPU Time-of-Day Interrupts

Basic Concept

Configuring a *time-of-day interrupt* is only required to have a part of your STEP 7 program execute at a specific date and time. When the interrupt is triggered, the main cyclical program in OB 1 will be interrupted at the configured time and date and the associated service routine (organization block) of the interrupt will be called for processing. Time-of-day interrupt can be configured such that upon reaching the target date and time, the associated OB is executed once and only once, or every minute, hourly, daily, weekly, monthly, every end of month, or yearly.

Essential Elements

The properties tab for time-of-day interrupts lists the Organization Blocks, OB10-to-OB17. All of the time-of-day interrupts are assigned a default priority 2 class. Up to eight time-of-day interrupts are supported by S7, but the exact number available is CPU-dependent. If one of these time-of-day interrupts is not supported by the opened CPU, then the OB and its parameters are grayed. Configuring a time-of-day interrupt requires activating the OB, selecting the execution interval, and setting the start date and time. Alternate priorities that may be assigned include priorities of 3-24.

Application Tips

Programming a time-of-day interrupt involves setting the start parameters, activating the OB, writing the code for the OB, and downloading the OB to the CPU. The OB is then ready to act as the service routine — called when the interrupt is generated. The parameters of time-of-day interrupts may be set here in the hardware configuration tool, or from within the STEP 7 program, using standard system functions. TOD interrupt application might include generating operator messages or production reports, or scheduled process readings.

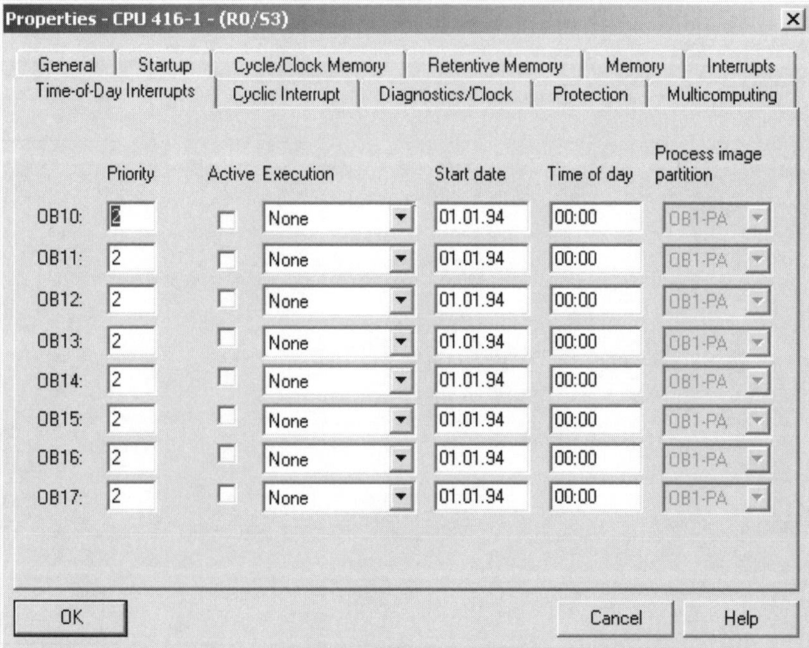

Figure 3-79. CPU Properties: Time-of-Day Interrupt parameters dialog.

Quick Steps: Configuring S7-400 CPU Time-of-Day Interrupts

STEP	ACTION
1	With the Hardware Configuration tool open to the desired project and station, from the central rack select the CPU module, right-click and select **Object Properties,** then select the *Time-of-Day Interrupts* tab.
2	Select the time-of-day interrupt OB whose parameters you wish to alter: to modify the **Priority** class click in the field and enter a valid alternate priority of 3-to-24.
3	Click inside the **Active** check box to activate the OB. Time-of-day interrupts may also be activated from the STEP 7 program using system function SFC 30 (ACT_TINT).
4	Click the drop arrow in the **Execution** field, and select whether the OB should be called once, every minute, every hour, every day, every week, every month, every month-end, or every year. Time-of-day interrupts may also be set from the STEP 7 program, using system function SFC 28 (SET_TINT).
5	Next, click in the **Start Date** field and enter the start date for the time-of-day interrupt. If the start date is not entered on the properties dialog, it can be set from the STEP 7 program using the system function SFC 28 (SET_TINT).
6	Click in the **Time of Day** field and enter the start time for the time-of-day interrupt. The start time can be entered here on the properties dialog, or from the STEP 7 program, using the system function SFC 28 (SET_TINT).
7	Click inside the **Active** check box to mark the OB as activated. If the time-of-day interrupt OB is not activated on the properties dialog, it can be activated on demand from the STEP 7 program. Use system function SFC 30 (ACT_TINT) to activate time-of-day interrupts.
8	Confirm the configuration parameters with the **OK** button.
9	Use the toolbar **Save and Compile** button to generate the System Data object before downloading to the CPU.

Configuring S7-400 CPU Cyclic Interrupts

Basic Concept

A *cyclic* or "*watchdog*" *interrupt*, is generated at a fixed periodic interval, and triggers execution of a specific cyclic interrupt OB. Cyclic interrupts are used to interrupt the CPU's normal processing at precise intervals (e.g., every 100 ms, every 500 ms, or every 5 sec) to execute code that must be processed on a regular basis. Typical uses of cyclic interrupts include communications, control loops, temperature control, and sampling analog inputs.

Essential Elements

The properties tab for cyclic interrupts lists Organization Blocks OB30 through OB38. These are the nine cyclic interrupt OBs supported by S7. Each OB is listed with its assigned priority (7-15), and the default cyclic interval at which the OB is processed. Which of these OBs are actually available, however, depends on the CPU. In the CPU 416-1, for example, all nine OBs are available. OBs that are grayed out are not supported by the opened CPU.

Configuring changes to default parameters is not necessary unless required. While the default priorities may be changed, only alternate priorities of 3-24 are allowed. To prevent loss of interrupts, however, care should be taken to not assign a priority twice. It is also possible to modify the period of execution and to enter a Phase offset value. If several cyclic interrupts will be activated, a phase offset will ensure that interrupts are distributed throughout the CPU cycle and are not triggered simultaneously.

Application Tips

Starting a cyclic interrupt simply involves determining which OB will be used, modifying the default parameters if required, creating and downloading the organization block code as part of your program. The OB will be processed according to the specified interval, which is timed from when the mode selector is moved from STOP to RUN. Finally, it is essential that the code programmed in a cyclic OB is executed in significantly less time than the call interval.

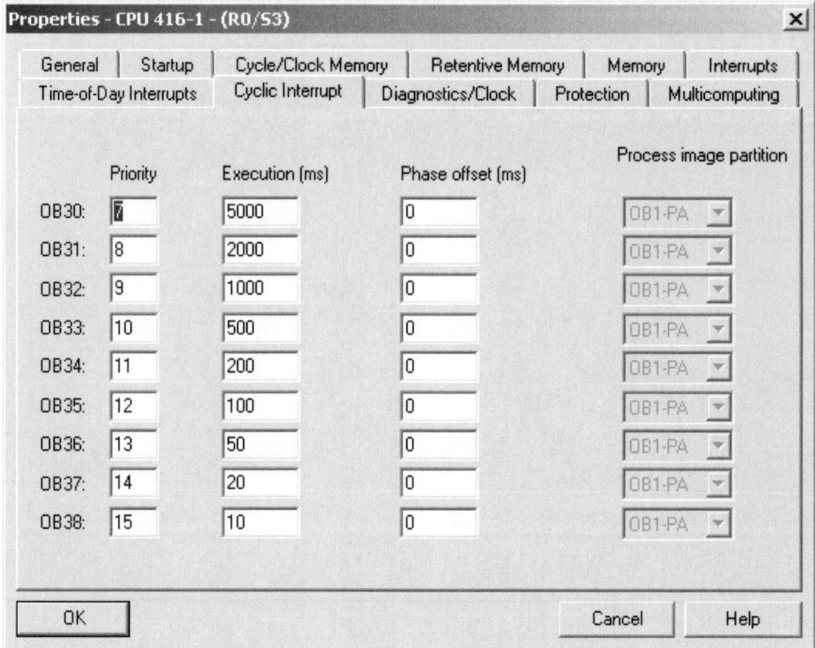

Figure 3-80. CPU Properties: Cyclic Interrupt parameters dialog.

Quick Steps: Configuring S7-400 CPU Cyclic Interrupts

STEP	ACTION
1	With the Hardware Configuration tool open to the desired project and station, from the central rack, select the CPU module, right-click and select **Object Properties,** then select the *Cyclic Interrupts* tab to modify the configuration parameters.
2	Select the cyclic OB whose parameters you wish to alter, and enter a new **Priority** class if you wish to modify the default priority; click in the field and enter a valid alternate (3-24). Assigning a priority twice should be avoided to prevent loss of interrupts due to multiple (>12) interrupts of the same priority occurring simultaneously.
3	Enter a new **Execution** value if you wish to modify the default processing interval for the OB; click in the field, and enter a valid value in the range of 10 milliseconds to 6 seconds (i.e., 10-to-60,000).
4	If multiple cyclic interrupt OBs will be running, then enter a **Phase Offset** value for each, in milliseconds, if you wish to ensure that processing of the OBs is distributed somewhat evenly throughout the total CPU cycle. A valid range for this field is (0 -to-60000). The default value of 0 represents no phase offset.
5	If the S7-400 CPU supports system update of **PI Partition** (process image partitions), you may enter the number of the process image segment for the interrupt OB in question, if required. The CPU will update the assigned process image segment, whenever the associated interrupt OB is called. A valid range for this field is (1 -to- 8). The default value of "0" represents no process image partitioning.
6	Confirm the configuration parameters with **OK**.
7	Use the toolbar **Save and Compile** button to generate the System Data object before downloading to the CPU.

Configuring S7-400 Digital Input Module Properties

Basic Concept

The properties of standard digital input modules in the S7-400 are viewed or edited from the Hardware Configuration. When inserted into the hardware configuration digital input modules may be started without any modifications to the default properties unless the module is an interrupt input type (Also see *Configuring S7-400 Interrupt Input Properties*). Basic information about the module is provided on the *General* tab. The addresses assigned to the module are provided on the *Addresses* tab (Also see *Configuring S7-400 Digital I/O Module Addresses*).

Essential Elements

The ***Short Description***, the same as given in the hardware catalog, gives important features of the digital input module such as number of input circuits, supply voltage, and number of circuits per group (i.e., per common). For example a 120 VAC input module may have 32-inputs with four groups of eight circuits (grouping of 8). The ***Order Number*** reflects the currently open module, which is assigned to a specific slot in the configuration — it should match the part number of the physically installed module. The ***Name*** field shows the default name of the module. Although the name can be changed, it is not recommended. The **Comment** field allows more details about the module or its application to be documented.

Application Tips

The writing area on a module connector limits what descriptive information that can be written for each digital input. Use the comment field to provide detailed information on the use of the digital inputs of a module.

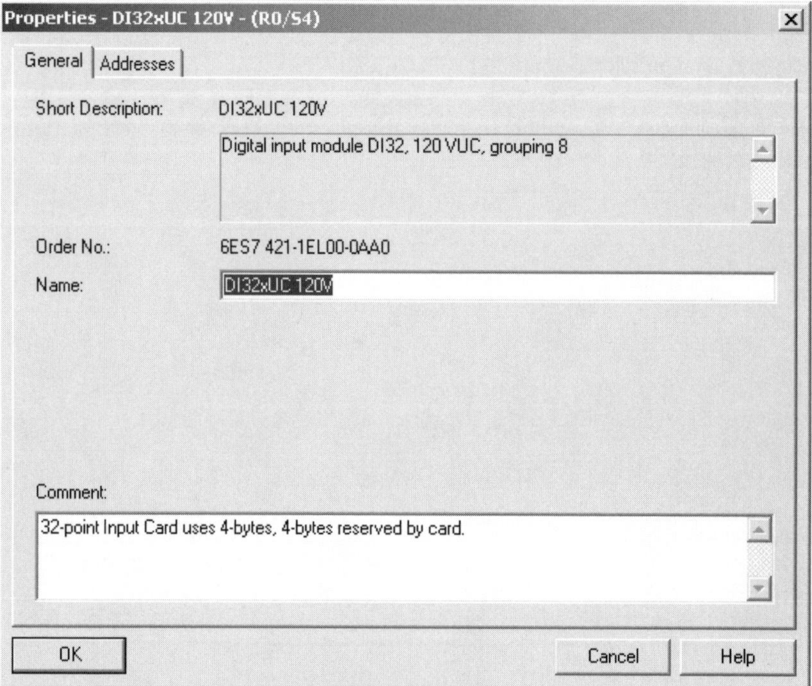

Figure 3-81. S7-400 Digital Input General Properties dialog.

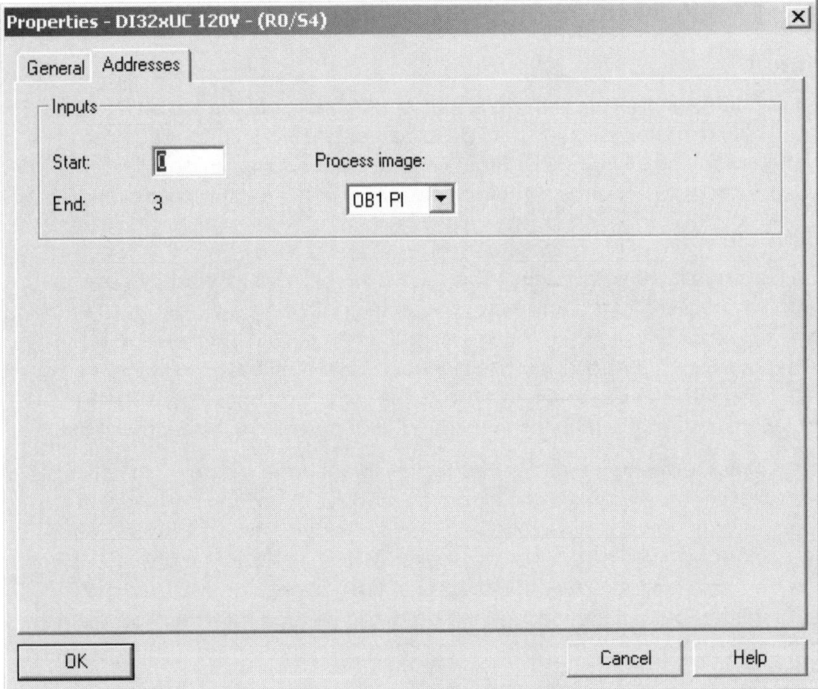

Figure 3-82. S7-400 Digital Input Address Properties dialog.

Quick Steps: Configuring S7-400 Digital Input Module Properties

STEP	ACTION
1	With a station open in the hardware configuration, select a rack to view its installed modules; double click on a digital input module whose properties you wish to view or edit. Not all properties may be edited (e.g., *Order Number, Short Description*).
2	On the *General* tab, the **Short description** of the module provides a brief summary of the module features (e.g., number of inputs, supply voltage, circuit grouping).
3	The **Order Number** is the part number of the open module in the configuration and should reflect the digital input module installed in the same slot in the physical rack.
4	Click in the **Name** field to enter a new name; the default name is based on the number of inputs and the supply voltage prefixed with **DI** for digital input.
5	Use the **Comment** field to enter a description of the module inputs or application.
6	Select the *Address* tab to view or modify the starting byte address of the module. To modify the start byte address, de-activate the **System Selection** check box, and enter a new start byte-address; confirm by pressing **OK**. See *Configuring S7-400 Digital I/O Module Addresses.*

Configuring S7-400 Interrupt Input Properties

Basic Concept

The general and address properties of the interrupt input module are the same as for the standard digital input module, however, a third dialog is provided for setting the interrupt module's interrupt properties. By default, hardware interrupts are disabled so if this module is used then it will be necessary to enable interrupt capability from the properties dialog.

Essential Elements

Activating the **hardware interrupt** parameter enables all of the interrupt inputs for use. An **input delay** can be specified for the entire module, based on the type of signals connected. The input delay should closely match the delay characteristics of the input device in order to protect against false input signals. Finally, each input channel may be set to trigger an interrupt on the **rising edge**, **falling edge**, or on both the rising and falling edge of the signal. If neither parameter is activated, then the associated channels will not trigger an interrupt.

Activating the **diagnostic interrupt** enables the module to generate a diagnostic interrupt. The diagnostic functions may vary depending on the module, but on the interrupt input module includes monitoring the supply voltage of the connected sensors and for a wire break in the connection between the input circuits and sensors. If monitoring is activated and the supply voltage fails, the group diagnostic event is entered into the diagnostic data storage area of the module. The data can be read from the module, in the STEP 7 program, using the system function to "read system status list" (i.e., SFC 51 RDSYSST) or evaluated in OB82.

Application Tips

Response to hardware interrupts in the user program is handled using organization blocks (OBs). In the S7-400 OB 40 through OB 47 are available and are assigned to the interrupt module on the *Addresses* tab. The interrupt OB must be programmed and downloaded to the CPU to be ready to handle an interrupt. When a hardware interrupt is triggered, the operating system interrupts the main program block (OB 1) and calls the interrupt OB.

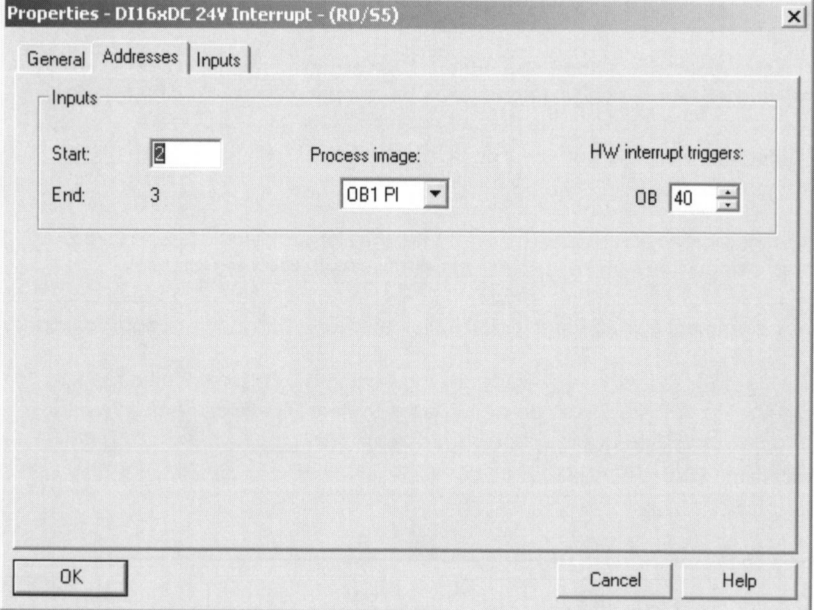

Figure 3-83. S7-400 Interrupt Input Module Properties: Addresses tab.

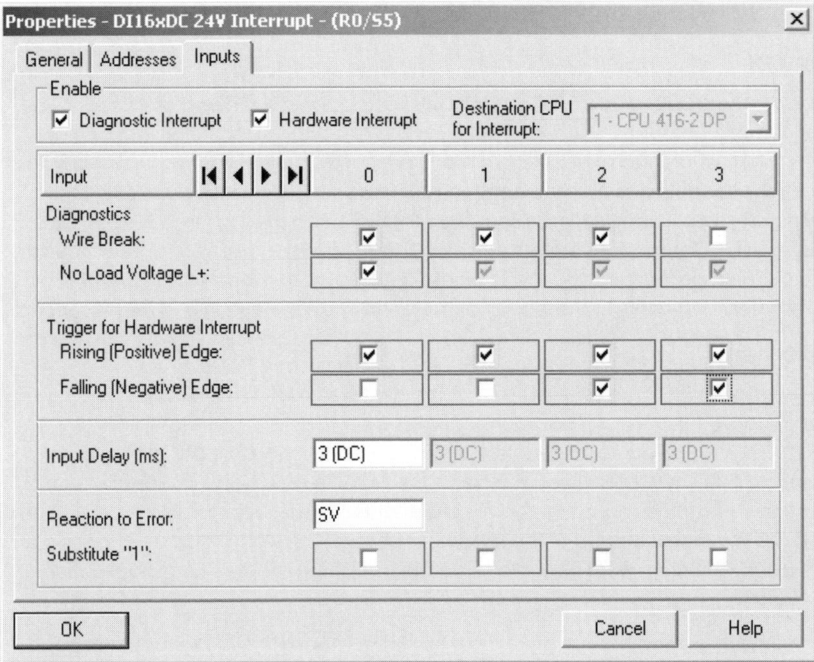

Figure 3-84. S7-400 Interrupt Input Module Properties: Input parameters tab.

Quick Steps: Configuring S7-400 Interrupt Input Properties

STEP	ACTION
1	With the required station open in the hardware configuration, select the rack where the module is installed; double click on the interrupt input module to open the properties dialog, then select the *Inputs* tab.
2	To use the module's diagnostic feature, activate the **Diagnostic Interrupt** check box and then activate the corresponding Wire Break check box for specific channels or the No Load Voltage check box for the module, causes the module to monitor and report on these conditions. A suitable program in OB82, can be used to evaluate which error has occurred and on which input the error occurred.
3	To use the module's hardware interrupt features, activate the **Hardware Interrupt** check box, then to enable use of the sixteen interrupt inputs. Activate the Rising Edge check box below the channel number if the input should trigger an interrupt on a 0-to-1 transition; or Activate the Falling Edge check box, if the input should trigger an interrupt on a 1-to-0 transition.
4	Click in the **Input Delay** field and select the typical signal delay time (in milliseconds) of the connected input signals. This parameter affects all inputs.
5	In the **Reaction to Error** field, select Keep Last Valid Value (KLV) (logic-0 or logic-1); or select Substitute Value (SV). If you select substitute value, for each output you wish to be set to logic 1, activate the check box for Substitute Value "1". Boxes not activated will be set to logic 0.
6	Confirm the configuration parameters with the **OK** button and use the toolbar **Save and Compile** button to generate the system data blocks.

Configuring S7-400 Digital Output Module Properties

Basic Concept

The properties of standard digital S7-300 digital output modules are viewed or edited from the Hardware Configuration. As well, when inserted into the configuration these modules may be started without any changes to the default properties, unless you wish to modify the start address or the module has output signal parameters you wish to influence. Like the digital input module, basic information about the module is provided on the *General* tab, addresses assigned to the module are provided on the *Addresses* tab (Also see *Configuring S7-400 Digital I/O Module Addresses*), and on some digital output modules an *Output* tab, contains parameters that allow the behavior of the outputs on the module to be influenced.

Essential Elements

The ***Short Description*** gives important features of the digital output module such as number of output circuits, supply voltage, and number of circuits per group (i.e., per common). For example a 120/230 VAC/2A output module may have 32-outputs with eight groups of four circuits (grouping of 4). The ***Order Number*** reflects the module assigned to a specific slot in the configuration — it should match the part number of the physically installed module. The ***Name*** field shows the default name of the module. Although the name can be changed, it is not recommended. The **Comment** field allows details about the module or its application to be documented.

A digital output module may include an *Output* tab if the module has **diagnostic interrupt** features that monitors and reports diagnostic events that have been enabled on the module. Diagnostic functions may include detection of *wire break*, *no load voltage*, *short circuit to ground*, and *short circuit to supply*, or *blown fuse*. This digital output may also have the ability to set a **substitute value** that determines the state that each output should be set to when the CPU goes to STOP.

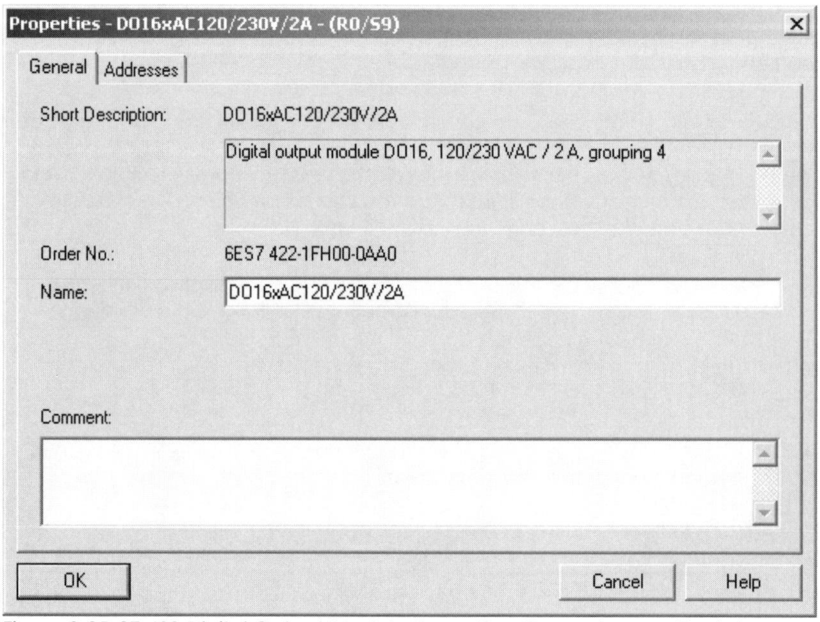

Figure 3-85. S7-400 Digital Output Module Properties: General properties tab.

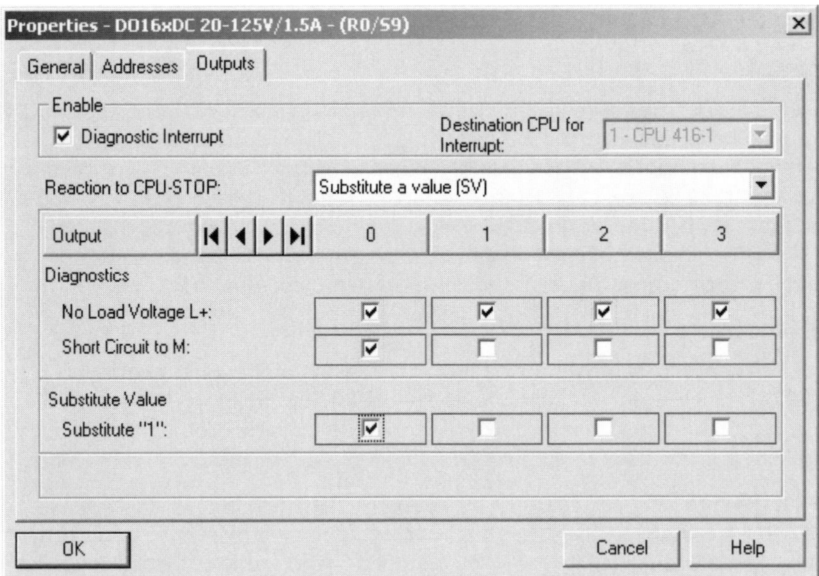

Figure 3-86. S7-400 Digital Output Properties: Output parameters settings.

Quick Steps: Configuring S7-400 Digital Output Module Properties

STEP	ACTION
1	With a station open in the hardware configuration, select a rack to view its installed modules; double click on a digital output module whose properties you wish to view or edit. Not all properties may be edited (e.g., *Order Number*, *Short Description*).
2	On the *General* tab, the **Short description** provides a brief summary of the module features (e.g., number of outputs, supply voltage, circuit grouping). The **Order Number** is the part number of the open module in the configuration and should reflect the digital output module installed in the same slot in the physical rack.
3	The default **Name**, which is modifiable, is based on the number of outputs, the supply voltage, and the output current, all prefixed with **DO** for digital output.
4	Use the **Comment** field to enter a description of the module outputs or application.
5	Select the *Output* tab to set operating parameters for the outputs of the module.
6	In the **Reaction to CPU STOP** field, select Keep Last Valid State (logic-0 or logic-1); or select Substitute a Value. If you select "substitute a value," then for each output you wish to be set to logic 1, activate the corresponding check box for Substitute "1". Boxes not activated will be set to logic 0.
7	Activate the **Diagnostic Interrupt** box to enable the module's ability to generate a diagnostic interrupt when any one of its diagnostic monitoring events occurs.
8	To activate diagnostic monitoring (e.g., Wire Break, No Load Voltage, Short Circuit to Ground (M), Short Circuit to Supply (L+), or Blown Fuse) for an output, activate the corresponding check box beneath each output circuit you wish to monitor.

Configuring S7-400 Digital I/O Addresses

Basic Concept

Just as in the S7-300, each digital input module, when inserted in the S7-400 configuration, reserves 4-bytes of the input image (**I**) memory; and each output module, when inserted, reserves 4-bytes of the output image (**Q**) memory. Reserving four bytes allows for a module with up to 32 inputs or 32 outputs to be inserted. Although default addressing in the S7-400 is similar to the S7-300, it is not fixed or slot-dependent. Also, full use of the input and output images is possible. This means that input and output modules may use the same starting byte addresses and that the assigned addresses of any module may be changed.

Essential Elements

As each module is inserted, four bytes are reserved in either the process image of inputs (input module) or process image of outputs (output module). The starting byte address for each successive module is based on the next available byte address following the previous input module (for inputs) or previous output module (for outputs), even if a slot is left empty.

If input modules are installed in the first four slots, as seen in the figure below, slot-4 would reserve bytes 0-3; slot-5 would reserve bytes 4-7; slot-6 would reserve bytes 8-11; and so on. If output modules are installed in the following four slots, slot-8 would reserve bytes 0-3; slot-9 would reserve bytes 4-7; slot-10 would reserve bytes 8-11; and so on. In the S7-400, assigned starting byte-addresses may be modified under the address properties of the module.

Application Tips

Digital inputs are accessed from the PII using the input memory identifier '**I**' prefixed to the address; digital outputs are accessed in the PIQ, using the output memory identifier '**Q**' prefixed to the address. Both inputs and outputs are referenced as "**byte.bit**". A 32-point output module in slot-4 would use addresses **Q 0.0** through **Q 3.7**; an input module would use **I 0.0** through **I 3.7**. Inputs/Outputs may also be accessed as bytes, words, or double words.

		I	I	I	I	Q	Q	Q	Q
		Slot-4	Slot-5	Slot-6	Slot-7	Slot-8	Slot-9	Slot-10	Slot-11
PS-407	S7-CPU	Byte-0	Byte-4	Byte-8	Byte-12	Byte-0	Byte-4	Byte-8	Byte-12
		0.0	4.0	8.0	12.0	0.0	4.0	8.0	12.0
		to	to	to	to	to	to	to	to
		0.7	4.7	8.7	12.7	0.7	4.7	8.7	12.7
		Byte-1	Byte-5	Byte-9	Byte-13	Byte-1	Byte-5	Byte-9	Byte-13
		1.0	5.0	9.0	13.0	1.0	5.0	9.0	13.0
		to	to	to	to	to	to	to	to
		1.7	5.7	9.7	13.7	1.7	5.7	9.7	13.7
		Byte-2	Byte-6	Byte-10	Byte-14	Byte-2	Byte-6	Byte-10	Byte-14
		2.0	6.0	10.0	14.0	2.0	6.0	10.0	14.0
		to	to	to	to	to	to	to	to
		2.7	6.7	10.7	14.7	2.7	6.7	10.7	14.7
		Byte-3	Byte-7	Byte-11	Byte-15	Byte-3	Byte-7	Byte-11	Byte-15
		3.0	7.0	11.0	15.0	3.0	7.0	11.0	15.0
		to	to	to	to	to	to	to	to
		3.7	7.7	11.7	15.7	3.7	7.7	11.7	15.7

Figure 3-87. Default digital I/O addressing in the S7-400 starts at byte-0, regardless of the slot, and reserves 4-bytes per slot. Input and output modules may use the same byte addresses.

Quick Steps: Configuring S7-400 Digital I/O Addresses

STEP	ACTION
1	From the SIMATIC Manager, open the required project and select the S7-400 Station for which addressing is required.
2	With the station selected, in the right pane of the station window double click on the Hardware object to open the station in the Hardware configuration tool.
3	Select a rack whose modules you wish to view the installed digital I/O modules or insert new digital I/O modules.
4	View the starting byte address of each input module in the **I-Address** column or each output module in the **Q-Address** column.
5	To modify the starting byte address of a module, select the module, right-click and select **Object Properties**, then select the Address tab.
6	To modify a starting byte address, de-activate the **System Selection** check box to disable automatic address determination by the system.
7	Enter a new start byte-address and confirm by pressing **OK**. Since in the S7-400 the complete PII and PIQ can be used, input and output modules may have the same addresses. Digital I/O start byte addresses, after byte-0, must be a multiple of 4 (e.g., 4, 8, 12, 16, 24, etc.).
8	The **Process Image** field shows the process image responsible for updating the module outputs. By default digital outputs are cyclically updated by the CPU.

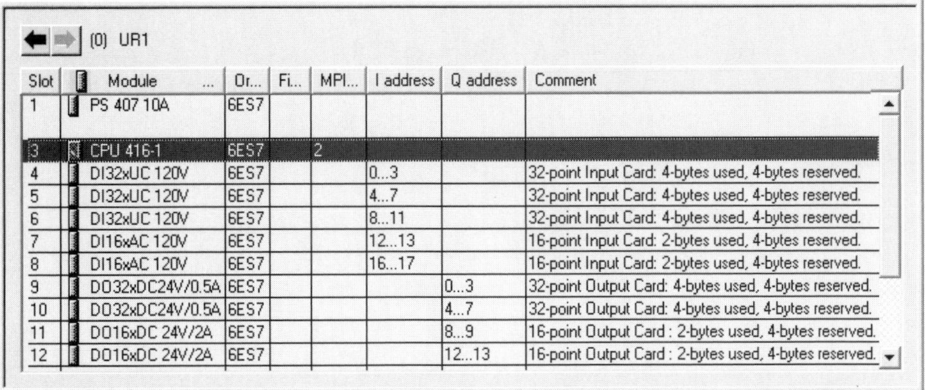

(0) UR1

Slot	Module ...	Or...	Fi...	MPI...	I address	Q address	Comment
1	PS 407 10A	6ES7					
3	CPU 416-1	6ES7		2			
4	DI32xUC 120V	6ES7			0...3		32-point Input Card: 4-bytes used, 4-bytes reserved.
5	DI32xUC 120V	6ES7			4...7		32-point Input Card: 4-bytes used, 4-bytes reserved.
6	DI32xUC 120V	6ES7			8...11		32-point Input Card: 4-bytes used, 4-bytes reserved.
7	DI16xAC 120V	6ES7			12...13		16-point Input Card: 2-bytes used, 4-bytes reserved.
8	DI16xAC 120V	6ES7			16...17		16-point Input Card: 2-bytes used, 4-bytes reserved.
9	DO32xDC24V/0.5A	6ES7				0...3	32-point Output Card: 4-bytes used, 4-bytes reserved.
10	DO32xDC24V/0.5A	6ES7				4...7	32-point Output Card: 4-bytes used, 4-bytes reserved.
11	DO16xDC 24V/2A	6ES7				8...9	16-point Output Card : 2-bytes used, 4-bytes reserved.
12	DO16xDC 24V/2A	6ES7				12...13	16-point Output Card : 2-bytes used, 4-bytes reserved.

Figure 3-88. Default digital addressing in S7-400. Note that although 4-bytes (32-bits) are always reserved, the actual byte usage of the module is displayed under the I-address and Q-address columns.

Configuring S7-400 Analog Input Module Properties

Basic Concept

The general properties of analog input modules in the S7-400 are viewed or edited from the Hardware Configuration. Basic information about the module, like part number and a brief description, is provided on the *General* tab. The addresses assigned to the module are provided on the *Addresses* tab (See *Configuring S7-400 Analog I/O Module Addresses*). Before analog input modules are started it may be necessary to set certain module operating parameters (See *Configuring S7-400 Analog Input Signal Parameters)*.

Essential Elements

The ***Short Description***, the same as given in the hardware catalog, gives important features of the analog input module such as acceptable input signals, bit resolution, and number of input channels. For example an eight channel analog input module may have 14-bit resolution and accept 0-20 mA/4-20 mA current signals and several voltage ranges. The ***Order Number*** reflects the currently open module, inserted in a specific slot in the configuration — it should match the part number of the actual installed module. The ***Name*** field shows the default name of the module. Although the name can be changed, it is not recommended. The **Comment** field allows details of the module's application to be entered.

Application Tips

The writing area on the module connector limits what descriptive information can be written for each input channel. Use the comment field to provide detailed information on the use of the analog inputs of a module.

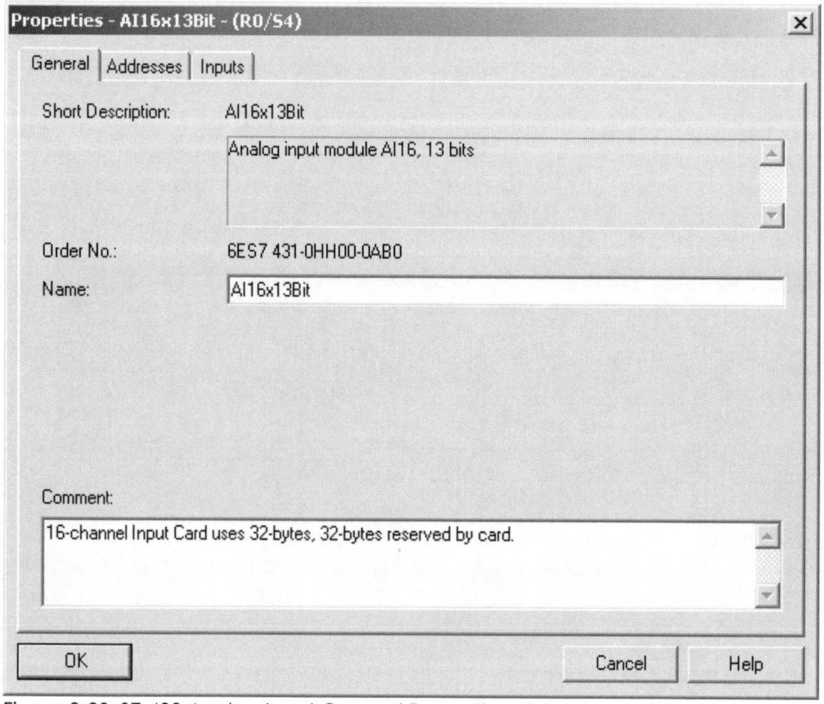

Properties - AI16x13Bit - (R0/S4)

General | Addresses | Inputs

Short Description: AI16x13Bit

Analog input module AI16, 13 bits

Order No.: 6ES7 431-0HH00-0AB0

Name: AI16x13Bit

Comment:

16-channel Input Card uses 32-bytes, 32-bytes reserved by card.

OK | Cancel | Help

Figure 3-89. S7-400 Analog Input General Properties dialog.

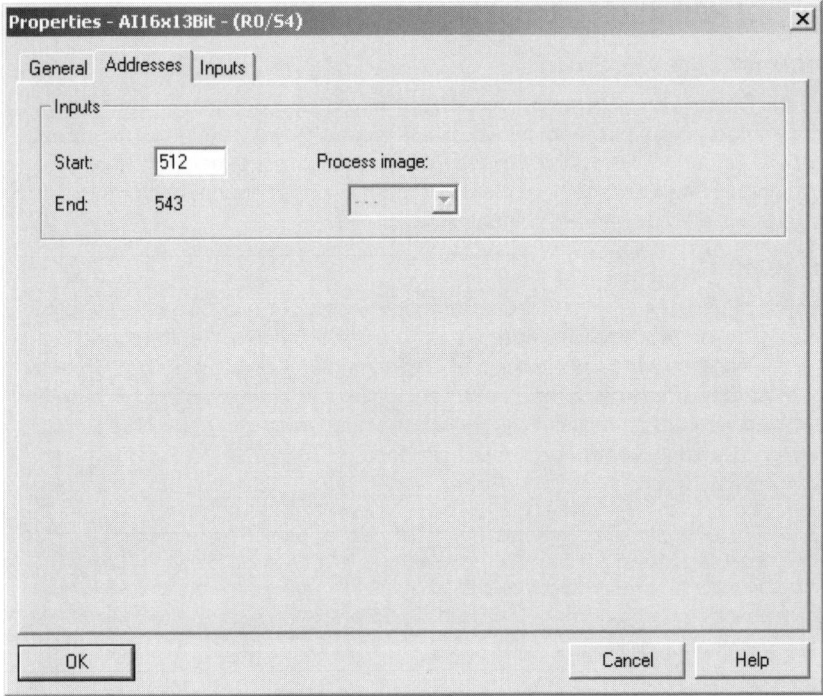

Figure 3-90. S7-400 Analog Input Address Properties dialog.

Quick Steps: Configuring S7-400 Analog Input Module Properties

STEP	ACTION
1	With a station open in the hardware configuration, select a rack to view its installed modules; double click on an analog input module whose properties you wish to view or edit. Not all properties may be edited (e.g., *Order Number, Short Description*).
2	On the *General* tab, the **Short description** provides a brief summary of the module features (e.g., module type, number of inputs, bit resolution).
3	The **Order Number** is the part number of the open module in the configuration and should reflect the analog input module installed in the same slot in the actual rack.
4	Click in the **Name** field to enter a new name; the default name is the number of analog input channels and the bit resolution prefixed with **AI** for analog input
5	Use the **Comment** field to enter a description of the module inputs or application.
6	Select the *Address* tab to view or modify the starting byte address of the module. To modify the start byte address, de-activate the **System Selection** check box and enter a new start byte-address; confirm by pressing **OK**. See *Configuring S7-400 Analog I/O Module Addresses*.

Configuring S7-400 Analog Input Signal Parameters

Basic Concept

In addition to the *General* and *Addresses* properties tabs used to configure S7-400 analog input modules, most analog input modules will also have one or two Input properties tabs. The *Input-Part 1* or *Input-Part 2* tabs allow you to view or configure the module's Hardware Interrupt or Diagnostic Interrupt features as well as determine input channel parameters that affect the type of signals may be used at each input.

Essential Elements

The configuration of analog input signal parameters involves two concerns. In the Enable Section you can activate the module's hardware or diagnostic interrupt features; and in the Input Section, you will determine the input signal characteristics for each channel. On most analog input modules a hardware interrupt can be triggered when user-defined signal limits are exceeded and when all of the input channels have been read and converted (i.e., whenever new measured values are at all input channels).

Depending on the module, diagnostic events may be activated for individual channels or channel groups and include diagnostics such as *wire break*, *Underflow*, *Overflow*, or *Short Circuit to Ground*. When a diagnostic interrupt occurs, on a channel or within a channel group, the module reports the event to the CPU (diagnostic buffer entry). Module-related diagnostic information as well as channel-specific diagnostic events can be read from the module using system functions (SFCs).

Application Tips

On modules that use a plug-in range card to allow the various input signal ranges, ensure that the range card is inserted for the required channels. Also ensure that the encoding position on the module matches the letter (e.g., A, B, C, D) displayed on the input dialog (below each channel) in the fields labeled *Position of Measuring Range Selection Module*.

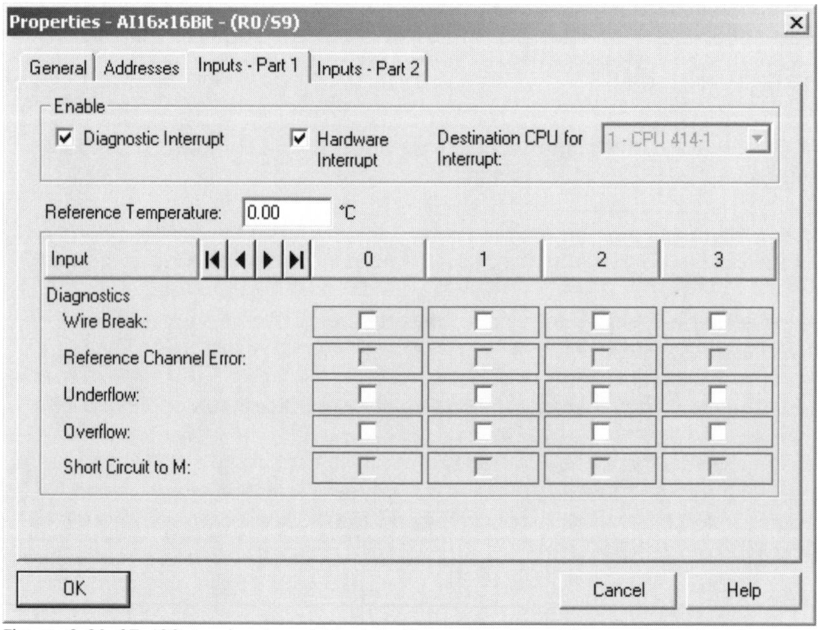

Figure 3-91. S7-400 Analog Input Module Properties: Input Parameters-Part 1.

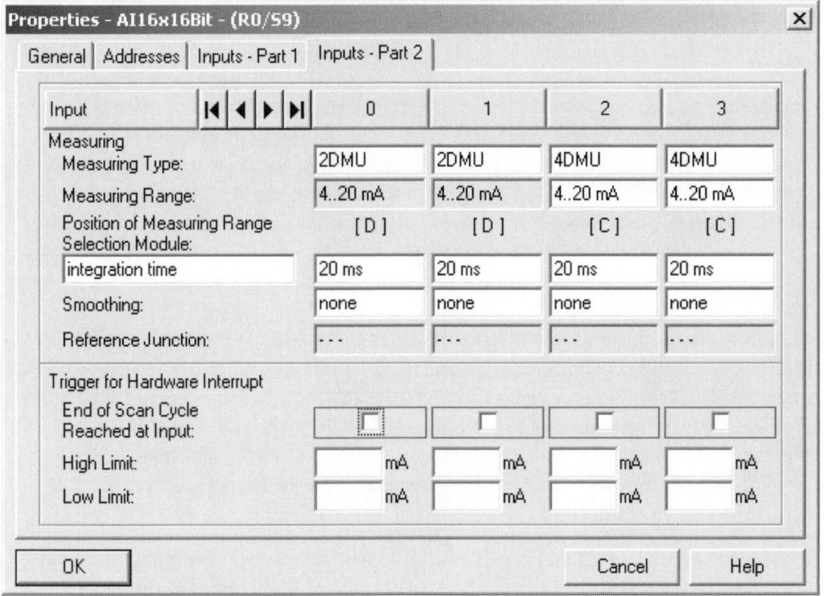

Figure 3-92. S7-400 Analog Input Module Properties: Input Parameters-Part 2.

Quick Steps: Configuring S7-400 Analog Input Signal Parameters

STEP	ACTION
1	Open to the required station configuration, double click on the analog input module to open the properties dialog, then select the *Inputs* tab.
2	Activate the check box **Diagnostic Interrupt** in order to use the module's ability to generate interrupt on certain module-specific diagnostic events.
3	Activate the check box **Hardware Interrupt when Limit Exceeded**, to allow inputs to generate an interrupt when the input signal exceeds (goes above or falls below).
4	Activate the check box **Hardware Interrupt at End of Scan Cycle** (if feature available) to allow the module to signal the CPU with a hardware interrupt when all of the input channels on the module have acquired new measured values.
5	If the "Diagnostic Interrupt" is enabled then for each channel, you may activate the check box for **Group Diagnostics** and for other channel diagnostics you wish to enable (e.g., Wire Break). If a diagnostic event occurs, a diagnostic interrupt is triggered and the module-dependent diagnostics are stored on the module.
6	Click in the **Measurement Type** field directly below each channel, and from the list of available measurement types (e.g., voltage (E), current, T/C, RTD) select one for the associated channel. If a channel is not connected, select "Deactivated."
7	Click in the **Measurement Range** field directly below each channel, and from the list of available measurement ranges select a range for the associated channel.
8	If the hardware interrupt is enabled for input signal limits exceeded, then for each channel, click in the corresponding fields and enter the **Upper limit** value and **Lower Limit** value. The module will trigger an interrupt when the input value exceeds the set "Upper limit," or when the input signal falls below the set "Lower Limit."

Configuring S7-400 Analog Output Module Properties

Basic Concept

The general properties of analog output modules in the S7-400 are viewed or edited from the Hardware Configuration. Basic information about the module, like part number and a brief description, is provided on the *General* tab. The addresses assigned to the module are provided on the *Addresses* tab (Also see *Configuring S7-400 Analog I/O Module Addresses*). Finally, there is the *Outputs* parameters tab, for viewing or setting the operating parameters of the module and individual output channels.

Essential Elements

The ***Short Description***, same as given in the hardware catalog, is a gives a shortlist of features of the analog output module such as output signal ranges, bit resolution, and number of output channels. The ***Order Number*** reflects the module assigned in the configuration tool and can be verified with part number of the physically installed module. The ***Name*** field shows the default name of the module. Changes to the module name are reflected in the configuration table. The **Comment** field allows more detail information about the module (e.g., how it is used) to be documented.

You will be able to configure, the *Outputs* parameters tab, for viewing or setting the signal type and range used by each output channel. The choices for the **Type of Output** are voltage and current;

Application Tips

The writing area on an analog output module connector limits what descriptive information that can be written for each output. Use the comment field to provide detailed information on the use of the analog output channels of a module.

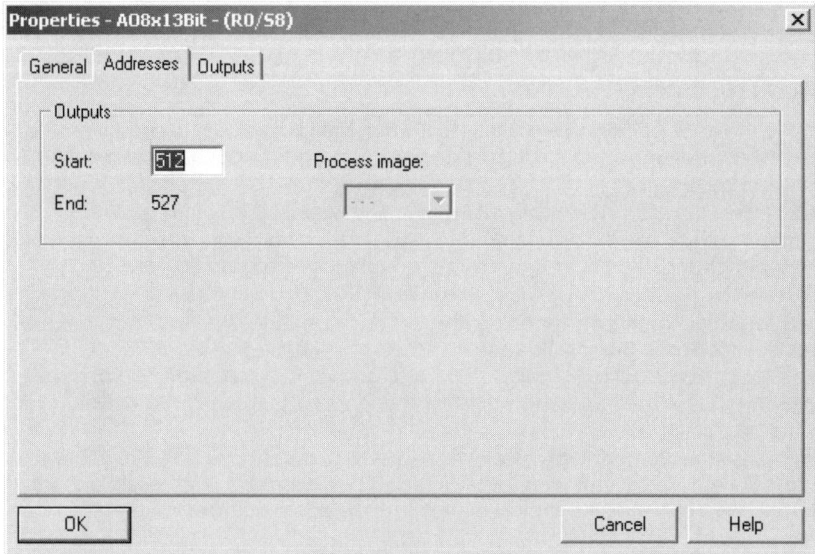

Figure 3-93. S7-400 Analog Output Module Properties: Addresses tab.

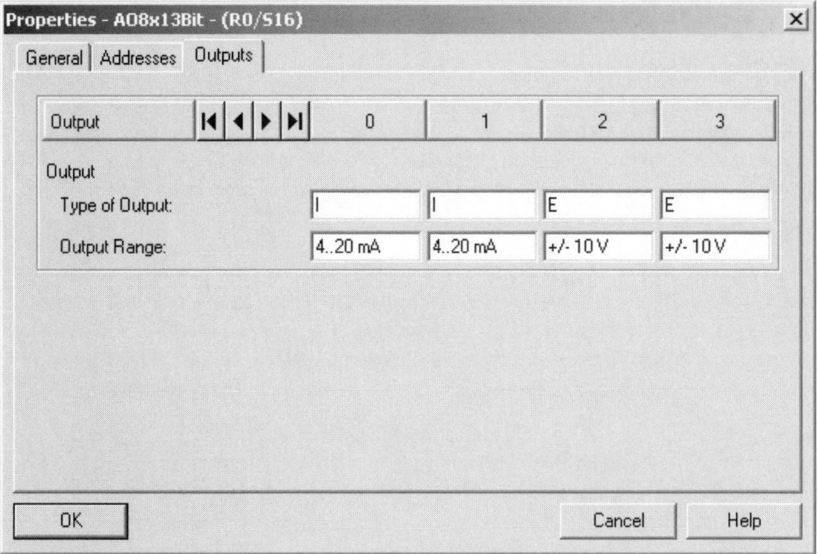

Figure 3-94. S7-400 Analog Output Module Properties: Output parameters tab.

Quick Steps: Configuring S7-400 Analog Output Module Properties

STEP	ACTION
1	With the required station open in the hardware configuration tool, select the central or expansion rack, or DP slave to display its installed I/O modules; select the analog output module and double click to open the properties dialog.
2	On the *General* tab, the **Short description** provides a brief summary of the module features (e.g., number of outputs, digital resolution). The **Order Number** is the part number of the open module in the configuration and should reflect the analog output module installed in the corresponding slot of the physical rack.
3	The default **Name**, which is modifiable, is based on the number of analog output channels and the bit resolution prefixed with **AO** for analog output.
4	Use the **Comment** field to enter a description of the module outputs or application.
5	Select the *Address* tab to view or modify the modules starting byte address.
6	Select the *Outputs* tab to view or modify the operating parameters for the module and for each analog output channel. Use the **Output** navigational buttons to see the output channels that are not in view.
7	Click in the **Measurement Type** field below each channel, and from the list of available measurement types [e.g., voltage (E), current (I)] select one for the associated output channel. If a channel is not connected, select "Deactivated."
8	Click in the **Measurement Range** field below each channel, and from the list of available ranges (e.g. ±10V, 1-5V, and 0-10V) select a range for the output channel.
9	Confirm the configuration parameters with the **OK** button and use the toolbar **Save and Compile** button to generate the *System Data* object.

Configuring S7-400 Analog I/O Addresses

Basic Concept

Each analog module inserted in an S7-400 configuration, depending on the number of channels will reserve sixteen or thirty-two bytes of the peripheral (P) memory area. Since each analog input or output channel requires a word (two bytes) to handle its digitized data, an 8-channel module reserves 16-bytes and a 16-channel module reserves 32-bytes.

Essential Elements

Default addressing for S7-400 analog I/O starts with byte-512 of peripheral memory, for both inputs and outputs. As each successive module is inserted, its starting byte address is based on the next available byte address following the previous analog input module (for inputs) or previous analog output module (for outputs), even if a slot is left empty. Like with digital modules analog input and output modules can have the same start bye addresses.

In the figure below, the S7- 400 has 16-channel analog input modules in slots 4-6 and 8-channel analog output modules in slots 7-8. Starting with byte-512, slot-4 reserves bytes 512 -to- 543; slot-5 reserves bytes 544 -to- 575; and slot-6 reserves bytes 576 -to- 607. Again from byte-512 for outputs, slot-7 reserves bytes 512 -to- 527; slot-8 reserves bytes 528 -to- 543. Each input address is prefixed with **PIW** (peripheral input word); each output address is prefixed with **PQW** (peripheral output word). The first analog input is **PIW 512**; the second input is **PIW 514**. The first analog output is **PQW 512**; the second analog output is **PQW 514**.

Application Tips

Since peripheral memory involves direct access to input or output modules, analog I/O must be handled via the program. In LAD and FBD an analog input is read by specifying the input address as the source in a move instruction; an analog output is written by specifying the output address as the destination of the move operation. In STL, a load operation reads an analog input by specifying the input address (e.g., L PW 512). A value is sent to an analog output with a transfer operation that specifies the output address (e.g., T PW 512).

		PIW/PQW	PIW/PQW	PIW/PQW	PIW/PQW	PIW/PQW	
		Slot-4	Slot-5	Slot-6	Slot-7	Slot-8	Slot-9
S7-PS407	S7-CPU	PIW512	PIW544	PIW576	PQW512	PQW528	IM
		PIW514	PIW546	PIW578			
		PIW516	PIW548	PIW580	PQW514	PQW530	
		PIW518	PIW550	PIW582			
		PIW520	PIW552	PIW584	PQW516	PQW532	
		PIW522	PIW554	PIW586			
		PIW524	PIW556	PIW588	PQW518	PQW534	
		PIW526	PIW558	PIW590			
		PIW528	PIW560	PIW592	PQW520	PQW536	
		PIW530	PIW562	PIW594			
		PIW532	PIW564	PIW596	PQW522	PQW538	
		PIW534	PIW566	PIW598			
		PIW536	PIW568	PIW600	PQW524	PQW540	
		PIW538	PIW570	PIW602			
		PIW540	PIW572	PIW604	PQW526	PQW542	
		PIW542	PIW574	PIW606			

Figure 3-95. S7-400 default analog addressing, starts at byte-512 for inputs and outputs.

Quick Steps: Configuring S7-400 Analog I/O Addresses

STEP	ACTION
1	From the SIMATIC Manager, open the required project and select the S7-400 Station for which addressing is required.
2	With the station selected, in the right pane of the station window double click on the Hardware object to open the station in the Hardware configuration tool.
3	Select a rack whose modules you wish to view the installed analog I/O modules or insert new analog I/O modules.
4	View the starting byte address of each input module in the **I-Address** column or each output module in the **Q-Address** column.
5	To modify the starting byte address of a module, select the module, right-click and select **Object Properties**, then select the Address tab.
6	To modify a starting byte address, first de-activate the **System Selection** check box to disable automatic address determination by the system.
7	Enter a new start byte-address and confirm by pressing **OK**. In the S7-400 analog input and analog output modules may have the same starting byte addresses (e.g., PIW 512, PQW 512). Analog I/O start byte addresses, after byte-512, must be at 16-byte boundaries.
8	The **Process Image** field has no meaning with analog modules and is grayed.

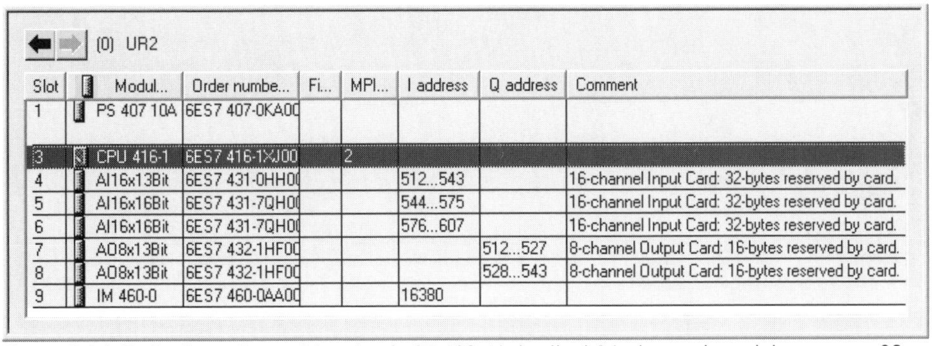

(0) UR2

Slot	Modul...	Order numbe...	Fi...	MPI...	I address	Q address	Comment
1	PS 407 10A	6ES7 407-0KA0					
3	CPU 416-1	6ES7 416-1XJ00		2			
4	AI16x13Bit	6ES7 431-0HH0			512...543		16-channel Input Card: 32-bytes reserved by card.
5	AI16x16Bit	6ES7 431-7QH0			544...575		16-channel Input Card: 32-bytes reserved by card.
6	AI16x16Bit	6ES7 431-7QH0			576...607		16-channel Input Card: 32-bytes reserved by card.
7	AO8x13Bit	6ES7 432-1HF0				512...527	8-channel Output Card: 16-bytes reserved by card.
8	AO8x13Bit	6ES7 432-1HF0				528...543	8-channel Output Card: 16-bytes reserved by card.
9	IM 460-0	6ES7 460-0AA0			16380		

Figure 3-96. Default analog addressing in S7-400. Note, that 16-channel module reserves 32-bytes and 8-channel modules reserves 16-bytes. Also note, under the **I**-address and **Q**-address columns, that analog inputs and outputs in the S7-400 may use the same addresses.

Configuring I/O Modules for Multi-Computing Operation

Basic Concept

All function modules (FMs), digital and analog signal modules (SMs), and communications processors (CPs) in a multi-computing configuration must be assigned to a specific CPU (i.e., CPU1, CPU2, CPU3, or CPU4) in the configuration. A module is assigned to a CPU by selecting the module and opening its object properties.

Application Tips

A module is assigned to a particular CPU by selecting the module and opening its object properties. Modules having interrupt capability will automatically indicate, under its object properties, the CPU to which it is assigned.

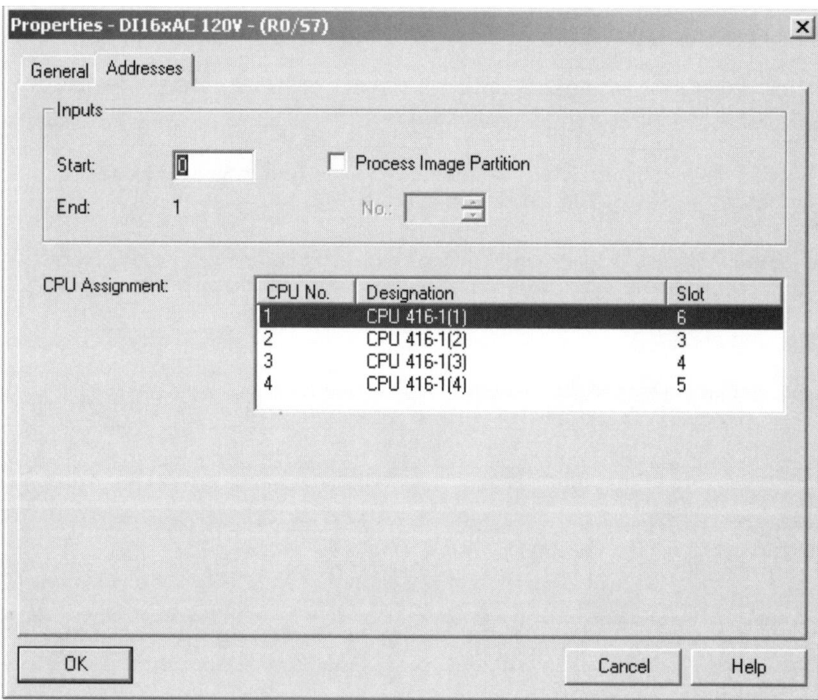

Figure 3-97. S7-400 I/O Module Addresses dialog in multiple CPU configuration. Each module must be assigned to a specific CPU.

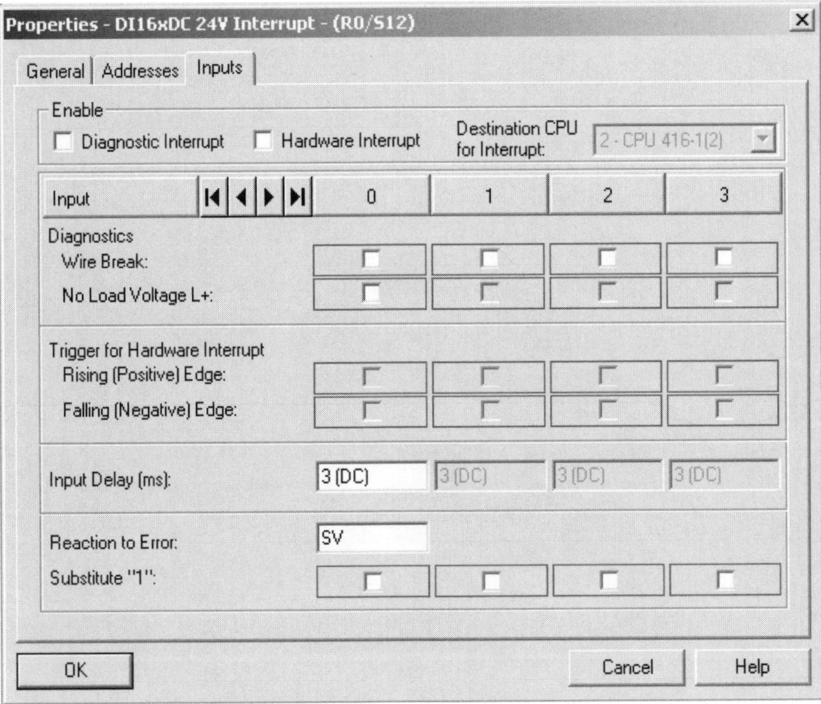

Figure 3-98. S7-400 Interrupt Input Module Properties: Input parameters tab, showing the *Destination CPU for interrupt* generated by this card.

Quick Steps: Configuring I/O Modules for Multi-Computing Operation

STEP	ACTION
1	Start the SIMATIC Manager and open the project with the multi-computing station.
2	Open the Hardware Configuration tool to display the multi-computing station and I/O racks and modules.
3	Select and double- click on a module and select its **Address** tab.
4	From the **CPU Assignment** window, displaying the multi-computing CPUs, select the CPU No. the module will be associated with.
5	Repeat the assignment process for all digital/analog signal modules (SMs), function modules (FMs), and communications processors (CPs) in both the CPU rack and all local and remote expansion racks.

Working with STEP 7 Programs and Data

Objectives

- Introduce STEP 7 Programming Principles
- Introduce STEP 7 Block Types
- Introduce S7 Program Processing
- Introduce Addressable S7 Memory Areas
- Introduce STEP 7 Data Types and Formats
- Introduce LAD/FBD Instruction Sets
- Introduce the LAD/FBD/STL Editor
- Opening, Editing, and Saving Blocks
- Create/Call Functions and Function Blocks

Introduction to STEP 7 Programming Principles

The STEP 7 programming environment supports the structuring of the control tasks of a machine or process and its corresponding control code into functional units. In such an approach, the system is broken down into logical or functional units and if necessary, each unit further broken down into subunits. Once broken down into units that reflect the major control tasks, the STEP 7 programming resources support development of the control code in modular sections called *blocks*. You may call these blocks to be processed as needed.

There are many advantages to creating your code in modular blocks. Each block is a working unit that may be called repetitively as needed, and may be stored in a library for later use. Since each block performs a specific function, the program is simplified, and each block may be tested individually. Code blocks may be copied and duplicated or slightly modified for reuse for the same or a similar function. Finally, each block may be called in the program for processing on a conditional or an unconditional basis.

Program Design Strategy

The program design strategy is largely concerned with the structure of the STEP 7 program. While the traditional linear approach to PLC programming is possible, a structured design, using modular blocks of code to perform different task, is the basis of the STEP 7 programming architecture. In either case, the design strategy is best developed in two steps. The first step involves partitioning the machine or process into logical units or segments, and each unit into subunits and operations. The second stage is to develop the control logic of the individual tasks and operations of each process unit. The standard STEP 7 package offers three languages which you may use separately or combined to develop your code blocks.

Defining Machine or Process Units

Any machine or process can be partitioned into its logical units. These units, which reflect the main subsystems, each performs a functional part of the whole system. By dividing the system into units, members of a development team will quickly grasp what has to be controlled. Well planned partitioning not only facilitates group development, but also allows for easier adaptation of the program for reuse, and for future extensions. Defining units is a first step in developing a program that is handled by STEP 7 code blocks.

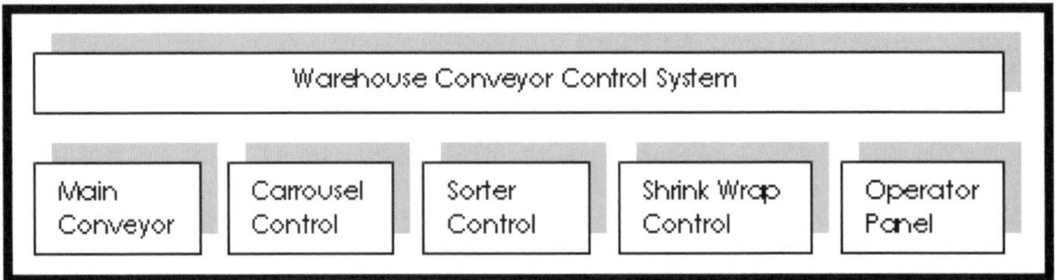

Figure 4-1. Typical division of machine/process into major functional or technological units.

Defining Subtasks and Operations

Machine or process units generally reflect the major systems involved in producing a finished product. Therefore, each can usually be further divided into functional *tasks,* which must be accomplished to complete the unit function. Each task will normally consist of several operations, which require control of individual devices, process loops, or other process elements for which control logic must be developed. This concept, illustrated in the following diagram, is one method of conveying the major procedures, subtasks, and operations for which a program must be developed. Non-hardware tasks such as communications, alarm processing, production accounting, and proprietary algorithms must also be considered.

Developing the Control Logic

The second stage of the program design strategy deals with developing control logic for the tasks of each process unit. In other words, the definition of how specific operations (*e.g., devices, process loops, etc.*) will be implemented. A given subtasks, for instance, may require control of several digital and analog actuators. Completion of the task may also require data processing, error checking, and perhaps communications functions. One-by-one, the control logic for each of these operations must be defined and then developed.

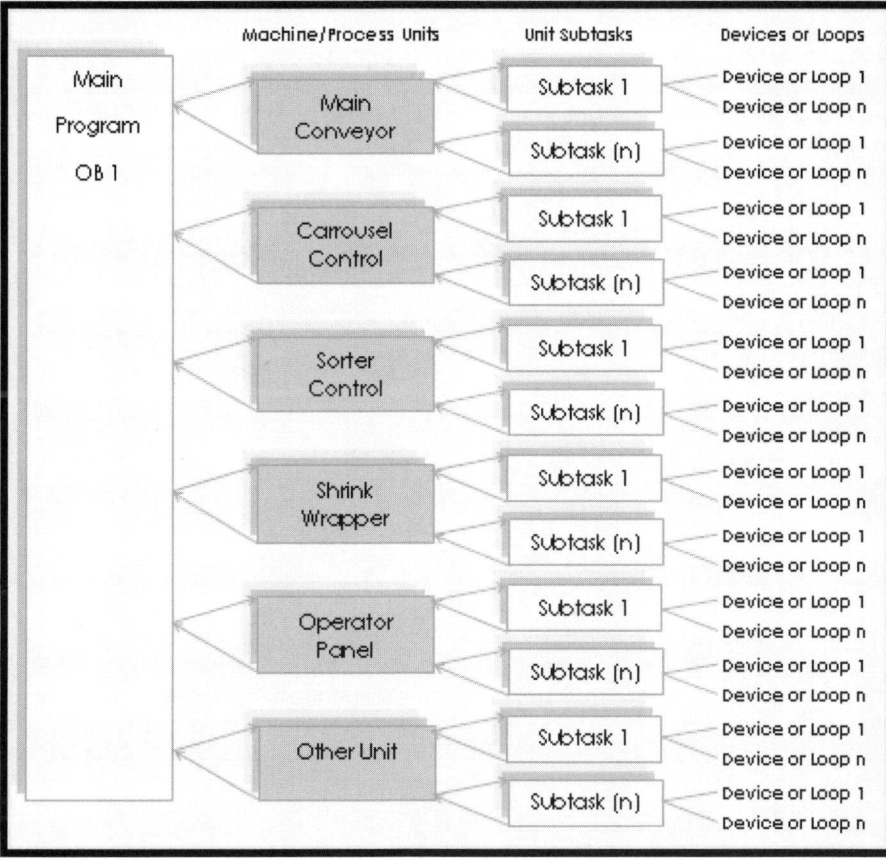

Figure 4-2. Illustration of program design based on division of control task into major functional or technological units, unit subtasks, and finally into device or loop operations.

Modular Functionality with Blocks

As you probably see by now, developing a STEP 7 program is about writing modular blocks of code, each of which perform a specific function in the overall control task. This concept, involves a program composed of a collection of procedures. In STEP 7, a procedure would be called a block. Each block might correspond to a subtask as shown in Figure 4-2. Here, the various tasks of each machine or process unit are divided into blocks, each of which would contain the control logic for the devices or operations that complete the subtask. Several blocks may be required to perform the subtasks that complete the unit control.

As you develop the program, you may subdivide it into as many sections or blocks as you like. To the degree that it is possible, each block should be self-contained and relatively independent of other blocks. With this approach, the block can be easily tested and debugged, and later problems can be pinpointed to a block based on the function.

STEP 7 Block Types

The STEP 7 programming environment provides three categories of blocks from which the user program may be developed. *User-blocks* are S7 resources that you are able to create and insert programs and data; *System blocks,* are blocks integrated as part of the S7 PLC operating system, yet may be called from the user program. *Standard Blocks* are off-the-shelf blocks that you may also incorporate in your program. Standard blocks support applications such as PID control and communications drivers. Several groups of standard blocks are supplied as libraries with STEP 7, while others may be purchased separately.

While all of the resource listed below may be used in your programs, the actual number of blocks that may be programmed (e.g., FBs, FCs, DBs), or that are available (e.g., OBs, SFCs and SFBs), is CPU-dependent. The size that a block can be (i.e., length in bytes) is also dependent on the CPU. You may view the performance characteristics for a CPU, from the online window of the SIMATIC Manager, after selecting the CPU then from the menu selecting **PLC ➢ Module Information**.

Table 4-1. STEP 7 Block Resources Overview.

Block Type	Identifier	Range	Category	Comment
System Blocks	SFB/SFC/SDB	-	System	Pre-Numbered
Organization Blocks	OB	OB 1 to OB n	User	Pre-Numbered
Function Blocks	FB	FB 1 to FC n	User	User Numbered
Functions	FC	FC 1 to FC n	User	User Numbered
Data Blocks	DB	DB 1 to DB n	User	User Numbered

System Blocks (SFB, SFC, SDB)

System blocks includes system functions (SFCs) and system function blocks (SFBs), both of which are integrated as part of the S7 CPU. These canned routines may be called in the user program. Since system blocks are canned functions, they have pre-defined block numbers and the code cannot be modified. System blocks handle many commonly required tasks that help reduce the overall development time. Numerous system blocks are available that allow the user program to access services of the CPU (e.g., enable and disable interrupts, set/read real-time clock) and perform services that include control, message-handling, and communications functions. A listing of SFCs and SFBs is provided in the appendix.

Finally, system blocks include system data blocks (SDBs), which contain data used by the PLC. SDBs contain the compiled configuration data generated by the hardware and network configuration tools, as well as function module configuration tools. Ultimately, the SDBs are downloaded to the CPU and to the associated modules (e.g., CPs, FMs, etc.).

Organization Blocks (OB)

Organization blocks (OBs) are one of three block types in which you may write parts of your STEP 7 code. The purpose of OBs is to provide a structured and simplified means for handling the various processing requirements of a total STEP 7 program. Organization blocks are important because they provide an interface between the S7 operating system and the user program — this is true since organization blocks are written by the user as part of the control program, but all OBs are called by the S7 operating system based on certain events or conditions. For example, on each start-up of the CPU, depending on the type of start-up either OB 100 or OB 101 (programmed with your start-up code) will be called and processed.

Just as OB 100 and OB 101 are assigned to a specific type of startup, each OB has a specific processing task, based on system-related event. When that specific condition or event occurs, the S7 operating system responds by calling the associated OB to be processed. The actual response, however, is based on the code that you will have written in the OB. By implementing code in the various organization blocks, you determine what code is processed at startup, during the normal cycle, in the event of timed or hardware interrupts, and system-related errors. An overview of organization blocks according to their processing function and the events or conditions that trigger their calling is shown in the following table.

Table 4-2. Overview of S7 Organization Blocks and their associated triggering event.

OB Type	OB No.	Call Event and Application Description
Normal Cyclical Processing	1	Called after each CPU start-up and thereafter, cyclically, to process the main user program.
Time-of-Day Interrupts	10-17	Called at user-configured time-of-day interrupt, to process user code that must be processed at a specific date and time.
Time-Delay Interrupts	20-23	Called after user-defined delay expires, to handle portions of the program that must be processed after the time delay expires.
Cyclic Interrupts	30-38	Called cyclically at defined intervals (e.g., 500 ms), to handle code (e.g., PID loops) that must be processed at precise intervals.
Hardware Interrupts	40-47	Called when any of the assigned process or module-generated interrupts occurs, to process the user-programmed service routine.
Multi-Computing Interrupt	60	Called on an interrupt generated by any of the CPUs in a multiple CPU configuration, to process a user-defined synchronized response.
Redundancy Error Interrupts	70-73	Called upon occurrence of a redundancy error in an S7-400 H-system, to process the user-programmed response.
Asynchronous Error Interrupts	80-87	Called at occurrence of a system-related error (e.g., power supply fault, or module failure), to process user-defined service routine.
Background Processing	90	Called to process non-critical code, when the actual cycle time of the current cycle is less than the user-defined minimum cycle.
CPU Startups Processing	100-101	Called at each CPU start-up and prior to calling OB1, to process initialization code that must be executed once at each startup.
Synchronous Error Interrupts	121-122	Called at occurrences of program-related runtime errors, to process the user-programmed service routine for synchronous errors.

Note: See *Programming Organization Block 1*, for guidelines on writing the code for OB 1.

Function Blocks (FB)

A Function Block (FB) is one of three block types (OB, FC, and FB) in which you may write code. FBs are intended for creating logic routines or algorithms where data that is either generated by or required by the block must be available from one call of the block to the next. To handle this memory requirement, you must assign a data block (DB) to each FB. The DB is opened for read/write access on each instance the FB is called, and is therefore called an *instance DB*. An FB can actually have several assigned instance DBs, allowing it to work with different sets of data. Each call of an FB/DB is referred to as an "instance." As seen in the template for programming and FB, like other blocks it contains a variable declaration section (top pane of window), for declaring variables; and a code section (bottom pane of window), for writing the program.

In an FB you may define *temporary **(TEMP)** local variables*, for those values that are only required by the FB while the block is processing; you may define *static **(STAT)** variables*, for those values that must be saved and available from one call of the FB to the next (I.e., memory). You may also define *formal parameters*, a special type of variable, which allows you to develop an FB as a black box function. When specifying formal parameters, you may define **IN**, **OUT**, or **IN_OUT** variables. **IN** variables are inputs to the function, read into the FB; **OUT** variables are outputs from the FB, written to by the FB; an **IN_OUT** variable is both read from and written to by the FB. Formal parameter variables to an FB act as substitute address holders. Each time you call the FB as a function in you program, you will supply a different set of actual addresses (*actual parameters*) in place of the formal parameters.

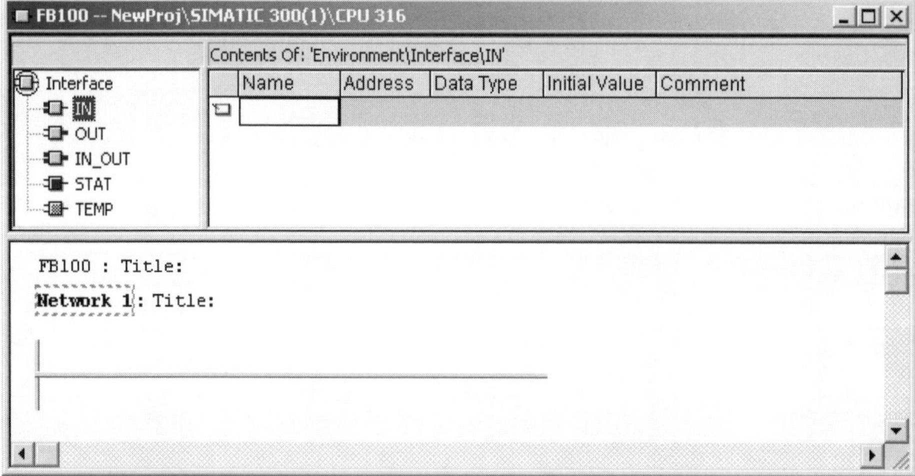

Figure 4-3. The LAD editor open to the template for programming a function block (FB).

Calling a Function Block

An FB can be called from any other block (e.g., OB, FB, or FC) in LAD, FBD, or in STL. If the FB is written without formal parameters, then in LAD/FBD you may drag the FB from the *FB Blocks* folder of the Program Elements and drop it into the current network. You must then specify the related instance DB just above the FB. You may call the FB with or without preceding logic conditions. If the FB has formal parameters, the same procedure is used but the address of each actual parameter must be supplied as required by each formal parameter.

In STL, an FB without formal parameters is called using the CALL instruction, to specify the FB number and the instance DB number (e.g., **CALL** FB 100, DB100). If the FB has formal parameters, each parameter is listed on a separate line beneath the CALL line, allowing you to specify an address for each actual parameter.

Functions (FC)

A Function (FC) is one of three user block types (OB, FB, and FC) in which you may write part of your program. Unlike the FB, the FC does not require an assigned data block (DB) as memory. However, an FC can access a global data block for reading or writing data that other blocks may access. FCs are suitable for simple logic to complex digital operations, and in many S7 applications, will account for as much as seventy to eighty percent of the total program. As seen in the template for programming and FC, like other blocks it contains a variable declaration section (top pane of block window), for declaring block variables; and a code section (bottom pane of block window), for writing the program.

In an FC, you may define *temporary (**TEMP**) variables*, for those values that are only required by the FC and are only available while the FC is processing. Each FC also allows you to define a **RETURN** variable that is returned to the calling block. You may also define *formal parameters*, a special type of variable that allows you to develop an FC as a black box function — similar to a box instruction. When specifying formal parameters, you may define **IN**, **OUT**, or **IN_OUT** variables. **IN** variables are inputs to the function and are read into the FC; **OUT** variables are outputs from the FC or written to by the FC; an **IN_OUT** variable is both read from and written to by the FC. Formal parameter variables to an FC act as substitute address holders. Each time you call the FC as a function in you program, you will supply a different set of actual addresses (*actual parameters*) in place of the formal parameters.

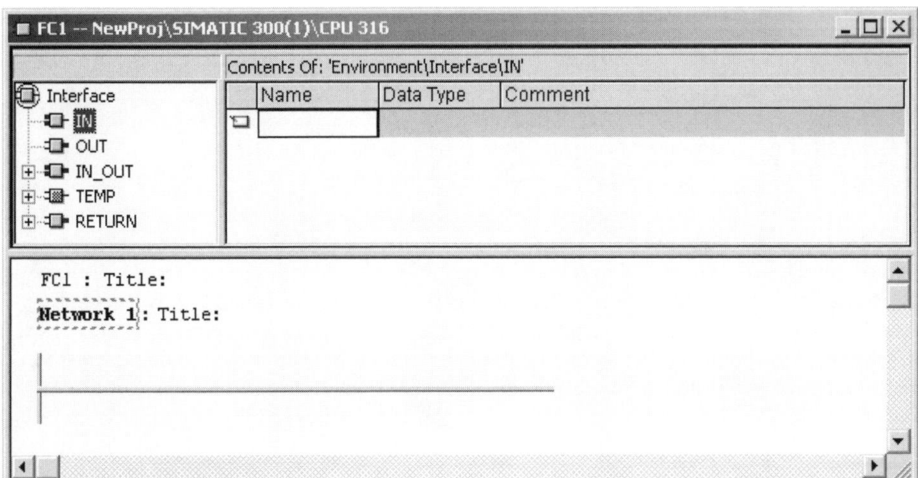

Figure 4-4. The LAD editor open to the template for programming a function (FC).

Calling a Function

An FC can be called from any other block (e.g., OB, FB, FC) in LAD, FBD, or in STL. If the FC is written without formal parameters, then in LAD/FBD you may call the FC using the CALL coil or output respectively; you must specify the FC number above the coil or output. You may also drag the function from the *FC Blocks* folder of the Program Elements and drop it into the current network. You may call the FC with or without preceding logic conditions. If the FC has formal parameters, the CALL coil or output cannot be used. Instead, you must take the FC box from the Program Elements browser and place it onto the network, where you must specify the address of each actual parameter as required by each formal parameter.

In STL, an FC without formal parameters is called using the CALL instruction to specify the FC number (e.g., **CALL** FC 1). If the FC has formal parameters, each is listed on a separate line beneath the CALL line, allowing you to specify an actual parameter (address) for each.

Data Blocks (DB)

Data Blocks (DBs) are the equivalent to the *Data Table* found in some PLCs. In STEP 7, the DB is a resource for organizing and storing the various constant and variable data required in your program. Typical constants might include presets for analog high and low limits, for PID loops; or recipe data values in a batch process. Variables might include analog input and output data, or perhaps various production totals.

DBs may be organized to your requirements and may contain variables of the same or of different data types. You also determine the name and order of each variable, and the total number of variables in each DB. Whereas intermediate results might be stored in Bit Memory (M) or in temporary local memory (L) of the block (i.e., OB, FC, FB), data blocks are used to store data stored that needs to be retained always.

DB10 -- New_prj\SIMATIC 300(1)\CPU 316

Address	Name	Type	Initial	Comment
0.0		STRUCT		
+0.0	TMP_PRESET_1	INT	350	Temperature Preset Zone1
+2.0	TMP_PRESET_2	INT	375	Temperature Preset Zone2
+4.0	TMP_PRESET_3	INT	400	Temperature Preset Zone3
+6.0	OVR_LMT_T1	BOOL	FALSE	ZONE 1 OVER Limit
+6.1	OVR_LMT_T2	BOOL	FALSE	ZONE 2 OVER Limit
+6.2	OVR_LMT_T3	BOOL	FALSE	ZONE 3 OVER Limit
=8.0		END_STRUCT		

Figure 4-5. Each data block may be defined according to your data requirements.

There are two data block types in STEP 7. The DB type, is defined when the data block is created, and is based on how the DB will be used in conjunction with logic blocks. *Shared DBs* (also called *global DBs*) are used to store data that may be read from or written to by any block within the complete program. An *instance DB*, on the other hand, is a data block assigned to a specific Function Block as required memory. The FB has read/write access to the instance DB and stores both its declared static (STAT) variables and formal parameter variables in the DB. When an instance DB is created, it must be explicitly assigned to an FB.

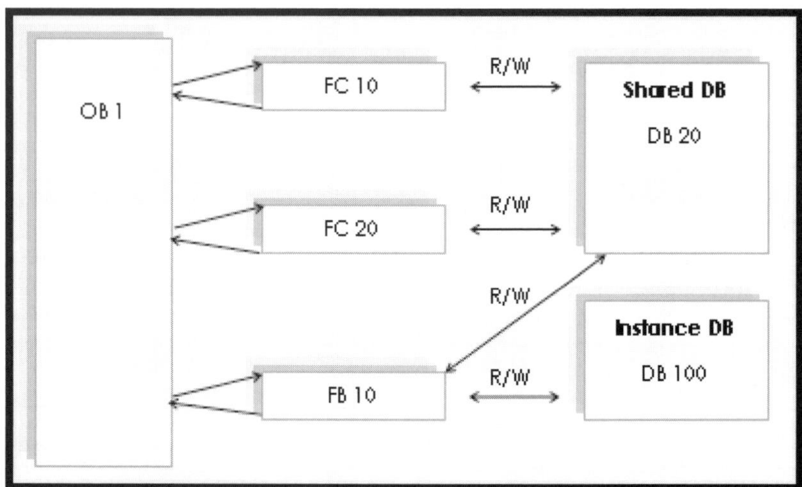

Figure 4-6. Illustration of Access to Instance and Shared (Global) Data Blocks.

Accessing Data Elements in Data Blocks

As data is entered into a data block, each element is stored in the order it is entered, starting from data byte zero. Data elements may use any valid data type (e.g., BOOL, INT, DINT, REAL, CHAR, and S5TIME). As each element is entered, an area is reserved that corresponds to the assigned data type. The start byte address of each element is shown in the 'Address' column of the data block (See Fig. 4-5). Whether a data element uses a bit, byte, word, or double word, the start byte is reflected in the absolute address of each element.

Before accessing a DB variable, you must open the data block first, using the appropriate instruction to reference the DB number (e.g., OPN DB n). Once a data block is open, any variable may be accessed, using the correct identifier and address (e.g., DBW4, *Table 4-3*). Each S7 CPU has two registers that it can use to open a DB. These registers, which the CPU uses to hold the number of the current DB, are called the DB-register and the DI-register. Although both global and instance DBs may be opened in either register, it is preferable that global DBs are opened with the DB-register and instance DBs with the DI-register.

Table 4-3. Partial Addresses used to access Shared DB or Instance DB variables.

Area	Identifier	Access Units	Example	Description of Example
Data Block (OPN DB n)	DB	**DBX** = Bit	DBX4.4	Bit 4 of Byte 4 of Shared DB (n)
		DBB = Byte	DBB4	Byte 4 of Shared DB (n)
		DBW = Word	DBW4	Word 4 of Shared DB (n)
		DBD = Double word	DBD4	Double word 4 of Shared DB (n)
Data Block (OPN DI n)	DI	**DIX** = Bit	DIX12.4	Bit 4 of Byte 12 of Instance DB (n)
		DIB = Byte	DIB12	Byte 12 of Instance DB (n)
		DIW = Word	DIW12	Word 12 of Instance DB (n)
		DID = Double word	DID12	Double word 12 of Instance DB (n)

Note: Data elements are accessed with partial addresses, using the DB or DI identifier, depending on which data block register (i.e., DB or DI) is used to open the data block.

While the CPU uses the DB and DI registers identically, some restrictions and guidelines help to minimize potential errors. In LAD/FBD, for example, you can only affect the DB register when you use the OPN coil or OPN output. Therefore, in LAD/FBD you may open a global data block. Any data element can then be accessed in the following networks until a new DB is opened. In LAD/FBD, the instance DB associated with an FB is, by design, opened with the DI-register whenever the FB is called. In STL, the open instruction is simply 'OPN DB n' (e.g., **OPN DB 22**), or OPN DI n (e.g., **OPN DI 23**). As shown in Table 4-3, a data word is addressed DBW if you used OPN DB and is addressed DIW if you used OPN DI.

Other methods for addressing the elements of a DB include *fully addressed access*, and *symbolic addressing*. Fully addressed access, only possible in LAD and FBD, does not require opening the DB. The DB is specified as part of the address (e.g., **DB 10.DBW 4**). Similar to full addressing, the symbolic address of each data location is derived by combining the assigned symbolic name of a data block with the assigned name of each element in the data block (e.g., **Zone_1.TMP_PRESET_3**). In this case, the DB name is Zone_1 and the variable is TMP_PRESET_3 (*See Figure 4-5*).

S7 Program Processing

An essential step in STEP 7 programming, is deciding how each block is called for processing. This step involves defining what should trigger the call of a block, and how often the block needs processing. Before any block can be processed, it must be downloaded to the CPU and called from somewhere within the program. How often a block needs processing and what triggers its call is generally based on the function the block performs. In any case, the processing needs of your program can be efficiently handled by the OBs listed in Table 4-4.

As you define how the blocks of your program will be processed, you will need to consider using OBs. Consider calling blocks that initialize values at each startup, from the *start-up OBs*. Call blocks that need processing at fixed intervals (e.g., every 500 ms), from a *cyclic OB (e.g., OB 30)*. Blocks that act as service routines, responding to process or module-generated interrupts, are called from an *interrupt OB (e.g., OB40)*. Blocks that need continuous processing, are called from OB1 (main program).

The Normal CPU Cycle

Normal cyclical processing of a STEP 7 program is handled by OB 1. After power is applied and the CPU is switched to RUN, OB 1 is called and is processed on each CPU cycle, until the CPU is stopped or power is removed. Since OB 1 is process continuously, most other blocks (i.e., FBs, FCs, SFCs, and SFBs) of your program will be called either directly or indirectly from OB 1. This is true since most blocks of the program will need evaluating on each cycle. During the normal cycle, OB 1 is processed from the first network until the last. If calls are made from OB 1 to other blocks, as each block completes its execution, control is passed back to OB 1.

Interrupting the Normal CPU Cycle

With the exception of the start-up cycle and background processing, all other processing tasks must interrupt the normal cycle of OB 1 to be processed. Start-up OBs are called just prior to the normal cycle, and background processing shares the normal cycle. When other events such as those that trigger hardware interrupts, timed interrupts, synchronous or asynchronous errors occur, the operating system responds by interrupting OB 1 and calling the appropriate OB. When the interrupting OB is done, control is returned to OB 1.

Table 4-4. Overview of S7 Organization Block Priority Classes.

OB Type	OB No.	Default Priority	Alternate Priorities
Cyclical Processing (Main Program)	1	23	None
Time-of-Day Interrupts	10, 11, 12, 13, 14, 15, 16, 17	2	2 -to- 24
Time-Delay Interrupts	20, 21, 22, 23	3, 4, 5, 6	2 -to- 24
Cyclic (Periodic) Interrupts	30, 31, 32, 33, 34, 35, 36, 37, 38	7, 8, 9, 10, 11, 12, 13, 14, 15	2 -to- 24
Hardware Interrupts	40, 41, 42, 43, 44, 45, 46, 47	16, 17, 18, 19, 20, 21, 22, 23	2 -to- 24
Multi-Computing Interrupt	60	25	None
Redundancy Error Interrupts	70, 71, 72, 73		
Asynchronous Error Interrupts	80, 81, 82, 83, 84, 85, 86, 87	26 (28 during startup)	24 -to- 26
Background Processing	90	.29	None
CPU Start-up Routine	100, 101, 102	27	None
Synchronous Error Interrupts	121, 122	-	-

The CPU uses a priority scheme to manage how other OBs may interrupt OB 1, as well as one another. With priority classes from 1- 28, priority 28 is the highest. If an OB is called while one of lower priority (e.g., OB 1) is being processed, the lower priority is interrupted. After the interrupting OB is done, the interrupted OB continues at the point immediately following the point of interruption. OBs that are assigned the same priority (e.g., OB 10-OB 17, priority 2) are processed in sequence, should they be triggered simultaneously. OB 90 has a priority of 29, although it is interpreted as 0.29, making it the lowest priority. In general, priorities are fixed in the S7-300; however, in the CPU 318 and in S7-400 CPUs, certain OB priorities may be altered.

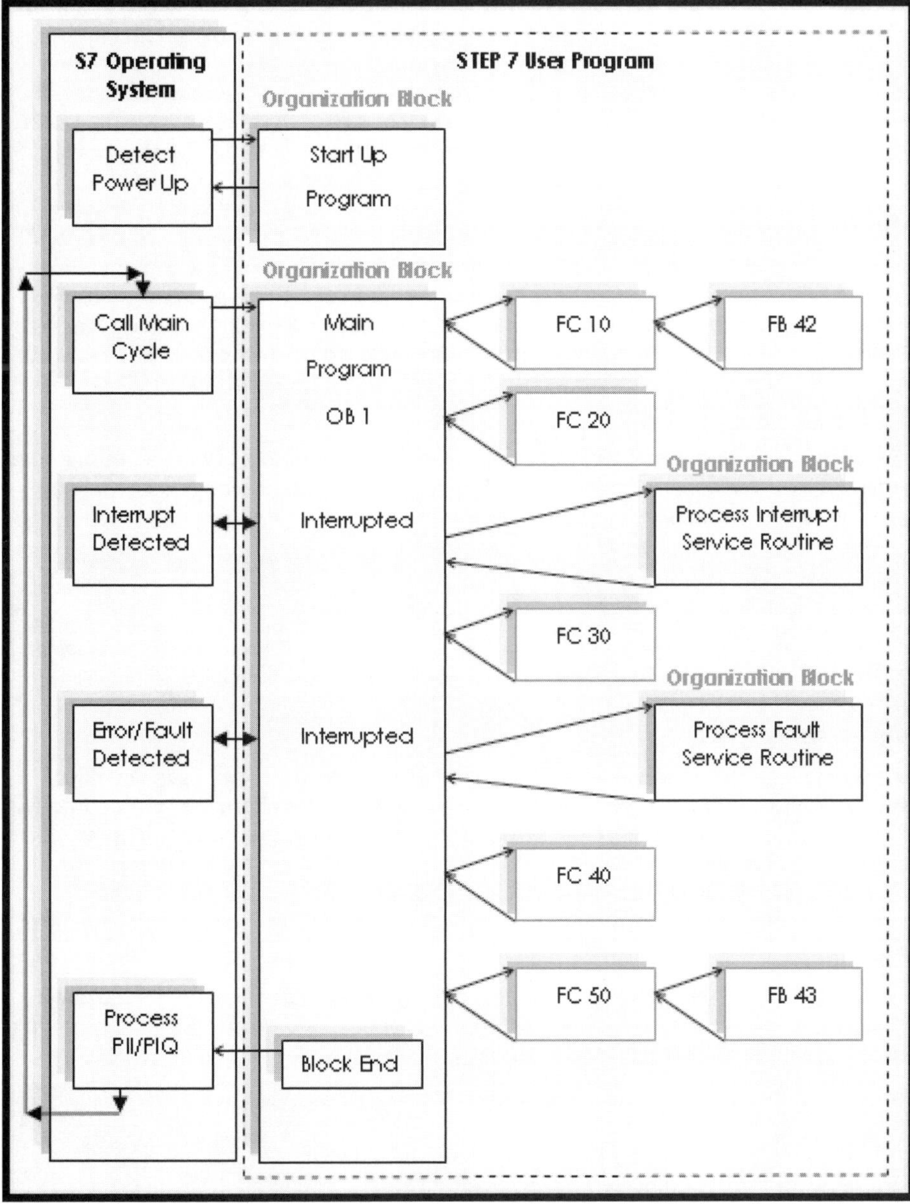

Figure 4-7. Illustration of S7 Program Processing; and Organization Blocks as the Interface between the S7 Operating System and the User Program.

Addressing S7 Memory Areas

S7 controllers, like other PLCs, contain memory areas that you may access in the user program. In the S7-300 and S7-400, these areas include *input image (I)*, *output image (Q)*, *bit memory (M)*, *peripheral input (PI)*, *peripheral outputs (PQ)*, *local (L)*, *timer (T)*, and *counter (C)* memory areas. In the following discussions we will look at how the locations in each of these areas may be addressed using the absolute address. The addresses of each of these areas may also be assigned a unique symbolic address using the *STEP 7 Symbol Editor*.

Each memory area is organized in bytes, starting from byte-0. S7 memory areas, allow access to a *bit*, *byte* (8-bits), *word* (16-bits), or a *double word* (32-bits). Referencing a word address accesses byte-n and n+1— a double word accesses byte-n, byte n+1, byte n+2, and byte n+3. Each absolute address, is prefixed by a unique identifier (e.g., I, Q, M, PI, PQ, L, T, and C) based on the area. A bit address is always referenced with the identifier followed by the byte and bit number separated by a period (e.g., **I 33.5**). A memory identifier is followed by a 'B', 'W', or 'D' to identify the location as a byte, word, or double word. All S7 memory areas, except for the local memory area may be referenced in any block.

Input Memory (I)

The *process image of inputs (PII)*, in S7 PLCs corresponds to the input image table in other PLCs. Prior to each program cycle the CPU reads the status of digital inputs and stores the results in the PII. The input circuits of each module are mapped to the PII based on the starting byte address that was assigned to the module during hardware configuration. Input addresses are accessed in the program via the address area for inputs, using the identifier letter '**I**'.

An input bit is addressed **I byte.bit**, a byte is addressed **IB n**, a word is addressed **IW n**, and a double word as **ID n** (e.g., I 9.7, IB 8, IW 8, or ID 8). See the *peripheral memory* area to learn how to obtain the instantaneous status of digital inputs.

ID 8	Input Byte-8 (IB 8)		Input Byte-9 (IB 9)		Input Byte-10 (IB 10)		Input Byte-11 (IB 11)	
Bit Positions	7	0	7	0	7	0	7	0

Figure 4-8. Illustration of input double word-8. Bit address I 9.7 is highlighted. On a 32-point module, with start address of input byte-8, byte-8 is the first group of eight input circuits.

Output Memory (Q)

The *process image of outputs (PIQ)*, in S7 PLCs corresponds to the output image in other PLCs. During each CPU cycle, the program determines the status of each output and stores it in the PIQ. At the end of each cycle, the CPU transfers the PIQ to the modules to reflect the program results. The output circuits of each module are mapped to the PIQ based on the starting byte address assigned to the module during hardware configuration. Output addresses are accessed in the program via the address area for outputs, using the identifier letter '**Q**'.

An output bit is addressed **Q byte.bit**, a byte is addressed **QB n**, a word is addressed **QW n**, and a double word as **QD n** (e.g., Q 5.7, QB 5, QW 4, or QD 4). See *peripheral memory* to learn how digital outputs are written to immediately.

QW 4	Output Byte-4 (QB 4)								Output Byte-5 (QB 5)							
Bit Positions	7	6	5	4	3	2	1	0	7	6	5	4	3	2	1	0

Figure 4-9. Illustration of output word-4. Bit address Q 5.2 is highlighted. On a 16-point module, with start address of byte-4, byte-4 is the second group of eight output circuits.

Bit Memory (M)

The bit memory area is the equivalent of internal storage bits in other PLCs. Bit memory is typically used when a bit, byte, word, or double word location is needed for an intermediate binary or digital operation. For example, for internal outputs that holds the bit status of internal logic operations, or the interlocking of logic circuits. Intermediate data results in arithmetic or other data operations might also use bit memory. Bit memory is typically used if the result does not need permanent storage, but is required throughout the entire program.

Bit memory addresses are accessed in the program using the identifier letter '**M**'. A bit location is addressed **M byte.bit**, a byte is addressed **MB n**, a word is addressed **MW n**, and a double word as **MD n** (e.g., M 101.2, MB 100, MW 100, MD 100). When configuring the CPU parameters, contiguous areas of bit memory can be defined as retentive. Otherwise, the contents of bit memory are lost when power is removed.

MW 100	Memory Byte-100 (MB 100)								Memory Byte-101 (MB 101)							
Bit Positions	7	6	5	4	3	2	1	0	7	6	5	4	3	2	1	0

Figure 4-10. Memory word-100: A bit address is written 'M byte.bit' (e.g., M 101.2).

Peripheral Memory (PI/PQ)

The peripheral area of memory allows the useable data of an input module to be read directly from the module, and the useable data of an output module to be written directly to the module. By accessing peripheral memory, the normal delay associated with the I/O image update at the end of each CPU cycle is eliminated. The identifier for accessing input modules is PI (for peripheral input), and for output modules is PQ (for peripheral output). A byte (PIB/PQB), word (PIW/PQW), or double word (PID/PQD) can be accessed.

Like other S7 memory areas, the peripheral memory starts at byte-0. In fact, the input image area (I) reflects data read directly from digital input modules by the CPU on each scan. The CPU reads the input modules starting with peripheral input byte-0 (PIB 0) and places the results in the input image area starting with input byte-0 (IB 0). You can also read a module from anywhere in your program by accessing the desired input byte from the PI area. Conversely, you may transfer data to a digital output module by accessing the desired peripheral output byte (e.g., PQB 0 updates the eight bits of the digital outputs at QB 0).

The peripheral memory area not only corresponds to the addresses assigned in the hardware configuration, for both digital modules, but also to analog modules. Unlike digital modules, analog input modules are not read cyclically to an input image; nor is an analog output image transferred cyclically to analog output modules. Each analog I/O must be read or written on demand in your program, via peripheral memory. Analog input addresses are prefixed with **PIW** and analog outputs are prefixed with **PQW**. In LAD/FBD, the Move operation is used to read and write analog I/O. In Statement List (STL), the load operation reads an analog input (e.g., **L** PIW 256) and the transfer operation writes to an analog output (e.g., **T** PQW 256).

L PIW 256	Peripheral Input Byte-256 (PIB 256)								Peripheral Input Byte-257 (PIB 257)							
First Analog Input	7	6	5	4	3	2	1	0	7	6	5	4	3	2	1	0

Figure 4-11. Accessing peripheral input word-256. The next analog input channel is PIW 258.

T PQW 256	Peripheral Output Byte-256 (PIB 256)								Peripheral Output Byte-257 (PIB 257)							
First Analog Output	7	6	5	4	3	2	1	0	7	6	5	4	3	2	1	0

Figure 4-12. Accessing peripheral output word-256. The next analog output is PQW 258.

Timer Memory (T)

The *Timer Memory* area of S7 memory contains the software timers of you program. Each timer is stored in one word location, which contains both the time preset value and the time base. Timer word locations are accessed by preceding the location number with the identifier 'T' to indicate timer. From the first location, timers are addressed **T0**, **T1**, **T2**, and so on. The number of available timers is CPU dependent. Timer addresses are used to designate the timer and is used in bit operations to check the timer status, and in word operations to check the remaining time. STEP 7 interprets the address appropriately, based on the operation. See *S7 Data Types and Formats,* and the *S5TIME* data type, for further details.

STEP 7 Timer Memory																
Time Base				Timer Preset Value in BCD (000-999)												Timer Word
15	14	13	12	11	10	09	08	07	06	05	04	03	02	01	00	
X	X	0	0	0	1	0	1	0	1	0	1	0	0	0	0	T0
X	X	0	1													T1
X	X	1	0													T2
X	X	1	1													T(n)

Figure 4-13. Illustration of Timer Word format. The preset value of T0 = 550 x .01, or 5.5 sec.

Counter Memory (C)

The *Counter Memory* area holds counter instructions. Like S7 timers, each counter reserves a word location, which is addressed in the program by preceding the location with the identifier 'C'. From the first location, counters are addressed **C0**, **C1**, **C2**, and so on. The number of available counters is CPU dependent. Counter addresses are used to designate the counter and is used in bit operations to check the counter status, and in word operations to check the current count value. STEP 7 interprets the address appropriately, based on the operation.

STEP 7 Counter Memory																
Bits 12-15 Unused				Counter Value in BCD (000-999)												Count Word
15	14	13	12	11	10	09	08	07	06	05	04	03	02	01	00	
X	X	X	X	0	1	0	1	0	1	0	1	0	0	0	0	C0
X	X	X	X													C1
X	X	X	X													C2
X	X	X	X													C(n)

Figure 4-14. Illustration of Counter Word format.

Local Memory (L)

Local memory (the L-stack) is memory allocated to a block (e.g., OB, FC, FB, SFC, SFB) for handling the temporary local variables declared in the block. The L-stack is made available to each block as it is called. When a block terminates, the L-stack is available to the next called block and the previous temporary data is overwritten. The identifier 'L' precedes local memory addresses. A local memory bit is addressed **L byte.bit**, a byte is addressed **LB n**, a word is addressed **LW n**, and a double word as **LD n** (e.g., L42.6, LB 2, LW 2, or LD 2).

The total size of the local memory is CPU dependent. The actual size of the local stack available to a block is actually the size allocated to the OB from which the block is called. By default, the local memory available is equally divided among each OB (e.g., 256) bytes). In S7-400 CPUs and in the CPU 318, the amount of local memory allocated to each priority can be modified. Since not all priorities require the same size local stack, a larger local stack can be assigned to priority classes needing more local memory. Furthermore, priority classes that are not required in the program may be deactivated, to free additional local memory.

Summary of S7 Memory Addressing

A summary of how each S7 memory area is addressed is provided in the following table. Each area may be referenced in the STEP 7 program or in the STEP 7 monitor/modify data utility, using either the absolute or symbolic address. The input (I), output (Q), and Bit Memory (M) areas all allow bit, byte, words, and double word access. Peripheral (P) memory does not support bit access. Counter and timer addresses may be used in bit operations to check the timer/counter status, and in word operations to check the remaining time/current count.

Table 4-5. Addressing Summary for Accessing S7 Memory Areas.

Area	Identifier	Access Units	Example	Description of Example
Input Image (PII)	I	I = Bit	I 30.7	Input Byte 30, bit 7
		IB = Byte	IB 30	Input Byte 30
		IW = Word	IW 30	Input Word 30; bytes 30-31
		ID = Double word	ID 30	Input Double word 30; bytes 30-33
Output Image (PIQ)	Q	Q = Bit	Q 44.7	Output Byte 44, bit 7
		QB = Byte	QB 44	Output Byte 44
		QW = Word	QW 44	Output Word 44; bytes 44-45
		QD = Double word	QD 44	Output Double word 44; bytes 44-47
Bit Memory	M	M = Bit	M 23.6	Memory Byte 23, bit 6
		MB = Byte	MB 23	Memory Byte 23
		MW = Word	MW 24	Memory Word 24; bytes 24-25
		MD = Double word	MD 8	Memory Double word 8; bytes 8-11
Counter	C	C = Counter	C 64	Counter 64
Timer	T	T = Timer	T 12	Timer 12
Local Stack	L	L = Bit	L 2.7	Local Memory Byte 2, bit 7
		LB = Byte	LB 2	Local Memory Byte 2
		LW = Word	LW 2	Local Memory Word 2; bytes 2-3
		LD = Double word	LD 2	Local Memory Double word 2
Peripheral Inputs & Peripheral Outputs	PI	PIB = Byte	PIB 44	Peripheral Input Byte 44
		PIW = Word	PIW 66	Peripheral Input word 66
		PID = Double word	PID 82	Peripheral Input Double word 82
	PQ	PQB = Byte	PQB 44	Peripheral Output Byte 44
		PQW = Word	PQW 66	Peripheral Output word 66
		PQD = Double word	PQD 82	Peripheral Output Double word 82

S7 Data Types and Formats

An understanding of the data types and formats used in STEP 7 is essential as you start to organize and define your data. As each variable required in a code block or data block is declared, you will explicitly define its data type. A variable's data type stipulate its width in bits (e.g., 1-bit, 8-bits, 16-bits, 32-bits, etc.), and how the binary data is represented and interpreted by STEP 7. How data is represented and interpreted is referred to as the format (e.g., signed integer, or character string). The range of values that a variable may use is also determined by data type. For example, data type INT, allows a range of -32768 to + 32767. The following discussions introduce *Elementary*, *Complex*, and *Parameter* data types.

Elementary Data Types

Elementary Data Types refer to the group of data types that represent variables that involve single data elements, all of which are 32-bits or less. Variables of elementary data type may be declared in all blocks, including data blocks, organization blocks, functions, and function blocks. Elementary data types include BOOL, BYTE, WORD, DWORD, INT, DINT, REAL, DATE, TIME, S5TIME, TIME-OF-DAY, and CHAR. Variables of elementary data type, may be used as direct input or output parameters of LAD both and FBD box instructions.

BOOL

Data type BOOL reserves a single bit, for a variable that can take on the permissible values of either TRUE or FALSE.

DB100 -- New_prj\SIMATIC 300(1)\CPU 316

Address	Name	Type	Initial value	Comment
0.0		STRUCT		
+0.0	OVR_LMT_T1	BOOL	FALSE	ZONE 1 OVER Limit
+0.1	OVR_LMT_T2	BOOL	FALSE	ZONE 2 OVER Limit

Figure 4-15. The first two bits of DB 100 are declared as BOOL variables.

BYTE

A variable of data type BYTE, reserves 8-bits. These bits are not evaluated individually, but as an 8-bit hex number. Each hex digit is represented in four-bits, allowing a two-digit Hex value. A constant of data type BYTE is specified using the identifier B#16#, where "B" is for byte, and '16' is for hex (e.g., B#16#0A). The permissible range of values is from B#16#00 to B#16#FF.

8	4	2	1	8	4	2	1	Weight
07	06	05	04	03	02	01	00	Bit No.
1	0	1	0	1	1	1	1	
A				F				B#16#AF

Figure 4-16. A BYTE variable

WORD

Variables of data type WORD reserve 16-bits. Data type WORD represents a 16-bit string, where the individual bits are not evaluated as a unit, therefore a WORD variable supports several bit-oriented uses. A WORD variable can simply be a 16-bit binary string of 1s and 0s; a 4-digit Hexadecimal number; a 3-digit count value in BCD; or two unsigned decimal values, each stored in one byte (e.g., 255, 255).

WORD As 16-bit String

A WORD variable is specified as a 16-bit string value, using the identifier 2# (e.g., 2#0000_1011_1111_1111). The range of this format then is 2#0000_0000_0000_0000 to 2#1111_1011_1111_1111. This option might be used, for example, to compare the 16-bits an input module to an expected 16-bit pattern; or to manipulate sixteen digital outputs.

WORD - As Four Digit Hexadecimal/3-Decade BCD

A variable of data type WORD is specified as a hexadecimal number using the identifier W#16#. 'W' is for word length, '16' is for hex (e.g., W#16#0FFF). Each hex digit is represented using four-bits, allowing a four-digit Hex number in a WORD variable. The permissible range of values for this format option is from W#16#0 to W#16#FFFF. This option might be used when up to 4-digit hex numbers are required, or as a shorter notation for indirectly specifying a binary pattern (e.g., W#16#FFFF instead of 2#1111_1111_1111_1111).

The 16-bit hex format can also support a 3-decade BCD value, simply by only entering hex digits 0-to- 9 in order to avoid invalid BCD numbers. BCD does not use a unique identifier, but instead uses the hexadecimal identifier. The range for a valid 16-bit BCD number is ±9,99, specified as W#16#F000 to W#16#0999. Bits 12-15 are used to indicate the sign. Negative is denoted by 1XXX, and positive by 0XXX, since only bit 15 is used for the sign.

8	4	2	1	8	4	2	1	8	4	2	1	8	4	2	1	Weight
15	14	13	12	11	10	09	08	07	06	05	04	03	02	01	00	Bit No.
1	1	1	1	1	1	1	1	1	1	1	1	1	1	1	1	
F				F				F				F				W#16#FFFF

Figure 4-17. WORD variable interpreted as 16-bit hexadecimal.

WORD - As a Count Value

A variable of data type WORD may be specified as a count value, which might be loaded to a counter on command. Recall, that each counter value reserves a single word and is interpreted as a 3-digit BCD value. When specified as a constant parameter input to an instruction or as an initial value of a WORD variable declaration, the count value is preceded by the identifier C# (e.g., C#500). The permissible range for this format is C#000 to C#999. This format could be used to compare the current value of a counter to a set point value. The compare for example, could compare C#100 to C6 (counter 6).

WORD - As Two Bytes Unsigned Decimal

A variable of data type WORD is specified as 2-bytes of unsigned decimal numbers, using the identifier B#. The two numbers are enclosed in parentheses and separated by a comma, for example, B#2(25, 175). This format maximizes 16-bit data storage if positive values from 0-255 are all that is required (e.g., indexing a loop to +10). The maximum range of each byte is from 0-255 or B# (0, 0,) to B#(255,255).

128	64	32	16	8	4	2	1	128	64	32	16	8	4	2	1	Weight
15	14	13	12	11	10	09	08	07	06	05	04	03	02	01	00	Bit No.
1	1	1	1	1	1	1	1	1	1	1	1	1	1	1	1	
255								255								B#(255,255)

Figure 4-18. A WORD variable interpreted as 2-bytes unsigned decimal.

DWORD

Variables of data type DWORD reserve 32-bits. Data type DWORD represents a 32-bit string, where the individual bits are not evaluated as a unit, thereby allowing a DWORD variable to be used for several bit-oriented purposes. A DWORD variable can simply be a 32-bit binary string of 1s and 0s; an 8-digit Hexadecimal number; or four unsigned decimal values, each stored in one byte (e.g., 255, 255, 255,1). Because the binary equivalent of Hex 0-9 is the same for BCD, DWORD variables may also be used in BCD operations.

DWORD - As 32-bit String

A DWORD variable is specified as a 32-bit binary string, using the identifier 2# (e.g., 2#0000_1011_1111_1111_0000_1011_1111_1111). The minimum value of this format is when all 32-bits are '0'; the maximum value is when all 32-bits are '1'. This option might be used, to compare the 32-bits of an input module to an expected 32-bit pattern; to manipulate thirty-two digital outputs simultaneously; or as a mask pattern in 32-bit word logic operations.

DWORD - As Eight Digit Hexadecimal/7-Decade BCD

A variable of data type DWORD is specified as a hexadecimal number using the identifier DW#16#. 'DW' is for double word length, '16' is for hex (e.g., DW#16#0FFF_A0FF). Each hex digit is represented using four-bits, allowing an 8-digit Hex number in a DWORD variable. The permissible range of values for 8-digit Hex is from DW#16#0000_0000 to DW#16#FFFF_FFFF.

The 32-bit hex format can also support a 7-decade BCD value, simply by only using the digits 0-to-9, in order to avoid invalid BCD numbers. The valid range for 32-bit BCD numbers is ±9,999,999. Specified using the 32-bit hex identifier, is DW#16#F9999999 to DW#16#09999999. Bits 28-31 are used to indicate the sign. Negative is denoted by 1XXX, and positive by 0XXX, since only bit 31 is used for the sign.

8421	8421	8421	8421	8421	8421	8421	8421	Weight
31 - 28	27 - 24	23 - 20	19 - 16	15 - 12	11 - 08	07 - 04	03 - 00	Bit No.
1111	1111	1111	1111	1111	1111	1111	1111	
F	F	F	F	F	F	F	F	DW#16#FFFFFFFF

Figure 4-19. DWORD variable interpreted as 32-bit Hexadecimal.

DWORD - As Four Bytes Unsigned Decimal

When a constant of data type DWORD is specified as 2-bytes of unsigned decimal numbers, the four numbers are separated by a comma, enclosed in parentheses, and preceded with the identifier B#; for example, B#(255, 255, 255, 1). This format maximizes 32-bit data storage if positive values from 0-255 are all that is required (e.g., indexing a loop to +10). The maximum range of each byte is from 0-255 or B#(0, 0, 0, 0) to B#(255, 255, 255, 255).

128 64 32 16 8 4 2 1		128 64 32 16 8 4 2 1		128 64 32 16 8 4 2 1		128 64 32 16 8 4 2 1		Weight
7	0	7	0	7	0	7	0	Bit No.
1 1 1 1 1 1 1 1		1 1 1 1 1 1 1 1		1 1 1 1 1 1 1 1		1 1 1 1 1 1 1 1		
255		255		255		255		

Figure 4-20. DWORD variable interpreted as 4-bytes unsigned decimal.

INT

Variables of data type INT represent signed decimal integers in a 16-bit word. The permissible range for data type INT variables is -32,768 to +32,767. A constant of type INT is entered in the declaration area of a code block, data block, or as an input value to an instruction, simply with the appropriate sign (e.g., +2575). No identifier is required.

S	-	-	-	-	-	1024	512	128	64	32	16	8	4	2	1	Weight
15	14	13	12	11	10	09	08	07	06	05	04	03	02	01	00	Bit No.
0	0	0	0	0	0	0	1	0	1	0	1	0	1	0	1	
							512	+	64	+	16	+	1	+	1	+597

Figure 4-21. Data type INT variable, interpreted as signed 16-bit integer.

Using the two's complement method of representing negative values, data type INT uses bit-15 to represent the sign. A '0' indicates a positive value; a '1' indicates negative. The remaining bits (0-14) hold the actual value. The two's complement value of any value is obtained by inverting each bit from right-to-left, but only after the first '1' is detected. This method is equivalent to inverting each bit (i.e., change 1s to 0s and 0s to 1s) then adding +1.

15	14	13	12	11	10	09	08	07	06	05	04	03	02	01	00	Bit No.
0	0	0	0	0	0	0	1	0	1	0	1	0	1	0	1	+597
1	1	1	1	1	1	1	0	1	0	1	0	1	0	1	1	-597

Figure 4-22. Two's Complement representation of data type INT negative number.

DINT

Variables of data type DINT represent signed decimal integers in an S7 double word (32-bits). The permissible range of DINT variables is -2,147,483,648 to +2,147,483,647. DINT variables should be used whenever the intended operations are expected to exceed the range of data type INT. An initial DINT constant is entered in a code block, data block, or as an input value to an instruction, using the *long integer* identifier 'L' (e.g., L#500000).

Using the two's complement method to represent negative values, DINT variables use bit-31 as the sign bit. A '0' in the sign bit indicates a positive value; a '1' indicates negative. The remaining 31 bits hold the value.

S								
31 - 28	27 - 24	23 - 20	19 - 16	15 - 12	11 - 08	07 - 04	03 - 00	Bit No.
0 1 1 1	1 1 1 1	1 1 1 1	1 1 1 1	1 1 1 1	1 1 1 1	1 1 1 1	1 1 1 1	+2147483647
1 0 0 0	0 0 0 0	0 0 0 0	0 0 0 0	0 0 0 0	0 0 0 0	0 0 0 0	0 0 0 1	-2147483647

Figure 4-23. Data type DINT variable, interpreted as signed 32-bit integer.

REAL

Variables of data type REAL represent floating-point numbers (e.g., 25.375). REAL variables are used to work with fractional numbers or to handle calculations or measured quantities that require floating-point accuracy. Internally, REAL variables are represented in the IEEE 32-bit floating-point format. In this format, bits 0-22 represent the *mantissa*, or the significant digits of the number; and bits 23-30 represent the *exponent*, a signed power of 10. Bit-31 represents the sign, where 0' indicates a positive value and '1' indicates a negative value.

A REAL constant is specified as the initial value for a declared variable or as a constant input to an instruction, using the REAL notation shown below. This shorthand notation places the first of the significant digits (non-zero) to the left of the decimal point and up to six additional significant digits to the right and preceding the exponent. An uppercase 'E' or lowercase 'e' is valid. The number in the exponent reflects the number of places the decimal point must be moved to the left or right of the most-significant digit, in order to represent the original value. A positive exponent reflects a move to the right — a negative exponent a move to the left.

Table 4-6. REAL (Floating-point) constants in scientific notation.

Original Value	REAL Notation	Original Value	REAL Notation
12345.67	1.234567 e +04	2000.0	2.0 E +03
.1234567	1.234567 e -01	20003.0	2.0003 E +04
-12.34567	-1.234567 e +01	.01	1.0 E -2
-0.0000023	-2.3 e -06	.001	1.0 E -3
+0.000000000675	6.75 e -10	.0001	1.0 E -4
1234567000000	1.234567e+12	10000	1.0 E +4

DATE

A variable of data type DATE reserves an S7 word, to represent a date value as an unsigned fixed-point number. Internally the time value is interpreted as a fixed-point number that represents the total days since January 01, 1990. This IEC format for specifying a date uses the identifier 'DATE#' or 'D', and the year, month and day separated by hyphens (e.g., DATE#2002-08-22). Data type DATE supports dates from January 1, 1990 through December 2168. This range of values is expressed DATE#1990-01-01 through DATE#2168-12-31.

TIME

Variables of data type TIME reserve an S7 double word. Constants of data type TIME are defined in units of days (D), hours (H), minutes (M), seconds (S), and milliseconds (MS), preceded by the identifier 'TIME' or 'T'. Unit identifiers may be entered as upper or lowercase, and unneeded units may be omitted (e.g., TIME#10H30M). This IEC format for specifying a time value is intended for specifying a delay (e.g., a 'wait time' in your STL code). Both positive and negative Time values are supported, with a range of TIME#-24d20h31m23s647ms to TIME#+24D20H31M23S647MS. The internal representation of the time value is as a signed 32-bit fixed-point number, interpreted by STEP 7 as total milliseconds.

TIME_OF_DAY

A variable of data type TIME_OF_DAY reserve an S7 double word. A TIME_OF_DAY constant is specified using the identifier 'TIME_OF_DAY#' or 'TOD#' preceding the hours, minutes, and seconds which are separated by colons (e.g., TOD#01:15:00). Milliseconds may be added to the time using a period as the separator, or omitting it altogether. The internal representation of the TIME_OF_DAY time value is as a signed 32-bit fixed-point number, interpreted by STEP 7 as total milliseconds.

CHAR

A variable of data type CHAR reserves one byte (8-bits) and represents a single ASCII character. All printable characters including upper and lower case alphabets, numbers 0-9, and some special characters are valid. An initial character value is specified for variables of data type CHAR, by enclosing the character using apostrophes (e.g., 'Y'). It is also possible with the MOVE instruction, to transfer either two or four characters enclosed in apostrophes, to a word or double word location. The special characters listed in the following table, are entered using a special notation. (See ASCII character table in Appendix C).

Table 4-7. Notation for Special CHAR Characters.

CHAR	ASCII Code (Hex)	ASCII Character
$$	24	Dollar Sign
$'	27	Apostrophe
$L or $l	0A	Line Feed Control (LF)
$P or $p	0C	New Page Control (FF)
$R or $r	0D	Carriage Return Control (CR)
$T or $t	09	Tab Control

S5TIME

A variable of data type S5TIME is specified to represent a preset value for a timer. This 16-bit format is the same as what was formerly used to store the STEP 5 timer preset. An initial S5TIME value is entered in units of hours (H), minutes (M), seconds (S), and milliseconds (MS), preceded by the data type identifier 'S5TIME#' or 'S5T#' (e.g., S5T#2H30M30S). The unit identifiers may be in upper or lowercase, and un-needed units may be omitted.

As described in detail under S7 Memory Areas, the internal format of the S5TIME data type involves a single word encoded in two BCD parts. The time value, represented in bits 00-11 of the timer word, allows a BCD value from 000-to-999; the time base, represented in bits 12-15, takes on encoded values of 0000, 0001, 0010, and 0011. These values are interpreted as .01 sec, .1 sec, 1.0 sec., or 10 sec., respectively.

Time Base (0000-0011)				Timer Value in 3-decades BCD (000-999)												
15	14	13	12	11	10	09	08	07	06	05	04	03	02	01	00	Timer Word
0	0	0	1	0	1	0	1	0	1	0	1	0	1	0	1	T0
X	X	0.1 sec		5				5				5				55.5 sec.

00 = .01 sec
01 = 0.1 sec
10 = 1.0 sec
11 = 10 sec

Figure 4-24. Illustration of Timer Word format. The preset value of T0 = 555 x 0.1, or 55.5 sec.

Table 4-8. Summary of S7 Elementary Data Types

Data Type	Bits	Format (units)	Constant Notation and Valid Range
BOOL	1	Boolean Text	TRUE/FALSE
BYTE	8	Hexadecimal	B#16#0 -to- B#16#FF
WORD	16	Binary........................ Hexadecimal........... BCD Count................ Two Unsigned Bytes..	2#0 -to- 2#1111_1111_1111_1111 W#16#0000 -to- W#16#FFFF C#000 -to- C#999 B#(0,0) -to- B#(255,255)
DWORD	32	Binary........................ Hexadecimal........... Four Unsigned Bytes.	2#0 -to- 2#1111_1111_1111_1111_1111_1111_1111_1111 DW#16#0000_0000 -to- DW#16#FFFF_FFFF B#(0,0,0,0) -to- B#(255,255,255,255)
INT	16	Signed Decimal	-32768 -to- 32767
DINT	32	Signed Decimal	L#-2147483648 -to- L#2147483647
REAL	32	IEEE Floating Point	±1.175495e-38 -to- ±3.402823e+38
DATE	16	IEC Date (1-day)	D#1990-1-1 -to- D#2168-12-31
TIME	32	IEC Time (1ms)	T#-24D_20H_31M_23S_648MS -to- T#24D_20H_31M_23S_647_MS
TIME_OF_DAY	32	Time of day (1ms)	TOD#0:0:0.0 -to- TOD#23:59:59.999
CHAR	8	ASCII Character	'A', 'B', 'C', 'D', and so on
S5TIME	16	S5 Time (10ms)	S5T#0H_0M_0S_10MS -to- S5T#2H_46M_30S_0MS

Note: Constants entered as binary, BCD count, or as unsigned bytes all revert to Hex values when declared, but may viewed in the desired format in the Monitor/Modify Variables tool.

Complex Data Types

Complex data types allow you to define variables that combine other data types and that is comprised of more than one data element. Complex data types, which include DATE-AND-TIME, STRING, ARRAY, STRUCT, and UDT, may only be used with the variables declared in data blocks, or those specified as temporary variables, or as formal block parameters.

STRING

A variable of data type STRING defines a string of up to 254 ASCII characters (CHAR). By default, a character-string reserves 256 bytes that allows for 254 characters and a 2 byte header. A string can be specified as STRING [**n**], where **n** defines the maximum number of characters (e.g., STRING[6]) the variable can accommodate. If no length is specified, 256 characters are reserved. An initial string is defined by enclosing the characters with apostrophes, or preceded by the dollar sign in the case of the special characters defined under data type CHAR. The actual storage requirement of a string is minimized by specifying the exact number of characters (e.g., string[12], with initial value 'System Ready') . If the initial string is shorter than the specified length then the actual length is reserved.

DATE_AND_TIME

Variables of data type DATE_AND_TIME are comprised of the date and time. When specifying this variable in a data block or as a static variable in a function block (FB), an initial value is entered using the identifier DATE_AND_TIME# or DT#, followed by the units of yyyy-mm-dd:hh:mm:ss:ms (e.g., DT#2004:12-31-23:59:59.999), where milliseconds may be omitted. Internally, this variable is stored in BCD format using 8-bytes. Starting with the byte (n), the variable contains the year (00-99), month (1-12,), day (1-31), hour (0-23), minute (0-59), second (0-59), millisecond (0-999), and finally the weekday (1-7), where Sunday is day 1.

ARRAY

An array variable represents a fixed number of data elements of the same data type (e.g., 10 values of data type INT), that may include all data types except parameter types. An array is declared as ARRAY [**x1**..**y1** }, where the dimension limits **x1** and **y1** define the size of the array. The index values may start at a negative integer, zero, or a positive integer (e.g., [-10.. 10], defines an array with 21 elements. Up to six dimensions may be declared by separating each dimension by a comma. ARRAY [**1..2**,**1..6**], defines a two-dimensional array with 12 elements. The data type of an array (e.g., INT, DINT, or REAL), is assigned on the declaration row immediately following the declaration of the array name and size (Fig. 4-25).

Initial values may be assigned to each array element by entering each value separated by commas. The same value can be assigned to consecutive elements by enclosing the value in parentheses preceded by a repetition factor (e.g., 5(100). Elements not assigned an initial value will default to zero. An array element in a single dimension array is addressed in the program as Name [index] (e.g., temp [4]). The full indices must be used when specifying an array element in a multi-dimensional array. Given the two-dimensional array 'ARRAY[1..3, 1..4],' the first element is addressed Name [1,1]; the third element is addressed Name [1,3]; the fourth element is addressed Name [2,1]; and the last element is addressed Name [3,4].

DB11 -- New_prj\SIMATIC 300(1)\CPU 316

Address	Name	Type	Initial value	Comment
0.0		STRUCT		
+0.0	Ex_1	ARRAY[1..10]	5 (100) , 5 (50)	Single dimension; 10 INT elements.
*2.0		INT		
+20.0	Ex_2	ARRAY[0..15,0..3]	64 (FALSE)	2-dimensional; bool elements.
*0.1		BOOL		
+36.0	Ex_3	ARRAY[-10..10]	10 (TOD#12:0:0.0)	Time-of-Day array; 21 elements.
*4.0		TIME_OF_DAY		
+120.0	Ex_4	ARRAY[1..2,3..4]	'Y', 'N', '0', '1'	2-dimensional array; 4 CHAR elements
*1.0		CHAR		

Figure 4-25. Declaring ARRAY variables.

STRUCT

A variable of data type STRUCT is represents a *structure* with a fixed number of elements that may involve any combination of other data types. In the declaration table, the STRUCT variable includes the data elements enclosed between the keywords STRUCT and END_STRUCT as shown in the figure below. Structures may be defined either in the variable declaration of the logic block or in a data block. The individual elements of the structure may be accessed using the structure name and the variable element name separated by a period (e.g., **Recipe_A.Sys_Rdy**).

DB12 -- New_prj\SIMATIC 300(1)\CPU 316

Address	Name		Type	Initial value	Comment
0.0			STRUCT		
+0.0	Recipe_A		STRUCT		Start Recipe A; Structure
+0.0		Sys_Rdy	BOOL	FALSE	Element == Recipe.Sys_Rdy
+2.0		Ingred_1	INT	1500	Element == Recipe.Ingred_1
+4.0		Ingred_2	REAL	1.502345e+002	Element == Recipe.Ingred_2
+8.0		Stops	ARRAY[1..10]	10 (FALSE)	End Recipe A; Structure
*0.1			BOOL		
=10.0			END_STRUCT		

Figure 4-26. Declaring ARRAY variables.

UDT (User Defined Type)

A UDT is a user-defined type that may include both elementary and complex variables. A UDT is essentially a structure, comprised of data elements of arbitrary data types. The intent of the UDT is to provide a user defined data structure that can be used as a data block template or as a predefined structure. Once defined, the UDT may be used as a template to define the data structure of several data blocks; or as a data type (essentially a structure).

The elements of the UDT are accessed like the elements of a data block or a structure. A specific data element is accessed using the name of the UDT and the variable name separated by a period. You can create a UDT in the Blocks folder of a program using the SIMATIC Manager or the standard editor. The UDT is assigned a number (absolute address) from UDT 0 to UDT 65,535, and can be assigned a symbolic address (e.g., PID Loop). A variable named PV, in a UDT named 'PID Loop' is addressed as '**PID Loop.PV**'.

UDT1 -- BrilliantTraining\S7 Program(1)

Address	Name	Type	Initial value	Comment
0.0		STRUCT		
+0.0	Loop_No	WORD	W#16#1	Loop Number
+2.0	PV	REAL	0.000000e+000	Process Input Variable
+6.0	SP	REAL	6.500000e+002	Setpoint
+10.0	UL	REAL	6.750000e+002	Upper Limit Alarm
+14.0	LL	REAL	6.550000e+002	Lower Limit Alarm
+18.0	CTRL	WORD	W#16#0	Control Word
=20.0		END_STRUCT		

Figure 4-27. A UDT (User-Defined Type) is created from the SIMATIC Manager.

Parameter Data Types

The *Parameter* data types refer to the group of data types that represent variables that allow timers and counters (e.g., T1, T2, C1, C2), as well as blocks (i.e., FCs, FBs, DBs) to be used as formal parameters in an FC or FB. With parameter data types, you may write an FC or FB in which a timer, counter, FC, FB, or DB number may be passed to the block on each call of the block. Parameter data types include TIMER, COUNTER, BLOCK_FC, BLOCK_FB, BLOCK_DB, BLOCK_SDB, POINTER, and ANY. The parameter type *POINTER* and *ANY* allow pointer addresses to be defined, for indirect addressing of specific addresses or address areas, with respect to transferring block parameters.

Overview of the LAD/FBD/STL Editor

In STEP 7, developing a program for the control system is to create a collection of code blocks and data blocks, each of which performs a specific task and that together form a complete program. The LAD/FBD/STL editor is the standard tool for STEP 7 program development. It alone is all that is required to build the program incrementally, using the blocks resources (e.g., OBs, FBs, FCs, SFCs, SFB, DBs) that have been previously presented.

Introduction

The LAD/FBD/STL editor may be started from the Windows Start Menu or from inside the SIMATIC Manager. After a block is generated in the offline blocks folder of a program, double clicking on the block launches the editor and opens the block in its own window. Once developed, a block may be copied to other STEP 7 programs or projects, and downloaded to the CPU. Each block can be checked for correct operation, using several monitoring and debugging tools (See following chapter). Once created and downloaded, the status of any block may be open and viewed online, showing the status of operating values.

Menus and Toolbar

Menu headings of the LAD/FBD/STL editor include *File*, *Edit*, *Insert*, *PLC*, *Debug*, *View*, *Options*, *Window*, and *Help*. File operations allow you to create, open, save, and compile blocks. Cut, Copy, and Paste operations from the Edit menu or from the right click allow you to copy program elements within a block or between blocks and programs. PLC operations support online access to CPU operations such as download, monitoring and diagnostic tools. View operations allows switching between the LAD, FBD, and STL representations, and for components of the editor to be displayed or hidden; while the Window menu allows the editor and block windows to be arranged to your convenience.

The toolbar buttons, listed below, represent some of the most frequently used menu operations or software utilities that you may launch while using the programming editor.

Table 4-9. LAD/FBD/STL Programming Editor core Toolbar Buttons.

Icon	Toolbar Function	Icon	Toolbar Function
	Create New Station in Project		Monitor Block Status Online
	Open Station Offline Window		Display Overview Window
	Open Offline/Online Partner		Display Details Window
	Save		Insert New Network
	Compile SCL Text (into block)		Insert Normally-Open Contact
	Cut Selected Object		Insert Normally-Closed Contact
	Copy Selected Object		Insert Coil Instruction
	Paste Object from Clipboard		Open Parallel Branch (LAD)
	Download to PLC		Close Parallel Branch (LAD)
	Toggle Symbol Addresses On/Off		

Program Elements/Call Structure Window

The *Program Elements/Call Structure window* contains the various LAD/FBD instruction folders on the Program Elements tab, and a tree-like structure of the call hierarchy of the currently open program. While the editor is opened, the Program Elements/Call Structure window can be hidden from view or displayed by selecting **View ➢ Overview** from the menu (toggled on/off). You may also "dock" or "undock" the window by double-clicking on the window's Control bar/Title bar. When undocked, the window can be resized and moved around to suit your convenience. You may dock the window on either the left-side or the right-side of the main window by dragging and dropping it on the left or right edge of the window.

The Program Elements tab, when selected, contains the New Network object, used to insert a new network; and the folders of the instruction categories, for whichever of the LAD/FBD languages is currently enabled. STL instructions are not listed. While programming, you may double-click on the New Network object to insert a new network, and you may open an instruction folder and drag the desired instruction onto the current network. The window also contains folders for any user FB, and FCs that have been created, and any SFCs and SFB that you may have copied to your program. You may also drag *FCs, FBs, SFCs*, from their folder and drop it into the current network.

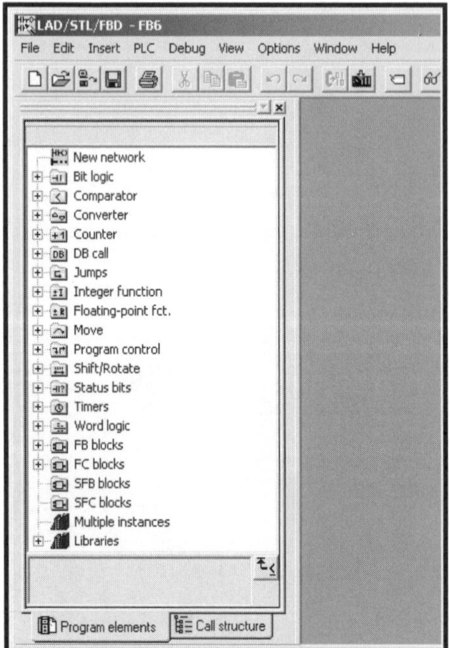

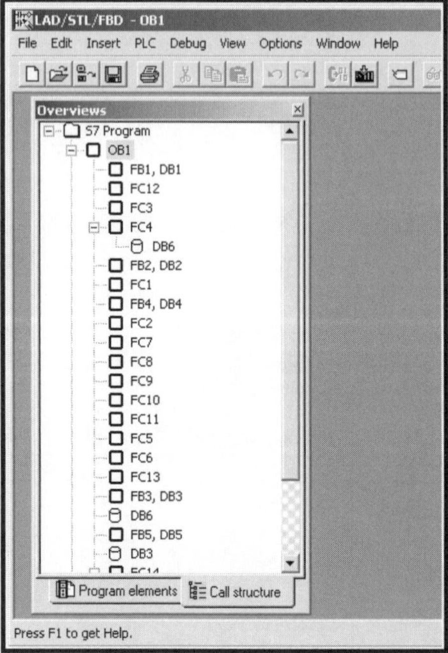

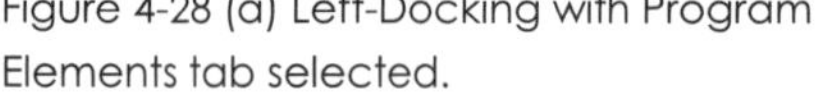
Figure 4-28 (a) Left-Docking with Program Elements tab selected.

(b) Un-Docked window with Call Structure tab selected.

The Call Structure tab, when selected, contains a tree-like structure of the calling hierarchy of the open program. The call structure shows the called blocks and the relationships between each and its calling block. Each block in the project is listed at the far left, with OBs being left-most. The blocks called from each OB are indented underneath it is a list of the blocks that it can call or use. Blocks which are not called from anywhere in the program are shown at the bottom of the call structure and marked with a black cross. When you click on the plus (+) symbol, the lower structures of the call hierarchy are expanded and displayed. When you click on the minus (-) symbol, the lower structures of the call hierarchy are collapsed and no longer displayed. With a block selected, from the right-click, you may navigate directly to the block or to the location from which the block is called.

Block Window

Each block is opened in its own window. You may arrange the open windows in a *horizontal*, *vertical* or *cascaded* layout. As seen in the figure below, each block window consists of a variable declaration pane and a code pane, where the actual program code is written. If the block is new, to begin writing code, simply click inside the pane at the first network; for existing blocks that you wish to edit, find and select the network you wish to edit. You may open as many blocks as you like, and must select the specific block you wish to work with. While a block is open, you may copy any block element (e.g., Network Title, Network Comment, Network Logic, etc.) and paste it to the same or to a different block.

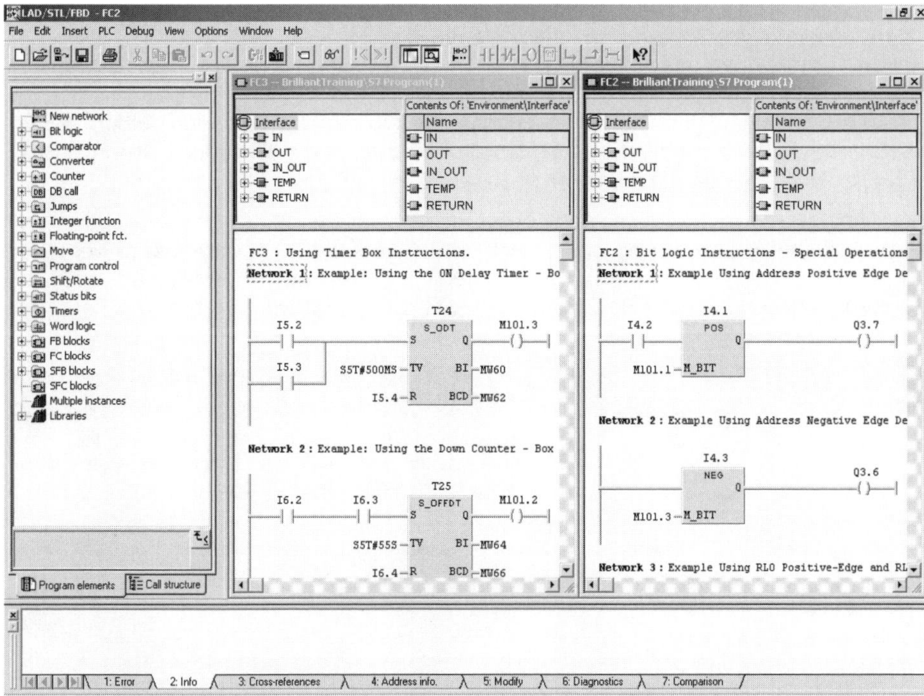

Figure 4-29. LAD/FBD/STL Editor tool with docked Program Elements/Call Structure window; Program window with two open block windows arranged vertically. The Details window has several tabs that provides tools for monitoring and diagnosing the control program and data.

Details Window

The *Details window* is opened from the menu by selecting View Details and hidden by again selecting the Details option from the View menu. The Details window allows you to select any one of seven views, by selecting the Error, Info, Cross-Reference, Address Info, Modify, Diagnostics, or Comparison tab.

Each tab provides a different service that enhances working with the open block. The cross-reference tab, for instance, lets you see in what other networks or blocks are the addresses of the current network located. The comparison tab is used in conjunction with the block compare function. When the results of a compare operation are determined, the compared blocks are shown in two windows and the found differences listed on the comparison tab.

STEP 7 Instruction Set Overview

The discussion following, will introduce each of the STEP 7 instruction categories for LAD (*Ladder*) and FBD (Function Block Diagrams). The purpose of each instruction category is briefly described, and then followed by a table that lists the operations it includes. The operation of each instruction is explained in the table. In the LAD/FBD/STL editor of STEP 7, each of these categories is represented as a folder object that contains the operations shown in the table. For the purposes of this discussion, the Bit logic category is divided into *Basic Operations*, and *Special Operations*.

Bit Logic Instructions — Basic Operations

The first group of STEP 7 bit-logic instructions, so-called *Basic Operations*, is used in developing relay equivalent control circuits and internal binary decision logic. With instructions that include normally-open and normally-closed contacts, and the standard output coil, these bit logic instructions allow development of series, parallel, and series-parallel logic combinations. Set and Reset output instructions, also known as latch/unlatch outputs, allow bit addresses to be latched ON until unlatched, regardless of the driving logic. The invert RLO instruction inverts the logic status at the point where the instruction is inserted.

These instructions, which examine and control the status of any addressable bit location, are the basis for creating the driving logic of externally connected discrete devices as well as for triggering internal operations such as timers, counters, conversions, arithmetic calculation, program flow control, word logic, shift-rotate and move operations.

Table 4-10. LAD/FBD Bit-Logic Instructions — Basic Operations

LAD	FBD	Brief Description
Output ??.? —( )—	Output ??.? ??.?–[=]	If the result of the logic operation (RLO) preceding the referenced output address = logic 1 then assign logic 1 to the output address, otherwise assign logic 0 to the output.
Normally-Open Contact ??.? —| |—	Input ??.? —|	Examine the referenced address for a logic 1 status. A status of logic 1 allows logic continuity; logic 0 interrupts continuity.
Normally-Closed Contact ??.? —|/|—	Negate Input ??.? —o|	Examine the referenced address for a logic 0 status. A status of logic 0 allows logic continuity; logic 1 interrupts continuity.
Set Output ??.? —(S)—	Set Output ??.? ??.?–[S]	If the result of the logic operation (RLO) preceding the referenced output transitions from logic 1 -to- logic 0, then assign logic 1 to the address, Remain at logic 1 until reset.
Reset Output ??.? —(R)—	Reset Output ??.? ??.?–[R]	When the result of logic operations (RLO) preceding the referenced output transitions from logic 1 -to- logic 0, then assign logic 0 to the address, Remain at logic 0 until set again.
Invert RLO —|NOT|—	Negate RLO —o	Invert the RLO (result of logic operation) status at the point at which the instruction is inserted (i.e., logic 0 -to- logic 1; logic 1 -to- logic 0).

Note: In FBD, the Negate operation performs the same function as the NC contact in LAD. The Negate in FBD also serves the same function as the NOT operation in LAD (i.e., to invert the RLO or power flow at the point of insertion.

Bit Logic Instructions — Special Operations

This second group of STEP 7 bit-logic operations includes edge detection operations for detecting positive and negative transitions. An edge can be detected in a specific address, or in the status that results from a logic operation (RLO), that precedes the instruction. These instructions are useful for simply detecting a discrete transition, or when triggering another operation (e.g., counter, arithmetic calculation, or move operation) that should only be triggered on the transition from 0-to-1 (positive edge), or from 1-to-0 (negative edge).

The Set/Reset flip-flop operations either allow the set or reset to have priority when both signals occur simultaneously. The mid-line output instruction is what it says; it allows an output operation to be placed at a point in the logic circuit other than at the end of the network. In this way, logic duplication can be minimized. Finally, the Save output copies the logic state that precedes the instruction to the BR status bit of the CPU status word.

Table 4-11. LAD/FBD Bit-Logic Instructions — Special Operations

LAD	FBD	Brief Description
Mid-Line Output ??.? —(#)—	Mid-Line Output ??.? ??.?— #	Stores intermediate result of logic operation (RLO) for the logic circuit up to and preceding the point of insertion.
SAVE RLO to BR —(SAVE)—	SAVE RLO to BR ??.?— SAVE	Saves result of the preceding logic operation (RLO), to the binary result bit of the status word.
Positive RLO Edge Detection ??.? —(P)—	Positive RLO Edge Detection ??.? ??.?— P	Detects 0-to-1 transition resulting from the preceding logic operation (RLO). An edge detection is signaled by the output going high for one CPU scan (single pulse).
Negative RLO Edge Detection ??.? —(N)—	Negative RLO Edge Detection ??.? ??.?— N	Detects 1-to-0 transition resulting from the preceding logic operation (RLO). An edge detection is signaled by the output going high for one CPU scan (single pulse).
Set/Reset Flip Flop ??.? SR —S Q— ...—R	Set/Reset Flip Flop ??.? SR ??.?—S ...—R Q—	Implements Set/Reset flip-flop, giving priority to the Reset function if both **S** and **R** input lines go TRUE simultaneously.
Reset/Set Flip Flop ??.? RS —R Q— ...—S	Reset/Set Flip Flop ??.? RS ??.?—R ...—S Q—	Implements Reset/Set flip-flop, giving priority to the Set function if both **R** and **S** input lines go TRUE simultaneously.
Address Negative Edge Detection ??.? NEG — Q— ??.?—M_BIT	Address Negative Edge Detection ??.? NEG ??.?—M_BIT Q—	Detect a negative-edge transition of a specific bit address. An edge detection is signaled by the output going high for one CPU scan (single pulse).
Address Positive Edge Detection ??.? POS — Q— ??.?—M_BIT	Address Positive Edge Detection ??.? POS ??.?—M_BIT Q—	Detects a positive-edge transition of a specific bit address. An edge detection is signaled by the output going high for one CPU scan (single pulse).

Counter Instructions

The counter instructions include the *Up counter*, the *Down counter*, and the *Up/Down counter*, which counts both up and down. The UP and Down counters are implemented both as box functions and as output instructions. Box counter functions support the full functionality of the counter in a single instruction, allowing the counter to be triggered, a preset value loaded, monitored, and reset. The UP and Down counter output operations simply support counting. Separate instructions would be used to load a preset, or to reset a counter output. The compare operation would be used to check the current count value.

Each counter you specify, numbered C0, C1, C2, and so on, will reference a single word in the counter memory area. The exact number of counters available is CPU-dependent.

Table 4-12. LAD/FBD Counter Instructions Summary

LAD	FBD	Brief Description
Set Counter Coil ??? —(SC)— ???	Set Counter Output ??? SC ??.?– ???– PV	Whenever the driving logic transitions from logic 0 to logic 1, load the specified preset value **PV** to the addressed counter.
UP Counter Coil ??? —(CU)—	UP Counter Output ??? CU ??.?–	Whenever the driving logic transitions from logic 0 -to- logic 1, increment the specified counter by one count.
DOWN Counter Coil ??? —(CD)—	DOWN Counter ??? CD ??.?–	Whenever the driving logic transitions from logic 0 -to- logic 1, decrement the specified counter by one count.
UP Counter ??? S_CU CU Q S CV PV CV_BCD R	UP Counter ??? S_CU CU S CV PVCV_BCD R Q	Whenever the logic input line **CU** transitions from logic 0 -to- logic 1, increment the specified counter by one count. Logic 1 on the reset input line **R** resets the counter to zero.
Down Counter ??? S_CD CD Q S CV PV CV_BCD R	Down Counter ??? S_CD CD S CV PVCV_BCD R Q	Whenever the logic input line **CD** transitions from logic 0 -to- logic 1, decrement the specified counter by one count. Logic 1 on the reset input line **R** resets the counter to zero.
UP-Down Counter ??? S_CUD CU Q CD CV S CV_BCD PV R	UP-Down Counter ??? S_CUD CU CD S CV PVCV_BCD R Q	For each logic 0 -to- logic 1 transition on the **CU** input line, increment the specified counter by +1; for each logic 0 -to- logic 1 transition on the **CD** input line, decrement the count by +1. Logic 1 on the reset input line **R** resets the counter to zero.

Timer Instructions

STEP 7 timers, are based upon the STEP 5 implementation of timers, and as such are referred to as S5 Timers. The instruction set includes five timer types, each of which is implemented as a box function and as a simple output instruction. Box timers support the full functionality of the timer in a single instruction, allowing the timer to be loaded with a preset value, started, reset, and monitored. Timer outputs simply allow a specific timer type to be started with a given preset value. Separate instructions are used to reset or monitor the timer's operation.

Each timer you specify, numbered T0, T1, T2, and so on, will reference a single word in the timer memory area. The exact number of timers available is CPU-dependent.

Table 4-13. LAD/FBD Box Timer Instructions Summary

LAD	FBD	Brief Description
ON-Delay Timer 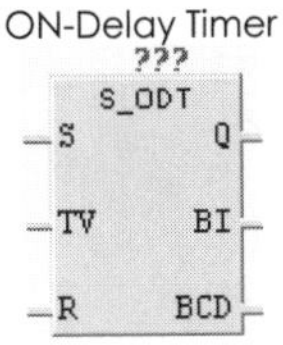 ??? S_ODT S Q TV BI R BCD	ON-Delay Timer 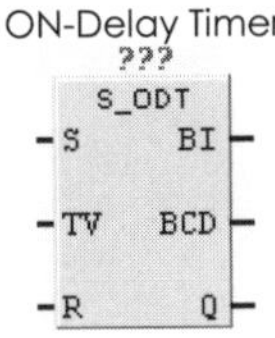 ??? S_ODT S BI TV BCD R Q	When the enable input **S** transitions from 0 -to- 1, the timer starts timing and continues to time unless the enable goes false or the reset input **R** transitions from 0 -to- 1. The output **Q** is activated after the preset time has elapsed, and stays energized until the enable signal transitions from 1 -to- 0 or the timer is reset.
OFF-Delay Timer 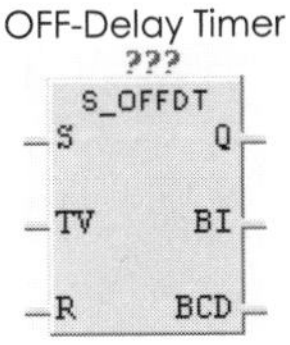 ??? S_OFFDT S Q TV BI R BCD	OFF-Delay Timer 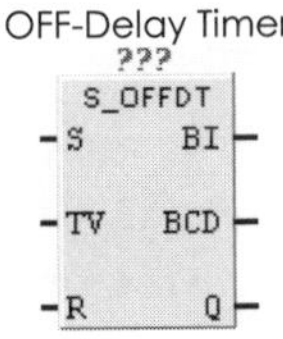 ??? S_OFFDT S BI TV BCD R Q	When the enable input **S** transitions from 0 -to- 1, the timer output **Q** is activated. The timer starts timing when the enable signal **S** transitions from 1-to- 0. The timer will run until the programmed delay expires, unless the reset signal **R** transitions from 0 -to- 1. After the timed delay expires, the timer output **Q** is de-activated. The output Q is also de-activated whenever the timer is reset.
Retentive ON-Delay Timer 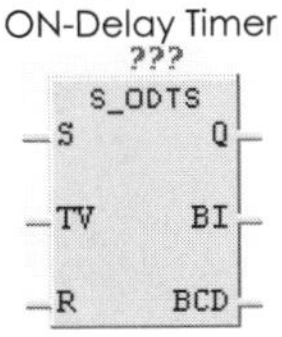 ??? S_ODTS S Q TV BI R BCD	Retentive ON-Delay Timer 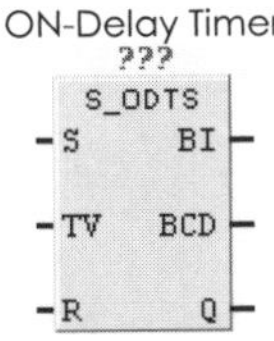 ??? S_ODTS S BI TV BCD R Q	When the enable input **S** transitions from 0 -to- 1, the timer starts timing and continues to time and continues to time even if the enable signal **S** transitions from 1-to- 0. If the enable signal changes back to '1' before the timer expires, the timer will restart. The output **Q** is activated after the preset time has elapsed, and remains activated until the timer is reset.
Pulse Timer 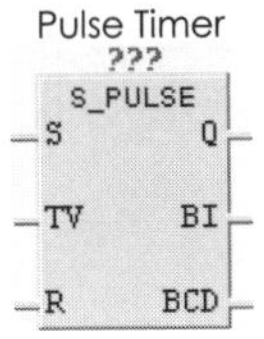 ??? S_PULSE S Q TV BI R BCD	Pulse Timer ??? S_PULSE S BI TV BCD R Q 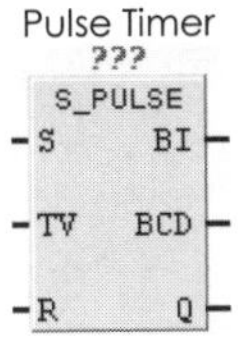	When the enable input **S** transitions from 0 -to- 1, the timer starts timing and continues to time for as long as the **S** input is '1', or until the programmed delay expires. While the timer is running, the output **Q** is activated for as long as the enable signal **S** is activated. If the enable signal **S** returns to '0' before the timer expires, the timer is stopped and the output Q is de-activated. While the timer is running, a transition from 0 -to- 1 on the reset line **R** resets the timer.
Extended Pulse Timer 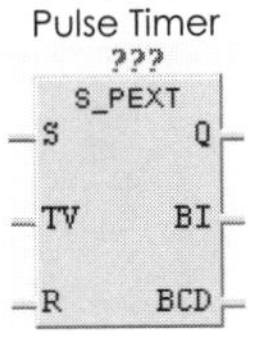 ??? S_PEXT S Q TV BI R BCD	Extended Pulse Timer 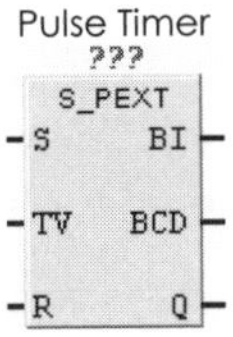 ??? S_PEXT S BI TV BCD R Q	When the enable input **S** transitions from 0 -to- 1, the timer starts timing and continues to time until the programmed delay expires, even if the input **S** returns to '0'. The output **Q** is activated whenever the timer is running. While the timer is running, a transition from 0 -to- 1 on the reset line **R** resets the timer.

Conversion Instructions

A conversion operation changes the format or data type of a variable to a new format or data type — generally before the data is further used. For instance, a value input as BCD must be converted before use in an integer arithmetic operation. Conversely, if the integer result must be output to a BCD display, then the value would need to be converted to BCD. If an arithmetic or comparison is to be performed on the data stored in two locations, the variables must be of the same data type or format to avoid erroneous results. Furthermore, since STEP 7 arithmetic operations are only performed on integer, double integer or real numbers, a variable not of these data types would require the appropriate conversion.

Conversion operations include conversions for variables of data type INT (integer) and DINT (double integer) variables, BCD numbers, and REAL numbers. There are also four operations for converting a REAL number to DINT format by rounding. These four operations differ in how the number is rounded with respect to the fractional part of the floating-point number.

Table 4-14. LAD/FBD Conversion Instructions Summary

LAD	FBD	Brief Description
BCD to Integer BCD_I EN ENO ???–IN OUT–???	BCD to Integer BCD_I ...–EN OUT–??? ???–IN ENO–	When **EN** is at logic 1, convert the 3-digit BCD value at supplied at **IN** to a 16-bit INT value. Put the result in the location specified at **OUT**.
Integer to BCD I_BCD EN ENO ???–IN OUT–???	Integer to BCD I_BCD ...–EN OUT–??? ???–IN ENO–	When **EN** is at logic 1, convert the 16-bit INT value supplied at **IN** to a 3-digit BCD value from 000-999. Put the result in the location specified at **OUT**.
Integer to Double Integer I_DI EN ENO ???–IN OUT–???	Integer to Double Integer I_DI ...–EN OUT–??? ???–IN ENO–	When **EN** is at logic 1, convert the 16-bit INT value supplied at **IN** to a 32-bit DINT value. Put the result in the location specified at **OUT**.
BCD to Double Integer BCD_DI EN ENO ???–IN OUT–???	BCD to Double Integer BCD_DI ...–EN OUT–??? ???–IN ENO–	When **EN** is at logic 1, convert the 7-digit BCD value supplied at **IN** to a 32-bit DINT value. Put the result in the location specified at **OUT**.
Double Integer to BCD DI_BCD EN ENO ???–IN OUT–???	Double Integer to BCD DI_BCD ...–EN OUT–??? ???–IN ENO–	When **EN** is at logic 1, convert the 32-bit DINT value supplied at **IN** to a 7-digit BCD value. Put the result in the location specified at **OUT**.
Double Integer to REAL DI_R EN ENO ???–IN OUT–???	Double Integer to REAL DI_R ...–EN OUT–??? ???–IN ENO–	When **EN** is at logic 1, convert the 32-bit DINT value supplied at **IN** to a REAL value (floating-point). Put the result in the location specified at **OUT**.

Note: When EN is '1' ENO will also be '1' except on error; then while EN is '1'; ENO is '0'.

Table 4-15. LAD/FBD Conversion Instructions Summary (Continued).

LAD	FBD	Brief Description
Ones Complement Integer INV_I EN ENO ???–IN OUT–???	Ones Complement Integer INV_I ... —EN OUT–??? ???—IN ENO–	When **EN** is at logic 1, complement or invert the integer value supplied at **IN** and **put** the result in the location specified at **OUT**.
Ones Complement Double Integer INV_DI EN ENO ???–IN OUT–???	Ones Complement Double Integer INV_DI ... —EN OUT–??? ???—IN ENO–	When **EN** is at logic 1, complement or invert the double integer value supplied at **IN** and **put** the result in the location specified at **OUT**.
Negate Integer NEG_I EN ENO ???–IN OUT–???	Negate Integer NEG_I ... —EN OUT–??? ???—IN ENO–	When **EN** is at logic 1, negate the integer value supplied at **IN** (change sign of value; positive-to-negative, negative-to-positive). Put the result in the location specified at **OUT**.
Negate Double Integer NEG_DI EN ENO ???–IN OUT–???	Negate Double Integer NEG_DI ... —EN OUT–??? ???—IN ENO–	When **EN** is at logic 1, negate the double integer value supplied at **IN** (positive-to-negative, negative-to-positive). Put the result in the location specified at **OUT**.
Negate REAL NEG_R EN ENO ???–IN OUT–???	Negate REAL NEG_R ... —EN OUT–??? ???—IN ENO–	When **EN** is at logic 1, negate the REAL value supplied at **IN** (positive-to-negative, negative-to-positive). Put the result in the location specified at **OUT**.
Round to Double Integer ROUND EN ENO ???–IN OUT–???	Round to Double Integer ROUND ... —EN OUT–??? ???—IN ENO–	When **EN** is at logic 1, convert the REAL value supplied at **IN** to Integer by Rounding to nearest Double Integer value. Put the result in the location specified at **OUT**.
Truncate Double Integer Part TRUNC EN ENO ???–IN OUT–???	Truncate Double Integer Part TRUNC ... —EN OUT–??? ???—IN ENO–	When **EN** is at logic 1, convert the REAL value supplied at **IN** to Integer by Truncating the fractional part of the REAL value. Put the result in the location specified at **OUT**.
Ceiling to Double Integer CEIL EN ENO ???–IN OUT–???	Ceiling to Double Integer CEIL ... —EN OUT–??? ???—IN ENO–	When **EN** is at logic 1, convert the REAL value supplied at **IN** to DINT value by Rounding to lowest DINT value greater-than-or-equal to **IN**. Put the result in the location specified at **OUT**.
Floor to Double Integer FLOOR EN ENO ???–IN OUT–???	Floor to Double Integer FLOOR ... —EN OUT–??? ???—IN ENO–	When **EN** is at logic 1, convert the REAL value supplied at **IN** to DINT value by Rounding to highest DINT value less-than-or-equal to **IN**. Put the result in the location specified at **OUT**.

Note: When EN is '1' ENO will also be '1' except on error; then while EN is '1'; ENO is '0'.

Integer and REAL Arithmetic Instructions

STEP arithmetic instructions support the four basic arithmetic operations of addition, subtraction multiplication, and division on numbers of data type INT (integer), DINT (double-integer), and REAL (floating-point). In STEP 7, real numbers are represented in IEEE floating-point format— therefore; the terms REAL and Floating-Point are often used interchangeably. These instructions allow the control program to perform basic production accounting, and the ability to solve simple algorithms for control, diagnostic, or other purposes. While the DIV_DI instruction produces the normal quotient result, the MOD_DI operation produces the remainder part of the quotient as the result.

Table 4-16. STEP 7 Integer and REAL Arithmetic Instructions

	LAD/FBD Arithmetic Instructions Using Data Type		
	INT	DINT	REAL
	Permissible Range	Permissible Range	Permissible Range
Operation	-32,768 to +32,767	-214,783,648 to +214,783,647	± 1.75495e-38 to ± 3.402823+e38
Addition	ADD_I	ADD_DI	ADD_R
Subtraction	SUB_I	SUB_DI	SUB_R
Multiplication	MUL_I	MUL_DI	MUL_R
Division with Quotient Result	DIV_I	DIV_DI	DIV_R
Division with Remainder	-	MOD_DI	-

Arithmetic operations are performed on the values specified at **IN1** and **IN2,** placing the result I the location specified at **OUT**. The arithmetic operation is always relative to IN1 (e.g., DIV_R; IN1divided by IN2). Acceptable values, as shown below, include constants within the valid range, the contents of absolute memory locations, or declared variables of the correct data type. If absolute memory locations are supplied, the correct memory width must be specified (i.e., word, double word). In any case, the contents of the specified memory locations or variables must be of the correct data type and fall within the permissible range.

Table 4-17. Example LAD/FBD Integer, Double Integer, and REAL Addition Instructions.

IN1/IN2/OUT	INT	DINT	REAL
	ADD_I EN ENO ???–IN1 OUT–??? ???–IN2	ADD_DI EN ENO ???–IN1 OUT–??? ???–IN2	ADD_R EN ENO ???–IN1 OUT–??? ???–IN2
Constants (in permissible range)	-32,768 to +32,767	-214,783,648 to +214,783,647	± 1.75495e-38 to ± 3.402823+e38
S7 Memory Locations (Absolute or Symbolic)	IW 28 QW 42 MW 54 DB6.DBW12	ID 28 QD 42 MD 54 DB6.DBD12	ID 28 QD 42 MD 54 DB6.DBD12
Declared Variables (of correct data type)	#Value_1 #Value_2 #Total	#Value_1 #Value_2 #Total	#Value_1 #Value_2 #Total

Note: When EN is '1' ENO will also be '1' except on error; then while EN is '1'; ENO is '0'.

In addition to the four basic arithmetic operations, the STEP 7 instruction sets include standard mathematical functions, as listed in the table below. These operations include functions for finding angle and radian measurement, absolute value, square root, square of a number, natural log of a number, and finally, a function to determine the exponent of a number with a radix of base e. In the STEP 7 programming software, these operations are located in the floating-point folder of the Program Elements tab.

Table 4-18. LAD/FBD: Standard Math Functions (REAL number operations).

LAD/FBD	Name	Brief Description
SIN EN ENO ???-IN OUT-???	Sine	When the **EN** signal is at logic 1, find the Sine of the REAL value (radian angle) supplied at **IN**. Put the result in location specified at **OUT**.
COS EN ENO ???-IN OUT-???	Cosine	When the **EN** signal is at logic 1, find the Cosine of the REAL value (radian angle) supplied at **IN**. Put result in location specified at **OUT**.
TAN EN ENO ???-IN OUT-???	Tangent	When the **EN** signal is at logic 1, find the Tangent of the REAL value (radian angle) supplied at **IN**. Put the result in location specified at **OUT**.
ASIN EN ENO ???-IN OUT-???	Arc Sine	When the **EN** signal is at logic 1, find the Arc Sine of the value supplied at **IN** (-1 to +1). Put angle result (radians) in location specified at **OUT**.
ACOS EN ENO ???-IN OUT-???	Arc Cosine	When the **EN** signal is at logic 1, find the Arc Cosine of the value supplied at **IN** (-1 to +1). Put angle result (radians) in location specified at **OUT**.
ATAN EN ENO ???-IN OUT-???	Arc Tangent	When the **EN** signal is at logic 1, find the Arc Tangent of the value supplied at **IN**. Put angle result (radians)in location specified at **OUT**.
ABS EN ENO ???-IN OUT-???	Absolute Value	When the **EN** signal is at logic 1, find the Absolute Value of the value supplied at **IN**. Put the result in location specified at **OUT**.
SQRT EN ENO ???-IN OUT-???	Square Root	When the **EN** signal is at logic 1, find the Square Root of the value supplied at **IN**. Put the result in location specified at **OUT**.
SQR EN ENO ???-IN OUT-???	Square	When the **EN** signal is at logic 1, find the Square of the value supplied at **IN**. Put the result in location specified at **OUT**.
LN EN ENO ???-IN OUT-???	Natural LOG	When the **EN** signal is at logic 1, find the Natural Log of the value supplied at **IN**. Put the result in location specified at **OUT**.
EXP EN ENO ???-IN OUT-???	Exponent	When the **EN** signal is at logic 1, find the Exponent of the value supplied at **IN**. Put the result in location specified at **OUT**.

Note: When EN is '1' ENO will also be '1' except on error; then while EN is '1'; ENO is '0'.

Compare Instructions

STEP 7 compare instructions allow comparison of integer (INT), double-integer (DINT), and REAL (floating-point) values. There are six compare types, each of which may be performed using each of the three numeric data types (i.e., INT, DINT, REAL). The compare tests include *equal, not equal, less than, greater than, less than-or-equal,* and *greater-than-or-equal*. Compare operations are often used for range checking, set point control, or limit checks on calculations.

Table 4-19. LAD/FBD Compare Instructions Summary.

	LAD/FBD Compare Instructions Using Data Type		
	INT	**DINT**	**REAL**
	Permissible Range	Permissible Range	Permissible Range
Compare Test	-32,768 to +32,767	-214,783,648 to +214,783,647	± 1.75495e-38 to ± 3.402823+e38
Equal	CMP==I	CMP==D	CMP==R
Not Equal	CMP<>I	CMP<>D	CMP<>R
Less Than	CMP <I	CMP <D	CMP <R
Greater Than	CMP >I	CMP >D	CMP >R
Less Than or Equal	CMP<=I	CMP<=D	CMP<=R
Greater Than or Equal	CMP>=I	CMP>=D	CMP>=R

Comparisons allow decisions based on the outcome of the specified comparison of the values input at **IN1** and **IN2**. The compare test is always relative to IN1 (e.g., CMP>I; IN1> IN2). Acceptable values, as shown below, include constants within the valid range of the data type, the contents of absolute memory locations, or local variables declared in a code block. If absolute memory locations are supplied as the input values, then the correct memory width must be specified (i.e., word, double word). In any case, the compared values must fall within the permissible number range of the data type being compared.

Table 4-20. Example LAD/FBD Integer, Double Integer, and REAL Compare Equal Instructions.

IN1/IN2	**INT**	**DINT**	**REAL**
	CMP ==I ???–IN1 ???–IN2	CMP ==D ???–IN1 ???–IN2	CMP ==R ???–IN1 ???–IN2
Constants (in permissible range)	-32,768 to +32,767	-214,783,648 to +214,783,647	± 1.75495e-38 to ± 3.402823+e38
S7 Memory Locations (Absolute or Symbolic)	IW 28 QW 42 MW 54 DB6.DBW12	ID 28 QD 42 MD 54 DB6.DBD12	ID 28 QD 42 MD 54 DB6.DBD12
Declared Variables (of correct data type)	#Value_1 #Value_2	#Value_1 #Value_2	#Value_1 #Value_2

Note: Symbolic addresses may be substituted for absolute addresses.

Program Flow Control Instructions

Program flow control instructions, provide one of the ways in which you can control how your program is processed. In short, you will be able to alter the way in which the CPU processes your program so that processing needs are serviced on an as-needed basis. Overall, these instructions can be used to create a more efficient and more organized program, reduce the processor scan time, and in some cases may be inserted to troubleshoot sections of the control program, and then later removed.

In both the LAD and FBD instruction browser, the Jump and Label instructions are grouped with so-called "Jumps" operations; Master Control Relay, Call, and Return instructions are all grouped with "Program Control" operations; and Open DB Coil, is under "DB Call." For the purposes of our discussions, all of these instructions are grouped together as "Program Flow Control Instructions."

Table 4-21. LAD/FBD Program Flow Control Instructions Summary

LAD	FBD	Brief Description
Label ???	Label ???	Four-character label defining a network to which a **JMP** or **JMPN** directs program execution. Combined with jumps, allows skipping of portions of logic.
Jump ??? —(JMP)—	Jump ??? JMP ...–	Causes internal block jump to labeled network based on conditional or unconditional logic. The jump destination is specified by label address specified above the **JMP** output.
Jump-if-NOT ??? —(JMPN)—	Jump-if-NOT ??? JMPN ??.?–	When the conditional logic is at logic 0 (RLO = 0), an internal block jump is directed to the network identified by the **label** address specified above the **JMPN** output.
MCR Activate —(MCRA)—	MCR Activate MCRA	This unconditional output, when encountered, enables the use of MCR zones up to the point of the next MCR De-activate instruction. Paired with MCRD instruction.
MCR De-Activate —(MCRD)—	MCR De-Activate MCRD	This unconditional output, when encountered, disables the use of MCR zones up to the point of the next MCR Activate instruction. Paired with MCRA instruction.
MCR ON —(MCR<)—	MCR ON MCR< ??.?–	This conditional output, when activated by the driving logic, begins an MCR zone. Paired with MCR-OFF (End) instruction.
MCR OFF —(MCR>)—	MCR OFF MCR>	This unconditional output, when encountered, ends an MCR zone. Paired with MCR-ON (Begin) instruction.
Call Block Coil ??? —(CALL)—	Call Block ??? CALL ...–	When the driving logic (conditional or unconditional) is at logic 1, the referenced Function (FC) or System Function (SFC), having no formal parameters, is called for processing.
Open Data Block ??? —(OPN)—	Open Data Block ??? OPN	When this unconditional instruction is encountered, the shared data block (DB) referenced above the instruction is opened, allowing data access to the data locations.
Return Coil —(RET)—	Return Output RET ??.?–	When the conditional logic driving this output is true, execution of the current block is terminated and control is returned to the calling block; otherwise continue with the following network.

Status Bit Instructions

S7 controllers generate a variety of status conditions that result from software operations. These status bits, which are combined into what is referred to as the *CPU status word*, are used by the CPU in controlling binary logic operations (e.g., AND/OR relay logic), and set by the CPU as a result of various digital operations (e.g., compares, arithmetic, conversions, shift-rotate, and word logic). Although the CPU status word is not directly available to the user program, it is made available indirectly using the status bit instructions of Table 4-21.

In LAD, these instructions are implemented as normally-open and normally-closed contacts that may be combined with other instructions. Just as the standard NC contact allows an inverse condition to be examined, use of the normally-closed status bit instructions allow you to check the inverse condition of each status bit. For example, when the normally-closed contact of the "result bit greater than zero," is used a zero status will allow power flow through the contact. In FBD these instructions are available with positive evaluation only. You may combine these instructions with the 'negate RLO' instruction to examine for the inverse status bit result. Status bit operations may also be combined with other bit operations.

Table 4-22. LAD: Status Bit (Result Bits) Instructions Summary.

LAD Instructions		**Name**	**Brief Description**
NO	**NC**		
BR	BR	Binary Result Bit	Use the NO contact to check the BR bit for logic 1 to allow power flow; use NC BR bit to check for logic 0 to allow power flow.
UO	UO	Unordered Bit	Check if previous floating-point operation involved at least one invalid (Unordered) floating-point number.
OV	OV	Overflow Bit	Check if previous math operation resulted in a value outside of permissible negative or positive range.
OS	OS	Overflow Bit Stored	Check if previous series of math operations resulted in a value outside of the permissible range.
>0	>0	Greater Than Zero Result Bit	Check result of previous math operation for greater than zero.
<0	<0	Less Than Zero Result Bit	Check result of previous math operation for less than zero. Combine in series or parallel with other contacts.
>=0	>=0	Greater Than or Equal Zero Result Bit	Check result of previous math operation for greater-than-or-equal zero.
<=0	<=0	Less Than or Equal Zero Result Bit	Check result of previous math operation for less than or equal zero.
==0	==0	Equal Zero Result Bit	Check result of previous math operation for equal zero.
<>0	<>0	Not Equal Zero Result Bit	Check result of previous math operation for not equal zero.

Note: These same instructions implemented in FBD are represented in a box. In STL, these operations are implemented as Jump instructions (i.e., jump to label), allowing internal block jumps based on the status result bits.

Word Logic Instructions

Word Logic operations allow you to perform standard Boolean operations of AND, OR and Exclusive OR on two word (16-bit) locations or two double-word (32-bit) locations. These instructions are quite useful in performing diagnostic checks on machine or process states. Through the use of these instructions, the desired states or the resulting word or double word location can be *masked in* (set to '1'), using the OR operation; or *masked out* (set to '0'), using the AND operation. It is also possible to invert multiple bits of the result or to determine what bits differ from an expected bit pattern, using the XOR operation.

Table 4-23. LAD/FBD Word Logic Instructions Summary

LAD	FBD	Brief Description
AND Word 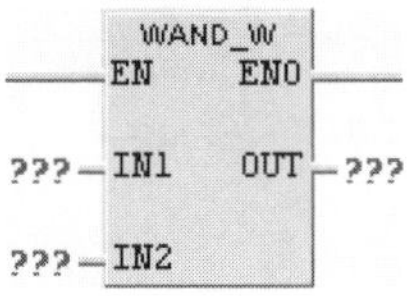 WAND_W EN ENO ???–IN1 OUT–??? ???–IN2	AND Word WAND_W ...–EN ???–IN1 OUT–??? ???–IN2 ENO– 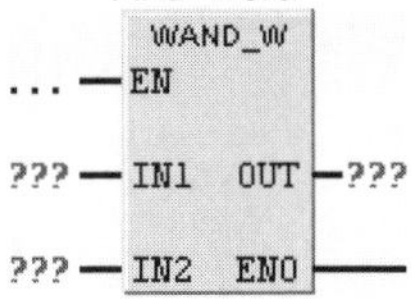	When the **EN** signal is at logic 1, perform a bit-by-bit logical AND on the two 16-bit values supplied at (**IN1**/**IN2**). Put result in the location specified at **OUT**.
OR Word 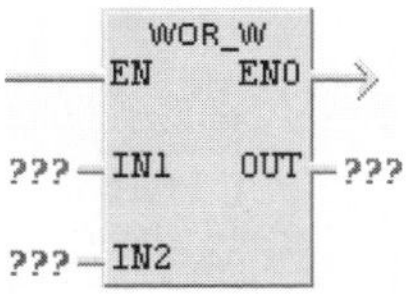 WOR_W EN ENO ???–IN1 OUT–??? ???–IN2	OR Word WOR_W ...–EN ???–IN1 OUT–??? ???–IN2 ENO– 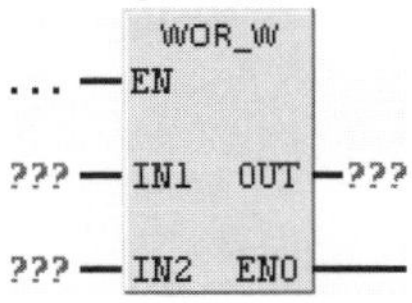	When the **EN** signal is at logic 1, perform a bit-by-bit logical OR on the two 16-bit values supplied at (**IN1**/**IN2**). Put result in the location specified at **OUT**.
Exclusive OR Word 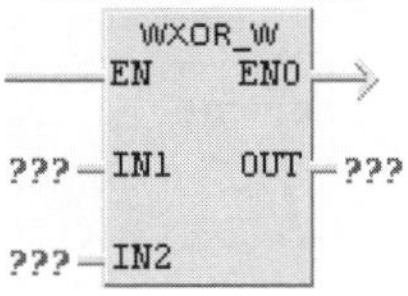 WXOR_W EN ENO ???–IN1 OUT–??? ???–IN2	Exclusive OR Word 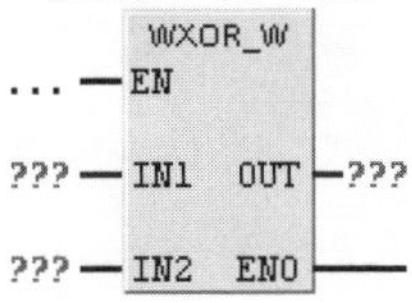 WXOR_W ...–EN ???–IN1 OUT–??? ???–IN2 ENO–	When the **EN** signal is at logic 1, perform a bit-by-bit logical XOR on the two 16-bit values supplied at (**IN1**/**IN2**). Put result in the location specified at **OUT**.
AND Double-Word 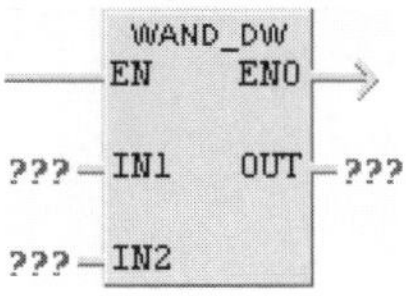 WAND_DW EN ENO ???–IN1 OUT–??? ???–IN2	AND Double-Word 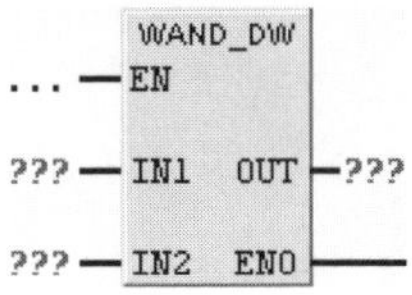 WAND_DW ...–EN ???–IN1 OUT–??? ???–IN2 ENO–	When the **EN** signal is at logic 1, perform a bit-by-bit logical AND on the two 32-bit values supplied at (**IN1**/**IN2**). Put result in the location specified at **OUT**.
OR Double-Word 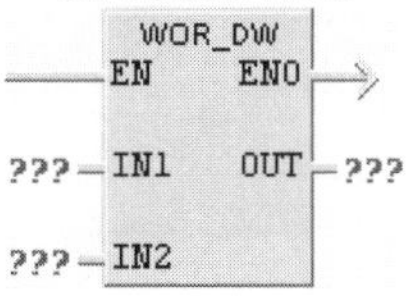 WOR_DW EN ENO ???–IN1 OUT–??? ???–IN2	OR Double-Word 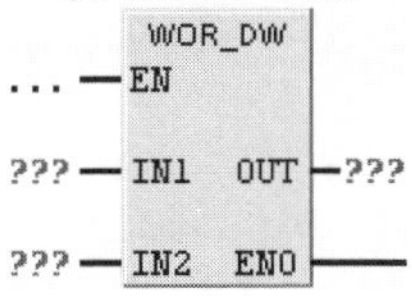 WOR_DW ...–EN ???–IN1 OUT–??? ???–IN2 ENO–	When the **EN** signal is at logic 1, perform a bit-by-bit logical OR on the two 32-bit values supplied at (**IN1**/**IN2**). Put result in the location specified at **OUT**.
Exclusive OR Double-Word 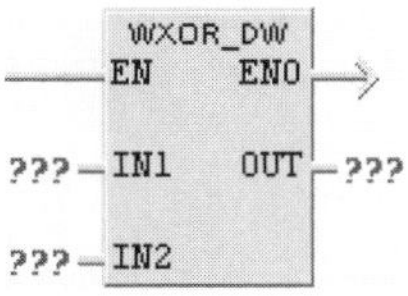 WXOR_DW EN ENO ???–IN1 OUT–??? ???–IN2	Exclusive OR Double-Word 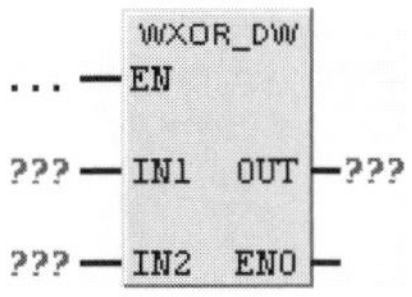 WXOR_DW ...–EN ???–IN1 OUT–??? ???–IN2 ENO–	When the **EN** signal is at logic 1, perform a bit-by-bit logical XOR on the two 32-bit values supplied at (**IN1**/**IN2**). Put result in the location specified at **OUT**.

Note: When EN is '1' ENO will also be '1' except on error; then while EN is '1'; ENO is '0'.

Shift-Rotate and Move Instructions

Shift and rotate operations move the bit data of a specified word or double word location to the left or right. These instructions are often used in materials conveying applications where information on discrete parts must be tracked for some distance, while the product is in motion, and usually prior to performing a secondary operation.

The move operation copies data from one memory location to another, leaving the source unchanged. Often, data is moved to a new location either prior to performing an operation or just after an operation has been performed, in order to preserve the data. Changing a timer preset, a process set point, or saving and stacking data for transfer to another device, all are cases where move operations might be used.

Table 4-24. LAD Shift-Rotate and Move Instructions

LAD/ FBD	Name	Brief Description
SHL_W EN ENO ???-IN OUT-??? ???-N	Shift Left Word	Shift the 16-bits of the Word specified at **IN**, by n-bits to the left.
SHR_W EN ENO ???-IN OUT-??? ???-N	Shift Right Word	Shift the 16-bits of the Word specified at **IN**, by n-bits to the right.
SHL_DW EN ENO ???-IN OUT-??? ???-N	Shift Left Double Word	Shift the 32-bits of the specified Double-Word, by n-bits to the left.
SHR_DW EN ENO ???-IN OUT-??? ???-N	Shift Right Double Word	Shift 32-bits of the Double-Word specified at **IN**, by n-bits to the right.
ROL_DW EN ENO ???-IN OUT-??? ???-N	Rotate Left Double Word	Rotate the 32-bits of the Double-Word specified at **IN**, by n-bits to the left.
ROR_DW EN ENO ???-IN OUT-??? ???-N	Rotate Right Double Word	Rotate the 32-bits of the Double-Word specified at **IN**, by n-bits bits to the right.
SHR_I EN ENO ???-IN OUT-??? ???-N	Shift Right Integer	Shift the 16-bits bits of the Integer value specified at **IN**, by n-bits to the right, while maintaining the sign bit.
SHR_DI EN ENO ???-IN OUT-??? ???-N	Shift Right Double Integer	Shift the 32-bits of the Double Integer value specified at **IN**, by n-bits to the right, while maintaining sign bit.
MOVE EN ENO ???-IN OUT-???	Move Word	Copy the variable specified at **IN** to the location specified at OUT.

Comments on Working with STEP 7 Programs and Data

The programming tools and software resources of STEP 7 are designed to facilitate the development of a well-planned and organized program — a program that can be easily started and later maintained. Tasks associated with developing your program with STEP 7 are truly straightforward and simple. If thoughtfully designed and developed, the program will process efficiently and correctly, and allow easy interpretation of block operations and flow.

In the remainder of this chapter, examples of developing blocks and using the STEP 7 programming instructions are presented in a step-by-step manner. Regardless of the size of your control task, you can accomplish the required tasks by breaking the job into smaller tasks. The following checklist will aid in this regard as well as in highlighting major development steps.

Checklist: Working with STEP 7 Programs and Data

- *Define major technological/functional units of the machine/process.*
- *Define subtasks of each machine/process unit, including inputs and outputs.*
- *Specify one or more blocks (i.e., FCs, FBs, etc.) to handle each subtask.*
- *Determine data block required by each block (i.e., global DBs, instance DBs).*
- *Define control logic for each device/operation of a logic block.*
- *Assign absolute addresses to Inputs, Outputs, Timers, and Counters.*
- *Assign symbolic addresses to Inputs, Outputs, logic and data blocks.*
- *Define preliminary allocation of bit memory for each logic block.*
- *Determine default language (i.e., LAD/FBD/STL) for creating blocks.*
- *Create and enter data for each global data block.*
- *Develop code blocks, creating control logic for each task (i.e., each device or operation).*
- *Determine the processing requirement (how often and from which OB) of each block (i.e., start-up only, cyclical, periodic interval, on interrupt, error reaction, etc.).*
- *Download and test each block separately by calling the block from OB 1.*

Viewing and Editing Symbolic Addresses

Basic Concept

Symbolic addresses are alphanumeric names assigned to the absolute addresses in your program, using the Symbol Editor. Whereas symbols created in the symbol table may be accessed anywhere in the program and are considered global, *Local Symbols*, refer to names given to the variables assigned in the declaration table of a block. The use of local symbols is restricted to the block where they were declared. When you use symbolic addresses in the program or in a monitoring tool, global symbols must be enclosed in double quotes (e.g., "Motor_1"), and local symbols are preceded by a pound sign (e.g., #Value_A).

Essential Elements

To create a symbol address is to assign a meaningful name (24-characters max.) as a substitute for an absolute address. You may assign global symbols to all S7 memory areas (*See Table 4-5*) including Inputs (I), Outputs (Q), Bit Memory (M), Timers (T), Counters (C), Peripheral Inputs (PI), and Peripheral Outputs (PQ). Addresses assigned from these areas may include bit, byte, word, and double word locations. Symbol addresses may also be assigned to code S7 blocks (e.g., FB, FC, OB, SFC, and DB). You may assign symbolic addresses to blocks in the symbol editor or in the block properties dialog when the block is first created or later.

Application Tips

Symbol addresses are an additional aid in viewing and understanding your program. The editor can be started from the SIMATIC Manager, by double-clicking on the Symbol Table object; from the LAD/FBD/STL editor menu; or from within the Monitor/Modify tool, while editing a Variable Table (VAT). For fast entry of symbolic names for I/O module addresses, consider entering these from the Hardware Configuration tool as you insert each module.

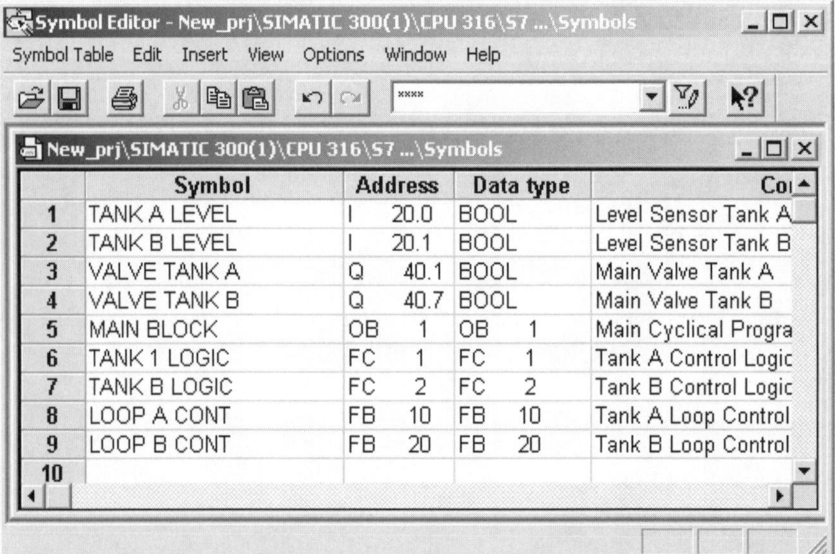

	Symbol	Address	Data type	Co
1	TANK A LEVEL	I 20.0	BOOL	Level Sensor Tank A
2	TANK B LEVEL	I 20.1	BOOL	Level Sensor Tank B
3	VALVE TANK A	Q 40.1	BOOL	Main Valve Tank A
4	VALVE TANK B	Q 40.7	BOOL	Main Valve Tank B
5	MAIN BLOCK	OB 1	OB 1	Main Cyclical Progra
6	TANK 1 LOGIC	FC 1	FC 1	Tank A Control Logic
7	TANK B LOGIC	FC 2	FC 2	Tank B Control Logic
8	LOOP A CONT	FB 10	FB 10	Tank A Loop Control
9	LOOP B CONT	FB 20	FB 20	Tank B Loop Control
10				

Figure 4-30. STEP 7 Symbol Editor, for assigning symbolic labels to S7 memory addresses.

Quick Steps: Viewing and Editing Symbolic Addresses

STEP	ACTION
1	From the SIMATIC Manager, open the correct the project and select the Program in which you wish to create new or edit existing global symbols.
2	With an S7 Program folder selected, from the right pane of the project window double click on the **Symbols** object, or right click on the object and select **Open Object**. The symbol table is presented like a spreadsheet, with a column for **Symbol**, **Address**, **Data Type**, and a **Comment**.
3	Click in the **Symbol** field and enter up to 24 characters for the symbolic name; click in the **Address** field and enter the associated absolute address; click in the **Comment** field and enter a descriptive comment of up to 80-characters. The data type of each absolute address is automatically inserted.
4	When done with entering symbolic addresses, from the menu select **Symbol Table ➢ Close**.

Creating a Data Block

Basic Concept

To create a new data block is to generate a new DB object. The new DB object is essentially a container used to store data for your process and program. Recall, that there are two types of data blocks — *Shared DBs*, which can be accessed by any of your program blocks; and *Instance DBs, each* associated with a specific Function Block (FB) as required memory.

Essential Elements

Data blocks are comprised of two sections: the *block header*, and *variable declaration*. When you generate a DB, the header dialog allows you to define attributes of the data block, including whether it should be an Instance or Shared DB. Once a shared DB is generated, you may open the block to define and initialize the data elements. When you create an instance DB you will simply assign it to an already existing FB.

Application Tips

Each FB you create will require an instance DB. You should create the FB and define the declaration table before creating the instance DB. When the instance DB is created and assigned to an existing FB, the structure of the declaration area of the FB (i.e., static variables, and formal parameters only), is written to the instance DB. In this way, the static variables and the actual parameters are stored in the DB on each instance the FB is called.

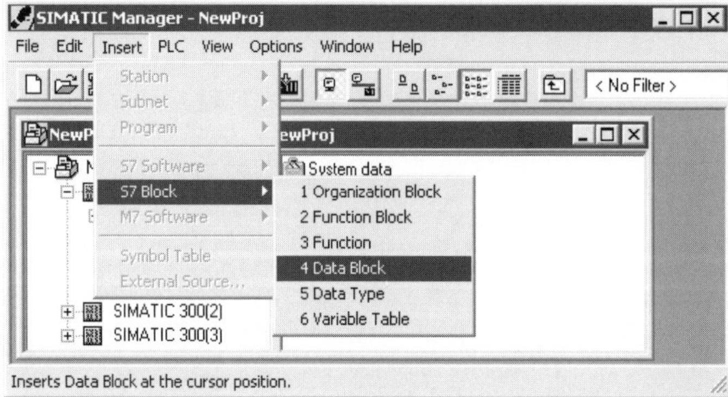

Figure 4-31. Generating a new S7 Data Block from the SIMATIC Manager.

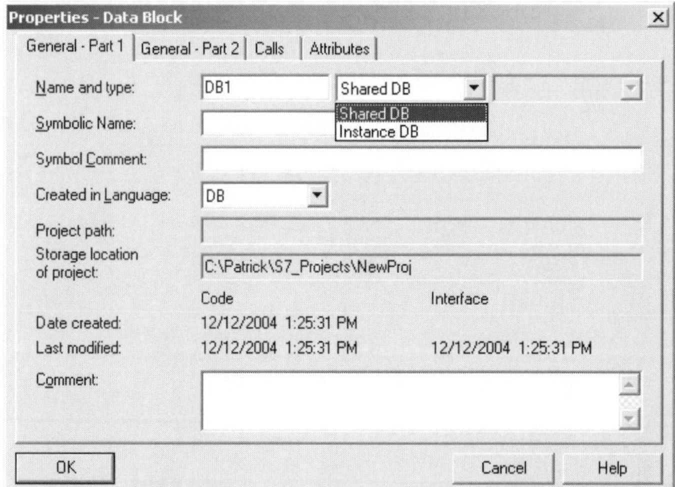

Figure 4-32. Data Block General Properties Part-1: Shared DB Selected.

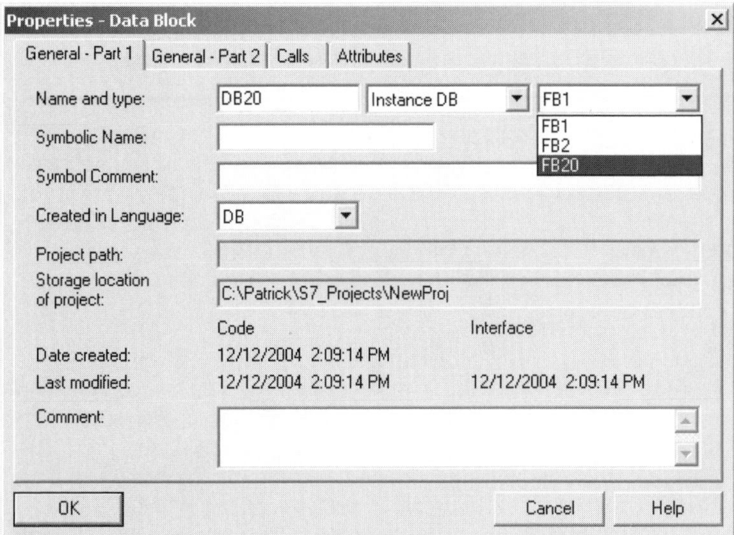

Figure 4-33. Data Block General Properties Part-1: Instance DB Selected.

Quick Steps: Creating a Data Block

STEP	ACTION
1	Open the project and select the Blocks folder of the program in which the new data block is to be inserted.
2	From the menu, select **Insert ➢ S7 Block ➢ Data Block,** to generate the new data block object. The block properties dialog is opened to the General Part-1 tab.
3	In the **Name and Type field,** the next available DB number is placed in the left field. Use either the next available number or type in a new DB number (e.g., DB 10).
	In the adjacent field, to the right of the DB number, select the drop arrow and choose Shared *DB* or *Instance DB*, based on the DB type you wish to create.
4	If Instance DB is chosen as the DB type, press the drop down arrow in the right-most field, and select the FB number to which the DB should be assigned.
5	Next, in the **Symbol Name** field, enter a label for the DB (up to 24-characters). The symbol name is entered into the Program's symbol table as a substitute address.
6	In the **Symbol Comment** field, enter a descriptive comment of up to 80-characters. The comment will appear with the symbol in the symbol table.
7	In the Comment field below, you may enter a descriptive comment for the entire data block, perhaps describing its main use in the application.
8	On the General Part 2 tab, an Instance DB will be assigned the same header (i.e., Version, Family, etc.) information of the FB to which it was assigned.
9	Press the **OK** button to confirm the DB header entries and to generate the new DB.

Editing a Data Block

Basic Concept

When you created data blocks for the first time, you may have created both shared DBs and instance DBs. The initial data structure and content of each instance DB you created was initialized by the editor when you assigned the DB to a specific function block. On the other hand, shared DBs will contain process data that will be available to any of the blocks of your program (for reading or writing data). You will determine the data organization and initial values for each data element of these shared data blocks. If the data blocks you wish to work with already exist, then you may insert the initial data or edit the current data contents.

Essential Elements

A DB, like other blocks, can be edited offline or online, and may be opened from the SIMATIC Manager or from the LAD/FBD/STL editor. In either case, when you are done editing the DB you can determine whether it is stored offline, online, in both places, or simply discarded. Variables are entered in the declaration table of a data block in the same fashion as the with logic blocks. This initial editing is done while in the *Declaration View* of the editor. For each variable, you must assign a **Name** and **Data Type**; an **Initial Value** and **Comment** are optional. Structures and arrays are also entered in the same fashion as in logic blocks.

The variable data in a data block may include both process constants and variables. Some values will be written to by your program while the system is in process. If you need to edit the actual values of DB variables at some later point, you will need to switch the editor to the *Data View*. Furthermore, at some point you may also wish to reinitialize the entire data block.

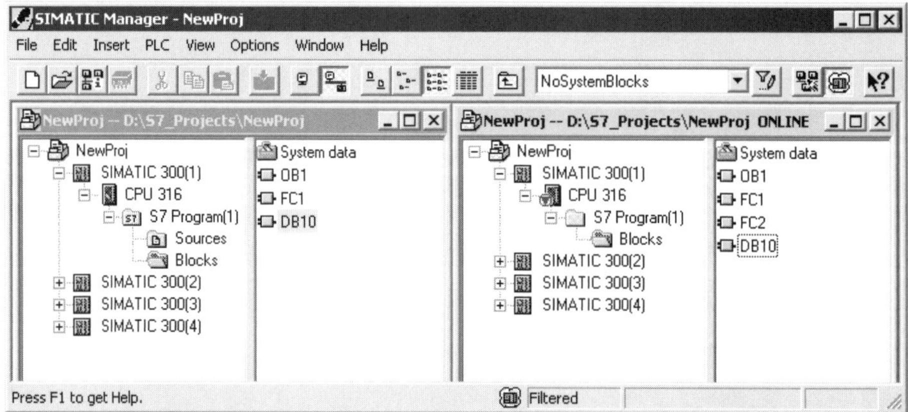

Figure 4-34. SIMATIC Manager open to "NewProj," with offline and online windows arranged vertically. The LAD/FBD/STL editor opens offline or online based on your selection.

LAD/STL/FBD - [DB10 -- New_prj\SIMATIC 300(1)\CPU 316]

File Edit Insert PLC Debug View Options Window Help

Address	Name	Type	Initial value	Comment
0.0		STRUCT		
+0.0	DB_VAR	INT	0	Temporary placeholder variable
=2.0		END_STRUCT		

Figure 4-35. Default screen when editing a Shared Data block for the first time.

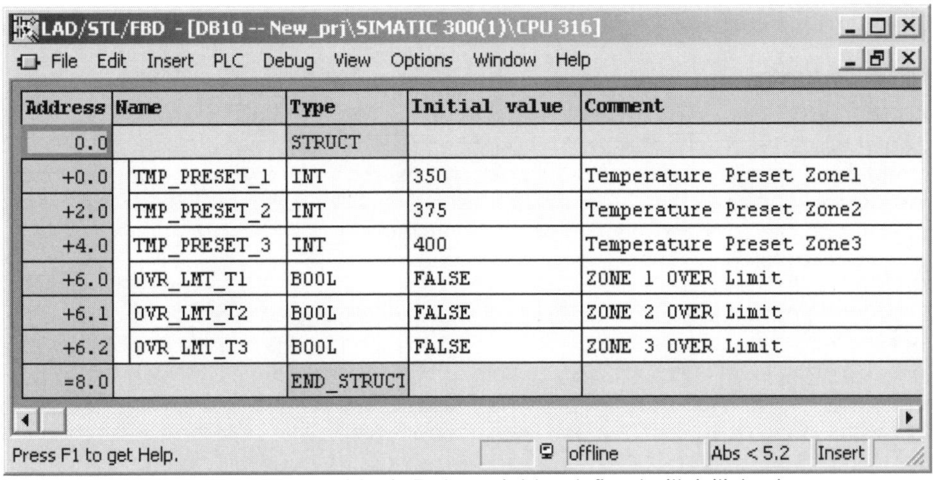

LAD/STL/FBD - [DB10 -- New_prj\SIMATIC 300(1)\CPU 316]

File Edit Insert PLC Debug View Options Window Help

Address	Name	Type	Initial value	Comment
0.0		STRUCT		
+0.0	TMP_PRESET_1	INT	350	Temperature Preset Zone1
+2.0	TMP_PRESET_2	INT	375	Temperature Preset Zone2
+4.0	TMP_PRESET_3	INT	400	Temperature Preset Zone3
+6.0	OVR_LMT_T1	BOOL	FALSE	ZONE 1 OVER Limit
+6.1	OVR_LMT_T2	BOOL	FALSE	ZONE 2 OVER Limit
+6.2	OVR_LMT_T3	BOOL	FALSE	ZONE 3 OVER Limit
=8.0		END_STRUCT		

Press F1 to get Help. offline Abs < 5.2 Insert

Figure 4-36. Editing a Shared Data block. Data variables defined with initial values.

Quick Steps: Editing a Data Block

STEP	ACTION
1	From the SIMATIC Manager, double-click on the data block you wish to edit.
2	From the menu, select **View ➢ Declaration View** in order to declare variables.
3	In the first data row, click in the **Name** field and enter a name; click in the **Type** field, right-click and select the data type from *Elementary Types* or *Complex Types*; click in the **Initial Value** field, and enter an initial value (optional); click in the **Comment** field and enter a comment (optional).
4	From the menu, select **Insert ➢ Declaration Line ➢ Before Selection** or **After Selection** to place the next variable either before or after the first row. Enter as many variables as required, and then from the toolbar press the **Save** button to save the DB offline; press the **Download** button to save the DB online in the PLC.
5	. Select **File ➢ Exit** to close the editor.
	Edit Actual Values of Shared or Instance DB data elements
2	With the data block open, from the menu select **View ➢ Data View** (if not already activated) to view the DB with an Actual Value column for editing the variables.
3	Click in an **Actual Value** field to modify any value as required.
4	If you simply wish to reset all of the variables of the DB to their initial values, from the menu, select the **Edit ➢ Initialize Data Block**.
5	Enter as many values as required, then press the **Save** button to save the DB offline; press the **Download** button to download the entire DB; for instance DBs, if you are using the DB parameter editor you will also have the option of only downloading the parameters values of the data block.
6	Select **File ➢ Exit** to close the editor.

Generating a New Code Block

Basic Concept

In STEP 7, any code blocks that you intend to write, you will first need to be created. To create a code block is to generate a block object of a specific type (i.e., OB, FB, or FC). The object, once created, is essentially a container in which you may open and then enter code as required. Although your STEP 7 program, may eventually consist of several user blocks that you develop, the complete program is built incrementally, block-by-block.

Essential Elements

Code blocks are comprised of the *block header*, *variable declaration area*, and *code section*. When you generate a block, you are actually creating the header, which allows you to define attributes of the block. Once generated, you ma open the block and begin writing your code. From the SIMATIC Manager, double-click on the block or right-click and select **Open Object**. From LAD/FBD/STL, open the block using **File ➢ Open** from the menu.

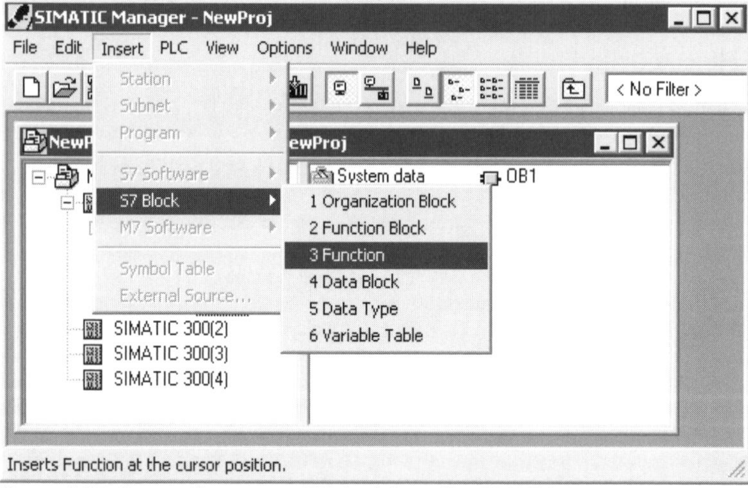

Figure 4-37. Generating a new S7 block from the SIMATIC Manager.

Properties - Function

General - Part 1 | General - Part 2 | Calls | Attributes

Name: FC19

Symbolic Name:

Symbol Comment:

Created in Language: LAD

Project path:

Storage location of project: C:\Patrick\S7_Projects\Brillian

	Code	Interface
Date created:	10/12/2004 5:27:57 PM	
Last modified:	10/12/2004 5:27:57 PM	10/12/2004 5:27:57 PM

Comment:

OK | Cancel | Help

Figure 4-38. Properties dialog: General Properties Part 1 - of Block Header.

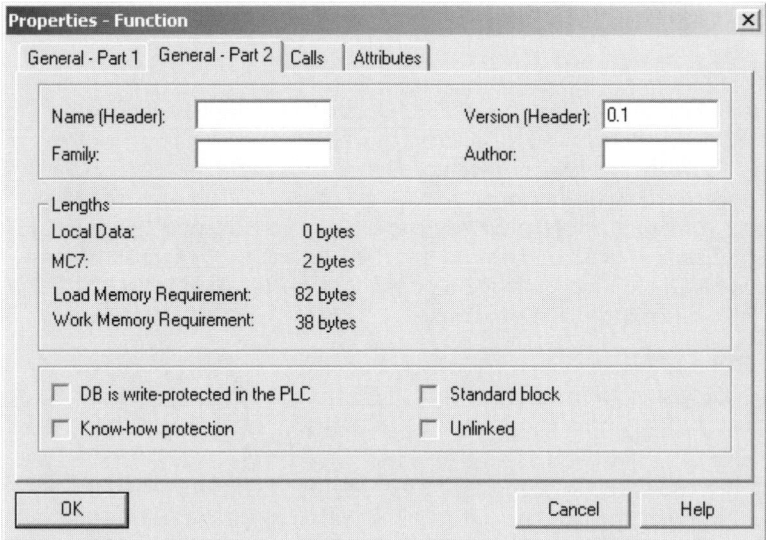

Figure 4-39. Properties dialog: General Properties Part 2 - of Block Header.

Quick Steps: Generating a New Code Block

STEP	ACTION
1	Open the project and select the Blocks folder of the program in which the new block is to be inserted.
2	From the menu, select **Insert ➢ S7 Block**, then select the block type (e.g., **Function**) to generate. The block properties dialog will be presented.
3	On the *General Part 1* tab, the **Name** field will contain the next available block number for the block type you have chosen. Either use the derived number or enter a new number (e.g., FC10, or FC 20). This name is the absolute address of the block.
4	Click in the **Symbol Name** field to enter a mnemonic name for the block. The symbolic name is actually a substitute for the absolute address.
5	Click in the **Symbol Comment** field to enter a descriptive comment. The comment, which may contain a maximum of 80-characters, will appear with the symbol name for the block when the symbol table is opened with the symbols editor.
6	Select the **Created in Language** drop down arrow, and choose the language in which the new block should be initially edited (e.g., LAD, FBD, or STL).
7	Select the *General Part 2* tab to enter additional properties for the block.
8	In the appropriate fields, enter a **Name** (not same as symbol), for the block; a **Family** name, that associates the block with a group of blocks; a **Version** number, that allows changes to the block to be managed; finally, an **Author** that identifies the user or perhaps department responsible for creating the block.
9	Press **OK** to confirm the configured block header and to generate the new block.

Navigating the LAD/FBD/STL Editor

Basic Concept

LAD/FBD/STL, the standard editor for creating STEP 7 programs, supports three language representations. The first language is LAD, a *Ladder Logic* representation; FBD or *Function Block Diagrams*, is a Boolean gate logic representation; and STL or *Statement List*, is a line-oriented text language similar to assembler or BASIC. With LAD/FBD/STL, you are able to develop a complete S7 program that combines user-developed organization blocks (OBs), function blocks (FBs), functions (FCs), and data blocks (DBs), as well as standard library and system blocks. LAD/FBD/STL is opened from the Windows Start menu — **Start ➢ SIMATIC ➢ STEP 7 ➢ LAD/FBD/STL**, or from within the SIMATIC Manager.

Essential Elements

In addition to menus and toolbars of the STEP 7 editor, there is the main window, the *Program Elements/Call Structure* window, and the *Details* window. The main window is a work area, which allows one or more blocks to be opened for editing in separate windows. The *Program Elements/Call Structure* allows you to access the STEP 7 LAD/FBD instruction folders, or view the overall calling structure of the open program. This window may be docked on either side of the main widow or may be hidden from view, using the View menu. The status bar, on the main window bottom, provides information on the status of the editor as well as PLC operations. Finally, the Details window, which may also be shown or hidden from view, gives you access to various views of program-related information.

Application Tips

You may copy block elements including *Block Title*, *Block Comment*, *Network Title*, *Network Comment*, and *Networks*, between blocks using Copy and Paste operations.

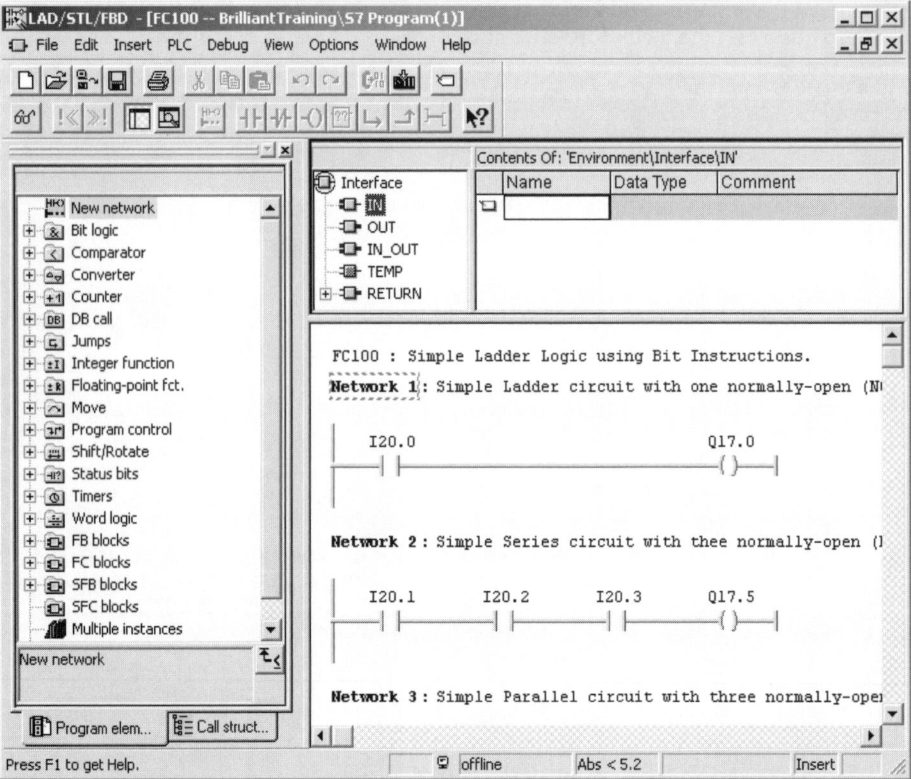

Figure 4-40. An opened block (FC 100) inside the LAD/FBD/STL programming editor.

Quick Steps: Navigating the LAD/FBD/STL Editor

STEP	ACTION
1	From the SIMATIC Manager, open the required project and S7 Program folder. Select the offline Blocks folder to display the *offline blocks* in the right pane. Alternatively, you can view the online Blocks folder by selecting **View** ➤ **Online** from the menu.
2	Double-click on a block or right-click on a block and select **Open** ➤ **Object** to open the block in the LAD/FBD/STL editor.
3	From the LAD/FBD/STL Editor menu select **View,** and then select **LAD,** to edit the block in *Ladder* format; select **FBD,** to edit the block in *Function Block Diagram* format; or select **STL,** to edit the block in *Statement List*.
4	Select **View** ➤ **Status Bar**, to display the editor status information on the bottom window edge. Select Status Bar again to hide the status bar.
5	Select **View** ➤ **Display With** ➤ **Symbol Representation** to toggle between the display of symbolic addresses or absolute addresses in the open program block.
6	Select **View** ➤ **Display With** ➤ **Symbol Information** to toggle the display of symbol information (i.e., symbol address, absolute address, comment) with each network.
7	Select **View** ➤ **Display With** ➤ **Comments** to display program comments or to allow entry of additional comments if required (toggles comments display on/off).
8	Select **View** ➤ **Overview** to display the *Program Elements/Call Structure* window. The *Program Elements* tab lists the instruction folders. Instructions may be dragged from a folder and dropped onto the current network. The *Call Structure* displays the hierarchy of how blocks are called in the open program; you may double click on any block to open the block.
9	From the menu, select **View** ➤ **Details** to display an information window at the bottom of the main window. The window contains seven tabs offering further access to related information for the open block.
10	From the *Program Elements* tab, expand the FC Blocks and FB blocks folders, to view user functions and function blocks that have been created; or expand the Libraries object and then the Standard Library folder, to access system block folders. FCs, FBs, SFCs, and SFBs may also be dragged and placed into a network.
11	To edit a program, find the network (e.g., Network 1) you wish to edit; click on a position in the network, then insert basic logic elements from the toolbar (e.g., normally-open and normally-closed contacts, coil, and branch instructions) or drag an instruction or block from the program elements window.
12	Use hotkeys for basic logic operations (e.g., **F2** ➤ NO contact; **F3** ➤ NC contact; **F7** ➤ Coil; **F8** ➤ branch open; and **F9** ➤ branch close). Use the **<Insert>** key to toggle between the insert mode and the overwrite mode while editing networks.
13	From the menu, select **File** ➤ **Exit** to close all open blocks and close the editor.

Opening, Editing, and Saving a Block

Basic Concept

In STEP 7, once a block object has been generated (i.e., OB, FB, FC, DB, etc.), it is essentially a container in which you may open and then enter and edit the code or data as required. A block can be edited offline or online, but in either case, it is actually temporarily held in a memory buffer on your PC until actually saved. When you are done editing a block, you may determine if it should be stored offline, online, in both places, or simply discarded.

Essential Elements

A block that already exists may be opened for editing from the SIMATIC Manager or from the STEP 7 LAD/FBD/STL editor. From the SIMATIC Manager you will open the offline blocks window of the required project, or to edit online you will need to open the online blocks window from the toolbar or from the menu. The block is selected and is opened by simply double-clicking on the block or with a right-click on the block and selecting **Open Object**.

Application Tips

As a matter of good practice and safety, consider making all program changes offline.

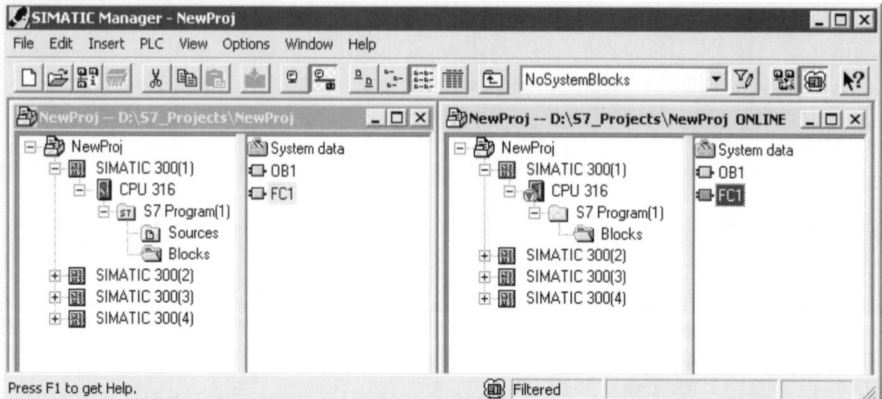

Figure 4-41. The project "NewProj" is opened with offline and online windows arranged vertically. The LAD/FBD/STL editor will open online or offline based on your selection.

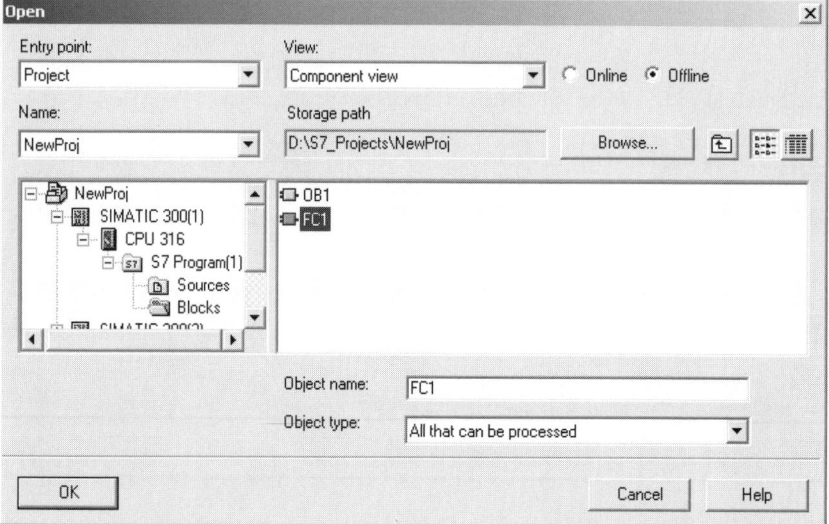

Figure 4-42. In the LAD/FBD/STL editor, File Open presents dialog for opening a block.

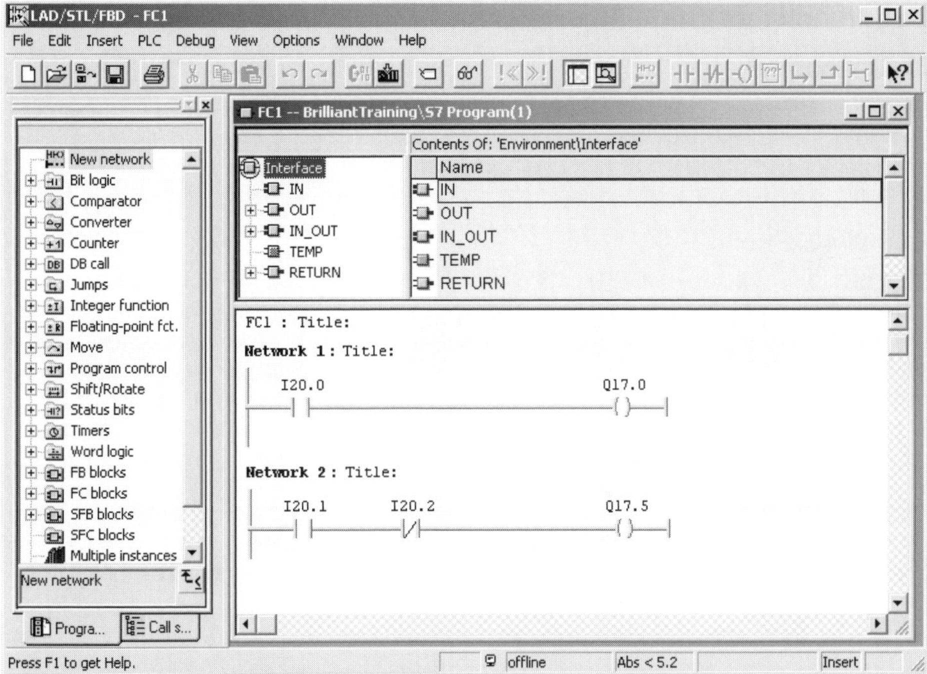

Figure 4-43. LAD/FBD/STL editor; FC1 block open in LAD; *Program Elements* window on left.

Quick Steps: Opening, Editing, and Saving a Block

STEP	ACTION
1	From the SIMATIC Manager open the required project, select the desired program, and then select the offline *Blocks* folder. Offline blocks are listed in the right pane.
2	Double-click on a block to edit, or right-click on the block and select **Open** ➢ **Object**. The LAD/FBD/STL editor opens with the offline block.
3	If the Program Elements tab is not in view, select **View** ➢ **Overview** from the menu.
4	From the menu, select **View,** then select the language of choice (e.g., **LAD, FBD,** and **STL**); in this case, select **LAD,** to edit the block in *Ladder* format. The Program Elements tab is refreshed to reflect a change in the choice of language.
5	Find the network you wish to edit, for example to add a normally-open contact to Network 1; click on a position in the network, then insert basic logic instructions from the toolbar (e.g., normally-open and normally-closed contacts, coil, and branch instructions) or drag an instruction from the instruction browser.
6	To insert a new network after a particular network, select the network and then from the menu select **Insert** ➢ **Network**, or from the toolbar press the **New Network** icon.
7	To save a block offline when done editing, from the menu select **File** ➢ **Save**; or from the toolbar, press the **Save** icon.
8	To save a block online when done editing, from the menu select **PLC** ➢ **Download**; or from the toolbar, press the **Download** icon. An online connection is required.

Documenting a Code Block

Basic Concept

Each user-written code block provides built-in mechanisms for easily documenting your code and data, thereby making your program easier to interpret and to troubleshoot. The documentation that you will enter for each block is stored in the offline program, but is available when either the offline block or its online equivalent is opened.

Essential Elements

Each block you create, whether in LAD, FBD, or STL, allows you to enter (1) **Variable Comments** of up to 80-characters, for each local variable declared in the declaration table; (2) a **Block Title** of up to 64-characters, to briefly title the block; (3) a **Block Comment**, of up to 64-K-bytes of text, to describe the purpose or operation of the block; (4) a **Network Title** of up to 64-characters, to title each logic Network; and (5) a **Network Comment** that allows text entry of up to 64-K-bytes per network, to describe the purpose or operation of each logic network. Blocks written in Statement List, also allow a **Line Comment** for each instruction line.

Application Tips

If a Symbol Table has been created for the program, symbolic addressing and symbol comments may be viewed by selecting **View ➤ Display with ➤ Symbol Representation** and **View ➤ Display with ➤ Symbol Information** from the menu.

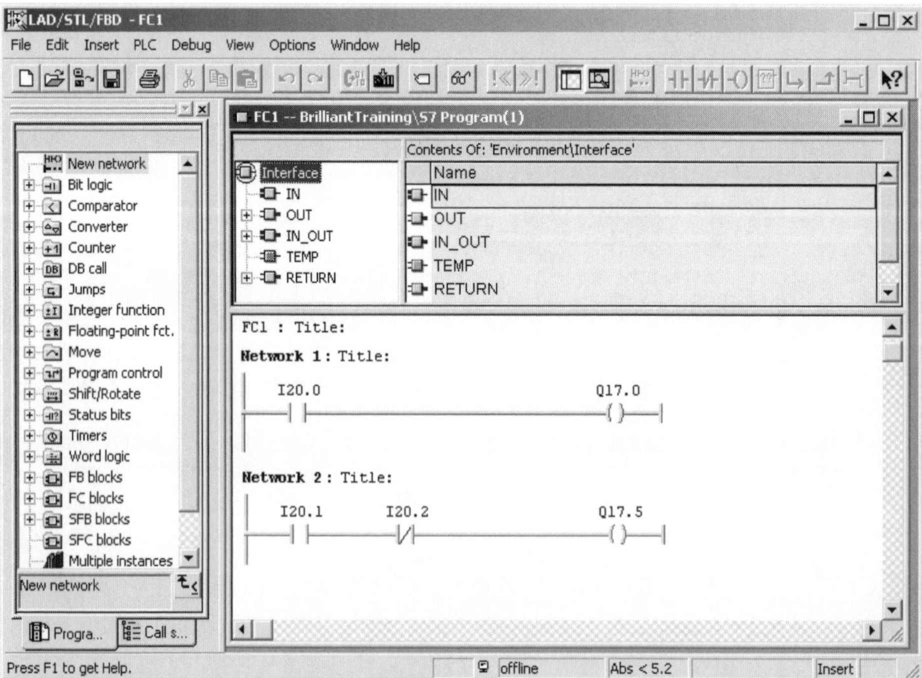

Figure 4-44. Block FC 1 opened in the LAD/FBD/STL editor, with LAD representation selected and with Display Comments not activated.

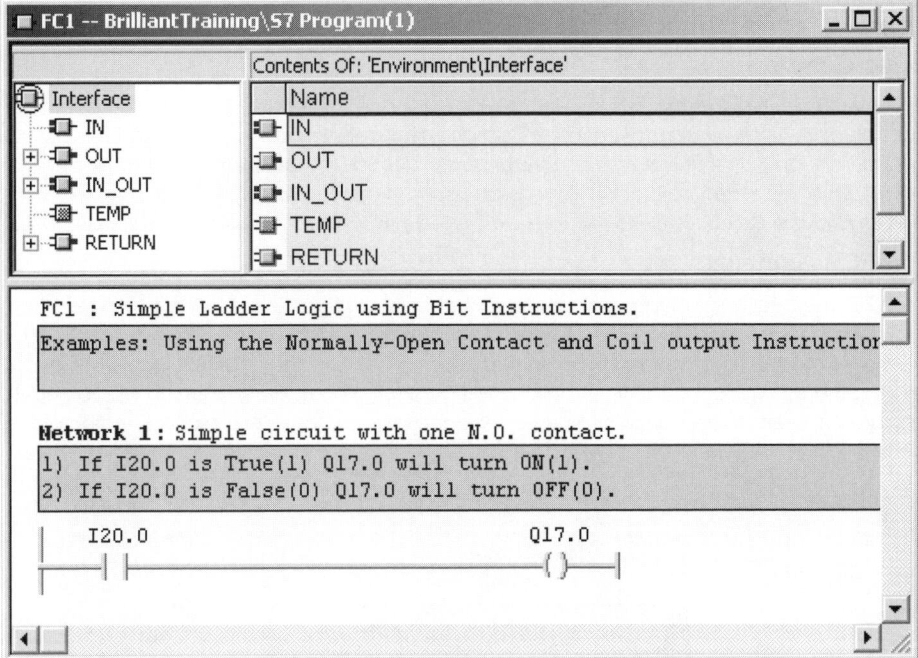

Figure 4-45. Block FC1 shown opened in the editor with comments displayed.

Quick Steps: Documenting a Code Block

STEP	ACTION
1	With the desired block open in the LAD/FBD/STL editor, from the menu, select **View ➢LAD** to choose Ladder as the language of choice for editing the block.
2	From the menu, selecting **View ➢ Display With ➢ Comments** activates comment entry or toggles the display of comments on and off. The check mark is shown when comments are displayed. No check mark is shown when comments are not active.
3	Click on the word 'Title' shown just to the right of the block name (e.g., **FC 1**: Title), to enter a **Block Title**. Enter up to a maximum of 64-characters for the title.
4	Click on the word 'Comment' in the gray field, below the block name, to enter a **Block Comment**. The field changes to white allowing entry of up to 64-K-bytes to describe the operation or purpose of the block. Click outside the box, when done.
5	Click on the word 'Title' shown just to the right of each network (e.g., Network 1: Title), in order to enter a **Network Title**. Enter up to a maximum of 64-characters.
6	Click on the word 'Comment' in the gray field, below each network, to enter a **Network Comment**. The field changes to white allowing entry of up to 64-K-bytes to describe the network operation or purpose. Click outside the box when done.
7	From the menu, select **File ➢ Save** to save the block and its documentation.

Programming an FC without Formal Parameters

Basic Concept

An FC written without formal parameters is typical for parts of your code involving basic logic operations in which no permanent data storage is required. Recall, that FCs may use temporary local variables, but these variables are handled in local memory, and are no longer available when the block is done processing. In applications consisting largely of discrete logic, most of the code will likely be written in FCs without use of formal parameters.

Essential Elements

In an FC, the declaration types include **IN**, **OUT**, and **IN_OUT**, which are used for declaring formal parameters; and **TEMP**, used to declare temporary local variables. Writing an FC without formal parameters mean you only need to define temporary (**TEMP**) local variables if they are needed. You may define as many temporary variables as required. Once you have declared the temporary variables, you will write the required logic, in the code section of the function.

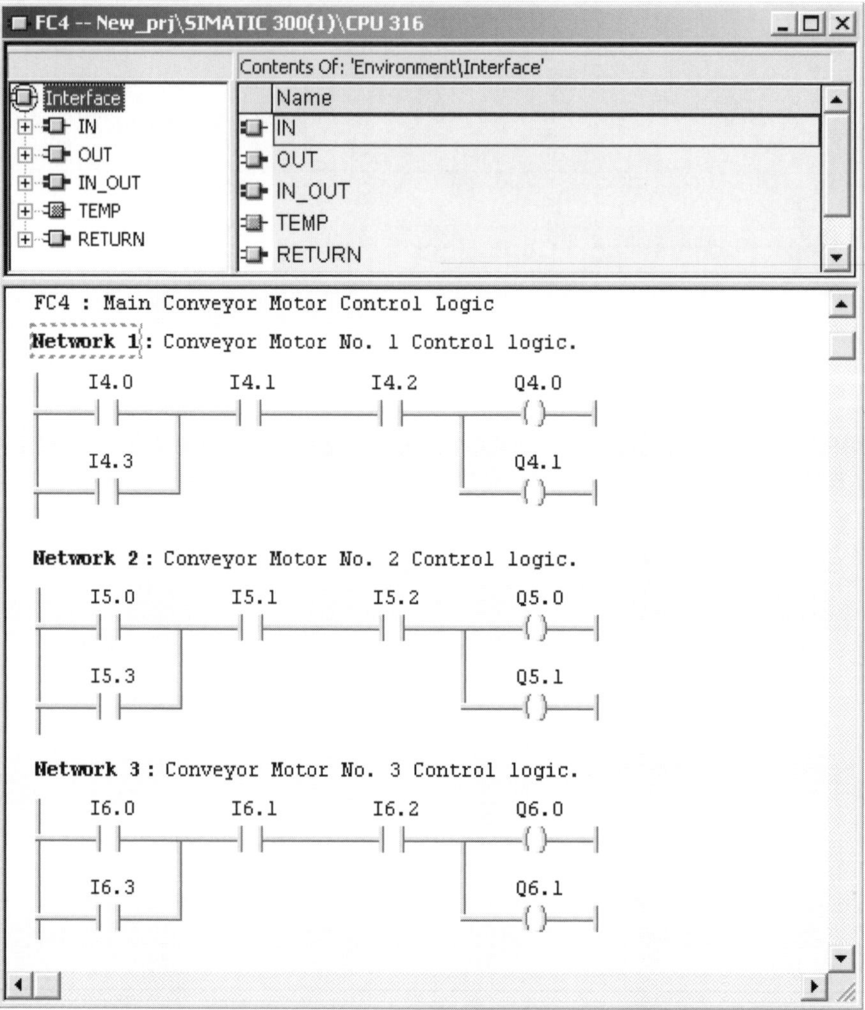

Figure 4-46. A Function (FC) written without use of formal parameters. See Figures 4-49 and 4-50 later, to see how an FC with formal parameters can replace repeated logic.

Quick Steps: Programming an FC without Parameters

STEP	ACTION
1	From the SIMATIC Manager, open the project and the required program *Blocks* folder; then double-click on the FC to open the block in the LAD/FBD/STL editor. Create the FC if necessary.
2	From the editor menu, select **View** ➢ then the language of choice (e.g., **LAD**).
3	Select **View** ➢ **Overview** from the menu if the *Program Elements/Call Structure* window is not in view (LAD/FBD instructions are in categorized folders).
4	To declare new temporary variables: First select the **TEMP** object in the left pane of the declaration table, then click in the **Name** field of the declaration row and enter the name for the variable. Click in the **Data Type** field and select the data type for the new variable from the drop list. Finally, click in the **Comment** field and enter a descriptive comment. Enter as many temporary local variables as required.
5	To insert a temporary variable before an existing variable: First select the existing variable row, then right click and select **New Declaration Row**. The new declaration row is placed above the row you selected. You may also select a declaration row and drag it to a new position.
6	To enter code for the block: Click inside the code section, starting with network 1, then write the code for each network operation, using either absolute addresses or symbolic addresses (if already defined). **Note**: Network 1 is automatically inserted.
7	To start a new network, from the menu select **Insert** ➢ **New Network**, or from the toolbar press the **New Network** button.
8	From the menu select **File** ➢ **Save** when done with the last network.
9	Select **File** ➢ **Exit** to close the editor.

Calling an FC without Formal Parameters

Basic Concept

With the exception of organization blocks, you must call all blocks to ensure that they are processed. A block is called by referencing its absolute address (e.g., FC 10) or symbolic address (e.g., "Sorter_1") with the appropriate call instruction. A call to the FC can be unconditional on every CPU cycle or conditional when specific logic conditions are met.

Essential Elements

In both LAD and FBD, an FC with no parameters can be called using the **CALL** coil or as a box function (**CALL FCn**). The Call coil instruction, however, may only be used when the FC has no formal parameters. If the call is programmed without any preceding logic then it is executed on every CPU scan (an unconditional call); if logic conditions precedes the call (a conditional call) then the FC is called when the conditions are met; if the logic conditions are not met, the instruction following the call is processed.

In Statement List (STL), an FC written without formal parameters can also be called conditionally or unconditionally. The conditional call (e.g., **CC** FC 10) is executed if the preceding result of logic operations is satisfied (RLO = logic 1). The unconditional call (e.g., **UC** FC 10) is executed on every CPU cycle.

Application Tip

Calls to FCs having no parameters, are typically conditional or unconditional calls from the main organization block (OB1) or from another block (e.g., OB, FC, FB). Based on the purpose of an FC, you will determine where it is called from and how often it is called. Conveyor control logic, for example, might be called from OB 1 whenever the conveyor system is enabled. On the other hand, an FC written to scale analog inputs might be called several times in one cycle, based on the number of input channels that need scaling.

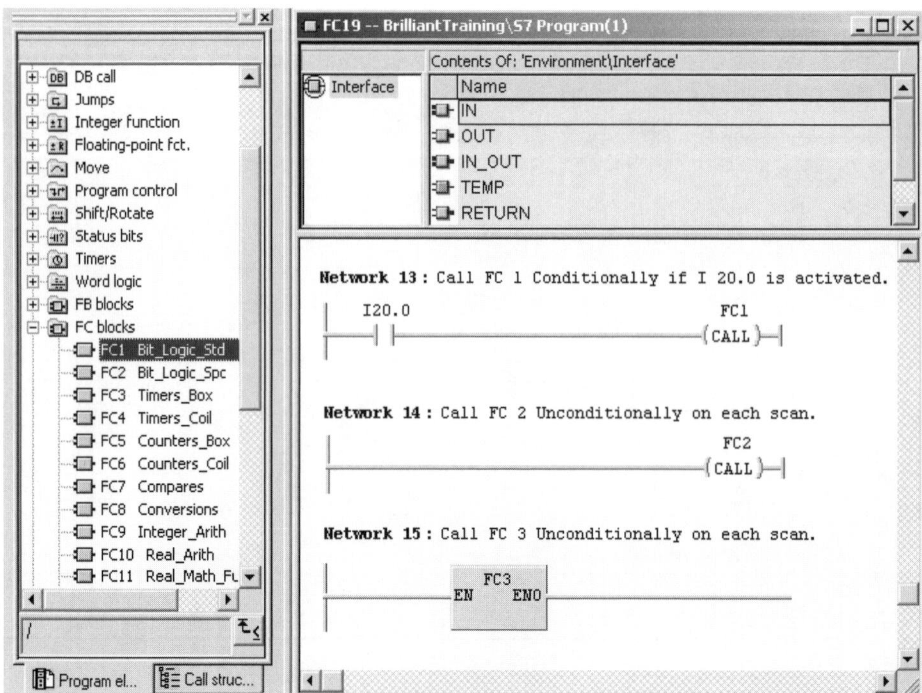

Figure 4-47. Calling FCs written without Formal Parameters in LAD, using Call Coil or Call Box.

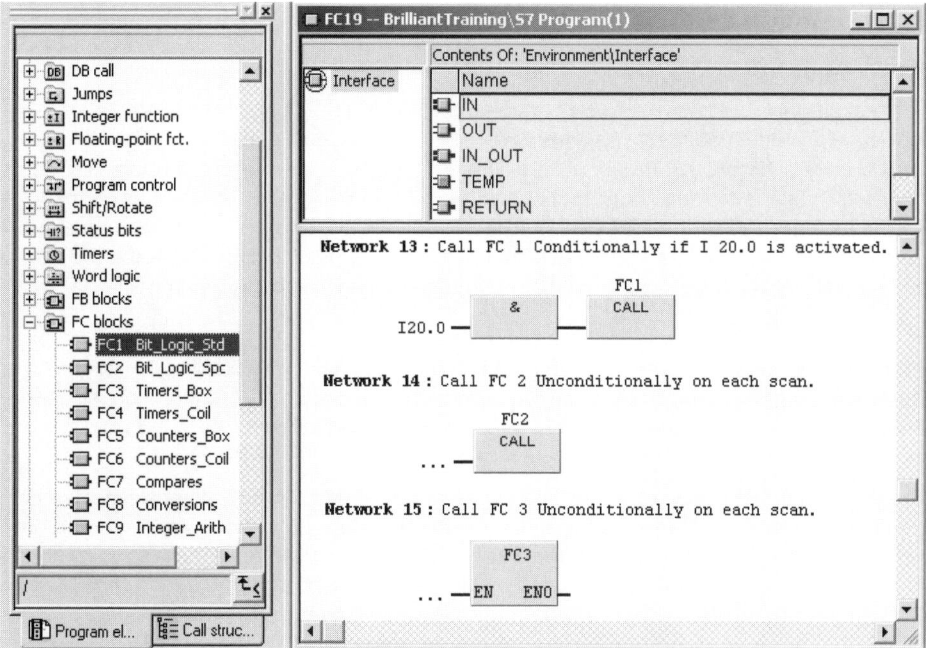

Figure 4-48. Calling FCs written without formal parameters in FBD.

Quick Steps: Calling an FC without Formal Parameters

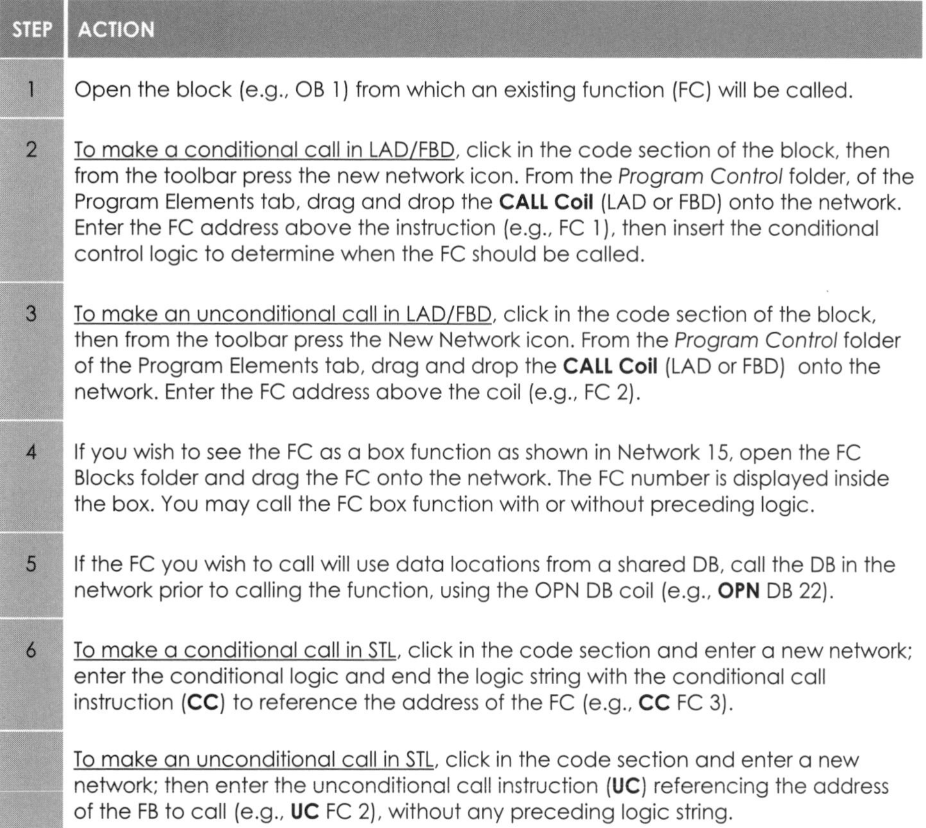

STEP	ACTION
1	Open the block (e.g., OB 1) from which an existing function (FC) will be called.
2	To make a conditional call in LAD/FBD, click in the code section of the block, then from the toolbar press the new network icon. From the *Program Control* folder, of the Program Elements tab, drag and drop the **CALL Coil** (LAD or FBD) onto the network. Enter the FC address above the instruction (e.g., FC 1), then insert the conditional control logic to determine when the FC should be called.
3	To make an unconditional call in LAD/FBD, click in the code section of the block, then from the toolbar press the New Network icon. From the *Program Control* folder of the Program Elements tab, drag and drop the **CALL Coil** (LAD or FBD) onto the network. Enter the FC address above the coil (e.g., FC 2).
4	If you wish to see the FC as a box function as shown in Network 15, open the FC Blocks folder and drag the FC onto the network. The FC number is displayed inside the box. You may call the FC box function with or without preceding logic.
5	If the FC you wish to call will use data locations from a shared DB, call the DB in the network prior to calling the function, using the OPN DB coil (e.g., **OPN** DB 22).
6	To make a conditional call in STL, click in the code section and enter a new network; enter the conditional logic and end the logic string with the conditional call instruction (**CC**) to reference the address of the FC (e.g., **CC** FC 3).
	To make an unconditional call in STL, click in the code section and enter a new network; then enter the unconditional call instruction (**UC**) referencing the address of the FB to call (e.g., **UC** FC 2), without any preceding logic string.

Programming an FC with Formal Parameters

Basic Concept

By using *formal I/O parameters*, you can develop FCs as a black-box operation, similar to a box addition instruction, that can be called whenever needed — supplying different values or addresses on each call. The formal parameters of the FC will represent the standard inputs and outputs of the operation and act as address placeholders. An FC written, for example, to control a specific AC motor type would require formal parameters of the standard I/O of such a motor. On each call of the FC in your program, the actual input and output addresses for a specific motor are plugged into the FC in place of the formal parameters.

Essential Elements

Formal parameters are specified using the **IN**, **OUT**, and **IN_OUT**, declarations. An **IN** declaration represents an input to the FC, a value to be read; an **OUT** declaration represents an output from the FC, a value to be written; the **IN_OUT** declaration represents both an input to and output from the FC, a value that is read and written by the FB.

The example AC motor FC might include inputs for the **Start PB**, **Stop PB**, **Emergency Stop PB**, and **Holding Coil** contacts, and outputs for the **Motor** starter, and a **Status** light. If temporary variables are required, they are specified using **TEMP** declarations. After defining as many formal parameters and temporary variables as required, the block code is then written using one or more networks just as you would normally, but instead of using actual input and output addresses, you will enter the formal parameters instead (See Figure 4-49).

Application Tips

Use of formal parameters is a great advantage in that while the actual code is written once, the FB may be called repeatedly. Memory use is also minimized and errors are reduced.

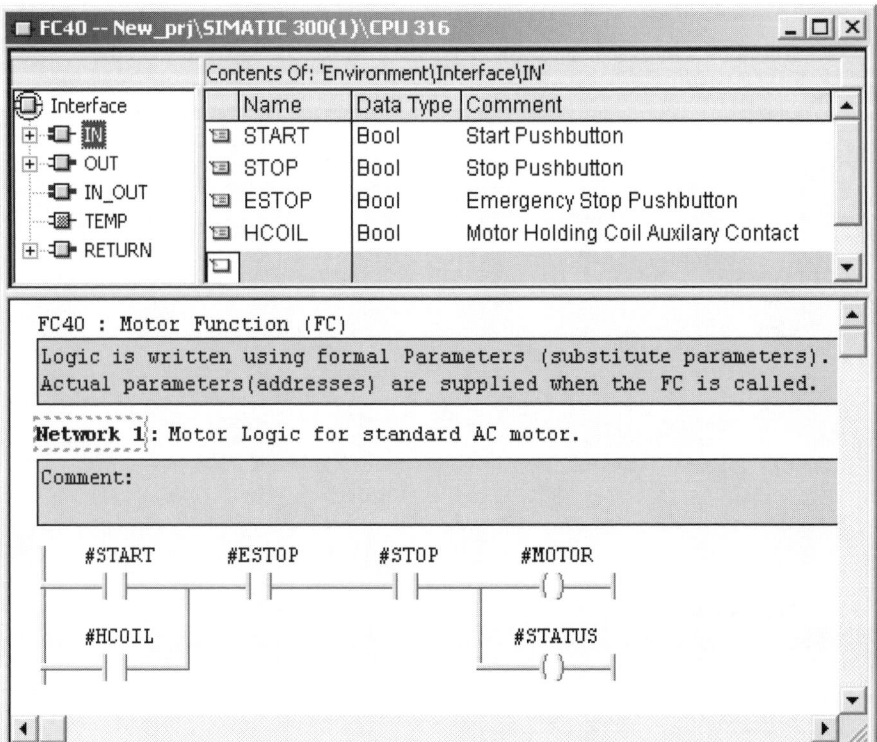

Figure 4-49. FC 40 Motor Control Logic: IN formal parameters shown in declaration table.

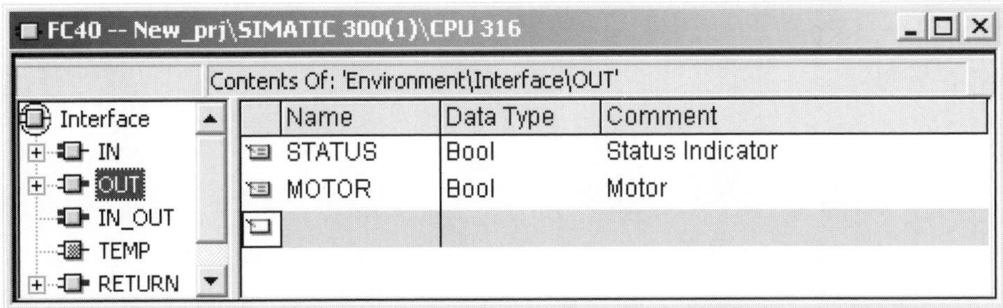

Figure 4-49a. FC 40 Motor Control Logic: OUT formal parameters shown in declaration table.

Quick Steps: Programming an FC with Formal Parameters

STEP	ACTION
1	From the SIMATIC Manager, open the project and the required program *Blocks* folder, and then double-click on the FC to open the in the LAD/FBD/STL editor. Create the FC if necessary.
2	From the menu, select **View** ➢ then the programming language to use (e.g., **LAD**); then select **View** ➢ **Overview** from the menu if the Program Elements/Call Structure window is not in view.
3	To declare a new formal parameter: From the left pane of the declaration table, select the parameter type you wish to declare (i.e., **IN**, **OUT**, or **IN_OUT**), then click in the **Name** field of the first declaration row and enter the name for the variable. Click in the **Data Type** field and select the data type from the drop list. Finally, click in the **Comment** field and enter a descriptive comment. Enter as many variables of a particular parameter type before moving to a new formal parameter type.
4	To declare a new temporary variable, first select the **TEMP** object in the left pane of the declaration table; then click in the **Name** field of the first declaration row and enter the name for the variable; click in the **Data Type** field and select the data type for from the drop list. Finally, click in the **Comment** field and enter a descriptive comment. Enter as many temporary local variables as required.
5	To insert a formal parameter or temporary variable before an existing variable: First select the TEMP object or the desired formal parameter type (i.e., **IN**, **OUT**, or **IN_OUT**) from the left pane, to display the current records of that type. Then, select an existing variable row, right click and select **New Declaration Row**. The new row is inserted before the row you selected. You may also select an existing variable and drag it to a new position.
6	To enter code for the block, click inside the code section; starting with network 1, write the code for the device or operation as you would normally, using the formal parameters where you would normally use absolute addresses (See Figure 4-49).
7	From the menu, select **File** ➢ **Save** when done with the last network.
8	Select **File** ➢ **Exit** to close the editor.

Calling an FC with Formal Parameters

Basic Concept

An FC you developed with *formal parameters* is meant to be called multiple times throughout your program. For example, the FC written in the previous task was developed to control AC motors of a specific type. Each time you call the FC to control a different motor, you must supply a new set of addresses (i.e., *actual parameters*) for that motor.

Essential Elements

When called in LAD or FBD, an FC with parameters appears as a box containing its input parameters on the left side of the box and output parameters on the right side. Each parameter has a name that reflects the actual input or output for which it is a placeholder. The AC motor control Function includes **START**, **STOP**, **ESTOP**, and **HCOIL** inputs; **MOTOR**, and **STATUS** are outputs. On each call of the FC, you must supply an absolute or symbolic address at each parameter. As you use an FC with formal parameters, you must ensure that the correct addresses are entered at the appropriate parameters and are of the same data type (e.g., BOOL, WORD, INT, and DINT). In this example, all of the parameters are of data type BOOL, and require that addresses of bit variables be specified (e.g., I 4.1, or Q 5.1).

Application Tips

A system function (SFC) with parameters is handled in the same manner as a user-developed function (FC) with parameters. In LAD/FBD, the EN/ENO signals allow the FC or SFC with parameters to be used in a network like any other box operation. To use an SFC or existing FC correctly, you will need appropriate documentation to know the purpose and operation of the block. It is essential to know the type of memory variable (e.g., I, Q, M, PI, PQ, and DBDW) and data type (e.g., BOOL, DINT, INT, and CHAR) expected at each parameter.

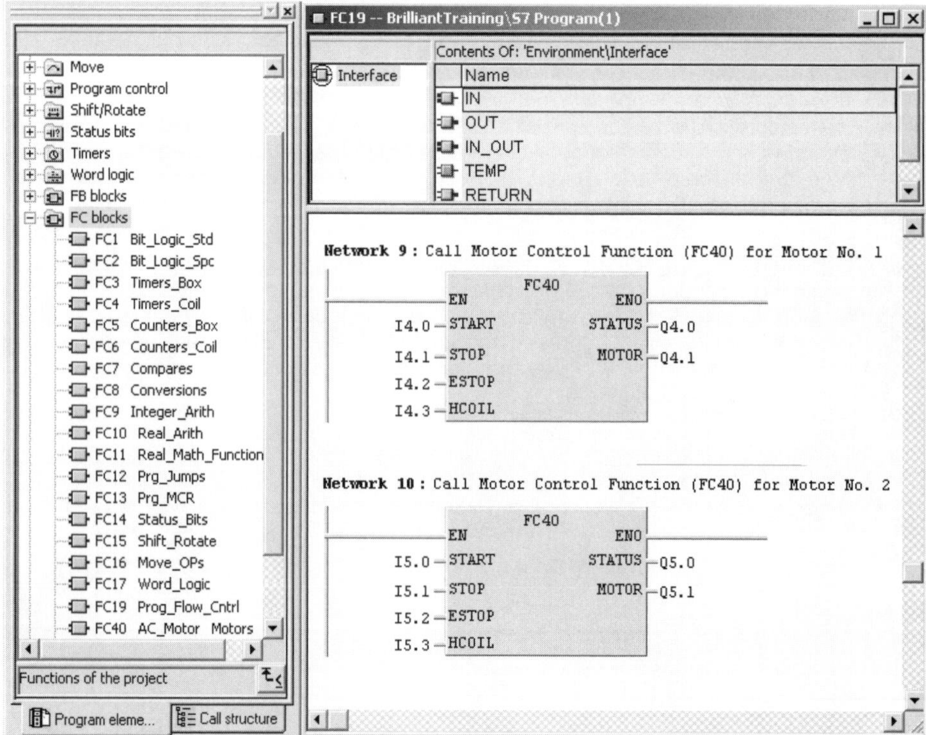

Figure 4-50. Call from FC19 to FC40 (Motor Control Logic) for two motors.

Quick Steps: Calling an FC with Formal Parameters

STEP	ACTION
1	Open the desired S7 program and the calling block (e.g., FC5) from where you will make a call to an existing function (FC) with parameters.
2	From the **View** menu, select **LAD** as the programming language to use.
3	If the instruction browser is not already in view, from the menu select **View** ➢ **Overview**.
4	To make first call in LAD, click in the code section of the block, then from the toolbar press the New Network icon to enter a new network in which to call the FC.
5	From the Program Elements tab, open the **FC Blocks** folder to display all user-created Functions (FCs) of the currently opened S7 program; locate the desired FC number (e.g., FC 40), then drag and drop the FC onto the network.
6	For each parameter, enter the actual addresses for the first call of the operation. Insert logic in the **EN** input line for a conditional call to the FC; insert the FC without logic conditions preceding the **EN** input line for an unconditional call.
7	To make second call in LAD, press the **New Network** button to insert a new network in which to call the function. Repeat the programming process, of steps 5 and 6, but supply the next set of actual addresses (absolute or symbolic) for the next device or operation.
8	From the menu, select **File** ➢ **Save** when done with the last network.
9	Select **File** ➢ **Exit** to close the editor.

Programming an FB without Formal Parameters

Basic Concept

An FB written without formal parameters is typical in subroutines involving data operations, where one or more values must be maintained, for use on the next call of the FB. Examples might include an FB that performs analog input filtering or that totals line production counts. On each call of a totalizer routine, the total value is either incremented or decremented by adding or subtracting one from the previous total value. In the filtering routine, consecutive readings of an analog input are added and then divided by the number of samples taken.

Essential Elements

In an FB without formal parameters, the **IN**, **OUT**, and **IN_OUT** declarations are not used. To define variables that must be maintained as memory in the instance data block, you must declare them as static (**STAT**) variables. To define variables that are needed only while the block is processing you must declare temporary (**TEMP**) variables. Once you have defined the required variables, you may write the block code in the code section of the block.

Application Tips

FB50, shown in Figures 4-51 and 4-51a, uses a routine that filters four analog inputs to illustrate an FB written without formal parameters. A static variable is required for each channel (e.g., **CH0_ACC**, **CH1_ACC**, **CH2_ACC**, **CH3_ACC**), to allow successive readings of each input to be added. A static variable (**SMP_ACC**) is used to count the number of readings (i.e., how many times the FB is called) . After four readings are taken, the sum of each input is divided by four, to obtain an average reading. Each filtered result is then saved to bit memory.

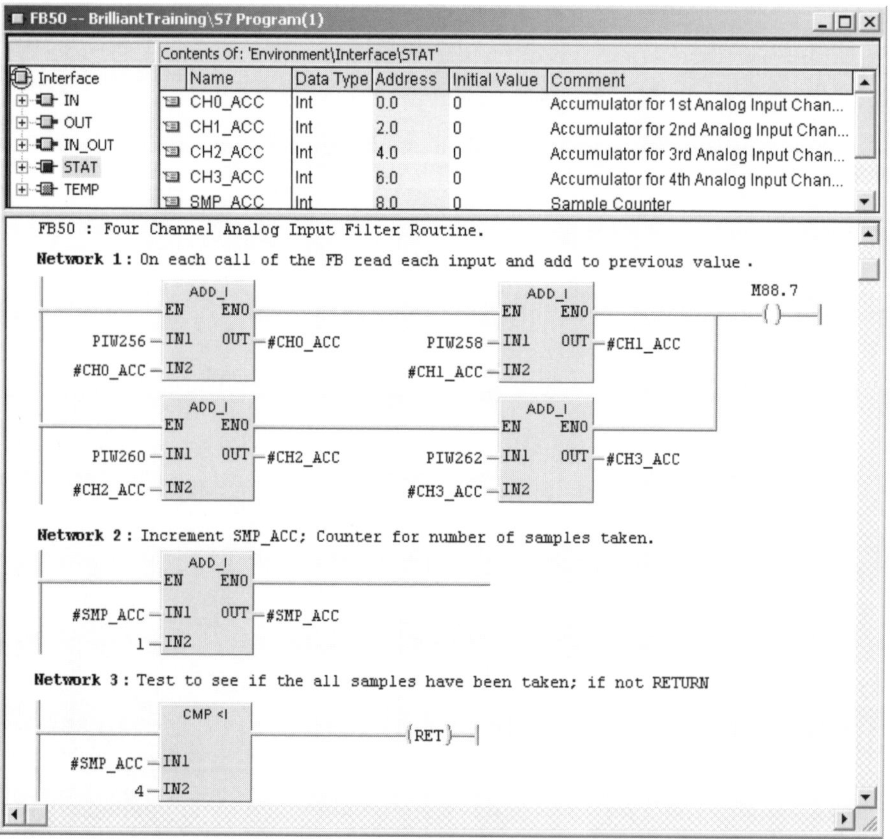

Figure 4-51. An FB without formal parameters, written as a 4-channel Analog Input Filter.

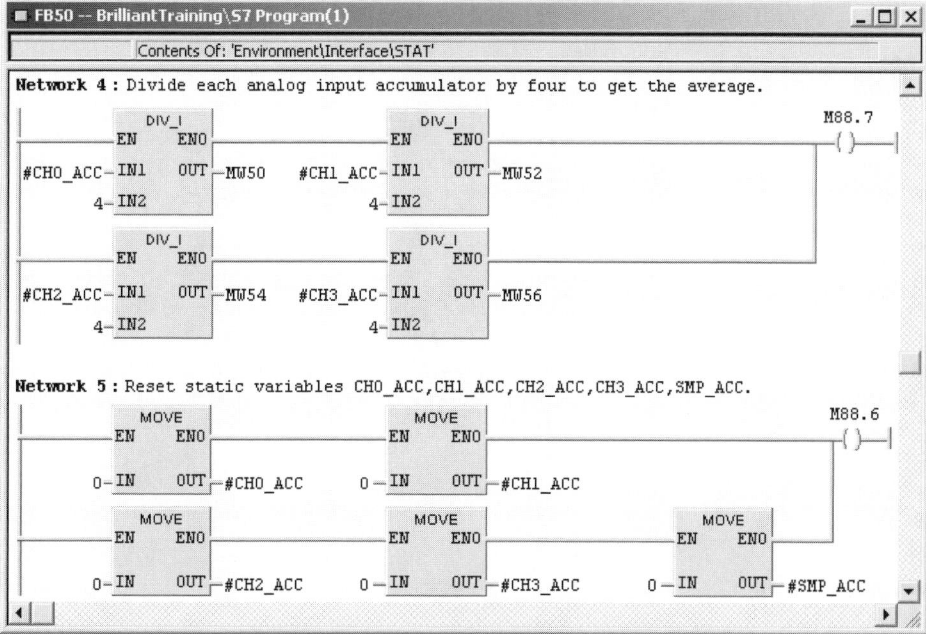

Figure 4-51a. Continuation of FB53: 4-Channel Analog Input Filter routine.

Quick Steps: Programming an FB without Formal Parameters

STEP	ACTION
1	From the SIMATIC Manager, open the project and the required program *Blocks* folder, and then double-click on the FB to open the in the LAD/FBD/STL editor.
2	From the menu, select **View** ➤ then the language of choice (e.g., **LAD** for ladder).
3	Select **View** ➤ **Overview** from the menu if the *Program Elements/Call Structure* window is not in view (LAD/FBD instructions are in categorized folders).
4	To declare a new temporary or static variable: First select the **TEMP** or **STAT** object in the left pane of the declaration table; then click in the **Name** field of the declaration row and enter the name for the variable. Click in the **Data Type** field and select the data type for from the drop list. Click in the **Comment** field and enter a descriptive comment. Enter as many temporary or static variables as required.
5	To insert a temporary or static variable before an existing variable: First select the **TEMP** or **STAT** object from the left pane of the declaration table, to display the current records of that type. Then, select an existing variable row, right click and select **New Declaration Row**. The new row is inserted before the row you selected.
6	To enter code for the block: Click inside the code section, starting with network 1, then write the code for each network operation, using either absolute addresses or symbolic addresses (if already defined). **Note**: Network 1 is automatically inserted.
7	To start a new network, press the **New Network** button from the toolbar.
8	From the menu select **File** ➤ **Save** when done with the last network; select **File** ➤ **Exit** to close the editor.

Calling an FB without Formal Parameters

Basic Concept

With the exception of organization blocks, all blocks must be called in order to be processed. A block is called by referencing its absolute address (e.g., FB 60) or symbolic address (e.g., "FILTER4CH") with the appropriate call instruction. A call to an FB can be unconditional, on every CPU cycle; or conditional, when specific logic conditions are met.

Essential Elements

In LAD and FBD, an FB with no parameters is called as a **CALL** box. All user-created function blocks, are located in the *FB Blocks* folder of the Program Elements tab. To call the FB, you must drag the box from the FB blocks folder and drop it onto the network. You must then specify the instance DB, belonging to the FB, above the box. The EN parameter allow you to call the FB with or without preceding logic. If the call is based on logic conditions, the FB is called when the conditions are satisfied; otherwise, the instruction immediately following the call is processed. In Statement List (STL), an FB without parameters and without any static variables can also be called conditionally, using the conditional call statement (e.g., **CC** FB 60); or unconditionally, using the unconditional call statement (e.g., **UC** FB 50). Since no static variables or parameters are used, no instance DB is used; nor does one need to be specified.

Application Tip

Calls to an FB may be from the main organization block (OB1) or from another block (e.g., OB, FC, FB). Based on the purpose of an FB, you will determine where it is called from and how often it is called. The analog filtering FB, for example, might be called from an FB responsible for processing analog I/O (e.g., reading, scaling, filtering), at periodic intervals. By supplying a different instance DB with each call, different data sets may be used.

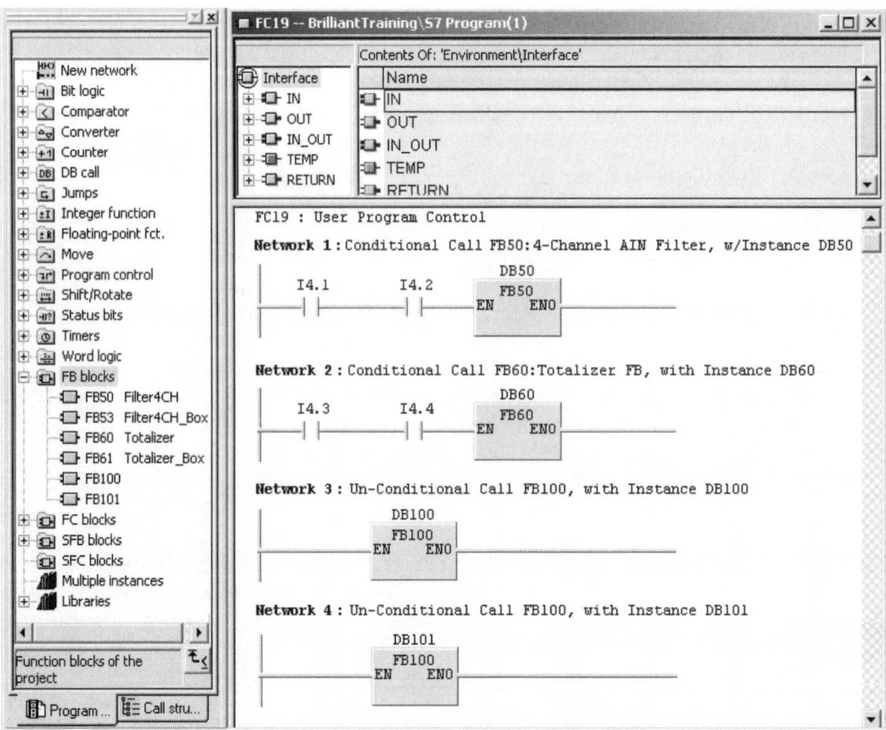

Figure 4-52. Calling an FB written without Formal Parameters in LAD/FBD, using CALL box.

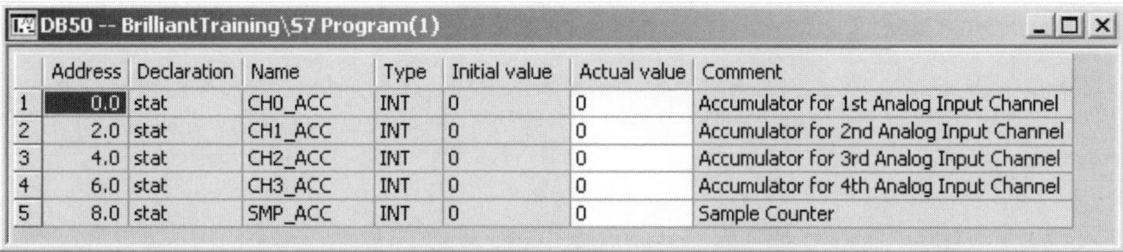

DB50 -- BrilliantTraining\S7 Program(1)

	Address	Declaration	Name	Type	Initial value	Actual value	Comment
1	0.0	stat	CH0_ACC	INT	0	0	Accumulator for 1st Analog Input Channel
2	2.0	stat	CH1_ACC	INT	0	0	Accumulator for 2nd Analog Input Channel
3	4.0	stat	CH2_ACC	INT	0	0	Accumulator for 3rd Analog Input Channel
4	6.0	stat	CH3_ACC	INT	0	0	Accumulator for 4th Analog Input Channel
5	8.0	stat	SMP_ACC	INT	0	0	Sample Counter

Figure 4-53. Instance DB 50, for FB50. The DB is opened automatically when the FB is called.

Quick Steps: Calling an FB without Formal Parameters

STEP	ACTION
1	Open the block (e.g., OB 1) from which an existing function (FB) will be called.
2	To make a conditional call in LAD, click in the code section of the block and enter a new network in which to call the function. To call the FB conditionally, enter the control logic to determine when the FB should be called, then from the *FB Blocks* folder of the instruction browser then drag and drop the **CALL COIL** at the end of the network. Enter the FB address above the coil (e.g., FB 1)
3	To make an unconditional call in LAD, click in the code section of the block and enter a new network in which to call the function. To call the FB unconditionally, simply open the *Program Flow Control* folder of the instruction browser then drag and drop the **CALL COIL** at the end of the network. Enter the FB address above the coil (e.g., FB 2)
	In Statement List (STL):
1	To make a conditional call in STL, click in the code section of the block and enter a new network in which to call the function; to call the FB conditionally enter the conditional logic and end the logic string with the conditional call instruction (**CC**) referencing the address of the FB to call (e.g., **CC** FB 1). No instance DB is needed.
2	To make an unconditional call in STL, click in the code section of the block and enter a new network in which to call the function; to call the FB unconditionally simply enter the unconditional call instruction (**UC**) referencing the address of the FB to call (e.g., **UC** FB 2), without any preceding logic string. No instance DB is needed.

Programming an FB with Formal Parameters

By using *formal I/O parameters*, you can develop FBs as a black-box operation, similar to a box addition instruction, that can be called whenever needed — supplying different values or addresses on each call. The formal parameters of the FB will represent the inputs and outputs of the operation and act as address placeholders. An FB, for example, to filter four analog input channels would require a formal input parameter to represent the address of each analog input, and an output parameter for each filtered result. On each call of the FB in your program, an actual address for each input and each output from the box is plugged into the FB in place of the formal parameters.

Essential Elements

Formal parameters are specified using the **IN**, **OUT**, and **IN_OUT**, declarations. The FILTER FB might include inputs for reading the analog input channels (e.g., **CH0_IN**, **CH1_IN**, **CH2_IN, and CH3_IN**); static variables for adding successive readings of each analog input (e.g., **CH0_ACC**, **CH1_ACC**, **CH2_ACC**, and **CH3_ACC**; outputs from the FB might include the filtered values (e.g., **CH0_AVG**, **CH1_AVG**, **CH2_AVG**, and **CH3_AVG**). If temporary variables are required, they are specified using **TEMP** declarations. After defining the formal parameters and temporary variables, the block code is then written using one or more networks just as you would normally, but instead of using actual input and output addresses, you will enter the formal parameters instead (See Figure 4-54).

Application Tips

Use of formal parameters is a great advantage in that while the actual code is written once, the FB may be called repeatedly. Memory use is also minimized and errors are reduced.

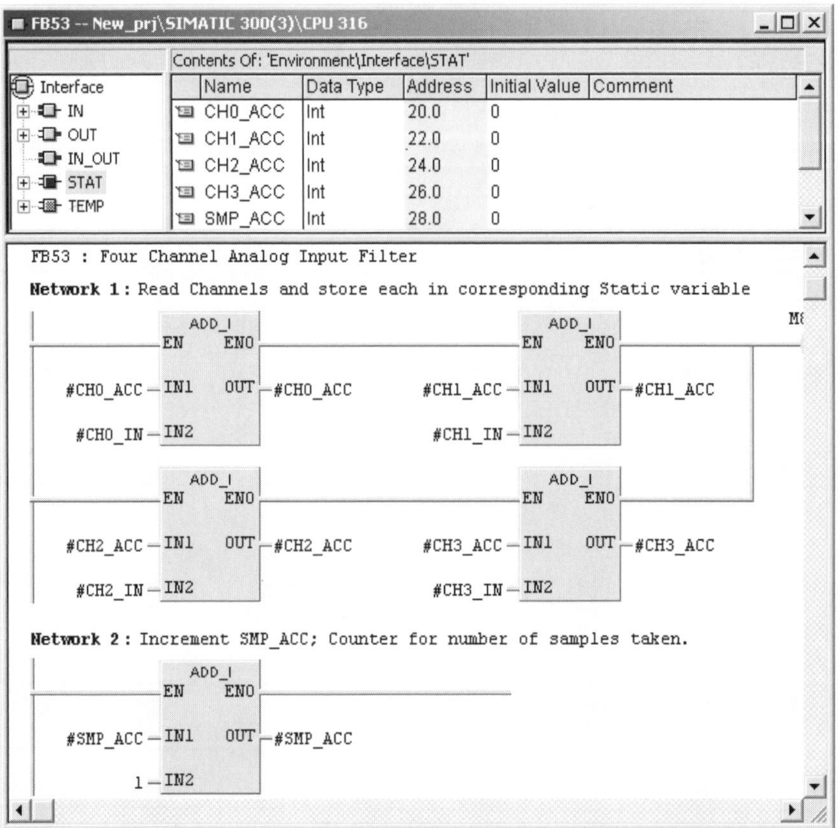

Figure 4-54. FB 53; 4-Channel Analog Input Filter Logic using Formal Parameters.

Quick Steps: Programming an FB with Formal Parameters

STEP	ACTION
1	From the SIMATIC Manager, open the project and the required program *Blocks* folder, and then double-click on the FC to open the in the LAD/FBD/STL editor.
2	From the menu, select **View** ➢ then the programming language to use (e.g., **LAD**); then select **View** ➢ **Overview** from the menu, if the Program Elements/Call Structure window is not in view.
3	To declare a new formal parameter: From the left pane of the declaration table, select the parameter type you wish to declare (i.e., **IN**, **OUT**, or **IN_OUT**), then click in the **Name** field of the first declaration row and enter the name for the variable. Click in the **Data Type** field and select the data type from the drop list. Finally, click in the **Comment** field and enter a descriptive comment. Enter as many variables of a particular parameter type before moving to a new formal parameter type.
	To insert a formal parameter before an existing variable: First select the desired formal parameter type (i.e., **IN**, **OUT**, or **IN_OUT**) from the left pane, to display the current records of that type. Then, select an existing variable row, right click and select **New Declaration Row**. The new row is inserted before the row you selected. You may also select an existing variable and drag it to a new position.
4	To declare a new temporary or static variable: First select the **TEMP** or **STAT** object in the left pane of the declaration table; then click in the **Name** field of the first declaration row and enter the name for the variable. Click in the **Data Type** field and select the data type for from the drop list. Finally, click in the **Comment** field and enter a descriptive comment.
5	To insert a temporary variable, or static variable before an existing variable: First select the **TEMP** or **STAT** object from the left pane, to display the current records of that type in the declaration table. Then, select an existing variable row, right click and select **New Declaration Row**. The new row is inserted before the row you selected. You may also select an existing variable and drag it to a new position.
6	To enter code for the block: Click inside the code section, starting with network 1, write the code for the device or operation as you would normally, using the formal parameters where you would normally use absolute addresses (See Figure 4-53).
7	From the menu, select **File** ➢ **Save** when done with the last network.
8	Select **File** ➢ **Exit** to close the editor.

Calling an FB with Formal Parameters

Basic Concept

An FC you developed with *formal parameters* is meant to be called multiple times throughout your program. For example, the FC written in the previous task was developed to control AC motors of a specific type. Each time you call the FC to control a different motor, you must supply a new set of addresses (i.e., *actual parameters*) for that motor.

Essential Elements

When called in LAD or FBD, an FC with parameters appears as a box containing its input parameters on the left side of the box and output parameters on the right side. Each parameter has a name that reflects the actual input or output for which it is a placeholder. The FB for filtering four analog inputs includes **CH0**, **CH1**, **CH2**, **CH3,** and **SAMPLE** inputs; **CH0_AVG**, **CH1_AVG**, **CH2_AVG**, **CH3_AVG,** and **DONE** are outputs. On each call of the FB, you must supply an absolute or symbolic address at each parameter. As you use an FB with formal parameters, you must ensure that the correct addresses are entered at the appropriate parameters and match the data type (e.g., BOOL, WORD, INT, and DINT) of the parameter. For example, the parameters that accept the analog inputs are of type INT. The filtered output is also of data type INT.

Application Tips

A system function block (SFB) with parameters is handled in the same manner as a user-developed function block (FB) with parameters. In LAD/FBD, the EN/ENO signals allow the FB or SFB with parameters to be used in a network like any other box operation. To use an SFB or existing FB correctly, you will need appropriate documentation to know the purpose and operation of the block. It is essential to know the type of memory variable (e.g., I, Q, M, PI, PQ, and DBDW) and data type (e.g., BOOL, DINT, INT, and CHAR) expected at each parameter.

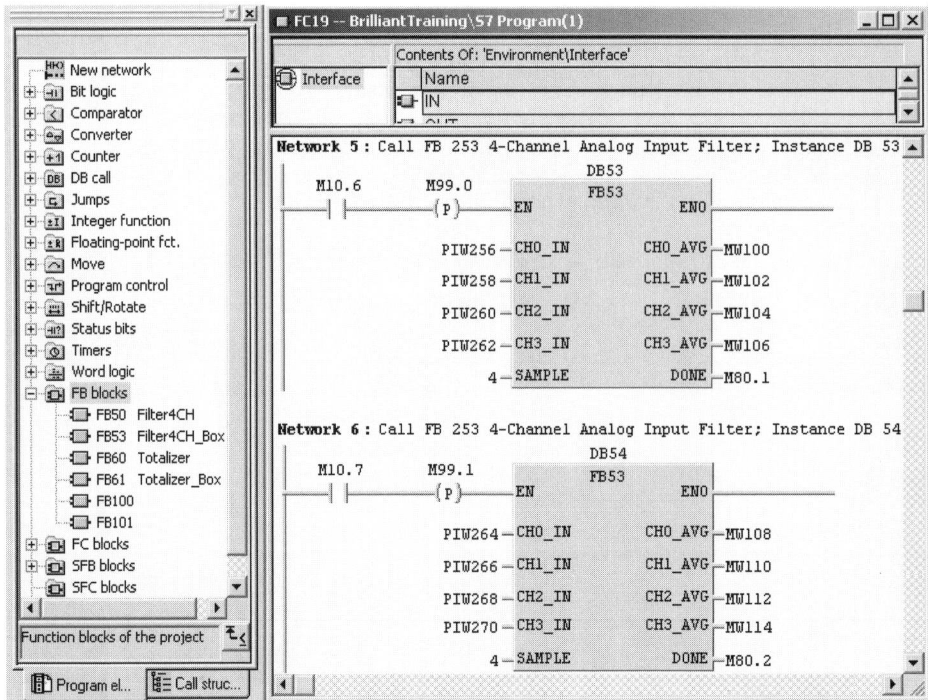

Figure 4-55. Call to FB 53 Filter; Four analog inputs are filtered on each call to the FB.

DB53 -- BrilliantTraining\S7 Program(1)

	Address	Declaration	Name	Type	Initial value	Actual value	Comment
1	0.0	in	CH0_IN	INT	0	0	Accumulator for 1st channel
2	2.0	in	CH1_IN	INT	0	0	Accumulator for 2nd channel
3	4.0	in	CH2_IN	INT	0	0	Accumulator for 3rd channel
4	6.0	in	CH3_IN	INT	0	0	Accumulator for 4th channel
5	8.0	in	SAMPLE	INT	4	4	Number of readings to take
6	10.0	out	CH0_AVG	INT	0	0	Filtered Result 1st Channel
7	12.0	out	CH1_AVG	INT	0	0	Filtered Result 2nd Channel
8	14.0	out	CH2_AVG	INT	0	0	Filtered Result 3rd Channel
9	16.0	out	CH3_AVG	INT	0	0	Filtered Result 4th Channel
10	18.0	out	DONE	BOOL	FALSE	FALSE	
11	20.0	stat	CH0_ACC	INT	0	0	Accumulator for 1st Channel
12	22.0	stat	CH1_ACC	INT	0	0	Accumulator for 2nd Channel
13	24.0	stat	CH2_ACC	INT	0	0	Accumulator for 3rd Channel
14	26.0	stat	CH3_ACC	INT	0	0	Accumulator for 4th Channel
15	28.0	stat	SMP_ACC	INT	0	0	Sample Counter

Figure 4-56. Instance DB 53, for FB53. The DB is opened automatically when the FB is called.

Quick Steps: Calling an FB with Formal Parameters

STEP	ACTION
1	Open the desired S7 program and the calling block (e.g., OB 1) from where you will make a call to an existing function block (FB) with parameters.
2	From the **View** menu, select **LAD** as the programming language to use.
3	From the **View** menu, select **Overview** if the instruction browser is not in view.
4	To make first call in LAD, click in the code section of the block, then from the toolbar press the New Network icon to enter a new network in which to call the function.
5	From the Program Elements tab, open the **FB Blocks** folder to display all user-created Function Blocks (FBs) of the currently opened S7 program; locate the desired FB number (e.g., FB 40), then drag and drop the FB onto the network.
6	For each parameter, enter the actual addresses for the first call of the operation. Insert logic in the **EN** input line for a conditional call to the FB; insert the FB without logic conditions preceding the **EN** input line for an unconditional call.
7	To make second call in LAD, insert a new network in which to call the function. Repeat the programming process, of steps 5 and 6, but supply the next set of actual addresses (absolute or symbolic) for the next device or operation.
8	From the menu, select **File** ➢ **Save** when done with the last network.
9	Select **File** ➢ **Exit** to close the editor.

Programming Basic Bit-Logic Operations

Basic Concept

Standard bit logic instructions provide instructions used in developing relay equivalent logic. By providing the ability to examine the ON/OFF status of any addressable bit variables, these instructions are combined to form series, parallel, and series-parallel combinations that determines how internal or external outputs are controlled. These instructions are also used in creating the driving logic to trigger operations like timers, counters, arithmetic calculations, program jumps, data move operations and communications.

Essential Elements

This group of operations allow the status of bit addressable variables or memory (e.g., I, Q, or M) to be examined using normally-open and normally-closed contacts; or controlled using the standard output or the set and reset outputs. Set and Reset outputs, provide the facility for latching and unlatching output signals. In essence, these operations allow you to develop logic to examine or influence the result of logic operations (RLO). For example, the NOT contact in LAD, and the Negate (NOT) input in FBD, allow the RLO to be inverted.

Application Tips

The following two figures illustrate basic logic combinations of series and parallel circuits. In FBD these combinations you will use the AND gate for series logic, and the OR gate for parallel logic. As many inputs as required may be input to a gate. Outputs may be placed in parallel in both LAD and FBD, however when working in LAD, the secondary outputs must not branch off the left power rail. Finally, if programming in STL, remember that when combining 'AND' and 'OR' circuits, that the OR must be performed first, which is ensured by placing the OR operation within parentheses.

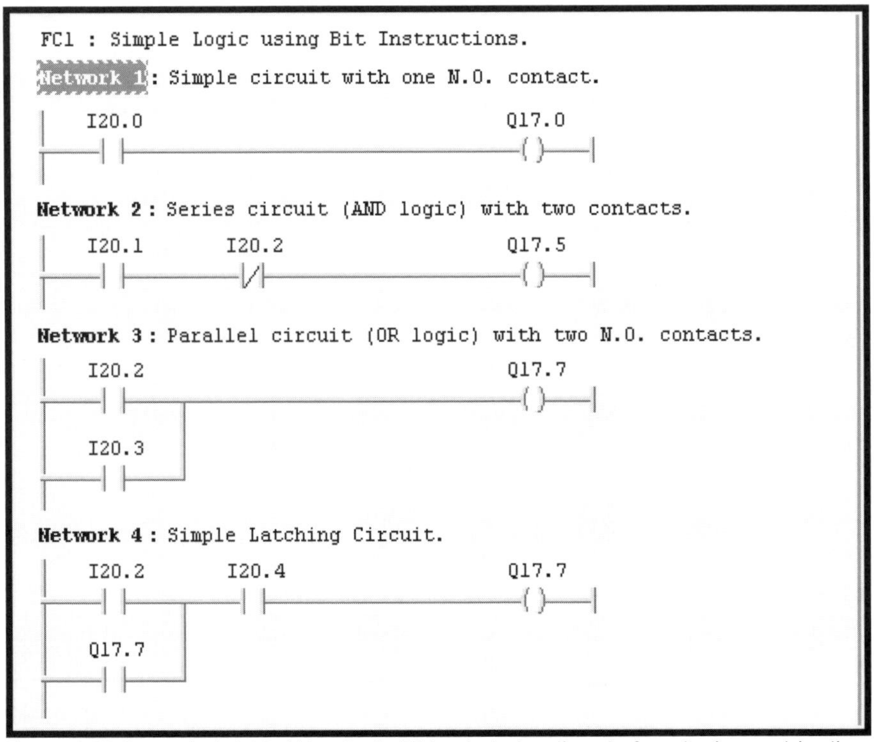

FC1 : Simple Logic using Bit Instructions.

Network 1: Simple circuit with one N.O. contact.

I20.0 Q17.0

Network 2 : Series circuit (AND logic) with two contacts.

I20.1 I20.2 Q17.5

Network 3 : Parallel circuit (OR logic) with two N.O. contacts.

I20.2 Q17.7

I20.3

Network 4 : Simple Latching Circuit.

I20.2 I20.4 Q17.7

Q17.7

Figure 4-57. Example in LAD: Using Basic Bit-Logic Instructions to form logic combinations.

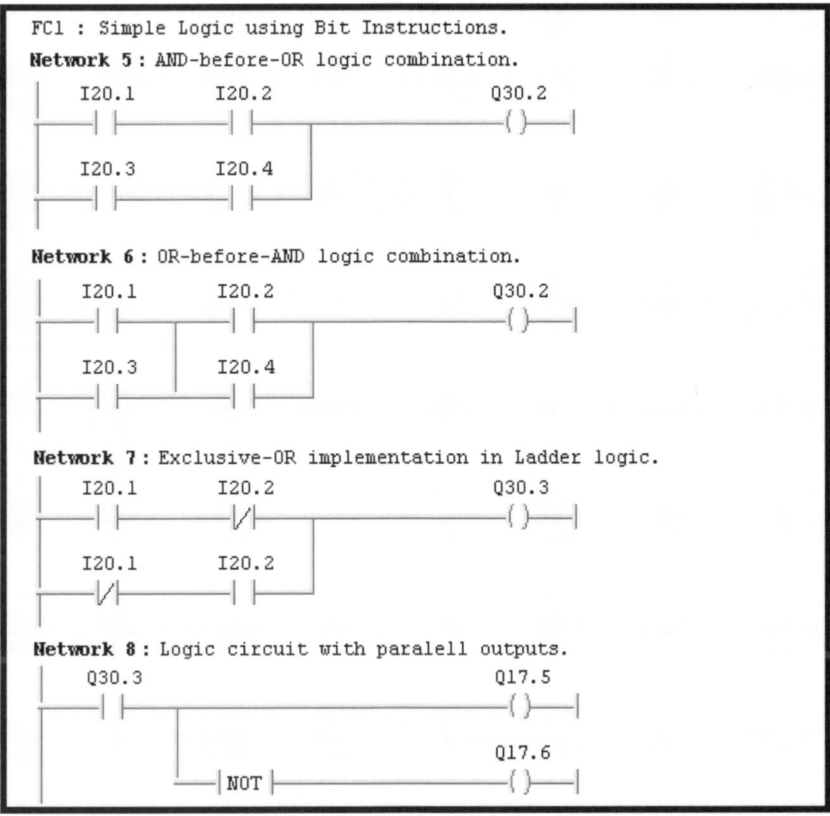

Figure 4-58. Example in LAD: Using Basic Bit-Logic Instructions to form logic combinations.

Guidelines: Programming Basic Bit-Logic Operations

No.	GUIDELINE
1	Open the *Bit Logic* folder of the Program Elements tab to access bit instructions like NO and NC contacts, NOT contact and coils.
2	Use toolbar buttons or hotkeys F2, F3, F7, F8, and F9 for frequently used operations like NO and NC contacts, branch open, branch close and coil.
3	Examine logic 0 or logic 1 status of memory operands including Inputs (I), Outputs (Q), memory bits (M), global DB bits (DBn.DBXy.x), temporary local data bits (Ln.y), static local data bits, timer status (Tn), counter status (Cn), or CPU status bits.
4	Use normally-open contact to examine for logic 1 to allow power flow; use normally-closed contact to examine for logic 0 to allow power flow.
5	Combine bit-logic operations to create series (AND), parallel (OR), and series-parallel (AND-before-OR, OR-before-AND) logic combinations (See Networks 1-7).
6	Combine multiple coils in parallel (See Network 8). Contacts may also be placed in the parallel coil branch, but only preceding the coil instruction (See Network 8).
7	Use the NOT contact to invert the result of logic operation at the point of insertion, or as the last contact to invert the normal operation of a coil (See Network 8).

Programming Set-Reset Operations

Basic Concept

The *Set* and *Reset coils* are used when a bit location must be set to logic 1 when the driving logic is logic 1, yet must remain set until reset to logic 0; even if the driving logic returns to logic 0. The set and reset operations depending on your requirements or preferences, can be programmed using two separate output instructions or using the Set/Reset or the Reset/Set flip-flop, which combine both operations in a single box function.

Essential Elements

The Set and Reset coils are generally programmed to operate as a pair, but are programmed in separate networks that each reference the same bit address. When power flows to the Set coil, the address is set to logic 1. When power flows to the Reset coil, the address is set to logic 0. No power flow to the Set or the Reset output, does not affect the signal state of the bit address. Whichever of the two operations is programmed last is considered the dominant of the two; this is true since when both coils are activated simultaneously, the one processed last overrides the status determined by the other.

The functionality of the *SR flip-flop* and the *RS flip-flop* is the same as the individual instructions except that both functions are handled in a single box instruction. The dominance of the set or reset in a flip-flop operation is based on which of the two input lines is processed last in the box. In the **RS** flip-flop, the **S** is processed last, and therefore is dominant should both occur simultaneously. In the **SR** flip-flop, the **R** is processed last, and therefore is dominant should both occur simultaneously.

Application Tips

Like standard outputs and other bit logic operations, set and reset coils and flip-flop instructions are generally driven by a combination of series-parallel logic that controls bit-addressable locations. Logic may be inserted preceding or following set/reset flip-flops. In FBD, an AND gate is inserted to create series logic; an OR gate is inserted for parallel logic. Set/reset coils may be paralleled in both LAD and FBD, but in LAD, parallel branches for secondary coils must not start at the left power rail.

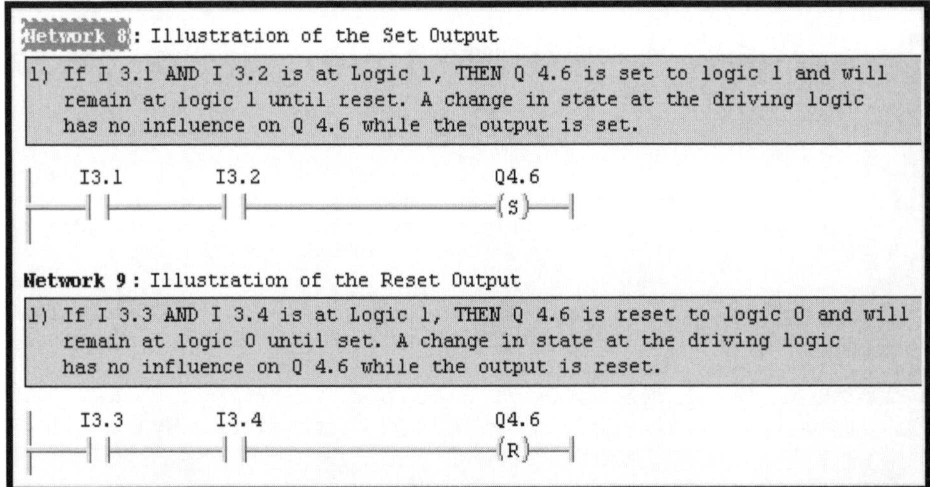

Figure 4-59. Example in LAD: Using Set and Reset Output Instructions.

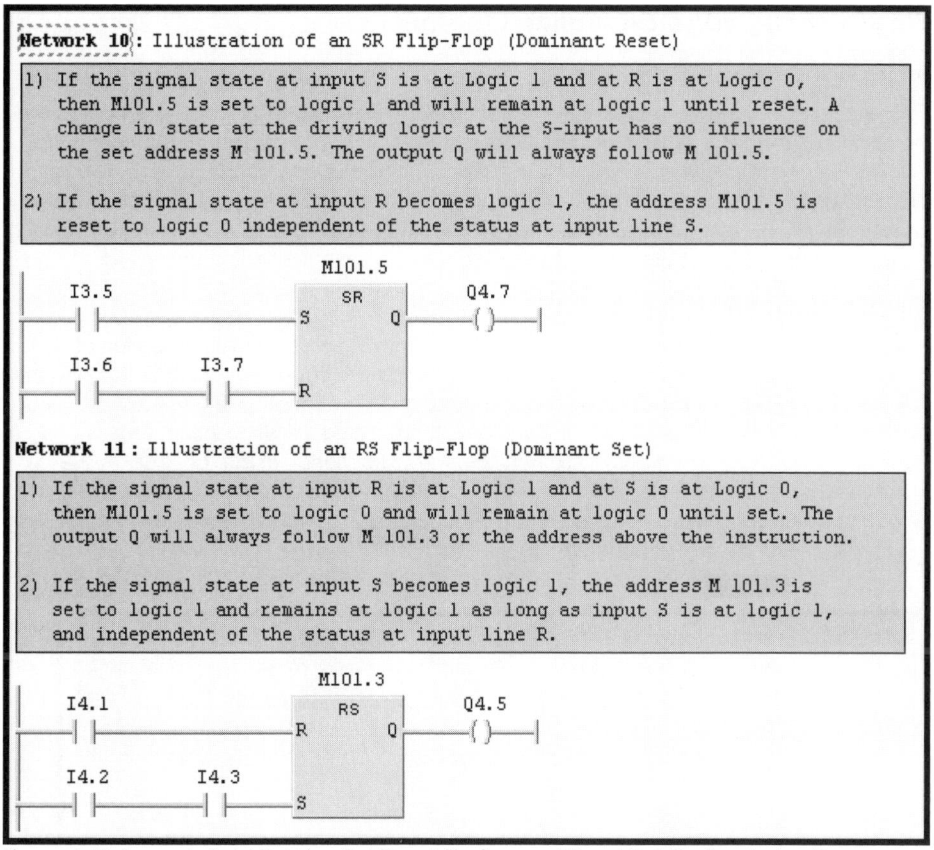

Network 10: Illustration of an SR Flip-Flop (Dominant Reset)

1) If the signal state at input S is at Logic 1 and at R is at Logic 0, then M101.5 is set to logic 1 and will remain at logic 1 until reset. A change in state at the driving logic at the S-input has no influence on the set address M 101.5. The output Q will always follow M 101.5.

2) If the signal state at input R becomes logic 1, the address M101.5 is reset to logic 0 independent of the status at input line S.

M101.5
I3.5 SR Q4.7
S Q
I3.6 I3.7
R

Network 11: Illustration of an RS Flip-Flop (Dominant Set)

1) If the signal state at input R is at Logic 1 and at S is at Logic 0, then M101.5 is set to logic 0 and will remain at logic 0 until set. The output Q will always follow M 101.3 or the address above the instruction.

2) If the signal state at input S becomes logic 1, the address M 101.3 is set to logic 1 and remains at logic 1 as long as input S is at logic 1, and independent of the status at input line R.

M101.3
I4.1 RS Q4.5
R Q
I4.2 I4.3
S

Figure 4-60. Example in LAD: Using Set/Reset flip-flop and Reset/Set flip-flop Instructions.

Guidelines: Programming Set-Reset Operations

No.	GUIDELINE
1	Set and reset operations are in the *Bit Logic* folder of the Program Elements tab.
2	Use the Set/Reset coils as a pair to provide the latch/unlatch functions. Both operations should reference the same address.
3	Use the SR or RS flip-flop box instruction to implement both the latching and unlatching function.
4	Combine series (AND) and parallel (OR) logic combinations, to drive the set/reset coils and SR or RS flip-flop operations.
5	SR or RS flip-flop operations may be placed in series and in parallel and combined with contact instructions placed before or after the flip-flop operation.
6	Set and reset coils and SR and RS flip-flop operations inside an MCR zone will maintain their last state when the MCR is not activated.
7	Like standard coils, set and reset coils may be combined as parallel coils and contacts may be placed in the branch preceding the set or reset coil.

Programming Edge Evaluation Operations

Basic Concept

Edge evaluation operations are mainly called upon to detect and indicate when a discrete transition occurs (e.g., motor switched off). These operations are also often used to ensure that certain operations (e.g., arithmetic calculation, or move operation) are triggered for processing only once on the transition of some event from logic 0 -to- logic 1(positive-edge), or from logic 1-to- logic 0 (negative edge). With the these operations, the need to develop complex binary logic to detect negative and positive edge transitions is eliminated.

Essential Elements

There are two types of edge detection operations, both of which support detection of negative and positive edge transitions. The *Address Positive-edge Detect* instruction and the *Address Negative-Edge Detect* instructions look for an edge transition in a specific bit address. The second edge evaluation type looks for an edge transition in the result of a logic operation (RLO) at any point in a logic network. The instructions include the *RLO Positive-Edge Detect* and the *RLO Negative-Edge Detect*. Both types of operations signal that an edge has been detected by causing a one-shot pulse at its output. The indicating signal is at logic 1 for one CPU scan, and then returns to logic 0.

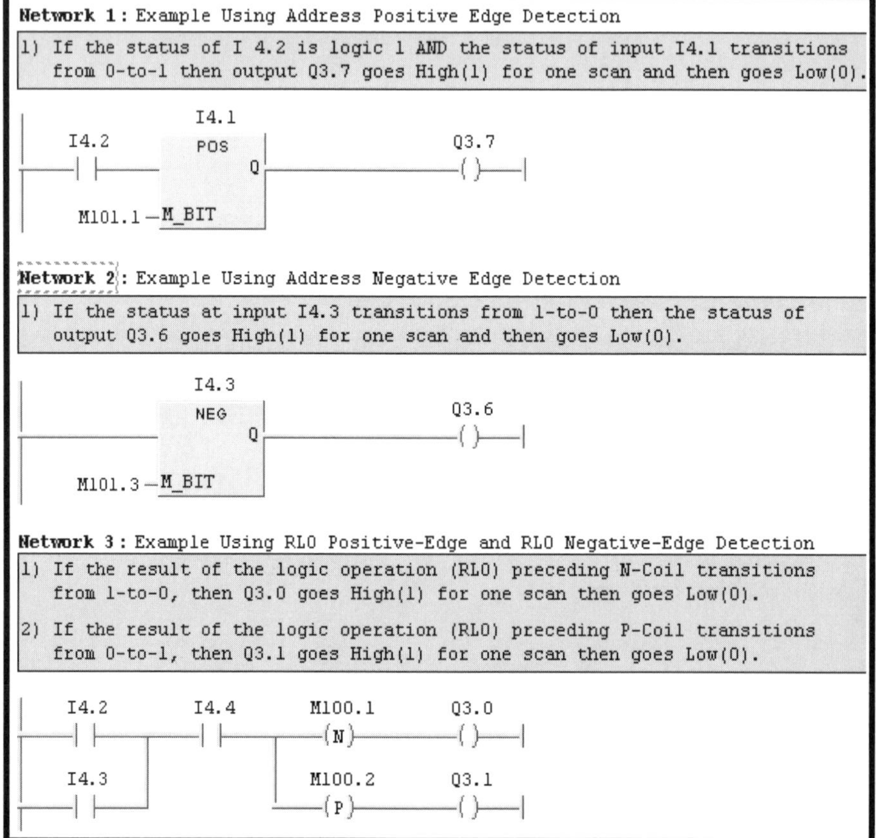

Figure 4-61. Example in LAD: Address and RLO Positive- and Negative-Edge Operations.

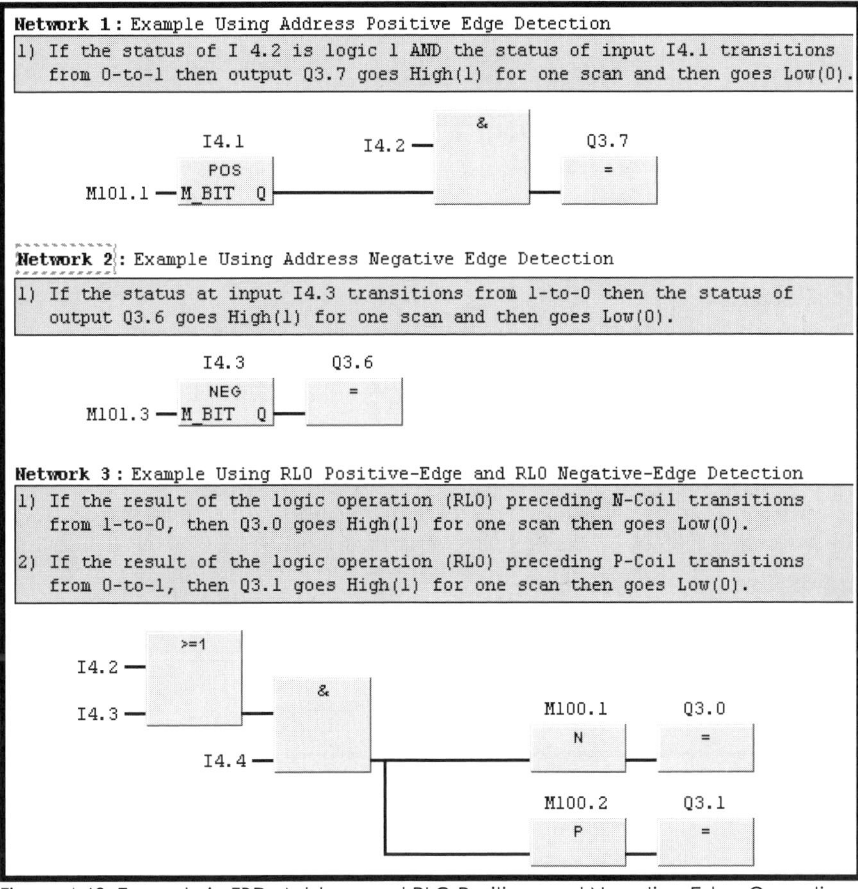

Figure 4-62. Example in FBD: Address and RLO Positive- and Negative-Edge Operations.

Guidelines: Programming Edge Evaluation Operations

No.	GUIDELINE
1	Access address edge detect and RLO edge detect edge evaluation instructions in the *Bit Logic* folder of the Program Elements tab.
2	Use the address edge detect operations to detect a signal change in a specific address; use the RLO edge detect operations to detect a signal change in the RLO at the point of insertion of the instruction.
3	The address edge detect operations may evaluate a specific address alone or in combination with other logic conditions(See Network 1 and Network 2); the instructions may be programmed in series and in parallel similar to a contact instruction.
4	The RLO edge detect operations must be placed after preceding logic conditions; the instructions may be placed in parallel branches, but only in a branch that starts at the left power rail.

Programming Counter Operations

Basic Concept

Programming a counter is first a matter of selecting the required counter type from the Program Elements tab, based on the name at the inside top of the box (e.g., **S_CU** is an UP counter), and placing it into the LAD or FBD network. The counter address must be entered just above the counter, using the absolute address (e.g., **C0**, **C1**, **C2**) or the symbolic address (e.g., AmTotal) if already defined. Inserting the driving logic for counter inputs (CU or CD, S, and R), and entering the output addresses for the current count is all that is required.

Essential Elements

A box counter has one count input line (**CU** or **CD**) or two (**CU** and **CD**) in the case of the UP/Down counter. A counter only increments or decrements on the logic 0 -to- logic 1 transition of CU or CD. When a logic 0 -to- logic 1 transition occurs on the set input **S**, the value specified at **PV** is loaded to the counter word as a start or new current count. PV is generally used as the start value for a down counter. Logic 1 at the reset line **R** resets the counter word to zero. On the output side, the status signal **Q**, is set to logic 1 whenever the count is greater than zero and is otherwise at logic 0. Finally, word locations specified at **CV** and at **CV_BCD**, allow the current count to be output in decimal and in BCD respectively.

Application Tips

A counter contains a WORD value in BCD and either counts from 000 to 999, or from 999 down to 000. The PV entry may be transferred to the counter at any time as a starting value, or as an override value. PV may be specified as a BCD constant (e.g., C#525), as a local variable of data type WORD, or as a word-width memory location (absolute or symbolic).

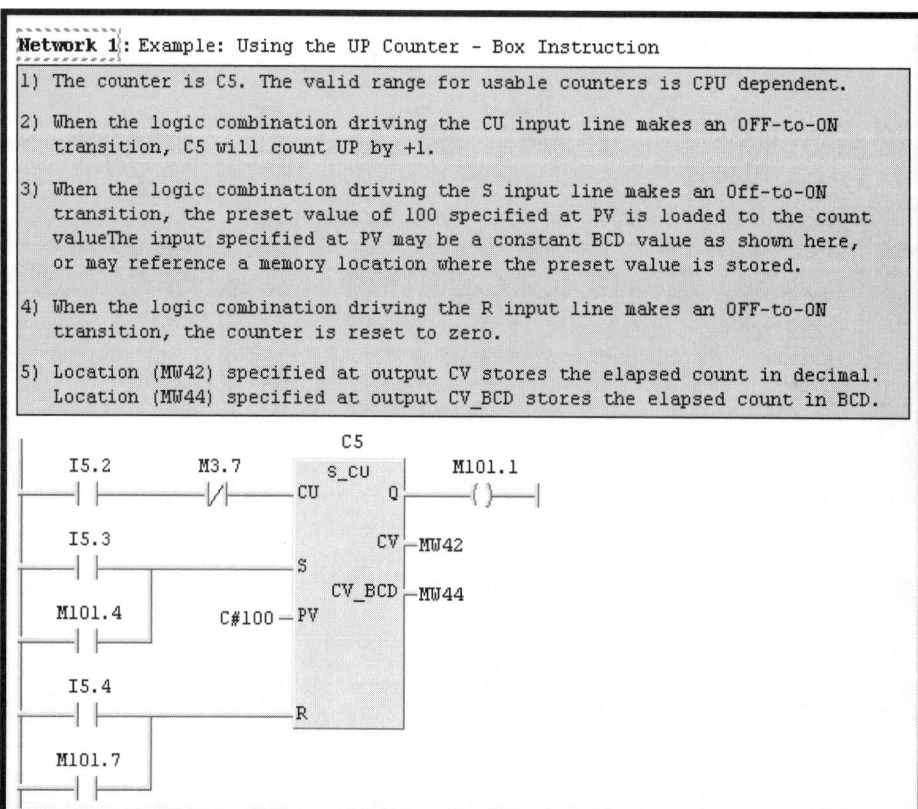

Figure 4-63. Example in LAD: Programming a Counter box instruction.

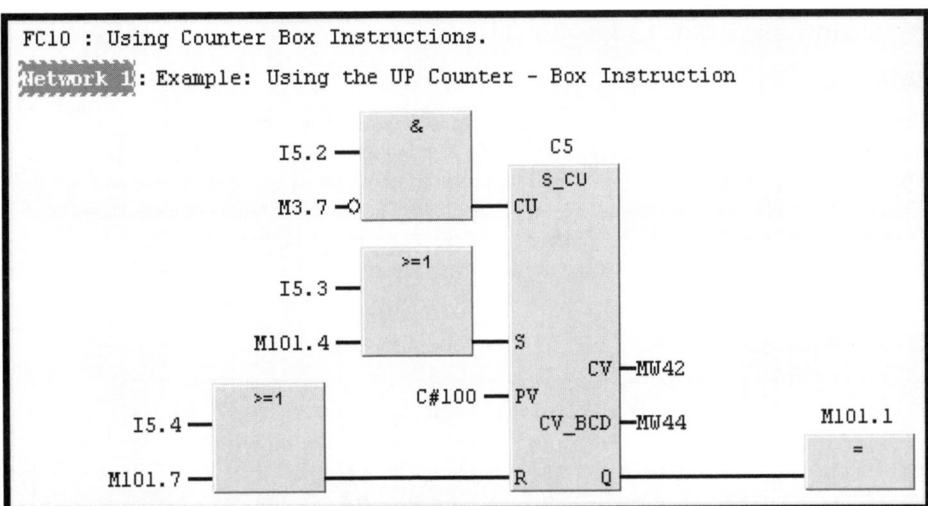

Figure 4-64. Example in FBD: Programming a Counter box instruction.

Guidelines: Programming Counter Operations

No.	GUIDELINE
1	Counter box and coil operations are in the *Counters* folder of the Program Elements tab.
2	Counters are addressed **C0, C1, C2**, and so on. Available number of timers is CPU dependent (e.g., C0 to C255)
3	Contacts may be arranged in series/parallel at each counter input (i.e., **CU**, **CD**, **S**, **R**), to achieve the desired control logic, as well as after the status signal (**Q**).
4	PV accepts valid BCD or Hex constants as preset count values (e.g., C#525 or W#16#0525). The maximum counter range and value for PV is 000 to 999.
5	PV may also be set as a variable by designating a valid word location (e.g., MW 40, IW 10) or using a local variable of data type WORD.
6	A counter is incremented (CU) or decremented (CD) by one, only when the RLO on the count input line or preceding the counter coil changes from logic 0 to logic 1.
7	You may check the status of a counter, using N.O. or N.C. contacts that reference the counter number (i.e., C4). The status Q is logic 1 whenever the counter value not equal to zero.
8	You can check the value of the counter at intermediate points for control or other purposes, using the compare operation and comparing the values at CV (decimal) or CV_BCD (BCD).

Programming Timer Operations

Basic Concept

Programming a timer is first a matter of selecting the required timer type from the Program Elements tab, based on the name at the inside top of the box (e.g., **S_ODT** is an on-delay timer), and placing it into the LAD or FBD network. The timer address must be entered just above the timer, using the absolute address (e.g., **T0, T1**, **T2**) or the symbolic address (e.g., Strt_Delay_1) if already defined. Inserting the timer start logic, specifying a preset value, inserting the reset logic, and entering output addresses for the remaining time value is all that remains.

Essential Elements

Inputs to a box timer include the timer address, specified above the box; a start logic **S** line, for entering the control logic that starts the timer; **TV**, for specifying a preset value to load to the timer; and a reset logic line **R**, to reset the timer. On the output side of the box, the timer status signal **Q**, signals the running status of the timer; and **BI** and **BCD**, are both for specifying where the current (or remaining) time is output. The remaining time is output as a decimal value at the location specified at BI and in BCD at the location specified at BCD.

Application Tips

The timer value **TV** may be specified as a constant using the S5TIME format; the range is S5T#0h_0m_0s_10ms to S5T#2h_46m_30s_0ms, where h = hour, m = minutes, s = seconds, and ms = milliseconds. Only the required units of the preset time need to be specified. **TV** may also reference a declared WORD variable, or a word-width memory location (e.g., MW 102) that contains the preset. The status signal **Q**, whose operation varies depending on the timer type, may drive a coil referenced with any unused bit (typically from bit memory). The status bit can be checked in your program using a NO or NC contact.

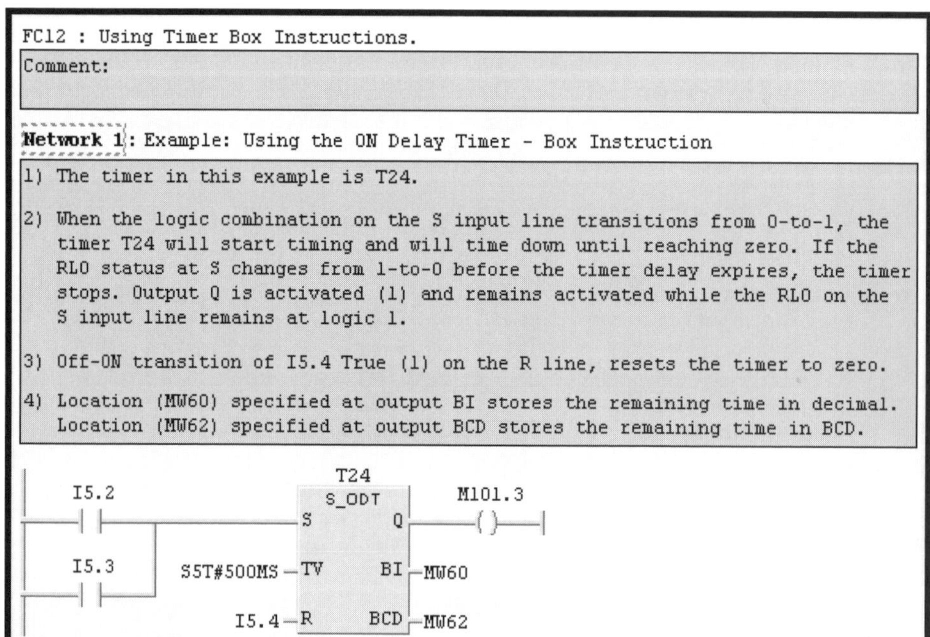

Figure 4-65. Example in LAD: Programming a Timer box instruction.

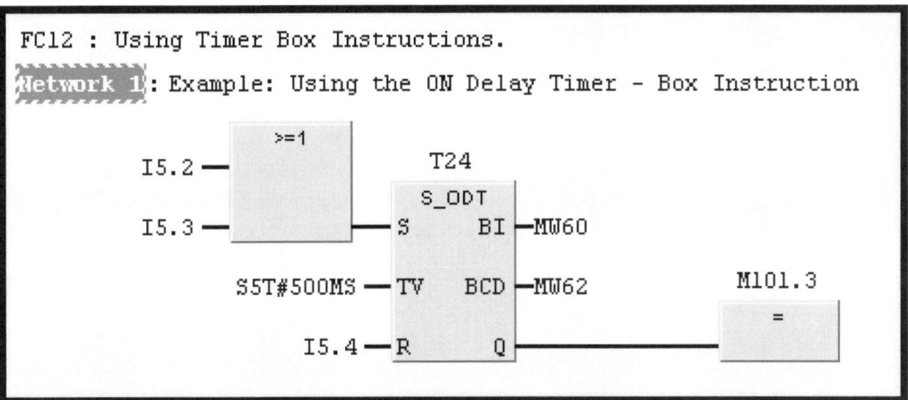

Figure 4-66. Example in FBD: Programming a Timer box instruction.

Guidelines: Programming Timer Operations

No.	GUIDELINE
1	Timer box and coil operations are in the *Timers* folder of the Program Elements tab.
2	Timers are addressed **T0, T1, T2**, and so on. Available number of timers is CPU dependent (e.g., T0 to T255)
3	A timer box may be placed in any branch that is connected to the left power rail.
4	Contacts may be arranged in series and in parallel at each binary input (i.e., **S**, **R**), of a timer, to achieve the desired control logic, as well as after the status signal (Q).
5	A valid preset time is S5T#10m_30s (10 MIN. 30 SEC.). Only the needed units of the timer value need be used (e.g., S5T#1500ms).
6	The range of a timer preset is 10 ms to 99990 seconds (= S5TIME#10ms to S5TIME#2h46m30s).
7	The operation of the timer status signal (**Q**) varies according to the timer type.
8	You may check the running status of a timer, using N.O. or N.C. contacts that reference the timer number.
9	You can check the remaining time at intermediate point for control or other purposes, using the compare operation and comparing the values at BI or BCD.
10	Time base units are binary encoded in bit-12 and bit-13 of the timer word of each timer. 00 = 0.01 seconds, 01 = 0.1 seconds, 10 = 1.0 seconds, and 11 = 10 seconds.

Programming Conversion Operations

Basic Concept

Programming a conversion operation is a matter of first selecting the required instruction, based on the name at the inside top of the box (e.g., **DI_BCD** is Convert Double Integer to BCD), and placing it into the LAD or FBD network. An input parameter must be specified as the value to be converted, and an output parameter as the location for storing the result. Finally, inserting the logic for enabling the conversion is all that is required.

Essential Elements

The enable input line **EN** allows control logic to determine if the conversion is executed. The parameter **IN**, which specifies the value to be converted, may be a constant or local variable of the correct data type, or a memory location of the correct width. When EN is at logic 1, the conversion is executed and the result is placed in the location specified at the parameter **OUT**. The variable specified at OUT may be a local variable of the correct data type, or a memory location of the correct width. The enable output line **ENO,** signals execution of the instruction and whether an error has occurred. If an error occurs while EN is at logic 1, ENO is set to logic 0; otherwise, the status of ENO follows the status of EN.

Application Tips

Conversions are used to change the format or data type of a variable. A BCD input value, for example, would require conversion before use in an integer arithmetic operation. The result would also need conversion, were it to be sent to a BCD display. STEP 7 arithmetic operations use integer, double integer or real numbers; values of any other data type would require conversion. Finally, when specifying the input and output variables of a conversion, remember that input and output variables may differ in data type and size, based on the operation. For instance, in the INT_ DINT conversion, the input is 16-bit and the output is 32-bit.

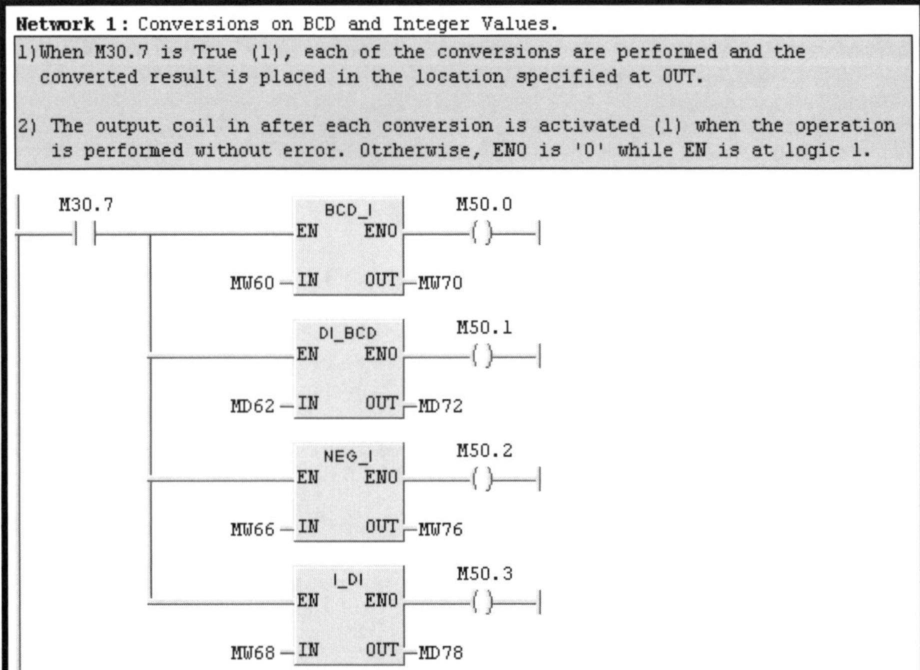

Figure 4-67. Example in LAD: Programming a Conversion instructions.

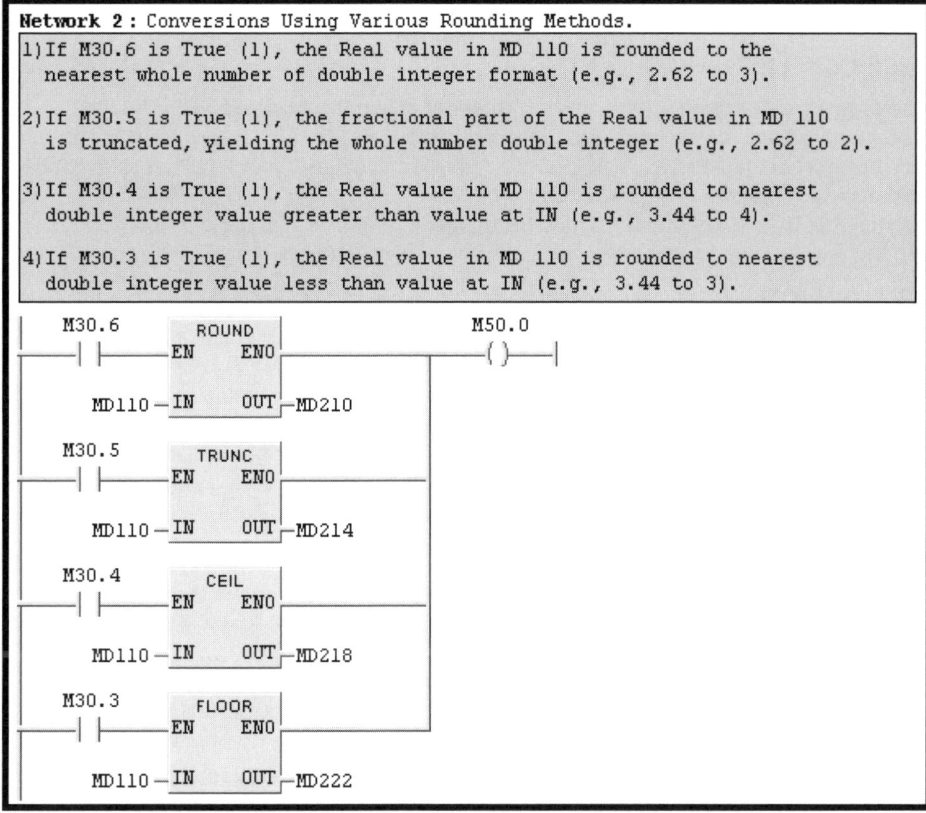

Figure 4-68. Example in LAD: Programming Conversion rounding instructions.

Guidelines: Programming Conversion Operations

No.	GUIDELINE
1	Conversion operations are in the *Converter* folder of the Program Elements tab.
2	Series and Parallel conditional logic may be inserted before the enable (EN) input and after the enable output (ENO).
3	Conversion boxes may be inserted in series by connecting the ENO line of one conversion box to the EN line of the following conversion box.
4	Conversion boxes may be inserted in parallel if placed in a branch that is directly connected to the left power rail of the network. When conversion boxes are placed in parallel, the network must be terminated with a coil.
5	If a network with two or more parallel branches contains conversion boxes in series, boxes are evaluated from left-to-right starting with the first branch.
6	Status bit instructions OV and OS may be used to check for errors resulting from invalid REAL numbers at IN, or values too large to convert to 3-digit or 7-digit BCD; status bit contacts may also be used to check if the result of a conversion is negative, positive, or zero.

Programming Compare Operations

Basic Concept

Programming a compare operation is a matter of first selecting the required instruction, based on the name at the inside top of the box (e.g., **CMP > I** is Compare Integer for greater than), and placing it into the LAD or FBD network. Each of the six compare types is available for comparing integers (**I**), double integers (**DI**), and real numbers (**R**), and is selected accordingly. The input parameters must be specified for the two values to be compared, and the enabling logic that causes the compare to be executed must be entered.

Essential Elements

The unlabeled input line to the compare, allows control logic to determine if the compare is executed. If no logic is inserted, the operation is always executed. The parameters **IN1** and **IN2** are where the two compare values are specified. The specified values, which must be of the same type, may include constants in the permissible range, local variables, or memory locations (e.g., MD 46) of the correct width. The unlabeled logic output line may drive a binary output to signal the logic-0 or logic-1 compare result, or may be used as direct input to other box instructions that accept binary inputs (e.g., the S-R Flip-flop, or UP Counter)

Application Tips

The compare operation can be placed in a network, much like a normally-open contact instruction, with preceding and following logic. When the compare is successful, power flows through the compare; if the compare test fails, power flow through the compare is interrupted. Also like a contact, multiple compare boxes may be placed in series or in parallel. If in parallel, a logic 1 result is produced for the branch when any compare is satisfied. If in series, a logic 1 result is produced only when all of the compares are satisfied.

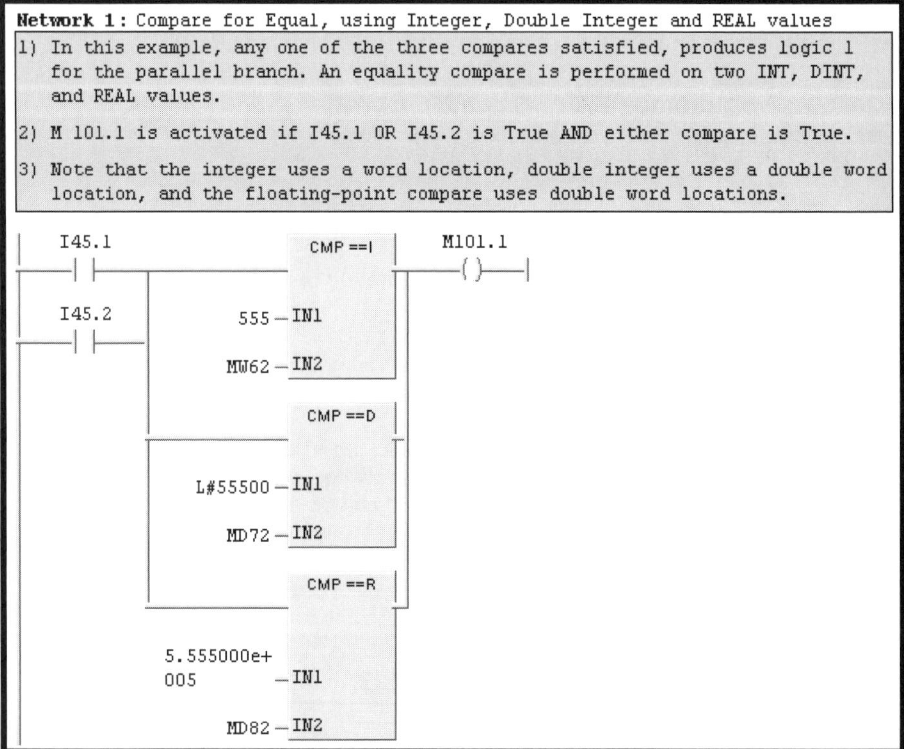

Figure 4-69. Compare Operations in LAD: Compare Equal for two INT, DINT, and REAL values.

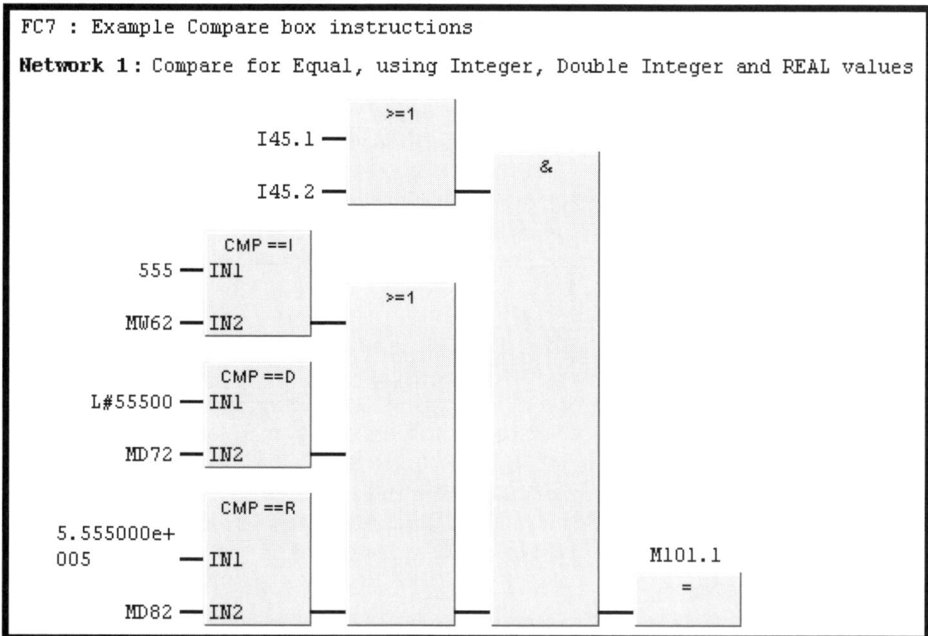

Figure 4-70. Compare Operations in FBD: Compare Equal for two INT, DINT, and REAL values.

Guidelines: Programming Compare Operations

No.	GUIDELINE
1	Compare operations are in the of the *Comparator* folder of the Program Elements tab.
2	Six compare types can be performed on variables of data type INT, DINT, and REAL.
2	The two values input at IN1 and IN2 must be of the same data type. For example, if the comparison is CMP≥ DINT, both IN1 and IN2 must be of type double integer or reference two double word memory locations.
3	To compare values other than INT, DINT, and REAL, the values must first be converted to one of these three data types. For example, BCD values are data type WORD or DWORD. Either a WORD –to- INT or DWORD –to- DINT conversion may be required in order to compare two BCD values.
4	Contact instructions may be arranged in series or parallel either preceding or following the compare box.
5	Compare boxes are like normally-open contacts and may be placed in series or in parallel. All compares in series must be satisfied to allow power flow. Compares in parallel require at least compare to be satisfied to produce logic 1 for the branch.
6	When REAL values are compared, the test fails if either of the values IN1 or IN2 are invalid (UO =1).

Programming Integer Arithmetic Operations

Basic Concept

Programming an integer arithmetic operation is a matter of selecting the instruction, from the Program Elements tab, based on the name at the inside top of the box (e.g., **MUL_DI** is Multiply Double Integers), and placing it into the LAD or FBD network. Two input parameters must be specified for the two integer (**I**) or two double integer (**DI**) values to be used in the calculation. An output parameter must be specified for the result. With the parameters defined, the enabling logic that causes the operation to be executed is all that is required.

Essential Elements

The two source values for the calculation, specified at **IN1** and **IN2**, must be of the same data type, and may include constants in the permissible range, local variables, or memory locations of the correct width (e.g., MW 42 for INT; or MD 42 for DINT). The variable specified at OUT may be a local variable of the correct data type, or a memory location of the correct width (e.g., MW 42 for INT; or MD 42 for DINT). The enable input line **EN** allows control logic to determine if the operation is executed. When **EN** is at logic 1, the operation is executed and the result is placed in the location specified at the parameter **OUT**. The enable output line **ENO,** signals execution of the operation and whether an error occurred.

Application Tips

If an error occurs during operation, ENO is set to logic 0 while EN is at logic 1; otherwise, the status of ENO follows the status of EN. After each arithmetic operation, you may use status bit instructions to check for overflow or if the result is, negative, positive, or zero. Use integer instructions, when whole number results are acceptable and your results are expected to fall within the permissible range (-32,768 to +32,767 for INT; -214,783,648 to +214,783,647 for DINT).

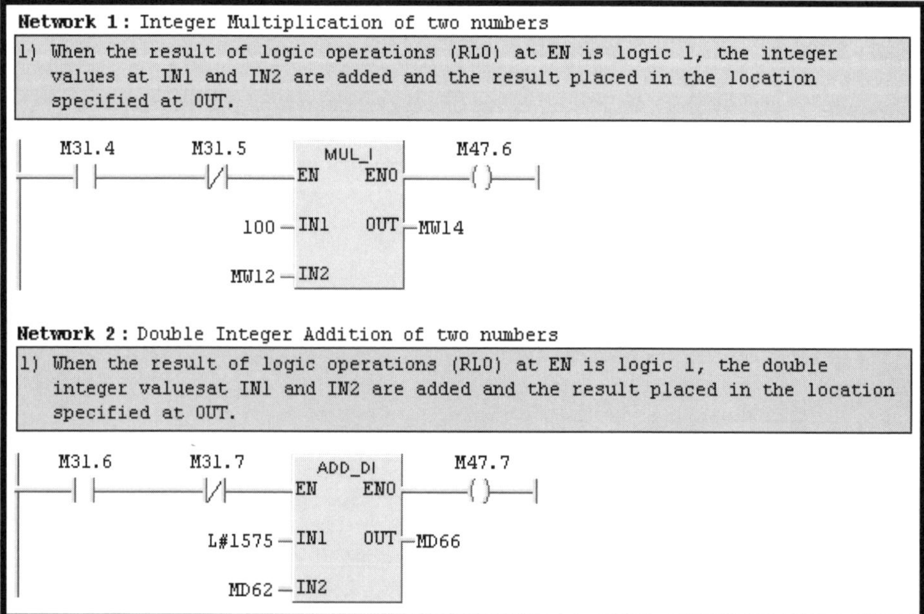

Figure 4-71. Integer Arithmetic in LAD: Integer Multiply; Double Integer Add.

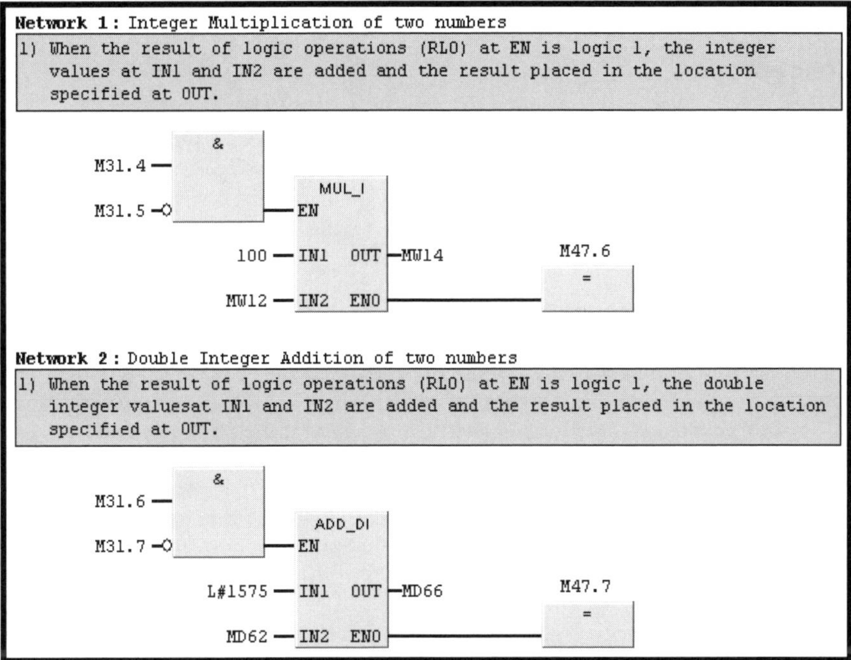

Figure 4-72. Integer Arithmetic in FBD: Integer Multiply; Double Integer Add.

Guidelines: Programming Integer Arithmetic Operations

No.	GUIDELINE
1	Integer (INT/DINT) operations are in the *Integer Functions* folder.
2	Use the INT boxes, for 16-bit operations; and the DINT boxes, for 32-bit operations.
3	Bit instructions may be combined in series or parallel before the enable EN input and after the enable output ENO output of the arithmetic box, to achieve the desired control logic that drives the integer box operations.
4	Integer arithmetic boxes may be inserted in series by connecting the ENO line of one operation to the EN line of the following operation. A following operation is only processed if the preceding operation is processed without error.
5	Use a temporary local variable as an intermediate buffer if you need to use the output value of a preceding operation in a following operation.
6	Integer arithmetic operations may be inserted in parallel if placed in a branch that begins directly connected to the left power rail of the network. In such a case, the network must be terminated with an output.
7	If a network has two or more parallel branches starting at the left rail, each of which has two or more integer arithmetic boxes in series, the order of processing is from left-to-right starting with the top branch, then the second branch, and so on.
8	Use status bit OV to check for an overflow after each operation, or after a series of operations use the OS status bit. Also use status bit instructions to determine if the result is negative, zero, or positive (e.g., <0, = = 0, or >0). See *Status Bit Instructions*.

Programming REAL Arithmetic Operations

Basic Concept

Programming a REAL arithmetic operation is a matter of selecting the instruction, from the catalog, based on the name at the inside top of the box (e.g., **DIV_R** is Divide REAL), and placing it into the LAD or FBD network. Two input parameters must be specified for the two floating-point or REAL (**R**) values to be used in the calculation, and an output parameter must be specified for the result. Finally, the enable logic for the operation may be inserted.

Essential Elements

Entry of each REAL arithmetic operation will require you to specify the source values for the calculation at parameters **IN1** and **IN2**; and the result location at the **OUT** parameter. Valid entries for IN1 and IN2 include a REAL constant in the permissible (± 1.75495e-38 to ± 3.402823e+38), local variables of data type REAL, or double word locations (e.g., DB22.DD12 or MD 44); the **OUT** parameter may be specified as a local variable of type REAL or a double word memory location (e.g., DB22.DD16 or MD 42). The enable input line **EN** allows control logic to determine if the operation is executed. When **EN** is at logic 1, the operation is executed and the result is placed in the specified location. The enable output line **ENO,** signals execution of the operation and whether an error occurred.

Application Tips

After a REAL calculation, you may use Status Bit instructions to determine if the result is negative, positive, or zero, or if one of the input values (IN1/IN2) is an invalid real number. If an error occurs while EN is at logic 1, ENO is set to logic 0; otherwise, the status of ENO follows the status of EN. Use REAL instructions, when whole number accuracy is not suitable.

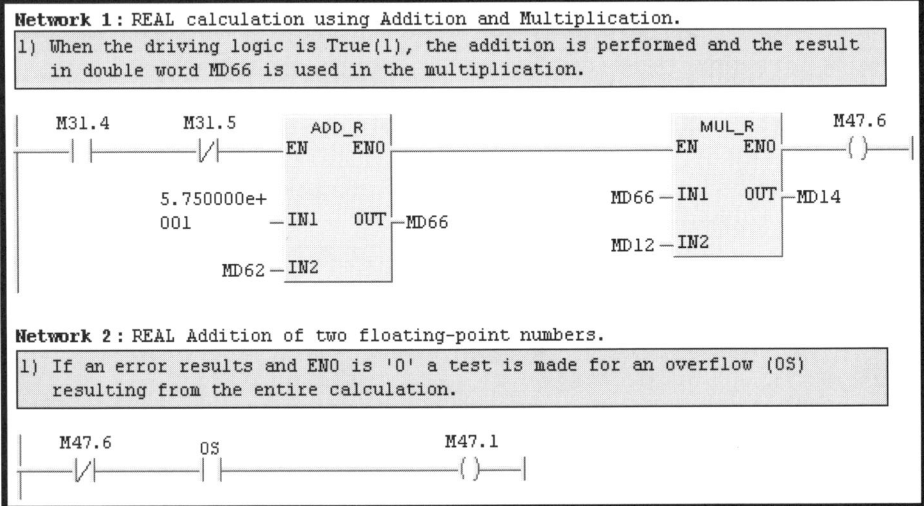

Figure 4-73. Example in LAD: REAL Multiply and REAL Addition.

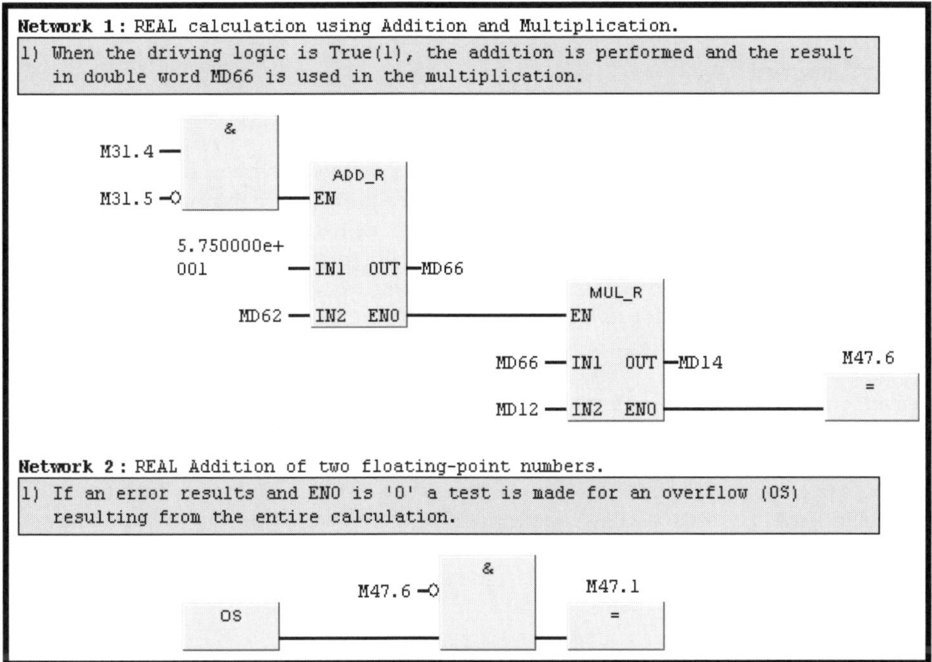

Figure 4-74. Example in FBD: REAL Multiply and REAL Addition.

Guidelines: Programming REAL Arithmetic Operations

No.	GUIDELINE
1	Real (REAL) arithmetic operations are in the *Floating-Point Functions* folder of the Program Elements tab. Use these instructions, for 32-bit floating-point operations.
2	Bit instructions may be combined in series or parallel before the enable EN input and after the enable output ENO output of the arithmetic box, to achieve the desired control logic that drives the integer box operations.
3	REAL arithmetic boxes may be inserted in series by connecting the ENO line of one operation to the EN line of the following operation. A following operation is only processed if the preceding operation is processed without error.
4	Use a temporary local variable as an intermediate buffer if you need to use the output value of a preceding operation in a following operation.
5	REAL arithmetic operations may be inserted in parallel if placed in a branch that begins directly connected to the left power rail of the network. In such a case, the network must be terminated with an output.
6	If a network has two or more parallel branches starting at the left rail, each of which has two or more REAL arithmetic boxes in series, the order of processing is from left-to-right starting with the top branch, then the second branch, and so on.
7	Use the OV status bit to check for an overflow after each operation, or after a series of operations use the OS status bit. Also use status bit instructions to determine if the result is negative, zero, or positive (e.g., <0, = = 0, or >0), or if either input value was an invalid REAL number. See *Status Bit Instructions*.

Programming Trigonometric and Other Math Functions

Basic Concept

STEP 7 REAL math functions include trigonometric, arc functions, logarithmic, and other standard math functions. While these instructions each operate on real numbers, there are input range restrictions on some of the operations. Programming these operations, is a matter of selecting the instruction, from the catalog, based on the name at the inside top of the box (e.g., **SQRT** is Square Root function), and placing it into the LAD or FBD network. Entry of each operation requires that a single input parameter be specified, as the value used in the operation; and a single output parameter, for storing the result. Finally, the enable logic for the operation may be inserted using conditional or unconditional logic.

Essential Elements

Entry of each REAL math function require the parameter **IN** be specified as the source value used in the function; and a parameter **OUT**, that designates where the result is stored. Valid entries for both IN and OUT include local variables of data type REAL, or double word locations (e.g., DB22.DD12 or MD 44). The **EN** line, which determines whether if the operation is executed, may be controlled by conditional logic or connected directly to the left power rail without conditions. When EN is at logic 1, the operation is performed using the floating-point value specified at IN and the result is placed in the location specified at OUT. If the operation executes without error ENO is set to '1'. If an error occurs, ENO is set to '0'.

Application Tips

Like with REAL arithmetic calculations, Status Bit instructions may be used after processing each math function, to determine if the input value (IN) is an invalid real number; or if the result is negative, positive, or zero. The OV and OS status bits are both set if the result is outside the permissible range. (See *Status Bit Instructions,* and *Programming Status Bit Operations*).

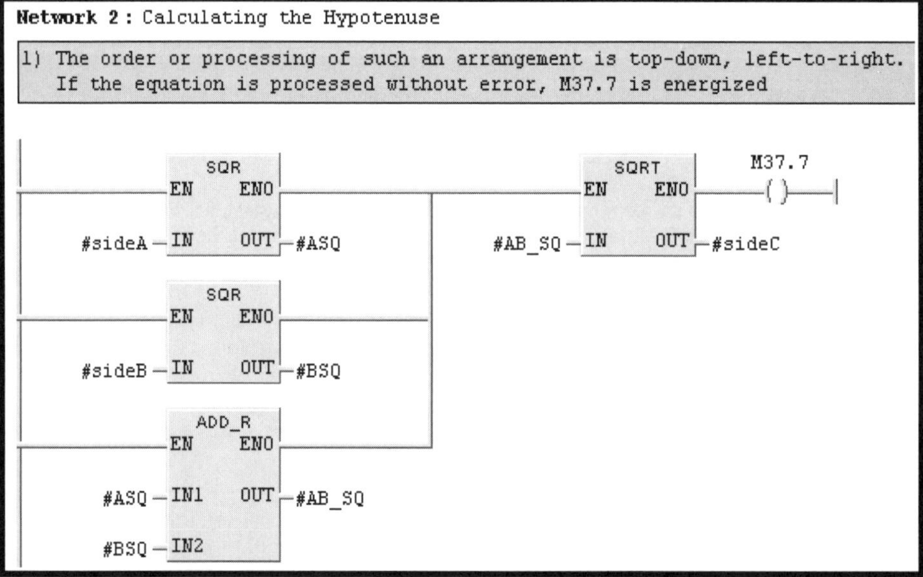

Figure 4-75. Example in LAD: Math Functions — REAL Addition, Square, and Square Root. In this example, first the square of side-a (#ASQ) and side-b (#BSQ) are found then the two are added, and finally, the square root of the sum (#AB_SQ) is determined.

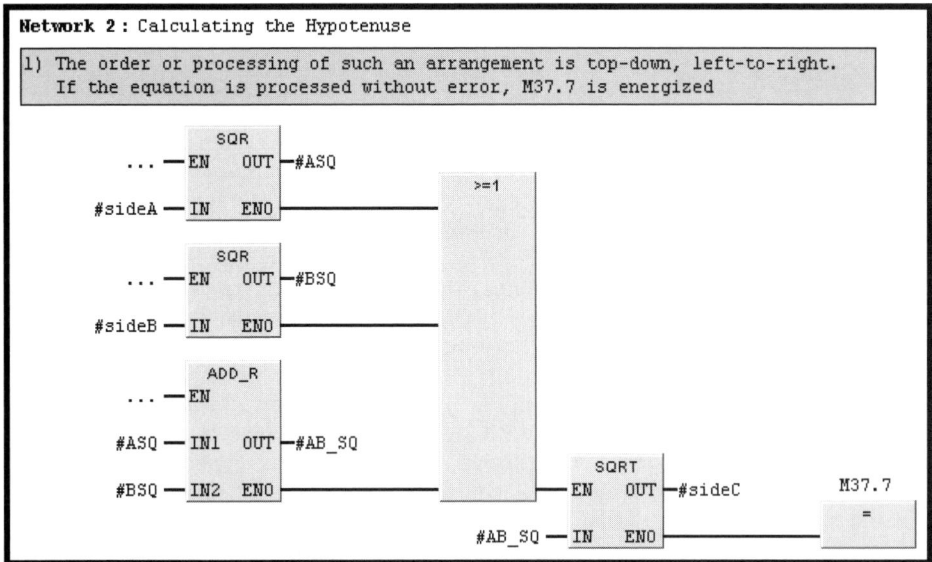

Figure 4-76. Example in FBD: Math Functions — REAL Addition, Square, and Square Root.

Guidelines: Programming Trigonometric and Math Functions

No.	GUIDELINE
1	Real math functions are in the *Floating-Point Functions* folder of the Program Elements tab.
2	Each of the trigonometric functions (SIN, COS, and TAN) assume an angle in radian measure (i.e., 0 to 2π, where π = +3.14593 e+00) as a REAL number input at IN.
3	Each of the arc functions (ASIN, ACOS, and ATAN) assume a REAL number that falls within a specific range (i.e., -1 to +1, ASIN; -1 to +1, ACOS; entire range, ATAN), as a REAL number input at IN.
4	Contact instructions may be inserted in series or parallel before the enable (EN) input, as well as after the enable output (ENO) of math functions.
5	Math functions may be inserted in series by connecting the ENO line of one function box to the EN line of the following function box.
6	Math function boxes may be inserted in parallel if placed in a branch that begins directly connected to the left power rail of the network. In such a case, the network must be terminated with an output.
7	If a network has two or more parallel branches starting at the left rail, each of which has two or more math function boxes in series, the order of processing is from left-to-right starting with the top branch, then the second branch, and so on.
8	Use the OV status bit to check for an overflow after each operation, or after a series of operations use the OS status bit. Furthermore, use status bit instructions to determine if the result is negative, zero, or positive (e.g., <0, = = 0, or >0), or if the input value was an invalid REAL number. See *Status Bit Instructions*.

Programming Jump, Label, and Return Operations

Basic Concept

Jump and label operations are used to facilitate jumps that take place internal to a block. When a Jump is executed the networks between the Jump and the Label are skipped. Processing resumes at the network following the Label. The Return allows the execution of a block to be terminated at an intermediate point in the block, prior to the normal block end.

Essential Elements

A **Label** instruction, which may consist of up to four characters, is always the destination to which a jump instruction is directed. Labels may consist of up to four alphanumeric characters, the first of which must be an alphabet. There are two LAD/FBD jump instructions, the **JMP** and **JMPN**. The JMP can be programmed with or without preceding logic conditions. With no conditions, the JMP is always executed. With logic conditions, the JMP is only executed when the result of the logic operation (RLO) is logic 1; if the RLO is logic 0, execution continues with the following network. The JMPN instruction must be programmed with preceding logic. If the result of the logic operation (RLO) is logic 0, the jump label is executed; if the RLO is logic 1, execution continues with the following network.

Application Tips

Jumps may be executed in the forward or backward direction. However, a jump made inside an MCR zone must be to a **Label** inside the same zone. Finally, while several jump networks may be directed to the same label, a label must appear only once in a block.

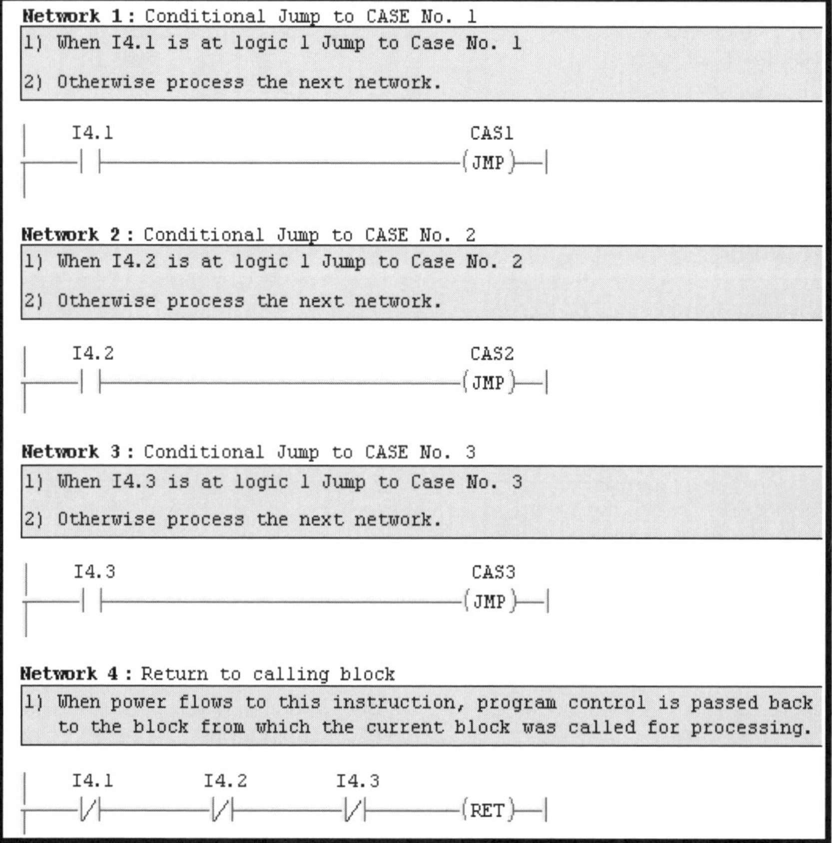

Figure 4-77. Jump to Label in LAD: Each JMP Label, passes control to a subroutine.

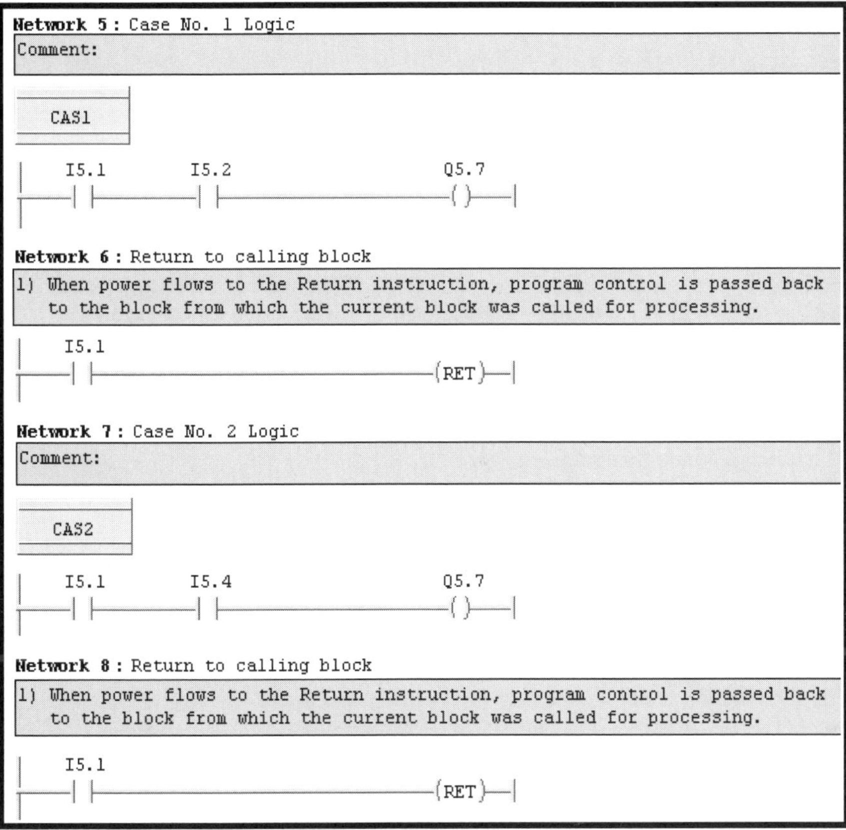

Figure 4-78. Jump to Label in LAD: Each Subroutines ends with a Return.

Guidelines: Programming Jump, Label, and Return Operations

No.	GUIDELINE
1	Jump and Label operations are in the *Jumps* folder of the Program Elements tab; the return operation is in the *Program Control* folder.
2	A label may consist of up to four alphanumeric characters, including the underscore. Labels must begin with a letter. Upper and lower case letters are interpreted as being different (e.g., CAS1 and cas1 are different).
3	Each label must be unique within a block; however, multiple jump coils from within a block may target the same label.
4	A jump is always to a label within the same block.
5	If a jump instruction is used inside an MCR zone, the target label must also be inside the same MCR zone and MCR area.
6	The JMP coil is used to jump on RLO = 1 and may be triggered conditionally (with preceding logic conditions) or unconditionally (without preceding logic conditions).
7	The JMPN coil is used to jump on RLO = 0 and may only be triggered conditionally (with preceding logic).

Programming MCR Operations

Basic Concept

MCR instructions allow logic zones to be created, in which the entire zone and the enclosed control logic to be enabled or disabled when the zone is enabled or disabled.

Essential Elements

There are four basic MCR operations. Each instruction, when programmed, is paired with one of the others. The MCR Activate (**MCRA**) and the MCR De-Activate (**MCRD**) instructions are paired to control activating and deactivating use of the S7 master control relay function. The *MCR Zone Start* (**MCR<**), and the *MCR Zone End* (**MCR>**), are paired to create a zone in which logic networks are enclosed. Networks inside an MCR zone are processed only if power flows (i.e., RLO = 1) to the MCR< instruction. If MCR zone dependency is switched ON (i.e., RLO to MCR< transitions to RLO = 0) then all outputs are reset to zero, except the Set/Reset outputs and flip-flops, which maintain their last states. A value of zero is written to all digital values or variables in box instructions.

Application Tips

MCR zones may be placed one inside another with up to eight nested MCR zones. If MCR zones are nested, the first MCR zone controls the dependency switching for all of the zones.

```
FC13 : Example of MCR Instructions.
Network 1: Activate use of S7 MCR function.
|
|----------------------------------------------(MCRA)--|

Network 2: Start of MCR zone 1.
|   I0.0
|---| |----------------------------------------(MCR<)--|

Network 3: Title:
|   I0.7        I4.3                              Q8.5
|---| |---------| |----+--------------------------( )--|
|                      |                          M0.6
|                      +--------------------------( )--|

Network 4: Title:
|   I0.4                                          Q16.0
|---| |-------------------------------------------(S)--|

Network 5: End of MCR zone 1.
|
|----------------------------------------------(MCR>)--|

Network 6: Title:
|   M5.5        I4.7                              M69.0
|---|/|---------|/|-------------------------------( )--|

Network 7: De-Activate use of S7 MCR function.
|
|----------------------------------------------(MCRD)--|
```

Figure 4-79. MCR Example: A single MCR zone.

Guidelines: Programming MCR Operations

No.	GUIDELINE
1	MCR operations are in the *Program Control* folder of the Program Elements tab.
2	To use the MCR capability in a block, you must define an MCR area within a block; MCRA (activate) defines the start of the area and MCRD (deactivate) defines the end of the area. MCRA and MCRD coils are connected directly to the left power rail without logic conditions.
3	An MCR zone is defined as the networks between a zone start instruction 'MCR<', and the zone end instruction 'MCR>'.
4	The MCR< coil, which starts a zone, must be driven by conditional logic.
5	The MCR> coil, which ends a zone, must be directly connected to the left power rail, without conditional logic.
6	MCR zones may be nested up to a depth of eight — that is, eight zones may be opened using the MCR< coil, before a zone is closed with the MCR> coil.
7	Overall MCR nesting depth is not affected when a new block is called from within an MCR zone. The current MCR zone remains open in the new block.
8	When a new block is called from within an MCR zone, the current MCR zone remains open in the new block although MCR use must first be re-activated with the MCR-A coil.

Programming Word Logic Operations

Basic Concept

Word logic instructions perform the specified Boolean logic operation on two source locations, based on the associated truth table, and place the result in a third location. Programming these operations is a matter of selecting the instruction, from the catalog, based on the name at the inside top of the box (e.g., **WOR_DW** is OR Double word), and placing it into the LAD or FBD network. Two input parameters must be specified as the variables used in the logic combination, and an output parameter for storing the result.

Essential Elements

Parameters **IN1** and **IN2** must be specified as the source values of each word logic operation and parameter **OUT** as the destination for storing the result. The enable line **EN** may be programmed with or without preceding control logic to determine if the operation is executed. When **EN** is at logic 1, the operation is performed bit-by-bit, using the two values at IN1 and IN2, and placing the result in the corresponding bit positions in the location specified at OUT. The enable output **ENO** is set to '1', signaling execution of the operation.

Application Tips

For word operations, valid entries include a WORD constant, a variable of data type WORD, or a word-width location (e.g., DB22.DW12 or MW 28) may be specified. For double-word operations, valid entries include a DWORD constant, a variable of data type DWORD, or a double word memory location (e.g., DB22.DD12 or MD 44) may be specified. The variable specified at OUT must be of the same data type or memory width as that specified at IN1 and IN2.

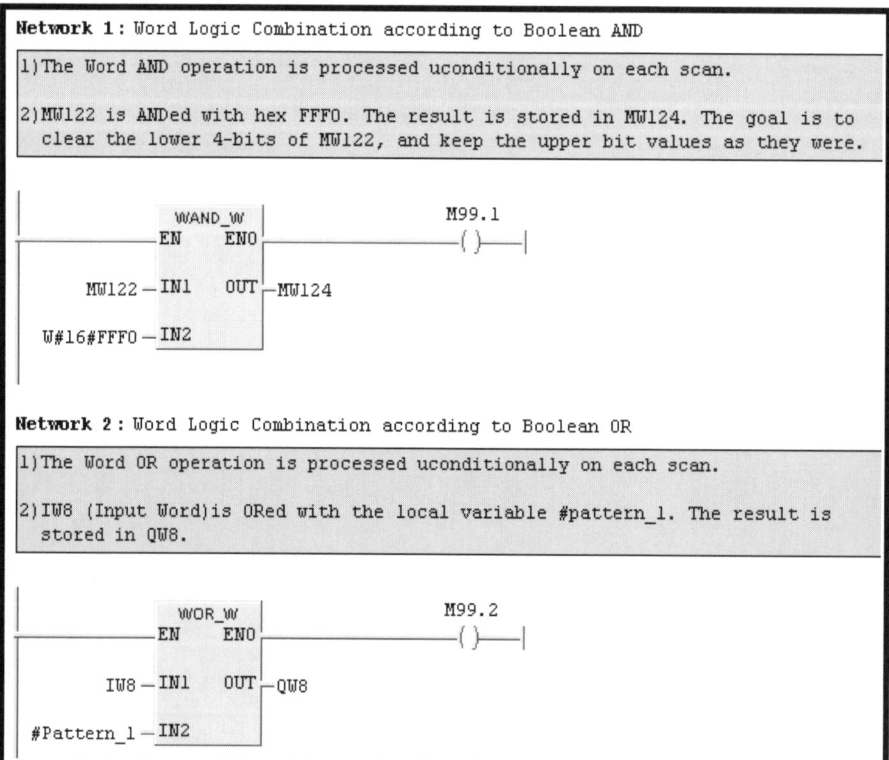

Figure 4-80. Example in LAD: Word AND, and word OR operations.

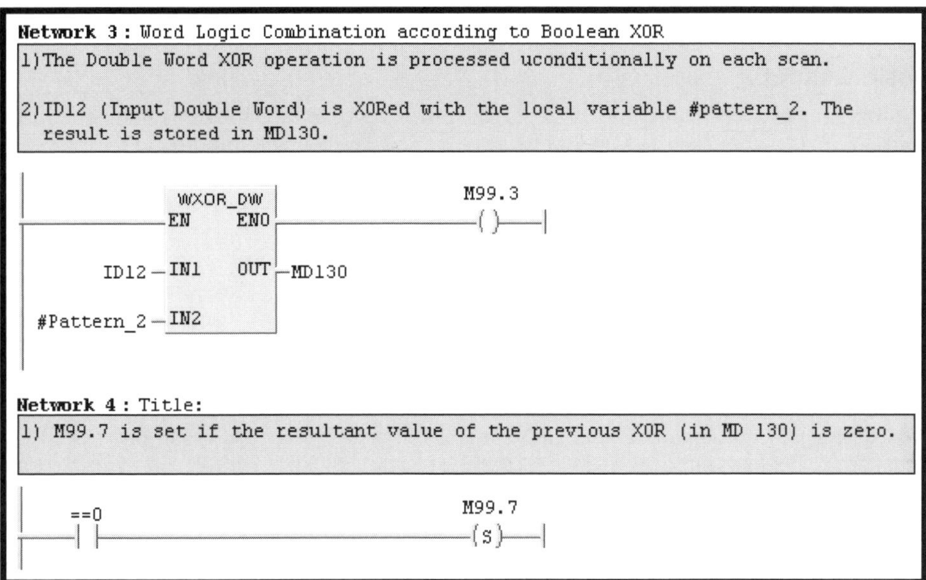

Figure 4-81. Example in LAD: Word XOR (exclusive OR) operation.

Guidelines: Programming Word Logic Operations

No.	GUIDELINE
1	Word logic operations are in the *Word Logic* folder of the Program Elements tab.
	You may insert contact instructions in series or parallel before the enable (EN) input and after the enable output (ENO), to create the desire control logic.
2	Word logic boxes may be inserted in series by connecting the ENO line of one logic box to the EN line of the following logic box. A box in series is only executed if the preceding box is processed without error.
3	Word logic boxes may be inserted in parallel if placed in a branch that begins directly connected to the left power rail of the network. In such a case, the network must be terminated with a coil.
4	If a network has two or more parallel branches starting at the left rail, each of which has two or more word logic boxes in series, the order of processing is from left-to-right starting with the top branch, then the second branch, and so on.
5	If the resultant bit-pattern at OUT is not equal to zero (all bits = 0), CC1 of the status word is set to '1'. You may use the >0 contact instruction to check for not equal '0'.
6	If the resultant bit-pattern at OUT is equal to zero (all bits = 0), CC1 of the status word is set to '0'. You may use the = = 0 contact instruction to check for equal '0'.
7	You may use status bit contacts instructions to check the result of a logic operation for all bits equal to zero or not all bits equal zero. See *Status Bit Instructions*.

Programming Shift and Rotate Operations

Basic Concept

Shift-rotate operations move the contents of a source variable by a specified number of positions to the left or right, and place the result in a destination variable. You would program these operations by selecting an instruction, from the catalog, based on the name at the inside top of the box (e.g., **SHL_W** is Shift Left Word), and placing it into the network.

Essential Elements

Parameter **IN** must be specified as the source value of each shift-rotate operation, and parameter **N** as the number of positions to shift or rotate the data. The enable line **EN**, which determines if the operation is executed, may be programmed with or without preceding logic. When EN is at logic 1, the contents of the WORD, DWORD, or integer (INT or DINT) variable are shifted (or rotated) by N –bit positions and the result is placed in the location specified at **OUT**. The enable output **ENO** is set to '1', signaling execution of the operation.

In a shift operation, the bits vacated on the shift end are filled with zeros and the bits shifted out on the opposite end are lost. In a rotate operation, bits shifted out on the far end are used to fill the bit positions vacated at the opposite end. Integer and double integer shift operations shift the bits of an integer variable, thereby changing the integer value, but while maintaining the original sign bit. The bit positions vacated by shift integer operations are always filled with the original sign bit (bit 15 for INT values, or Bit 31 for DINT values).

Application Tips

A typical application of the shift operation is to multiply or divide a variable by any binary multiple of two (i.e., 2, 4, 8, 16, etc.). The variable at IN is multiplied by 2, simply by shifting the variable contents to the left by one. Shifting two positions to the left multiplies by 4, and so on; conversely, shifting by 1 position to the right, results in division by 2; shifting 2 places to the right results in division by 4, and so on. The shift operation is often used in this manner, to index a value that controls an operation or when indexing a memory pointer value.

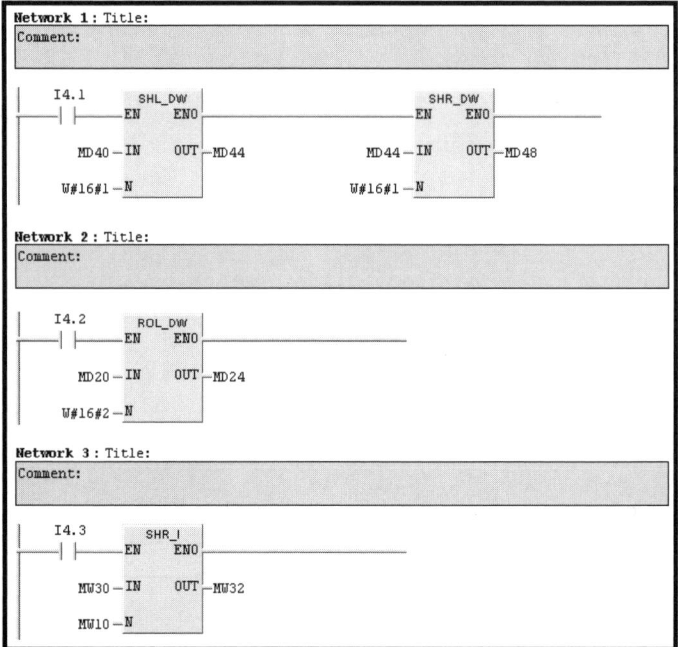

Figure 4-82. Example in LAD: Shift and Rotate operations.

Guidelines: Programming Shift and Rotate Operations

No.	Guidelines
1	Shift-Rotate operations are in the *Shift-Rotate* folder of the Program Elements tab.
2	Contact instructions may be inserted in series or parallel before the enable (EN) input and after the enable output (ENO) of shift and rotate instructions.
3	Shift-rotate boxes may be inserted in series by connecting the ENO line of one box to the EN line of the following box.
4	Shift-rotate boxes may be inserted in parallel if placed in a branch that begins directly connected to the left power rail of the network.
5	If a network arrangement involves two or more parallel branches, each of which has two or more Shift-rotate boxes in series, boxes are evaluated from left-to-right starting with the first branch, and so on.
6	In shift functions bit positions vacated after the shift are filled with zeros, except in the case of shift integer and shift double integer, where vacated positions are filled with the value of the sign bit of the original number (i.e., plus = '0' minus = '1').
7	In shift-rotate functions, the shift parameter (**N**), may be a variable or constant; if equal to zero, the operation is not executed; is greater than the highest bit position (15 in the case of word variables, 31 in the case of double word variables, then zero is placed in the result location OUT.
8	Shift integer operations always cause the variable to be divided by a binary multiple of 2 (i.e., 2, 4, 8, 16, etc). Shift by 1 divides by 2, shift by 2 divides by 4, shift by 3 divides by 8, and so on. The result is always the whole number after rounding down.

Programming Status Bit Operations

Basic Concept

S7 controllers generate a variety of status conditions that result from software operations. These status bits, which are combined into what is referred to as the *CPU status word*, are used by the CPU in controlling binary logic operations (e.g., AND/OR logic), and set by the CPU as a result of various digital operations (e.g., compares, arithmetic, and word logic). Although the CPU status word is not directly accessible to the user in LAD and FBD, it is made available indirectly using the status bit instructions. You may combine these instructions with other operations to evaluate the result of digital operations or as the basis of control logic.

Essential Elements

The overflow (**OV**) and overflow stored (**OS**) status bits are both set at the first time in a block where an arithmetic or math operation exceed the permissible range. The OV bit is reset at the next occurrence of an error-free arithmetic operation, and therefore can be checked immediately after each operation. The OV bit remains set until a new block call, therefore can be used to check for overflow after a series of calculations. The unordered (**UO**) status bit indicates that at least one of the values input to a previous REAL operation was invalid.

Application Tips

Condition code bits CC0 and CC1 are bits of the status word that report various results after integer and REAL arithmetic operations, as well as after compare, conversion, word logic, and shift operations. The various results indicated by CC0 and CC1 are made available to the user program via the so-called Results Bits which where briefly described back in Table 4-21. In this table, contact instructions are listed for checking the results of any arithmetic or math function relative to zero (e.g., equal zero, nonzero, positive, negative, and overflow). Additional results, which can be evaluated in your program, are provided in Appendix E.

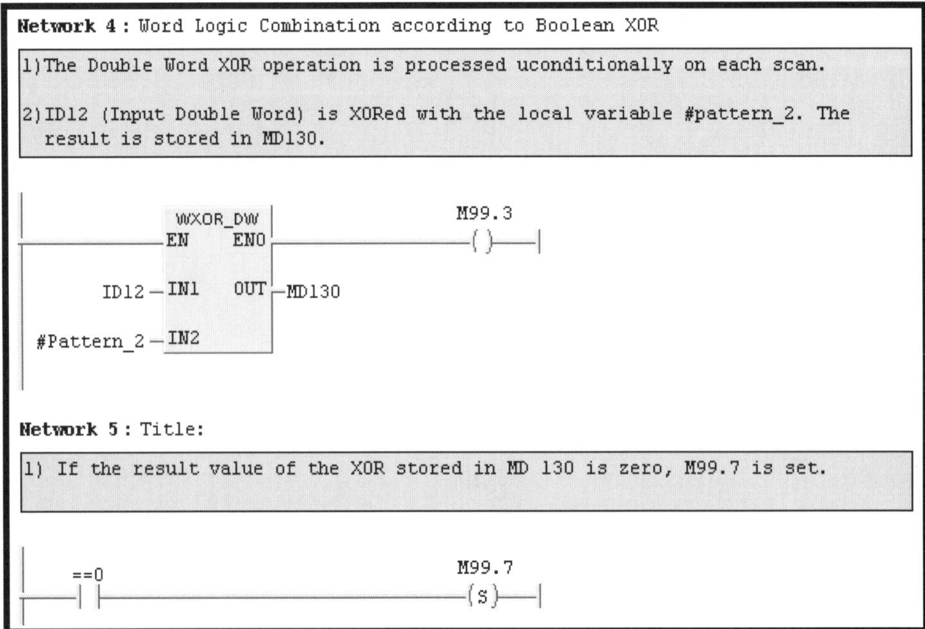

Figure 4-83. Example in LAD: Status Bit operation, using a Result bit Equal Zero contact.

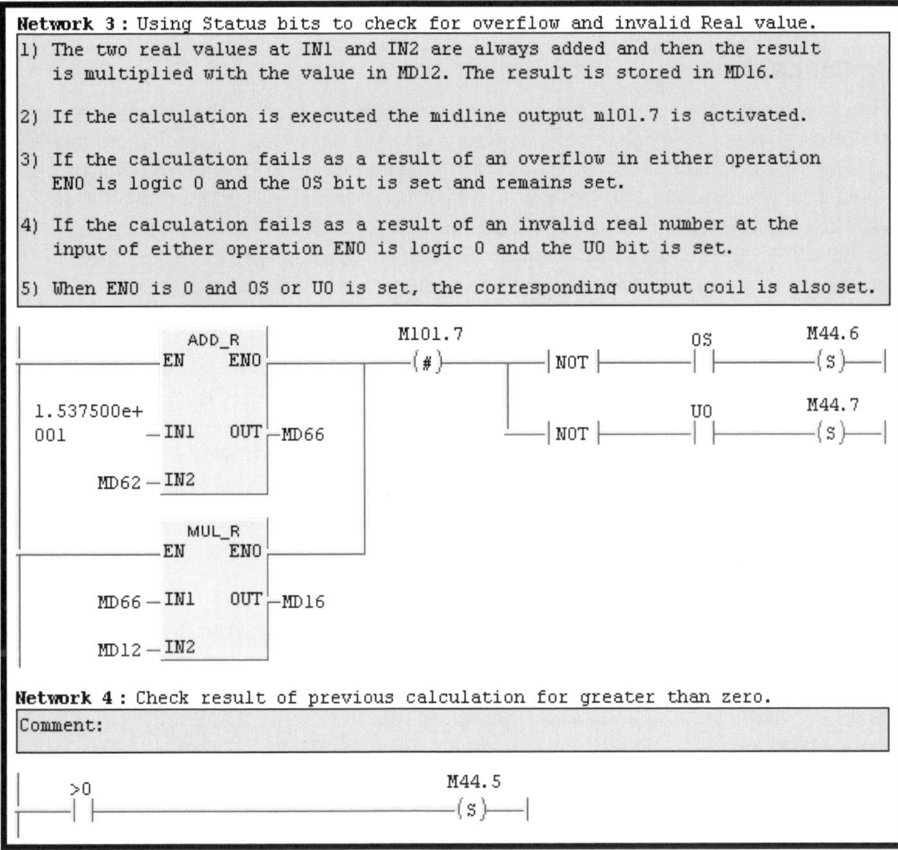

Figure 4-84. Example in LAD: Status Bit operations.

Guidelines: Programming Status Bit Operations

No.	GUIDELINE
1	Status bits operations are in the *Status Bits* folder of the Program Elements tab.
2	Use results bit instructions (e.g., **< 0**, **= = 0**, **> 0**) immediately after any arithmetic or math function, to check the result relative to zero. These instructions may be combined in series or parallel logic with other bit instructions.
3	Use the **OV** (overflow) status bit instruction immediately after a arithmetic or math function to check if the valid range is exceeded; use the **OS** (overflow stored) status bit immediately after a series of calculations, to check if an overflow occurred with any one of the instructions. The **OS** status bit, is set at the first overflow in a block, and is not reset until a new block is called.
4	Use the **UO** (unordered) status bit instruction immediately after a REAL arithmetic or math function, to check if either of the input values to the operation was invalid.
5	Additional result information is available for arithmetic and math functions, as well as for digital operations (e.g., word logic, comparison, shift, and conversions), using status bits CC0 and CC1 (condition code bits). Combinations of CC0/CC1 (e.g., 0/0, 0/1, 1/0, and 1/1) may be evaluated using the results bit contact instructions. See Appendix E.

Programming the Move Operation to Read and Write Data

Basic Concept

With the Move operation you can copy the contents of a memory location or variable to a new location or variable. Programming a Move operation is a matter of placing the move box into the LAD or FBD network, specifying an input parameter as the value to be moved, and an output parameter as the location to which the value is copied. The move instruction in LAD/FBD only handles elementary data elements of 8, 16, or 32 bits. You may however, use system blocks SFC 20 (BLCKMOV) and SFC21 (FILL) for data transfers involving multiple locations or complex data types (e.g., arrays, structures, user defined types).

Essential Elements

The enable input line **EN** allows control logic to determine if the MOVE is executed. You may specify the parameter **IN**, as a constant, a local variable of elementary data type (except for BOOL), or a valid byte, word, or double word memory location. When EN is at logic 1, the value at IN is copied to the location specified at the parameter **OUT**. You may specify the parameter **OUT**, as a local variable of elementary data type (except for BOOL), or a valid byte, word, or double word memory location. You do not have to specify the same data type or memory width at the IN and OUT parameters.

Application Tips

Valid memory areas at **IN** include **I**, **Q**, **M**, **L**, **T**, **C**, **PI**, and **DB** memory; valid memory areas at **OUT** include **I**, **Q**, **M**, **L**, **PQ**, **DB** memory. In the S7-300, the addresses of the entire input/output image areas (**I/Q**) may be specified at the IN or OUT parameter of the MOVE operation, regardless of whether or not the address is assigned to an installed module. In the S7-400, however, an address should only be accessed if it is assigned to an installed module.

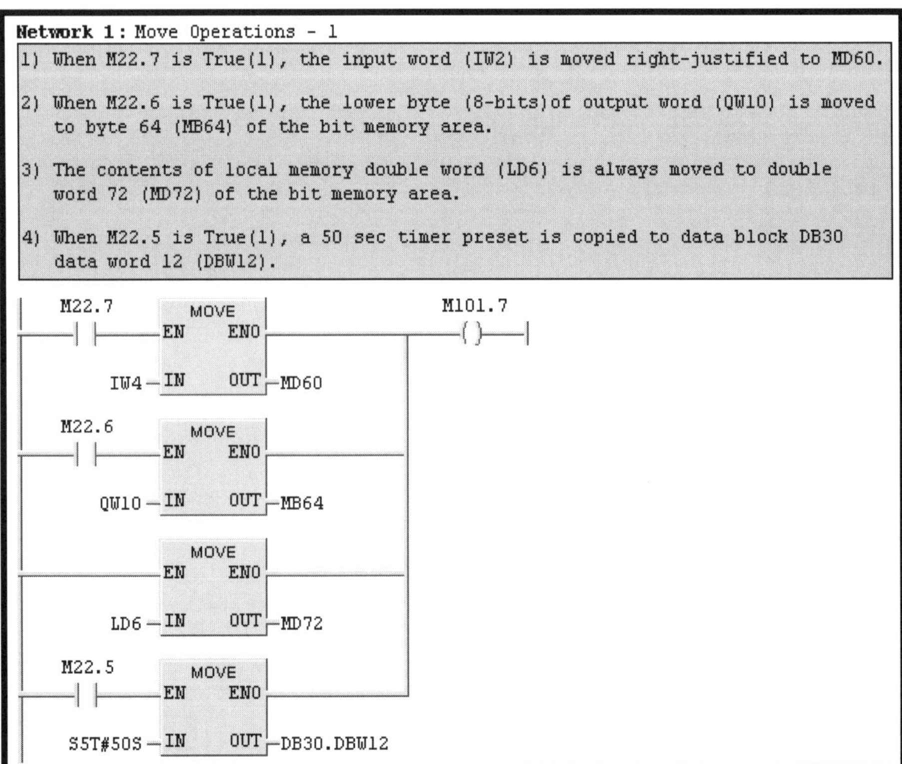

Figure 4-85. Example in LAD: Move operations using different source/destination areas.

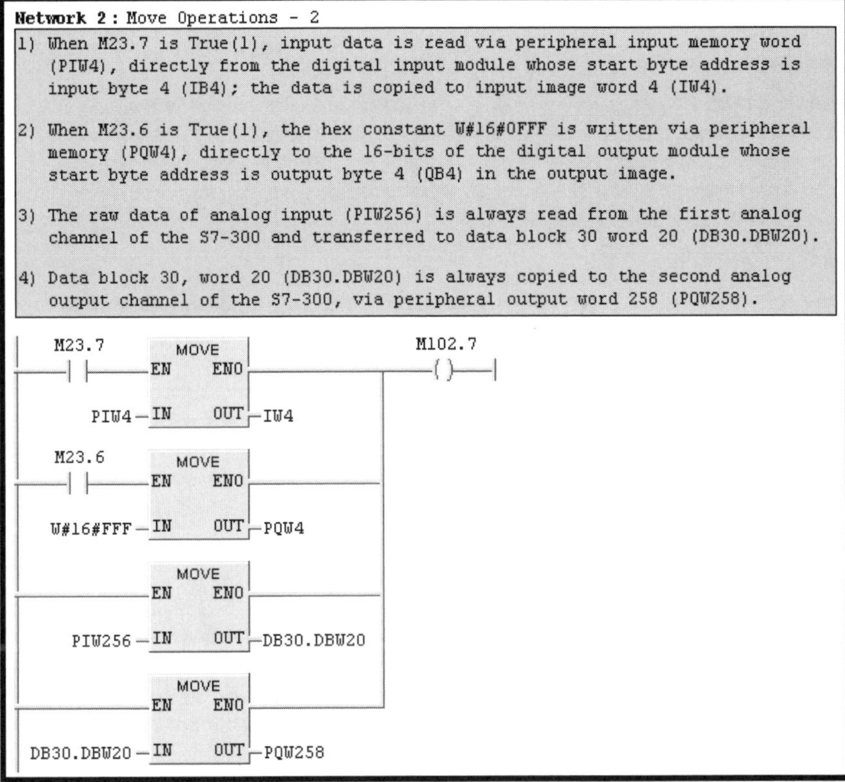

Figure 4-86. Example in LAD: Move operations using different source/destination areas.

Guidelines: Programming the Move Operation to Read and Write Data

No.	GUIDELINE
1	The Move operation is in the *Move* folder of the Program Elements tab.
2	If the size of the variable at IN is less than the size of the variable at OUT, the data is moved into OUT right-justified; the bits to the left are filled with zero. For example, if a WORD variable is moved to a DWORD variable, then the word is moved to the right-most 16 bits of OUT; the left-most bits are filled with zero (*Network 1, branch 1*).
3	If the size of the variable at IN is greater than the size of the variable at OUT, the least significant bytes of the variable at IN, that fit into the variable at OUT, are moved into OUT. For example, if a WORD variable is moved to a byte variable, then the least significant byte of the WORD variable is moved to OUT (*See Network 1, branch 2*).
4	Use the MOVE box to read digital inputs directly from the module; reference the input start byte address at **IN**. Use the MOVE box to write digital outputs directly; reference the digital output start byte address at **OUT** (*See Network 2, branch 1 and branch 2*).
5	Use the MOVE box to read raw data from an analog input module; reference the analog input channel address at **IN**. Use MOVE to write a value to an analog output module; reference the analog output channel address at **OUT** (*See Network 2, branch 3 and branch 4*).

Accessing Data in a Data Block

Basic Concept

Before accessing a DB variable, you must open the data block first, using the appropriate instruction to reference the DB number (e.g., OPN DB n). Once a data block is open, any variable may be accessed, using the correct identifier and address. Data block variables may be addressed in three methods – *partially-addressed operands, fully addressed operands*, and symbolic addressed operands.

Essential Elements

With partial addressing, you must first use an open data block instruction that references a global or instance DB. In STL, the open instruction is simply 'OPN DB n' (e.g., **OPN DB 22**), or OPN DI n (e.g., **OPN DI 22**). In LAD/FBD the open instruction is only valid for opening a global DB and is in the form of a coil instruction (OPN) placed in a network without preceding logic. After the open instruction, any data element can be accessed in the following networks (in LAD/FBD) or in the following statements (in STL), until a new data block is opened. Example partial addresses after using OPN DB are DBX8.7, a data bit; DBB10, a data byte; DBW20, a data word; and DBD12, a double data word. If you had used OPN DI, these locations would be addressed DIX8.7, DIB10, DIW20, and DID12.

With fully-addressed operands, you will specify the DB number as part of the bit, byte, word, or double word address of the location (e.g., DB22.DBX8.7, DB22.DBB10, DB22.DBW20, or DB22.DBD12). With fully-addressed operands, only possible in LAD/FBD, and only usable to access global data locations, you do not have to open the DB first. Symbolic addressing can also be applied with full addressing. The symbolic addresses are derived by combining the assigned symbolic name of the DB with the assigned name of each element in the data block (e.g., **Zone_1.TMP_PRESET_3**). The DB name is Zone_1 and the variable is TMP_PRESET_3.

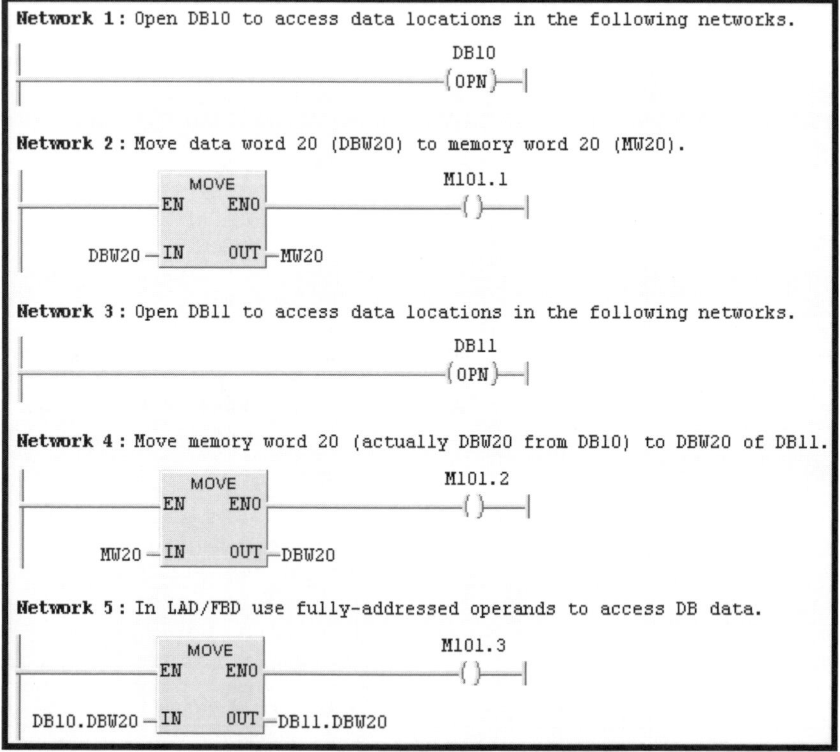

Figure 4-87. Accessing data locations using partial and using fully-addressed operands.

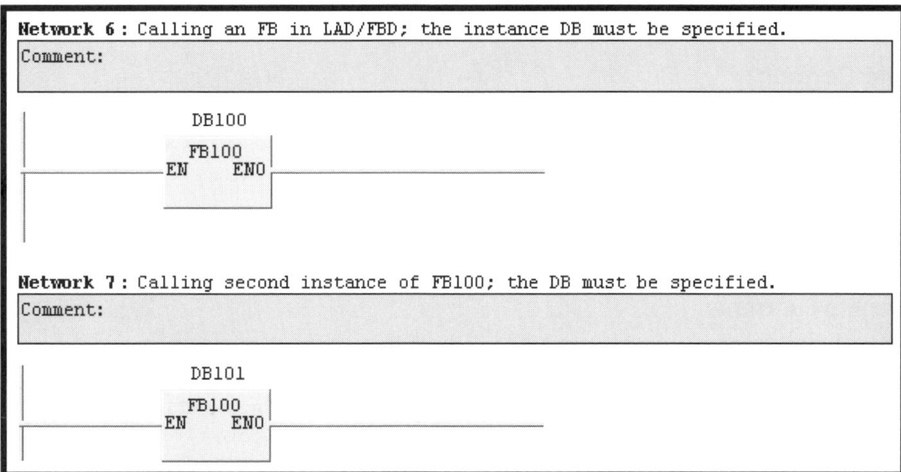

Figure 4-88. Specifying the Instance DB; the DB is opened when the FB is called

Guidelines: Accessing Data in a Data Block

No.	GUIDELINE
1	The OPN DB coil is in the *DB Call* folder of the Program Elements tab.
	Accessing a Global DB with partial and full addressing (LAD/FBD/STL)
2	The open DB coil can only be used to open a global data block, and may only be used in LAD and in FBD.
3	Only one OPN DB coil may be programmed in a network and it must be without conditional logic or other parallel outputs.
4	Following the OPN DB instruction in LAD, FBD, or STL, you may reference data locations of the specified DB, using *partially addressed* operands for data bits, bytes, words, or double words (e.g., DBX8.1, DBB4, DBW20, DBD4). The open data block may be accessed until a new data block is opened.
5	In LAD, FBD, and STL you may using *fully addressed operands* for bits, bytes, words, and double word locations of global data blocks. Full addressing means that you specify the DB number and location. For example DB10.DBW20, is DB10 word 20; DB10.DBX8.1, is DB10 byte 8 bit 1.). The OPN instruction is not required in this case.
	Accessing an Instance DB with partial addressing (STL)
1	An instance DB may only be referenced in LAD/FBD when the associated function block is called. The instance DB is opened when the FB is called.
2	Following the OPN DI instruction in the STL (Statement List) editor only, you may reference data locations of the specified instance DB, using *partially addressed operands* for data bits, bytes, words, or double words (e.g., DIX8.1, DIB4, DIW20, DID4). The open data block may be accessed until a new data block is opened.
3	The instance DB is also opened in STL in the call statement for calling the FB (e.g., Call FB100, DB100).

Programming Organization Block 1 (OB1)

Basic Concept

Organization Block 1 (OB 1) is the main program block of a STEP 7 program — responsible for the normal cyclical processing of your program. As such, most other blocks (e.g., FBs, and FCs), which collectively makeup the control program, are called for processing from OB 1. Properly written, OB 1 will facilitate a well organized program whose processing tasks are easily identified and followed. This task will introduce some basic guidelines and recommendations for writing the code for OB 1.

Essential Elements

Generally, most logic blocks of a STEP 7 program will be called either directly from OB1, or from another block (e.g., FC, or FB) that is called from OB1. The calls from OB1 will be largely dependent on how you have divided the functional control tasks into blocks and the processing requirements of each of these blocks. Blocks should be called from OB1 if they must be processed on each CPU cycle, or at least evaluated on each cycle but processed only if certain enabling conditions are met. When OB1 is used strictly for calling code blocks, the task of testing, startup, and diagnosing problems in the program, will be much easier.

Application Tips

As you develop new blocks, especially after the original startup, be sure that each newly created block is called either directly from OB 1, from another block that is called from OB1, or either directly or indirectly from another organization block.

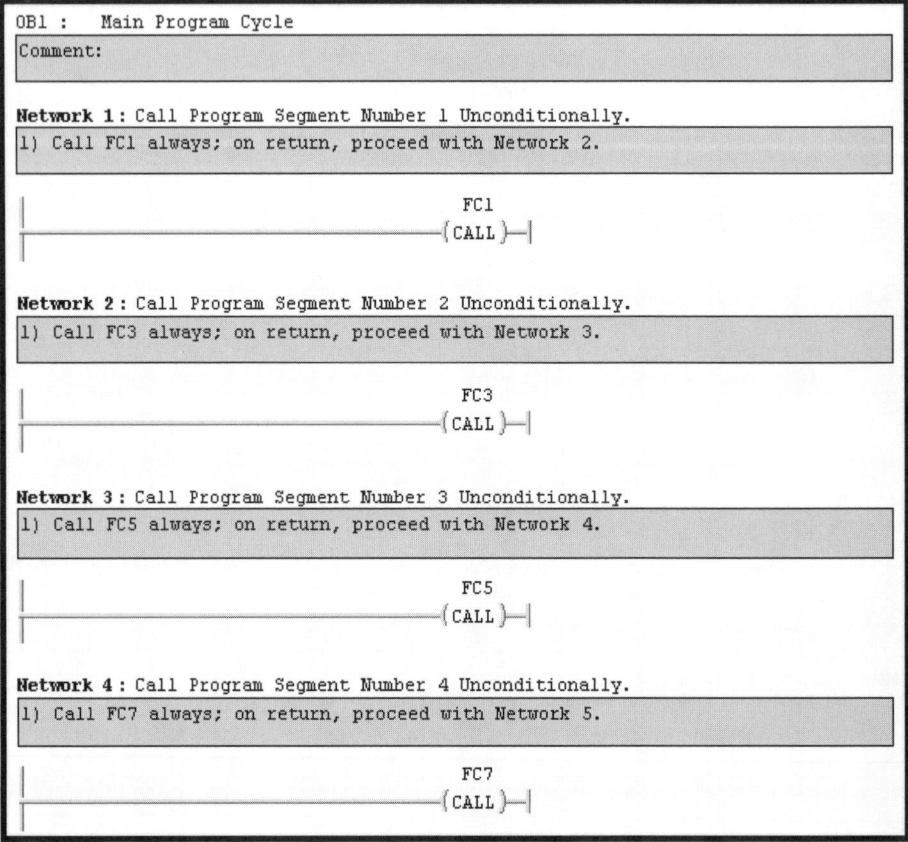

Figure 4-89. Unconditional calls to blocks from OB1 (organization block 1).

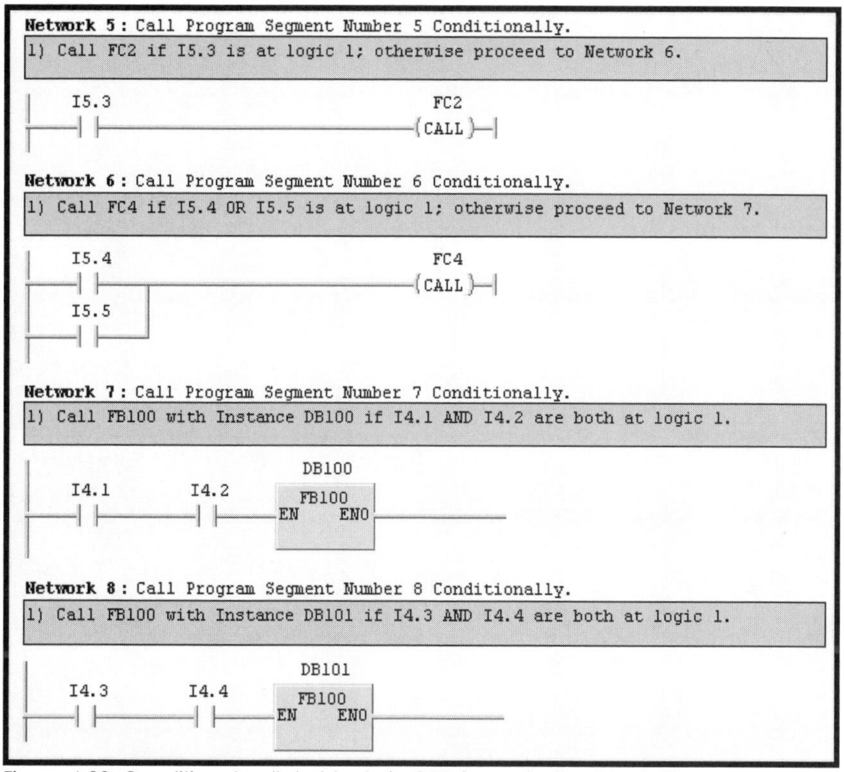

Network 5 : Call Program Segment Number 5 Conditionally.

1) Call FC2 if I5.3 is at logic 1; otherwise proceed to Network 6.

I5.3 FC2 (CALL)

Network 6 : Call Program Segment Number 6 Conditionally.

1) Call FC4 if I5.4 OR I5.5 is at logic 1; otherwise proceed to Network 7.

I5.4 FC4 (CALL)
I5.5

Network 7 : Call Program Segment Number 7 Conditionally.

1) Call FB100 with Instance DB100 if I4.1 AND I4.2 are both at logic 1.

DB100
I4.1 I4.2 FB100 EN ENO

Network 8 : Call Program Segment Number 8 Conditionally.

1) Call FB100 with Instance DB101 if I4.3 AND I4.4 are both at logic 1.

DB101
I4.3 I4.4 FB100 EN ENO

Figure 4-90. Conditional calls to blocks in OB1 (organization block 1).

Guidelines: Programming Organization Block 1 (OB1)

No.	GUIDELINE
1	Write a brief Block Title of OB 1 (e.g., Main Cyclical Program Block).
2	Write a Block Comment describing the function and purpose of OB 1.
3	OB 1 should only contain unconditional or conditional calls to FCs, FBs, SFCs, or SFBs; control logic for driving outputs or other purposes should not be placed in OB 1.
4	Program all unconditional block calls first, using a separate network for each call; make unconditional calls to blocks that must be processed on every CPU cycle.
5	Program all conditional block calls following the unconditional calls, using a separate network for each call. The entire conditional logic for a block call should be placed in one network.
6	Make conditional calls to blocks that must be processed whenever the driving logic conditions are met. If the RLO =1 the block is processed; if the RLO =0 the block is not processed and the following network is evaluated.
7	Except for instance DBs, opened with associated function blocks, avoid opening data blocks in OB1. Global data blocks (DBs) should be opened in the blocks where used.

Step 5

Managing Online Operations with S7 CPUs

Objectives

- Access CPU Operations without a Project
- Verify CPU Resources and Performance
- Change CPU Operating Modes
- Perform Reset on CPU Memory
- Access a Password-Protected CPU
- Set CPU Date and Time
- Compress CPU Memory
- Upload and Download a Program
- Upload and Download Selected Blocks
- Compare Online/Offline Programs
- Compare any Two Programs

Establishing Online Connections using STEP 7

As the development phase of your STEP 7 project progresses, more of your work will involve online interactions with the S7-300 or S7-400 CPU. With STEP 7 in the online mode, your will be able to perform activities such as downloading and uploading the user program, comparing programs blocks, resetting the CPU memory, accessing CPU information, viewing and switching the operating modes of the CPU, and many others which will are introduced in this chapter. Other online activities, associated with the use of monitoring and diagnostic tools, and generally required during the debugging process and later during the start-up and commissioning phases of the project, are covered in the next chapter.

The Standard Physical Connection

A *direct online connection* is established using the appropriate cable between the MPI interface of a programming system (PG/PC) and the MPI port on the S7 CPU. If your programming system is a PG, then a standard cable connects directly between the two MPI ports. If a PC is your programming system, then the serial cable is connected from the serial port of your PC, then to an MPI/RS232 adapter, and then to the MPI port on the CPU. You can then access the PLC via the *Online Project Window* or the *Accessible Nodes Window*.

If there is an MPI subnet with several CPUs connected, then each CPU must have a unique address. Initially, and before parameters are downloaded, each CPU is has the default MPI address of '2.' You can set a new MPI address for each CPU as you assign the CPU parameters for the stations of a project. Before connecting all of the CPUs to the subnet, you must make direct a connection between the PG/PC and each CPU, to download the hardware configuration. By downloading the *System Data* object from the *Offline Blocks* folder of the associated S7 Program, the new MPI address assigned in the hardware configuration will take affect immediately.

Using the Online Project Window

Using the *Online Project Window* is the preferred method of accessing a CPU online, if the PLC you wish to access has been configured in a project on your PG/PC. If a direct or subnet connection is established, then the Online Project Window is opened from the SIMATIC Manager menu by selecting **View ➢ Online**, or by selecting the Online Window icon from the toolbar. The online project window will display all Stations of the project, and will display the online data from the selected station. From the online window you may open the object properties for each online object, access the hardware configuration online, perform online operations of the PLC menu, and open program blocks online in the LAD/FBD/STL editor.

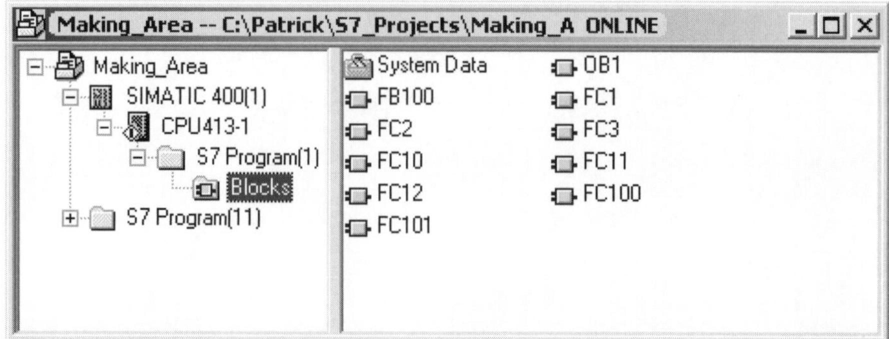

Figure 5-1. Online Project Window: Blocks folder selected.

Using the Accessible Nodes Window

You may open the *Accessible Nodes Window* from the SIMATIC Manager menu by selecting **PLC ➢ Display Accessible Nodes**, or using the Accessible Nodes toolbar button. From the Accessible Nodes Window, you can access a CPU online and perform the online operations of the PLC menu without having to open the associated project. This method is often used when the project is not yet created, or simply is not available on the PG/PC currently in use.

When the Accessible Nodes Window is opened, an MPI folder is displayed for each CPU module that can be reached over the MPI connection. The folder, which is identified by the CPU's MPI address, contains the online blocks of the CPU. In addition to allowing access to the online program of a CPU, it is also possible to access the online PLC menu operations. An important consideration when using the accessible nodes window is that you are operating without the offline project data, and conveniences such as symbolic addresses, comments, and program documentation will not be available. Making changes are not recommended.

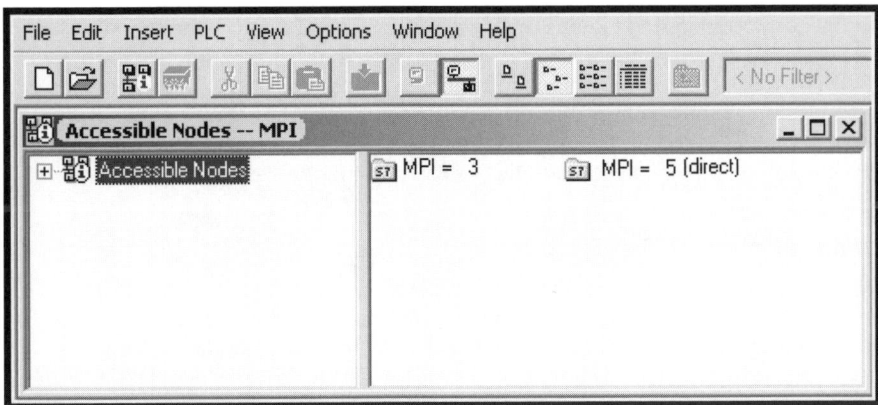

Figure 5-2. Accessible Nodes Window with two stations found.

Standard Online Operations with STEP 7

At some point, you will want to start testing parts of your work or the completed hardware configuration and user program. The first step, however, is to establish an online connection to the CPU, in order to download the required components. The user program, when completed, will consist of the hardware configuration, the user program, the network configuration and the CPU communication connection table (data).

All of the hardware configuration data for each station is stored in the *System Data* object, which is located in the offline blocks folder of the associated CPU program. When these components are completed, you may then download individual blocks, or the complete user program, which includes the *System Data* object. If you want to download modifications to the Hardware Configuration only, then only the System Data needs to be downloaded. Changes or modifications to the connection table can be downloaded from the Network Configuration tool, or from the CPU objects folder. Care must be taken to download changes to all communications partners.

Downloading the User Program to the CPU

Downloading to the CPU requires the CPU to be in either the STOP or RUN-P operating mode. Although you may perform a download in the RUN-P mode, switching to the STOP mode is recommended. When downloading, you may download all blocks or selected blocks. If the download is performed in RUN-P mode, you must consider the order in which blocks are downloaded, to ensure that blocks that are called are always transferred before the calling block. Before downloading the complete program, you should clear CPU memory to avoid having old blocks remaining in the CPU. To download hardware configuration changes only, simply select and download the System Data object from the offline blocks folder.

To test individual blocks, you must download the block (e.g., FC, FB), and the block that call the block to be tested (e.g., OB 1). Required data blocks must also be downloaded.

Uploading the User Program from the CPU

Like the download operation, the upload is an online operation that is performed from the SIMATIC Manager. You can only initiate the upload blocks operation from the Online Project Window. This restriction ensures that the uploaded blocks are directed to the offline blocks folder to which the CPU program is linked. Uploading blocks to the programming device is often useful for making a backup copy of the program currently in the CPU and perhaps for offline diagnostic purposes. To maintain an unmodified copy of the offline blocks, you might create a module-independent S7 Program and copy the offline Blocks folder to the new program before performing the upload. This backup can then be restored, for example, if alterations made to the uploaded program proved undesirable.

Comparing Online/Offline Programs

The online compare operation allows you to compare the online blocks residing in the CPU, to the offline blocks of the S7 Program associated with the CPU in the project. The online compare is initiated from the SIMATIC Manager menu by first selecting **View** ➢ **Online** to open the online project window. With the online window open, you may select the block folder or the individual blocks you want to compare, and then from the menu select **Options** ➢ **Compare Blocks**. By default, the Online/Offline compare option is selected and will compare the CPU blocks of the selected CPU to its offline blocks folder.

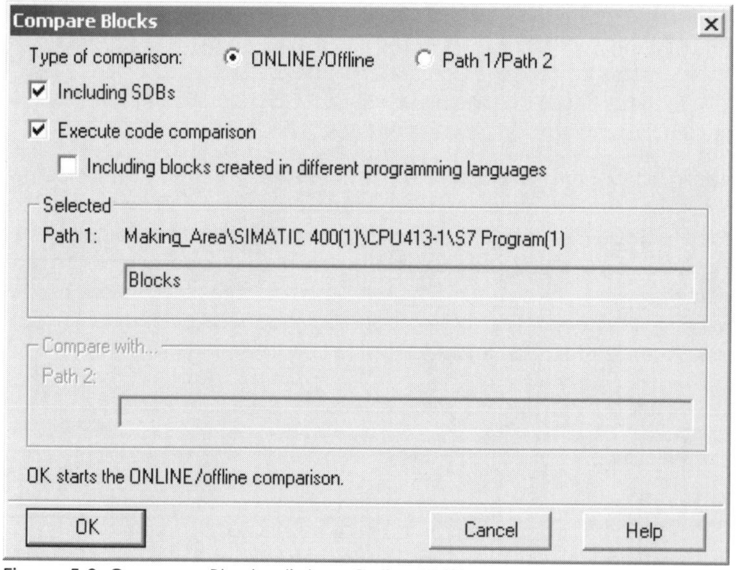

Figure 5-3. Compare Blocks dialog: Online/Offline compare selected.

Comparing Offline Programs – Path 1/Path 2

You may also compare blocks from any two programs by selecting the Path 1/Path 2 option as the type of comparison. This option is typically used to compare two offline programs but may also be used to compare the blocks of two online programs. Using the Path 1/Path 2 option, you may select any offline blocks folder or you may select online blocks while using the online project or accessible nodes window. Path 1 and Path 2, for example may be used to compare the online blocks of a CPU to the blocks of a CPU-independent program as shown in the figure below.

To configure the Path 1/Path 2 compare, simply select the block folder or the individual blocks you want to compare. Your first selection is entered into the dialog box as Path 1. You may then select a second block folder or selected blocks, which will be entered into the dialog as Path 2.

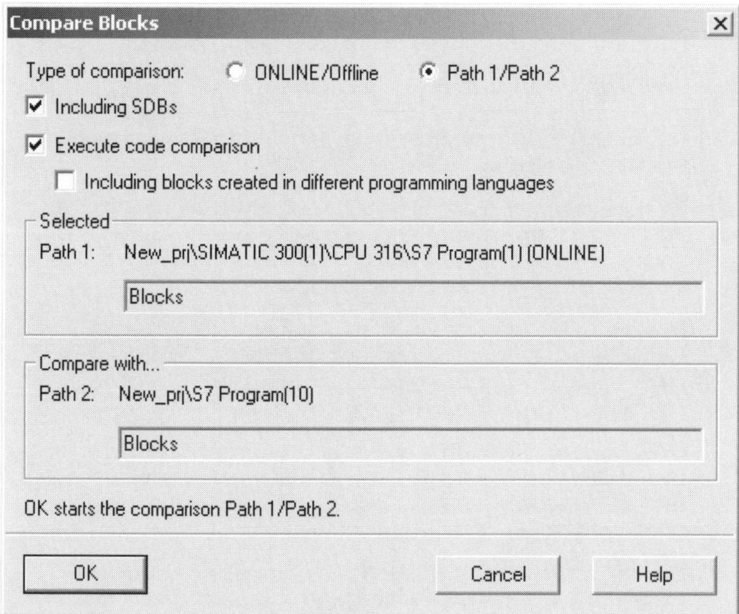

Figure 5-4. Compare Blocks dialog: Path 1/Path 2 compare selected.

Providing CPU Access Protection

Some S7 CPUs support password protection to limit access to the CPU's control program and online operations. This protection is in addition to the protection provided by the key-switch on the CPU face-plate. With password protection, the control program and its data, and online operations, such as upload/download blocks, are protected from unauthorized access (i.e., write access protection) regardless of the key-switch position.

Using CPU access protection involves setting the protection level and if required, defining a password. Once configured, as part of the CPU parameters, the protection level and password is downloaded to the module with the configuration data. Complete read and write access is possible for all password holders, regardless of the key-switch position, but three levels of protection may apply to non-password holders. **Level 1** is the default setting in which standard key-switch access is in operation. **Level 2** provides *write-protection*; read access is possible, but write access is denied without the password. **Level 3** provides *read/write protection*, which means that both read and write access are denied, independent of the key-switch position.

Accessing CPU Information and Operating Characteristics

While in the online mode, using the online project window or the accessible nodes window, you will be able to access information pertinent to any CPU that can be accessed via an online connection. This information, which is resident in the CPU and briefly described in the table below, can only be viewed from an online connection.

Table 5-1. Overview of S7 CPU Module Information.

Information Type	Typical Information
General	Basic information such as the CPU type, order number, firmware, and status information.
Time System	Provides CPU clock information, currently set time and date, status of CPU run-time meters, and other time-related information.
Diagnostic Buffer	A CPU diagnostic tool that tracks and displays the last 100 diagnostic events of the CPU (e.g., Stops, Startups, program faults, I/O faults, diagnostic interrupts, memory resets, power off, etc.).
Memory	Provides information related to CPU Load memory and Work memory, showing total amounts, amount used, and amount free. Also allows the CPU memory to be compressed (elimination of memory gaps to free up space).
Communication	Provides information concerning the communication resources of the CPU (e.g., transmission rates of the CPU interfaces, maximum connections, and percent of scan cycle set for communications load), and how they are currently used.
Scan Cycle Time	Provides CPU cycle time information, displaying the shortest scan, longest scan, and last scan time in milliseconds. In addition, set scan cycle parameters including scan monitoring time (watchdog), and set minimum cycle time are also shown.
Stacks	Provides detail information on the location and cause of the most recent CPU stop. The stacks also relate information prior to the stop, such as the sequence of block calls leading up to the stop, last open data blocks, and the last contents of the CPU accumulators.
Performance Data	Provides information on CPU features such as memory and I/O expansion limits, the total number and address range for digital I/O, timers, counters, and bit memory. Also reports, the number of each user block type that is supported (e.g., FB, FCs, OBs, DBs), what system blocks (i.e., SFCs, SFBs) are available, and the length of the local data stack.
Clear/Reset Memory	Allows you to clear the CPU memory if in the STOP mode.
Set Date and Time	Allows you to set the CPU's internal clock (e.g., Date and Time).
Operating Modes	Provides a view of the current CPU operating mode and allows you to switch the operating mode (e.g., RUN, RUN-P, and STOP).

Comments on Managing Online Interaction with S7 CPUs

The STEP 7 programming system offers many online tools and utilities that aid in managing the various components of your program, as well as for diagnosing problems in either the program or in the control system. In this chapter we will present standard online operations, typically utilized in the handling of the control program, ascertaining and protecting important information related to the CPU. The following chapter continues with online operations by presenting S7 online monitoring and diagnostic tools.

Like in preceding chapters several brief tutorials have been presented here, to offer a quick overview of handling online operations. In the remainder of this chapter examples of standard online operations such as accessing the CPU online without a project, uploading and downloading blocks, comparing online and offline blocks, clearing CPU memory, switching CPU operating modes, accessing a password protected CPU are all presented in a step-by-step manner. The following checklist highlights some of the key points in working online with the S7 CPU.

Checklist: Managing Online Interaction with S7 CPUs

- *Ensure a direct connection between CPU and programming system, using a standard serial cable (and PC Adapter for PC) or an MPI subnet connection.*
- *Go online with a specific CPU using the associated project and the Online Project Window or without a project, using the Accessible Nodes Window.*
- *Verify the resources of a CPU: using an online connection, select the CPU then right-click and select Module Information, then select the Performance Data tab.*
- *Verify that the CPU Date and Time are set correctly to ensure accurate time stamping of messages sent to the CPU's diagnostic buffer and the triggering of Time-of-Day interrupts.*
- *If password protection is enabled for your CPU, consider entering the password once from the SIMATIC Manager to avoid repeated interruptions when attempting online operations.*
- *Before downloading all blocks to the CPU, clear memory first to ensure that all old blocks are removed from the CPU and that the CPU is re-initialized.*
- *If changes are made to any part of the hardware configuration, you only need to download the System Data object of the associated station. The System Data object is located in the offline Blocks folder of the associated CPU.*
- *If memory is insufficient when a download is attempted, compress the memory to free up additional memory by eliminating gaps.*
- *When downloading individual blocks, with the CPU is in the RUN-P mode, you must ensure that called blocks are transferred to the CPU before calling blocks.*

Accessing Online Operations without a Project

Basic Concept

The Accessible Nodes utility allows you to quickly access an S7 CPU for monitoring and debugging, as well as other online operations which involve CPU interaction. Using Accessible Nodes you can connect online to all the S7 CPUs and other programmable modules that can be accessed over the MPI network. Often this method is used when the programming (PG/PC) system does not have a copy of the project, or you simply want to go online quickly without the delay of having to find and open the associated project.

Essential Elements

With a direct or networked MPI link to the CPU, using the Accessible Nodes utility you can access CPU module information, set the CPU clock, clear and reset memory, switch the CPU operating modes, and transfer blocks to and from the CPU. As shown in the figure below, the Accessible Nodes window will contain an S7 folder for each of the active S7 stations that can be reached. Each folder, numbered with the MPI address assigned to the CPU, contains the online blocks of the CPU. The PLC operations are accessed by selecting the desired folder, then from the menu or the right click selecting PLC menu.

Application Tips

You can also access monitoring and diagnostic tools including the *Monitor/Modify Variables*, *Hardware Diagnostics*, and the *CPU Diagnostic Buffer*. Monitoring and diagnostic tasks are discussed in the following chapter. Although you may gain access to the program and data for any CPU opened through the Accessible Nodes utility, this method should be reserved for monitoring and troubleshooting. Editing CPU programs and data without the project is not recommended, since it could result in the loss or distorting of critical project information.

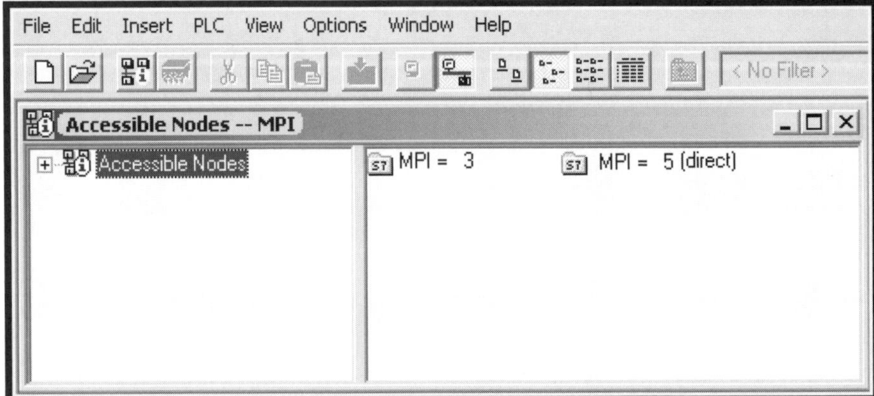

Figure 5-5. Displaying active MPI stations found using the *Accessible Nodes* utility.

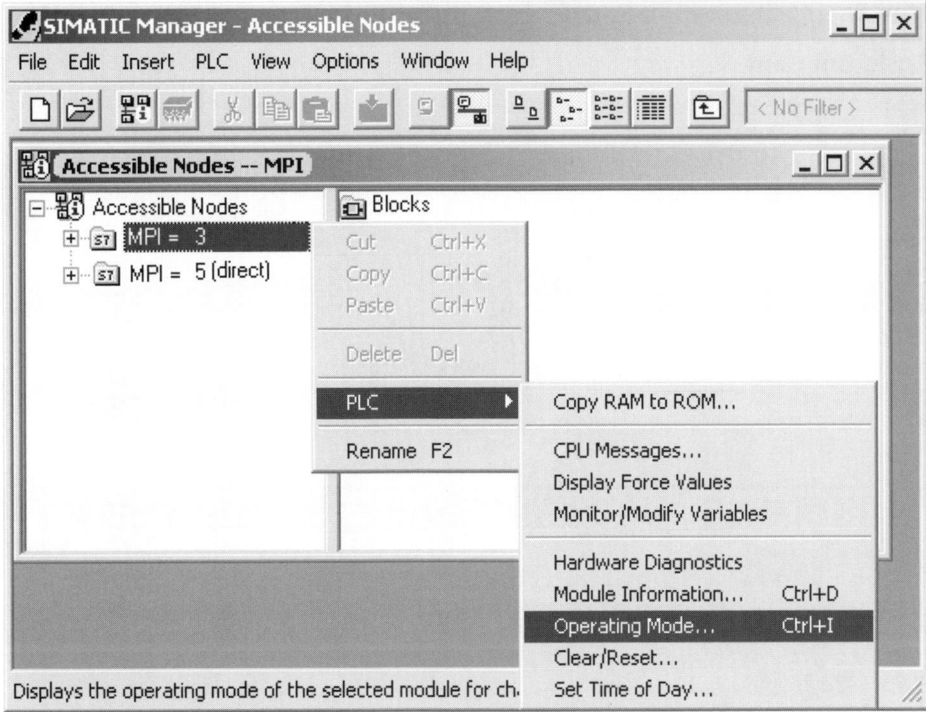

Figure 5-6. PLC menu online operations using accessible nodes utility.

Quick Steps: Accessing Online Operations without a Project

STEP	ACTION
1	Launch the SIMATIC Manager application from the Windows **Start** button.
2	From the menu, select **PLC ➢ Display Accessible Nodes**, or from the toolbar press the *Accessible Nodes* button. Each CPU that can be reached on the MPI subnet is displayed as a folder identified by the MPI address assigned to the CPU.
3	Select the *Accessible Nodes* folder and then from the menu select **View ➢ Expand All** to view the online Blocks folder of each CPU.
4	Select the Blocks folder of the desired station, then from the right pane double-click on the block you wish to open or right-click on the block and select **Open Object**.
5	To view performance characteristics of a CPU, right click on folder and select **PLC ➢ Module Information**. See task "*Viewing CPU Resources and Performance Data*."
6	To view or change the operating mode of a CPU, select **PLC ➢ Operating Mode**. See task "*Viewing and Switching CPU Operating Mode*."
7	To open the dialog for erasing the CPU memory, select **PLC ➢ Clear/Reset**. See task "*Memory Reset from the SIMATIC Manager*."
8	To open the dialog for adjusting the CPU clock, select **PLC ➢ Set Time of Day**. See task "*Setting the CPU Date and Time*."

Viewing CPU Resources and Performance Data

Basic Concept

Whether you are working with an already installed CPU or are planning to select and install a new one, you will need to be aware of the CPU performance and resources data. If, for example, you are planning on a particular program implementation or copying blocks from a library or other program, you will need to verify the compatibility of the resources.

Essential Elements

CPU resource information is resident in the CPU and is accessed via an online connection from either the SIMATIC Manager or the LAD/FBD/STL editor. **Module Information** for the CPU defines features such as memory and I/O expansion limits, the total number and address range for timers, counters, and bit memory, and for digital I/O. This information also reports, the number of each user block type is supported (e.g., FB, FCs, OBs, DBs), what system blocks (i.e., SFCs, SFBs) are available, and the length of the local data stack.

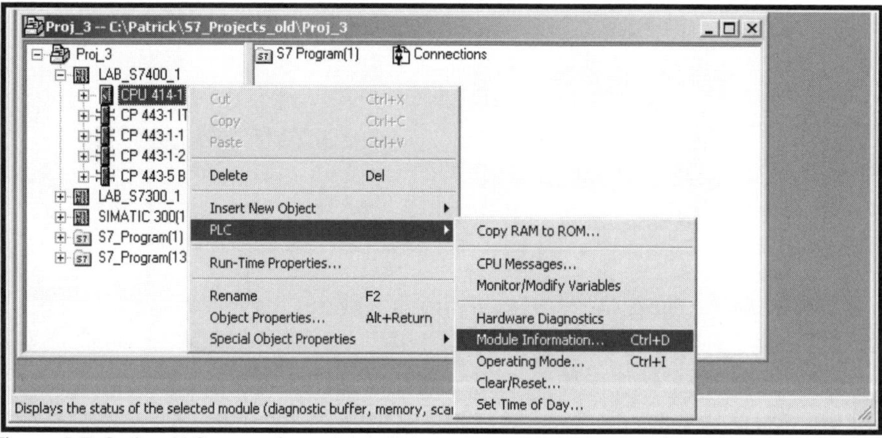

Figure 5-7. Online PLC menu from right click on CPU object.

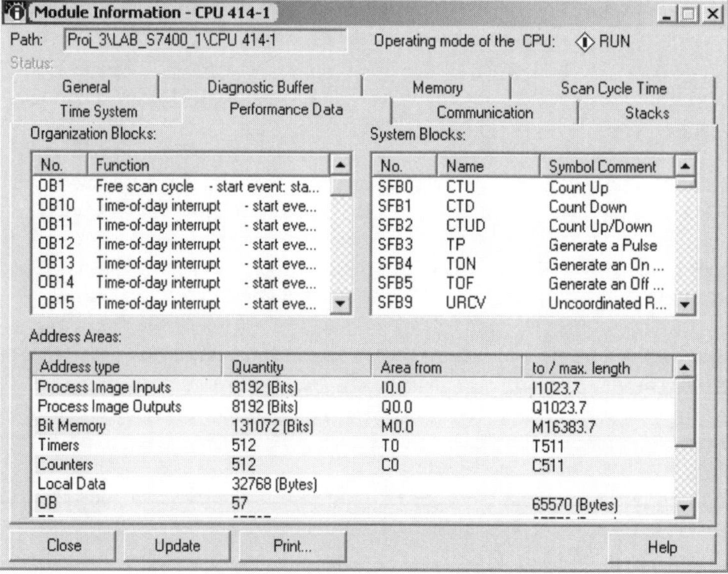

Figure 5-8. CPU Module Information: Performance Data tab.

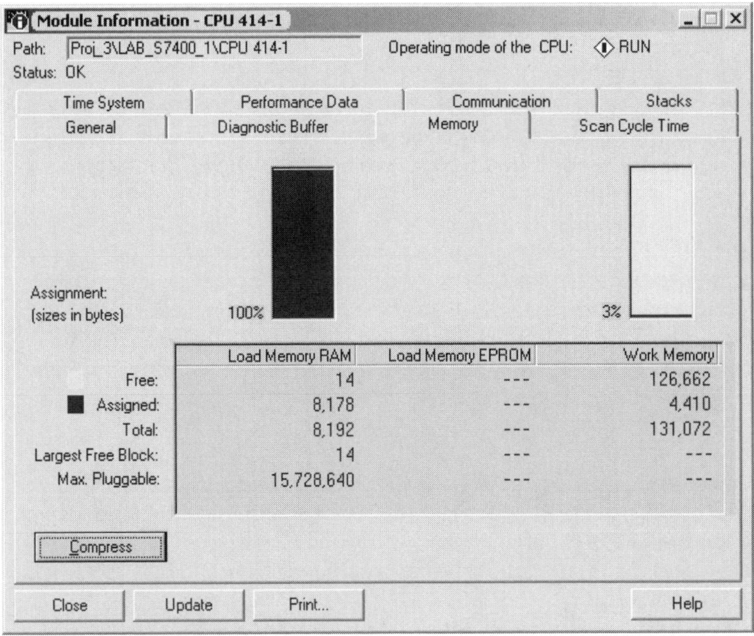

Figure 5-9. CPU Module Information: Memory overview tab.

Quick Steps: Viewing CPU Resources and Performance Data

STEP	ACTION
1	Open the SIMATIC Manager and select the desired CPU from the online project window or from the Accessible Nodes utility to select the CPU.
2	From the menu, select **PLC ➢ Diagnostic/Setting ➢ Module Information** to view CPU resources and performance data online (also from right click with CPU selected).
3	Select the *General* tab to view basic information of the open CPU, including its part number, firmware version, and the rack and slot in which the module is installed. The Status window displays information on the current operational status of the CPU.
4	Select the *Performance Data* tab to view a summary of the CPU resources (e.g., number of each block type, available system blocks, address range of input/output image tables, memory range for timers, counters, bit memory, and local memory).
5	Select the *Memory* tab to view memory resource data; the dialog shows total bytes of assigned and free memory for both work memory and load memory areas.
6	Select the *Communication* tab, to display information on the supported number of communications connections as well as the number presently available.
7	Select the *Scan Cycle Time* tab, to view the longest, the shortest, and the most recent (or last) CPU cycle time, since the last change from STOP to RUN.
8	Select the *Time System* tab to capture and display CPU time and date information, run-time meters, and other time related data for the open CPU.

Viewing and Changing CPU Operating Modes

Basic Concept

The CPU operating mode (e.g., RUN, RUN-P, STOP), may be switched from the CPU face plate (physically) or from the *Operating Mode* dialog (logically) of the STEP 7 software. The operating mode dialog can be opened and viewed with or without the associated project.

Essential Elements

In the *STOP mode*, the CPU program is not processed, and unless access protection is in place, complete read/write access to the CPU program and data is possible. In the *RUN-P mode* the CPU program is being processed, but may be modified. When the CPU is in the *RUN mode*, the main program block (OB1) is being called cyclically by the operating system to process the user program. In this mode, write access to the CPU is denied and program downloads or other write operations are not possible.

Application Tips

It is also possible to affect the operating modes of a CPU through by calling system functions (SFCs) from the control program. With the appropriate SFC, for example, it is possible for one CPU to stop, and restart the CPU of a communications partner. System function blocks (SFBs) for controlling the operating modes of a communications partner include: SFB 19 (START), for executing a complete restart in a partner; SFB 20 (STOP), for switching a partner to STOP; and SFB 21 (RESUME), for triggering a warm restart in the partner CPU. See Appendix B.

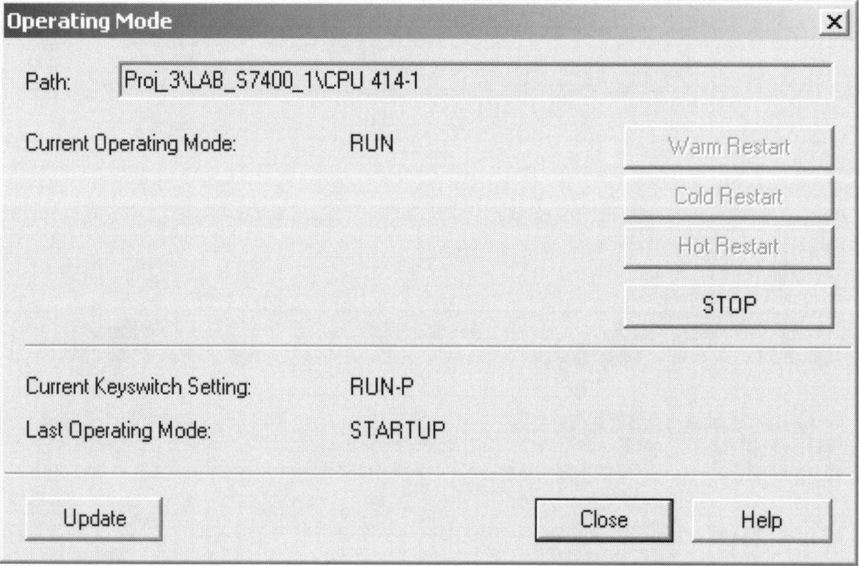

Figure 5-10. CPU Operating Modes dialog: Current operating mode is RUN.

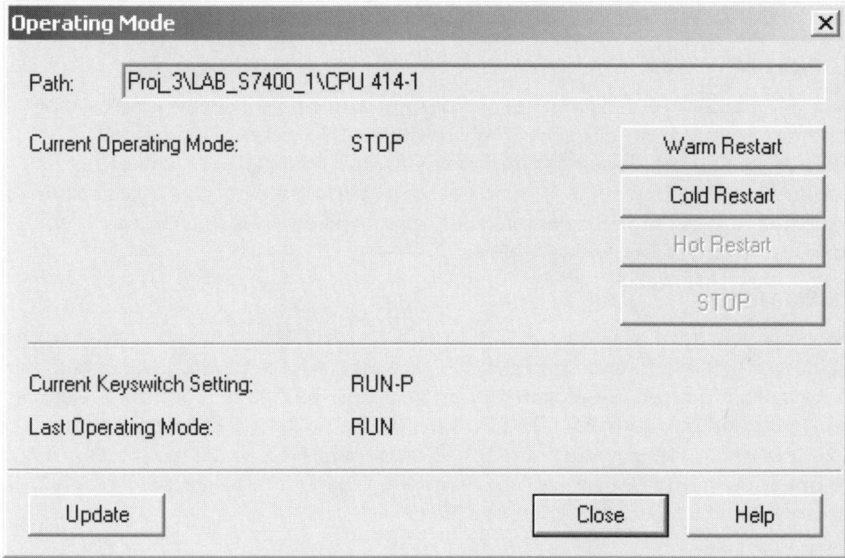

Figure 5-11. CPU Operating Modes dialog: Current mode is STOP, CPU ready for restart.

Quick Steps: Viewing and Changing CPU Operating Modes

STEP	ACTION
1	*To open a CPU without a project*, start the SIMATIC Manager and from the menu select **PLC ➢ Display Accessible Nodes** or from the toolbar press the Accessible Nodes icon; then, select a specific node address (e.g., MPI = 5), right click and select **PLC ➢ Operating Modes**.
2	*To open a CPU with a project*, start the SIMATIC Manager, open the desired project and then from the from the toolbar press the online window icon; then open the station folder, select the CPU module, right click and select **PLC ➢ Operating Modes**.
3	If the current operating mode is **RUN** or **RUN-P**, you may switch the CPU to **STOP**.
4	If the current operating mode is **STOP**, you may switch the CPU to **RUN** or **RUN-P**.
5	To restart, select **Warm Restart** to retain all DB data, retentive timers, retentive counters, and retentive bit memory; non-retentive data are reset. Triggers OB 100.
6	To restart, select **Cold Restart** to delete system generated DBs, reinitialize user DBs from load memory, and reset all other memory areas, including retentive. Triggers OB 102.
7	To restart, select **Hot Restart** (S7-400 only) to retain all data areas, and restart program at the point at which it was interrupted at the last stop. Triggers OB 101.
8	Press the **Update** button to refresh the dialog after changes have been made.

Resetting Memory from the SIMATIC Manager

Basic Concept

In the S7 world a memory reset, also called overall reset, is to delete both the load memory and work memory areas of the CPU and to set the CPU to its initial state (i.e., default parameters). A memory reset operation may be performed from the PLC menu of the SIMATIC Manager in STEP 7, or manually using the mode selector switch, on the front-plate of the CPU. You can access the STEP7 memory reset operation from either the Online Project Window or from the Accessible Nodes Window.

Essential Elements

To clear memory, you must first switch the CPU to STOP mode. Clearing memory, deletes user program blocks from the load and work memory. Address areas are all cleared, connections with communications partners are broken and cleared, and the CPU and other modules are re-initialized to their default state. Both the CPU's diagnostic buffer and MPI address are retained. In CPUs where the program is also stored on a Flash EPROM card the program is not erased. At power up the EPROM contents are transferred to the CPU's work memory. The MPI address is overwritten by the address found on the EPROM.

Application Tips

Memory reset is generally recommended before downloading your program for the first time or whenever reloading the entire program. Such a case may exist, for example, if you suspect faulty program operation or that the program residing in the CPU is not the most current. The CPU may also request a memory reset after determining that the program is corrupt or just after a PLC operating system update is performed. A CPU request for memory reset is indicated at the CPU by a slow flashing (1/2 sec.) of the STOP indicator.

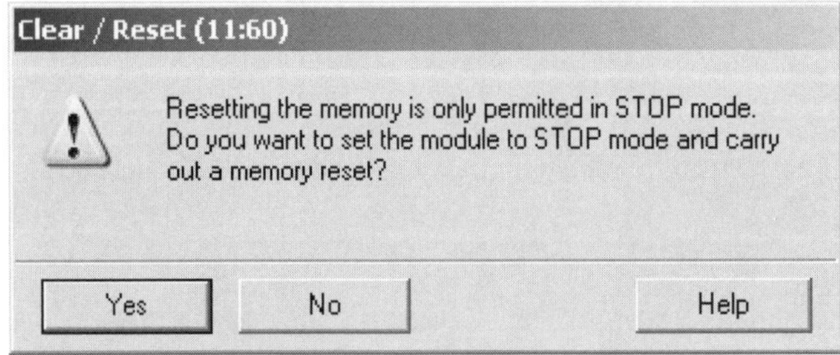

Figure 5-12. S7 CPU operating mode dialog.

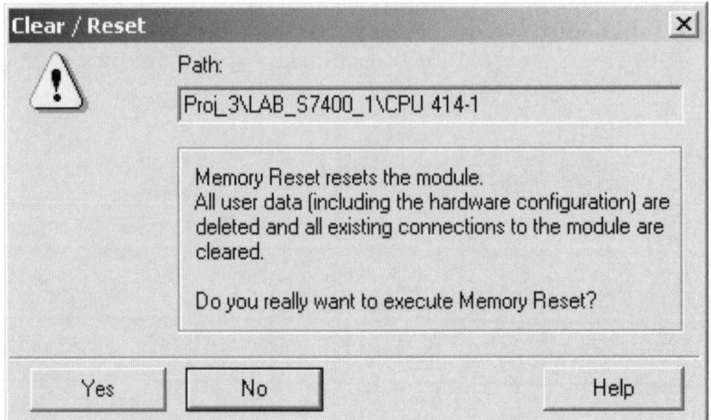

Figure 5-13. Clear memory confirmation dialog.

Quick Steps: Resetting Memory from the SIMATIC Manager

STEP	ACTION
1	Ensure that the CPU mode selector switch is switched to the RUN-P or STOP positions. This operation is not possible if the key-switch is in the RUN position.
2	Start the SIMATIC Manager, open the desired project and select **View ➢ Online** or press the Online Window icon to open the online project window. Alternatively, you may open the Accessible Nodes Window; press the Online Window toolbar icon.
3	From the project window, expand the desired Station folder and select the CPU module whose memory will be cleared; or from the Accessible Nodes window select the CPU according to the MPI address.
4	Select the CPU object, right-click and select **PLC ➢ Clear/Reset**; Confirm your intention to perform the memory reset by pressing the **YES** button. You will be prompted to place the CPU in the STOP mode if necessary.
5	After a memory reset, you may download your program and then restart the from the CPU mode selector switch or from the SIMATIC Manager (i.e., select **PLC Diagnostic/ Setting ➢ Operating Mode**, and then select the *Warm Restart* button.
	Manual Memory Reset from the CPU Mode Selector Switch.
1	Switch the CPU mode selector switch to the STOP position (The STOP LED will light).
2	Turn and hold the mode switch in the (memory reset) position, until the yellow STOP LED blinks at least twice at a 1 second period, after which it will remain illuminated.
3	Release the mode selector switch to the STOP position; and immediately return to the MRES position and hold for 4-5 seconds; the yellow LED will flash quickly while memory is being cleared and will flash slowly when done.
4	Release the switch back to STOP position and the overall reset should be complete.
5	Open the online blocks folder to verify the memory reset (System blocks should list).

Accessing a Password Protected CPU

Basic Concept

If password protection is enabled for a CPU or other module, then depending on the set level of protection and the specific module, entry of the correct password is required before certain online operations are possible. Read and write access is possible for password holders, independent of the key-switch position or the set level of protection. However, operations, involving CPU programs and data, are denied to non-password holders. The idea is to prohibit unauthorized online access that could result in adverse operation of the program or process.

Essential Elements

During normal operations, with level 2 or level 3 protections, upon each attempt to perform certain online functions you will be prompted to enter the required password. For example, Monitor/Modify Variables or Upload/Download blocks, are affected by protection level 2. If the correct password is entered, access is permitted to modules for which a particular protection level was set during parameter assignment. You can then establish online connections to the protected module and execute the online functions belonging to that protection level.

Application Tips

To avoid being interrupted at each attempt to go online with a protected CPU, you may set the access rights once from the SIMATIC Manager. When the access rights are set from the SIMATIC Manager, by entering a correct password, the access will apply until the last S7 application is closed, unless the access rights are canceled first. Using the menu command **PLC ➢ Access Rights ➢ Setup**, you can call the "Enter Password" dialog box directly. The current password access is cancelled from the SIMATIC Manager menu command **PLC ➢ Access Rights ➢ Cancel.**

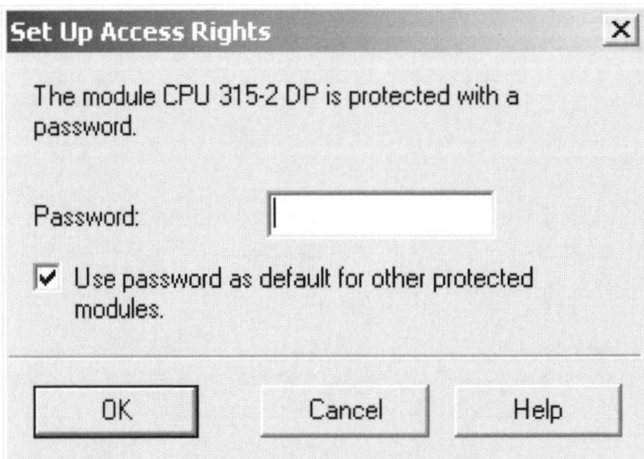

Figure 5-14. Password access rights dialog.

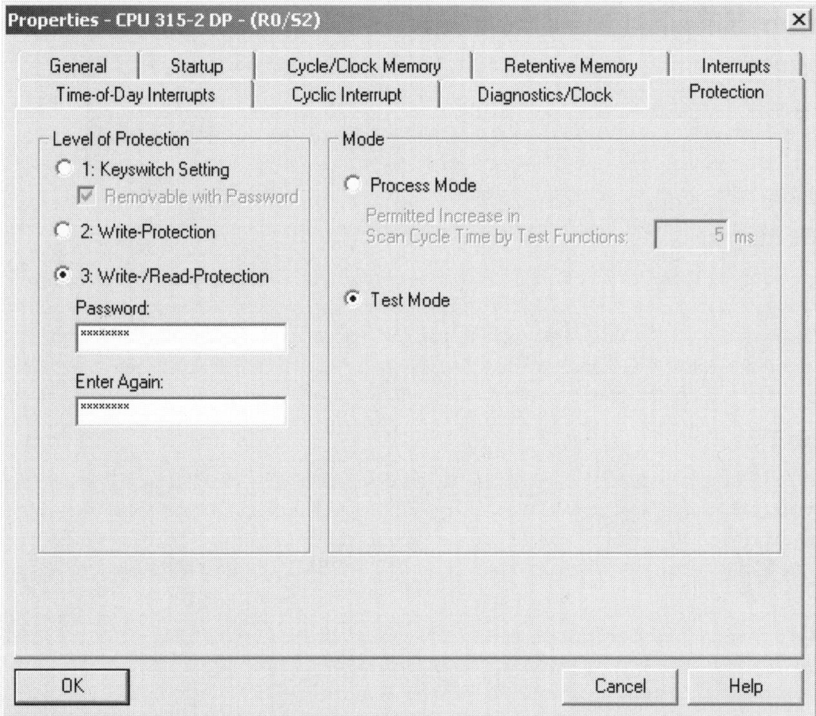

Figure 5-15. CPU Hardware Configuration dialog for setting CPU Protection.

Quick Steps: Accessing a Password Protected CPU

STEP	ACTION
1	Start the SIMATIC Manager and open to the desired project; use the Online Window icon to open the project online.
2	Expand the desired Station folder and select the CPU module you wish to access.
3	From the SIMATIC Manager menu, select **PLC** ➢ **Access Rights** ➢ **Setup** or with the CPU selected right click and select **PLC** ➢ **Access Rights** ➢ **Setup,** then enter the correct password when the access rights dialog box appears.
4	Activate the **Use Password as Default** checkbox in order to use the same password for any other installed modules that are password protected.
5	To cancel the access rights for the current session you may select **PLC** ➢ **Access Rights** ➢ **Cancel,** at any time from the SIMATIC Manager menu. Or if you are done, closing the SIMATIC Manager will terminate the current CPU access rights.

Setting the CPU Date and Time

Basic Concept

Most S7-300 and S7-400 CPUs incorporate a real-time clock which you may adjust from the STEP 7 software. Setting the date and time are important if your application will use functions such as the run-time meter, or interrupt service routines that are triggered on a specific date and at a specific time.

Essential Elements

The CPU clock may be manually adjusted to whatever time you choose, or synchronized with the system time and date of the STEP 7 PC or PG programming system. Although real-time may not be required in your application, setting the CPU clock with the correct time and date is still essential since the CPU clock is used to assign the date and time stamp to all events logged by the CPU Diagnostic buffer.

Application Tips

Not only can the CPU clock be adjusted manually as discussed here, but through the use of system functions (SFCs) within your program, it is also possible to set, stop, and read the current date and time. The system function for setting the date and time is SFC 0 (SET_CLK); the system function for reading the date and time is SFC 1 (READ_CLK); and the system function for synchronizing all CPU clocks is SFC 48 (SNC_RTCB). See Appendix B.

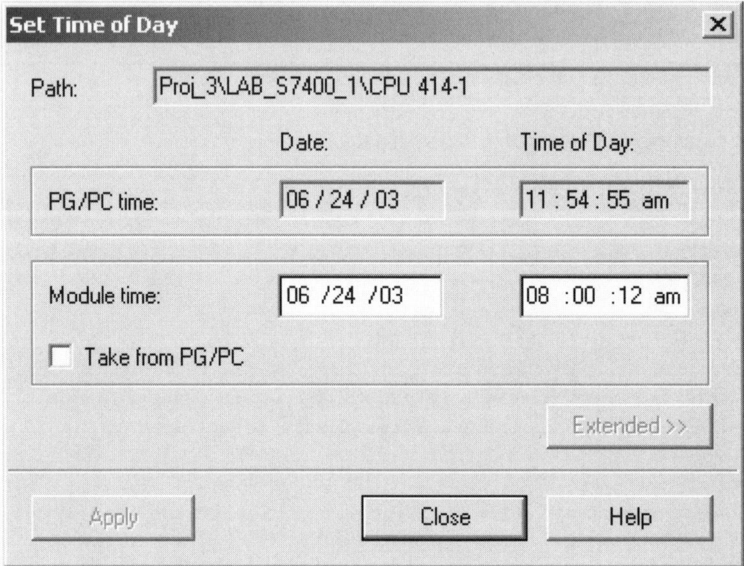

Figure 5-16. Set S7-300/S7-400 CPU Date/Time dialog.

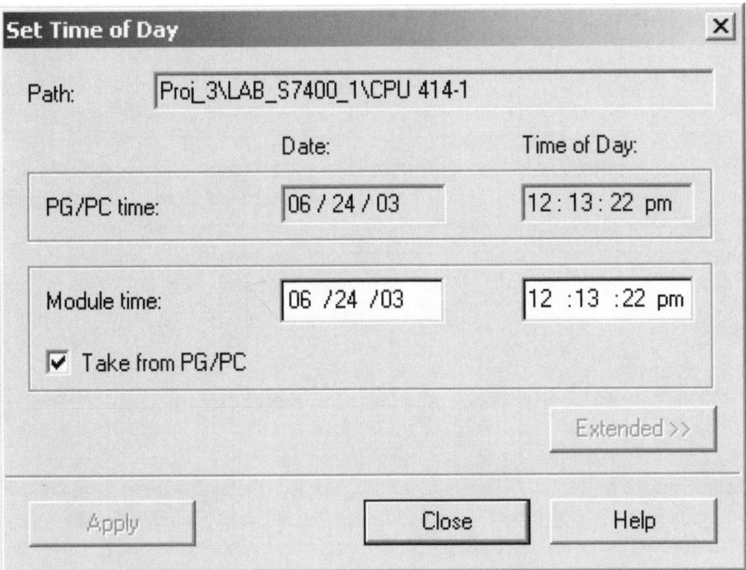

Figure 5-17. Synchronize CPU Date/Time with PC System Date/Time.

Quick Steps: Setting the CPU Date and Time

STEP	ACTION
1	Start the SIMATIC Manager and open to the desired project.
2	Open the station folder and select the CPU module for which you intend to set the time and date.
3	With the CPU selected right-click and select **PLC ➢ Set Time of Day**; or from the menu select **PLC ➢ Diagnostic/Setting ➢ Set Time of Day** to open the CPU clock dialog.
4	In the **Time** field under **Module**, enter the CPU time as hour: minute: second, in the format hh:mm:ss followed by am or pm.
5	In the **Date** field under **Module**, enter the CPU date as month/day/year, in the format mm/dd/yy.
6	Activate the "**Take from PG/PC**" check box and then press the **Apply** button, to set CPU time and date to match the operating system time and date of the PG/PC.

Compressing CPU Memory

Basic Concept

Similar to your PC hard drive, the work memory of an S7 CPU will sometimes require de-fragmenting. In this case, however, the process is referred to as "*Compressing*." During program development, many corrections and deletions take place online, that result in a fragmented memory. When an existing CPU block is edited and resaved, in actuality the new block is appended to the end of memory while the existing block is marked inactive. By pushing the blocks together, compression eliminates memory gaps formed by inactive blocks.

Essential Elements

The memory compress operation is activated from the Memory tab of the CPU Module Information dialog. With an online connection established you may open the project window or the accessible nodes window. With the desired CPU selected in the online project window or the desired MPI address folder selected in the accessible nodes window, right click and select **PLC ➢ Module Information**, then select the Memory tab. The best possible compression is achieved when you compress memory while the CPU is in the STOP mode. Compression is possible in the RUN-P operating mode — however blocks that are being executed cannot be shifted in memory while they are open. The compress function is not possible in RUN mode, since memory is then write-protected.

Application Tips

The memory compress operation may also be executed on demand from your S7 program, using the system function COMPRESS (SFC25). When called from SFC 25, the complete operation is performed over a number of CPU cycles. While still in progress, the SFC BUSY output reports logic 1; the DONE output reports logic 1 when the operation is completed. As mentioned before, the memory gaps are completely removed only when you perform the compress operation with the CPU in STOP mode.

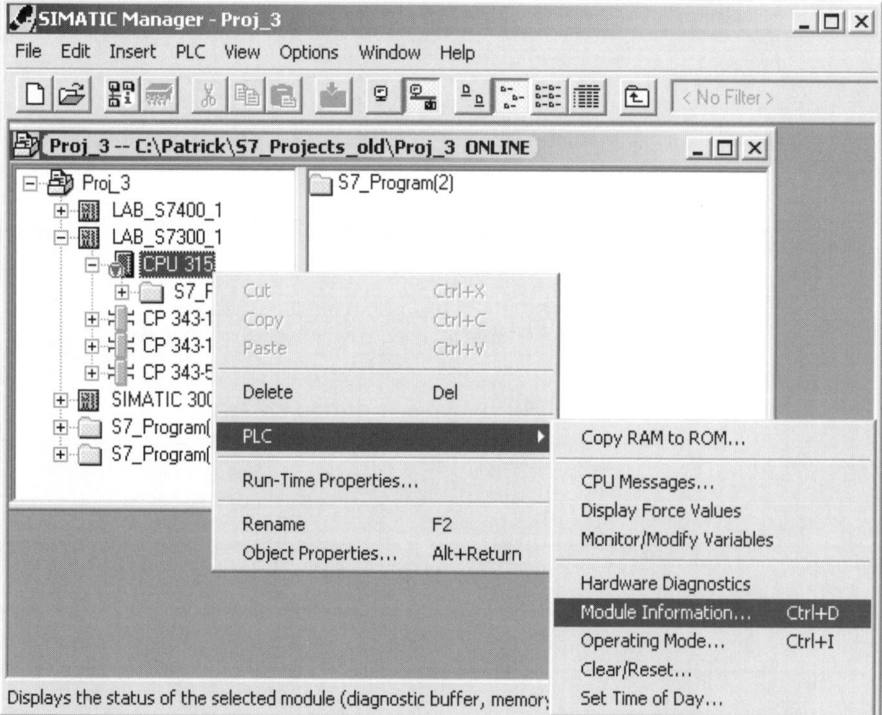

Figure 5-18. Online PLC menu from right click on CPU object in online project window.

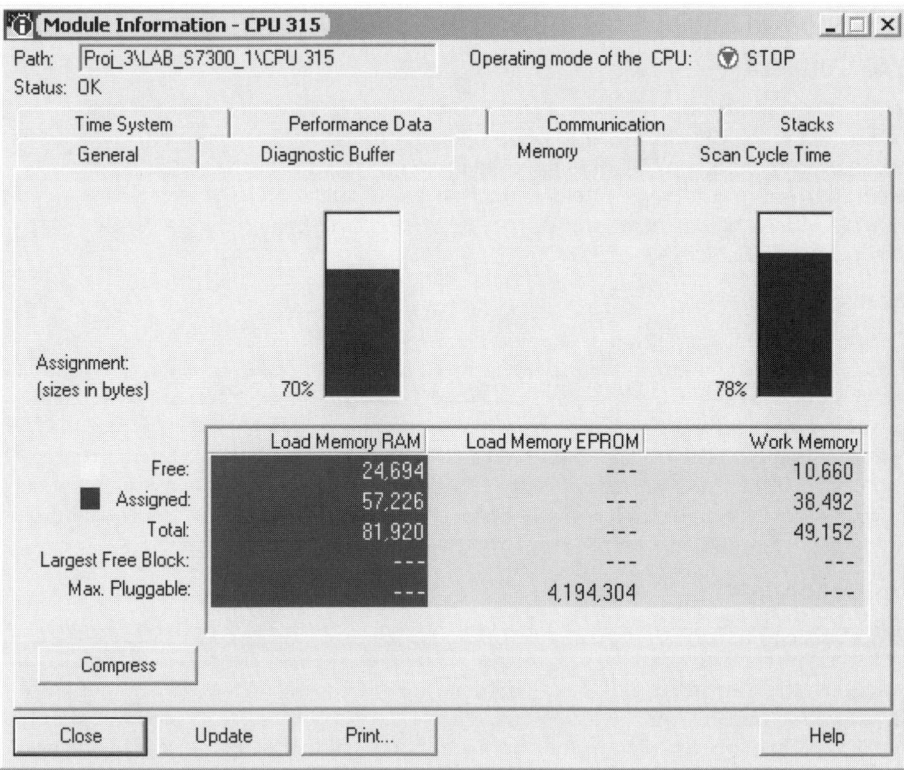

Figure 5-19. CPU Module Information: Memory tab.

Quick Steps: Compressing CPU Memory

STEP	ACTION
1	Start the SIMATIC Manager; open the desired project and *Station*.
2	Under the station, select the CPU whose memory you intend to compress.
3	With the CPU selected right-click select **PLC** ➢ **Module Information**.
4	Select the Memory tab from the Module Information dialog.
5	Press the **Compress** button to eliminate gaps from the CPU memory.
6	Press the **Update** button to refresh the dialog and view the compression results.

Downloading the S7 Program or Selected Blocks

Basic Concept

In order to be processed, a STEP 7 program (i.e., OBs, FCs, FBs, and DBs) must first be downloaded to the CPU. In a project, each CPU has an associated S7 Program folder that is located directly under the CPU object. Downloading a program involves selecting the correct CPU, opening its Program folder, and then selecting the offline *Blocks* folder. Only offline blocks may be downloaded. You may download the Blocks folder (i.e., entire program), or one or more selected blocks.

Essential Elements

To download an entire program or selected blocks, you must first ensure that there is an established online connection to the associated CPU. To download blocks, the CPU mode selector switch must be in either the STOP or RUN-P operating mode. If you are operating in the RUN-P mode during a download, you will be prompted to switch the CPU to the STOP mode prior to the download and to restart the CPU after the download. If you attempt top download all blocks, you will also be asked if the *System Data* should be transferred. Recall that the System Data contains the station hardware configuration. To download hardware configuration changes only, simply select and download the System Data object.

Application Tips

When downloading blocks, if the available load memory is too small you will be asked by STEP 7 if you wish to compress the CPU memory. Compressing the memory will make a larger contiguous space in memory to complete the download, by eliminating memory gaps. An alternate download method is to open both the offline and online project windows, select the block folder of each program, then from the offline window select the objects you want to download and drag them to the online window.

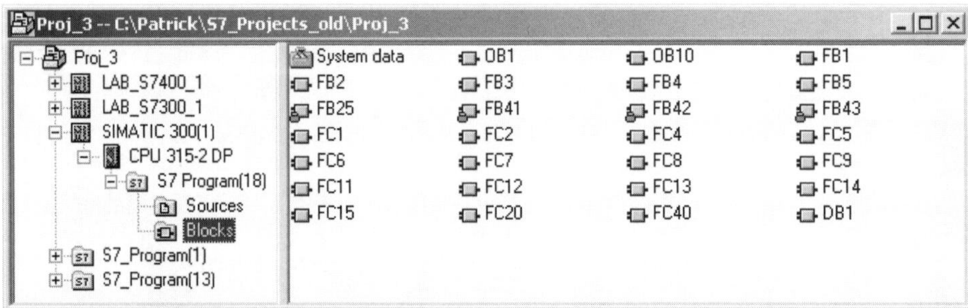

Figure 5-20. Offline Blocks Window: Downloading offline blocks folder (all blocks) to the CPU.

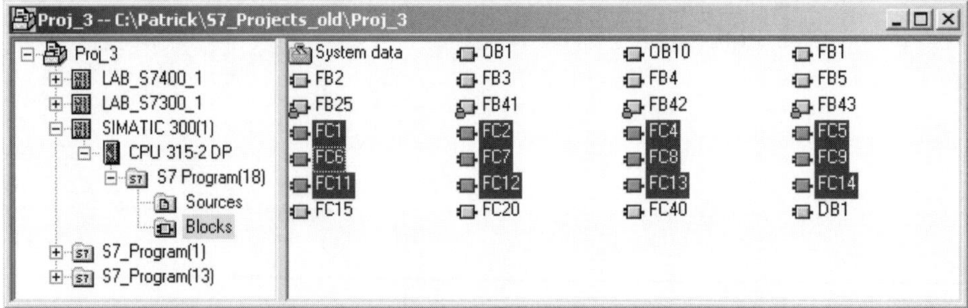

Figure 5-21. Offline Blocks Window: Downloading selected offline blocks to the CPU.

Quick Steps: Downloading the S7 Program or Selected Blocks

STEP	ACTION
	Download Program: With the offline Bocks folder selected, download entire program
1	If your CPU is in the RUN mode, you will need to switch to STOP before attempting the download.
2	Although it is not required, performing an overall reset on the CPU is recommended when downloading the entire program.
3	With the SIMATIC Manager open to the required project, from the offline blocks window, double click on the program folder under the required CPU, and then select the blocks folder under the program.
4	With the program folder selected, click on the Download icon from the toolbar; or from the menu select **PLC ➢ Download**, to start the download operation. When prompted by STEP 7 as to whether a block already in the CPU should be overwritten. You may respond **Yes**, **All**, **None**, **No**, or **Cancel**.
5	Choose **Yes** to overwrite the current block; **All**, to overwrite all blocks that already exist in the CPU; **None**, to not overwrite any existing blocks; **No**, to skip the current block; and **Cancel**, to abort the Download.
6	After a complete download, switch to the CPU operating modes dialog and perform a Warm Restart of the CPU (STOP to RUN).
	Download Blocks: With the offline Blocks folder open, download one or more blocks
1	If your CPU is in the RUN mode, you will need to switch to STOP before attempting the download. RUN P is allowed.
2	With the SIMATIC Manager open to the required project and the desired CPU folder open, select the offline *Blocks* folder. The blocks are displayed in the right pane.
3	To select two or more non-adjacent blocks to download, press and hold the CTRL-key down, and click on as many non-adjacent blocks as required.
4	To select a group of adjacent blocks in a row or column, press and hold the SHIFT-key, then click on the first then the last block of the row or column; press and hold the CTRL-key down, and click on non-adjacent blocks to include additional blocks to the already selected blocks.
5	With one or more blocks selected, click on the Download icon from the toolbar; or from the menu select **PLC ➢ Download**, to start the download operation. When prompted by STEP 7 as to whether a block already in the CPU should be overwritten. You may respond **Yes**, **All**, **None**, **No**, or **Cancel**.
6	Choose **Yes** to overwrite the current block; **All**, to overwrite all blocks that already exist in the CPU; **None**, to not overwrite any existing blocks; **No**, to skip the current block; and **Cancel**, to abort the Download.
7	Return the CPU to RUN mode if it was stopped.

Uploading the S7 Program or Selected Blocks

Basic Concept

To upload an S7 program is to transfer the online blocks of a CPU back to the offline blocks folder of the associated S7 Program. The blocks displayed in the online blocks window, are the blocks contained in the CPU's memory and can only be directed to the offline blocks folder of the program associated with the CPU. You may upload the entire online Blocks folder (i.e., complete S7 program), or one or more selected blocks.

Essential Elements

To upload the entire S7 program or selected blocks of a CPU, you must first ensure that there is an established online connection to the associated CPU. There are no other requirements for performing the upload unless the CPU is password protected. In this case, you will be prompted to enter the correct password before the operation can be completed. If you attempt to upload all blocks, you will also be asked if the *System Data* should be transferred — this process will overwrite your offline System Data. Recall that the System Data contains the hardware configuration for the selected PLC Station.

Application Tips

You may also use Copy and Paste to transfer the online blocks of a CPU to a program not associated with a particular CPU (i.e., CPU independent program). Simply open the online blocks window of the CPU, select the desired blocks, open the target offline blocks folder and perform the paste.

When you perform an upload of the online blocks folder, the complete offline blocks folder is overwritten. Prior to the upload, you may consider copying the offline blocks folder to a CPU independent program, in case for some reason you may need to revert to these blocks.

Figure 5-22. Online Blocks Window: Uploading all CPU blocks to the offline blocks folder.

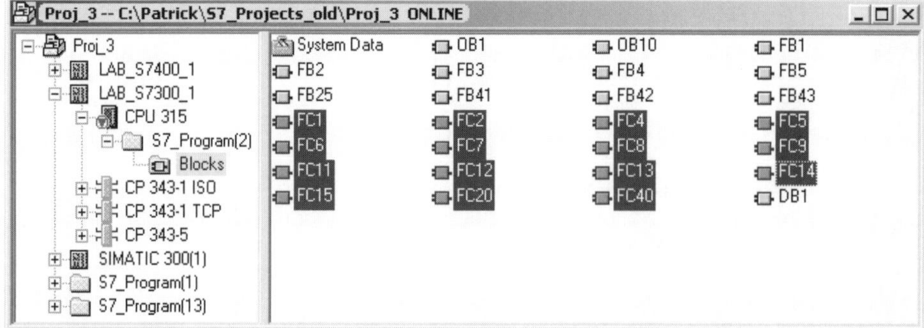

Figure 5-23. Online Blocks Window: Uploading selected CPU blocks to offline blocks folder.

Quick Steps: Uploading the S7 Program or Selected Blocks

STEP	ACTION
	Upload Program: With the online Blocks folder selected, upload the entire program
1	With the SIMATIC Manager open to the desired project, press the online window icon and then navigate to the correct *Station* and *CPU*.
2	Expand the CPU folder, then *S7 Program* folder, and then select the *Blocks* folder.
3	With the online blocks folder selected, click on the upload icon on the toolbar; or from the menu select **PLC ➢ Upload** to start the upload operation.
4	When prompted by STEP 7 as to whether an uploaded block that already exists in the offline folder should be overwritten, respond **Yes**, **All**, **None**, **No**, or **Cancel**.
5	Choose **Yes** to overwrite the current block; **All**, to overwrite all blocks that already exist in the offline blocks folder; **None**, to not overwrite any existing blocks; **No**, to skip the current block; and **Cancel**, to abort the Upload.
	Upload Blocks: With the online Blocks folder selected, upload one or more blocks.
1	With the SIMATIC Manager open to the desired project, press the online window icon and then navigate to the correct *Station* and *CPU*.
2	Expand the CPU folder, then *S7 Program* folder, and then select the *Blocks* folder.
3	With the online blocks folder selected, the CPU blocks are listed in the right pane.
4	In the online blocks window, hold the SHIFT-key and use the mouse to select a group of adjacent blocks in a row or column; to select two or more non-adjacent blocks, select each block while holding CTRL-key.
5	With the highlighted blocks selected, either click on the upload icon on the toolbar; or from the STEP 7 menu select **PLC ➢ Upload**, to start the upload operation.
6	When prompted by STEP 7 as to whether an uploaded block that already exists in the offline folder should be overwritten, respond **Yes**, **All**, **None**, **No**, or **Cancel**.
7	Choose **Yes** to overwrite the current block; **All**, to overwrite all blocks that already exist in the offline blocks folder; **None**, to not overwrite any existing blocks; **No**, to skip the current block; and **Cancel**, to abort the Upload.

Comparing Offline/Online Programs

Basic Concept

One of the options of the compare operation lets you compare the offline blocks of a specific program to the corresponding online blocks of the associated CPU. This compare operation can be executed from the SIMATIC Manager or from the LAD/FBD/STL editor.

Essential Elements

When defining the online/offline compare dialog, you must select a CPU program; you may select the offline blocks folder to compare all blocks or one or more selected blocks. You may include or exclude the System Data, and include or exclude blocks created in different languages. The compare outcome is presented in a *Results* window that lists the blocks with differences and allows you to go directly to the LAD/FBD/STL editor to view any differences.

Enable the "Compare code" check box to compare code blocks of the program. Once this function is enabled, you can also specify in the next check box whether blocks generated in different programming languages should be compared (e.g. OB1 in STL with OB1 in LAD).

Application Tips

When an off-line block folder is compared with an on-line block folder, only downloadable user blocks (e.g., OBs, FBs, FCs, and DBs) are compared. Variable tables (VATs), user defined types (UDTs) SFBs, and SFCs, are not compared.

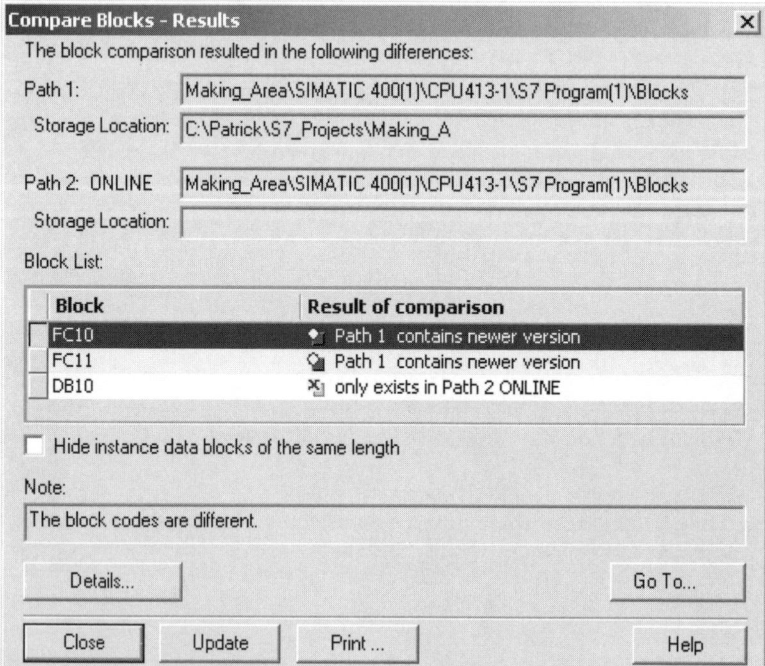

Figure 5-24. Compare Blocks Results dialog lists differing blocks.

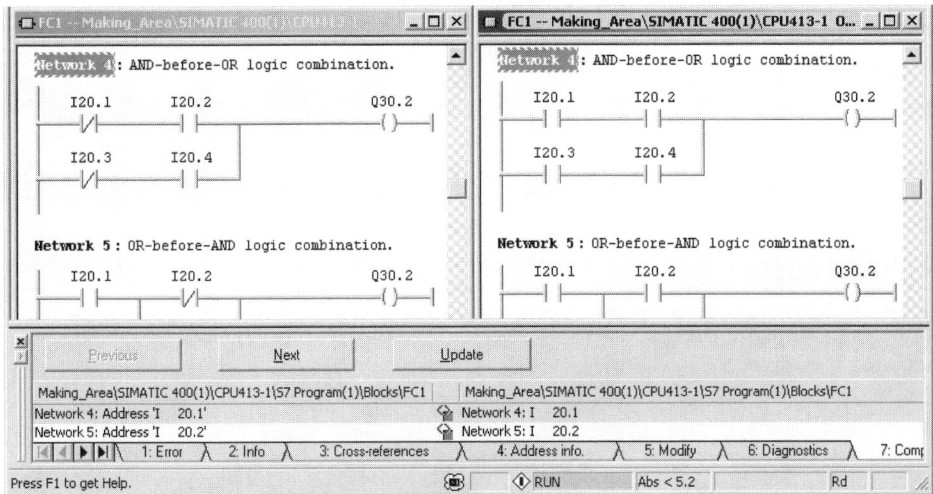

Figure 5-25. LAD/FBD/STL editor with Online/Offline windows open to first difference found.

Quick Steps: Comparing Offline/Online Programs

STEP	ACTION
1	Start the SIMATIC Manager with the desired project; Open the CPU program folder and select the offline blocks folder or a selected group of blocks to compare.
2	With the offline blocks folder or one or more blocks selected, from the menu select **Options ➢ Compare Blocks**; or right-click and select **Compare Blocks**.
3	When the Compare Blocks dialog is presented, activate the **Online/Offline** radio button, if not already selected. The selected offline block folder or the group of blocks you selected should already appear in the Path 1 field.
4	Activate or deactivate the **Including SDBs** check box to include or exclude the system hardware configuration in the compare operation.
5	Activate or deactivate **Execute Code Compare** to compare actual code or simply compare by date-time stamp of last time each offline/online block was saved.
6	Confirming your selection with **OK** starts the compare operation.
7	When, the **Compare Blocks Results** dialog is presented; it lists each block that differs in the online (Path 1) and offline (Path 2) programs. The result indicates which block was the *newer version*, and whether a block *only exists in the online* or *offline folder*.
8	Press the **Go To...** button opens the LAD/FBD/STL editor with online and offline windows of the block; the first found executable code difference is displayed.
9	Use the **Next** and **Previous** buttons to navigate forward to the next network with difference or backward to the previous network with a found code difference.
10	As corrections are made and the changes are saved, use the **Update** button to refresh the difference list to reflect the remaining networks with a code difference.

Comparing Two Programs (Path1/Path2)

Basic Concept

The STEP 7 compare operation allows blocks of any two offline programs to be compared for code differences. The operation may involve CPU or CPU-independent programs from the same or different projects. Outcome of the compare will present a windowed list of compared blocks having differences. This compare operation can be executed from the SIMATIC Manager or from the LAD/FBD/STL editor.

Essential Elements

When defining the offline compare dialog, you may use any two offline programs, including programs not associated with a CPU. You may select an offline blocks folder to compare the entire program, or selected blocks. You may include or exclude the System Data (the station configuration), and you may include or exclude blocks that were created in different languages. The compare outcome is presented in a *Results* window that lists the blocks with differences and allows you to go directly to view the differences in the LAD/FBD/STL editor. When the editor is open, ensure that the Detail view is activated from the menu.

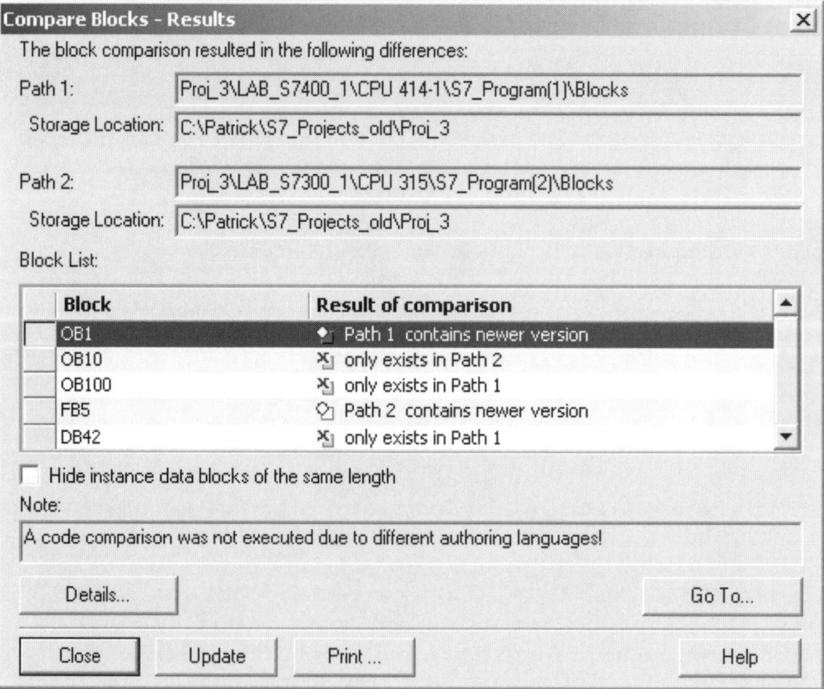

Figure 5-26. Compare Blocks Offline compare result, lists blocks found different.

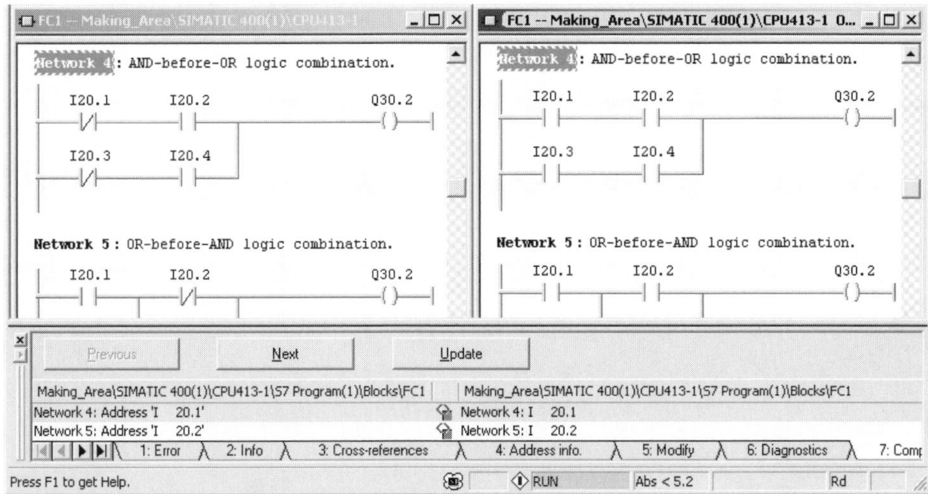

Figure 5-27. Compare dialog set to compare offline blocks folders of two programs.

Quick Steps: Comparing Two Programs (Path1/Path2)

STEP	ACTION
1	From the SIMATIC Manager, open the desired project and program, then select the offline blocks folder or a group of selected blocks (**Path 1**) to compare; then from the menu select **Options ➢ Compare Blocks**; or right-click and select **Compare Blocks**.
2	When the Compare dialog is presented, activate the **Path1/Path2** radio button, as the "**Type of Comparison**". Your first selection will already appear in the Path-1 field; navigate to the project and the second program folder to select the offline blocks folder or a group of selected blocks to be compared (**Path 2**).
3	Activate or deactivate the **Including SDBs** check box to include or exclude the system hardware configuration in the compare operation.
4	Activate or deactivate **Execute Code Compare** to compare actual code or simply compare by date-time stamp of last time each offline/online block was saved.
5	Confirm your selection with **OK** to start the compare operation.
6	When, the **Compare Blocks Results** dialog is presented; it displays a list of each block that differs in the Path 1 and Path 2 programs. The result indicates which offline block was the *newer version*, and whether a block *only exists in Path 1 or in Path 2.*
7	Press the **Go To...** button opens the LAD/FBD/STL editor with online and offline windows of the block; the first found executable code difference is displayed.
8	Use the **Next** and **Previous** buttons to navigate forward to the next network with difference or backward to the previous network with a found code difference. As you press Next or Previous, both editor windows are positioned at the selected network.
9	As corrections are made and the changes are saved, use the **Update** button to refresh the difference list to reflect the remaining networks with a code difference.

Working with Monitoring and Diagnostic Tools

Objectives

- Use Debug Monitor to Evaluate S7 Program
- Create and Use Variable Tables (VATs)
- Introduce Monitor/Modify Variables Utility
- Introduce and Apply Force Variable Utility
- Introduce the Diagnose Hardware Utility
- Introduce the CPU Diagnostic Buffer
- Introduce the CPU Stacks as Diagnostic Tools
- Introduce the Program Reference Data Utility

Tools for Monitoring Programs and Data

Using the Debug Monitor to Evaluate the Program Status

The STEP 7 *Debug Monitor* is an online debug utility, available in the LAD/FBD/STL editor, when an online connection to a CPU is established and the CPU is in either the RUN or RUN-P operating mode. Displaying a block in this mode allows you to view and to evaluate the code of a block as it is being processed. Combined with the Monitor/Modify Variables utility you are able to monitor and evaluate both your program and data online.

Using Variable Tables

A variable table (VAT) is a table of specific memory addresses that can be created and used to assist you in monitoring, modifying, or forcing memory and process variables in your program. VATs are generated from the Simatic Manager, like other blocks, and are placed stored in the offline Blocks folder. Since VATs are not code or data blocks, they are not downloaded to the PLC. Once created, the VAT can be recalled to view and manipulate a particular variable set in the Monitor/Modify Variables or Force Variables utilities. Up to 255 VATs may be generated, numbered from VAT 1 to VAT 255.

Monitoring and Modifying Variables

The Monitor/Modify Variables utility allow you to both monitor and modify the defined constants and variables of your process and S7 program. Using this utility, you will be able to view and manipulate, in real time, the various S7 memory areas (e.g., bit memory, timers, counters, inputs, outputs, peripheral inputs, peripheral outputs, and data block variables), that you have defined in a variable table (VAT). Using the monitor icons (eyeglasses in Fig. 6-1) you can start the monitor utility and view the screened variables as they update. This operation can be performed using a direct or network connection to the appropriate CPU. As you modify variables, you can use the appropriate toolbar icons, to trigger updates that cause new values to be written once or continuously.

Figure 6-1. Debug Monitor toolbar.

Forcing I/O and Memory Variables

Forcing is a useful program and I/O diagnostic tool available in all S7-400s, and with limited scope in some S7-300s. This feature allows you to selectively override the normal program influence on memory variables of your S7 program, as well as the influence of the normal I/O update on the actual status of input and output devices. Using this utility, you will be able to force the desired statuses of I/O devices or the values of memory variables, until they are released by your command. With forcing, output wiring can be checked, and control logic can be fully tested without having to go out and manually open or close switch contacts or to wait until a temperature or speed setpoint is reached.

Using STEP 7 Program Reference Data

STEP 7 provides several types of program reference data that can be displayed or print as reports. When you decide to use these reports, available from the Options menu, you have the option of displaying, filtering, or deleting the current program reference data (e.g., **Options ➢ Reference Data ➢ Display**). From this same menu, you may filter the reference data to determine what is reported, and regenerate it to process newly created blocks, or blocks that may have been edited. From the LAD/FBD/STL editor, you can also ensure that the reference data is automatically generated for each newly created block. Select **Options ➢ Customize** then on the *Blocks* tab, activate the "*Create Reference Data*" checkbox.

Program Reference data reports include the *Assignment Lists, Program Structure, Cross References, Unused Symbols,* and *Addresses without Symbols.*

Assignment List

The assignment list is a memory usage overview of the inputs (I), outputs (Q), bit memory (M), timers (T), and counters (C). The report is actually divided into two parts — the so-called I/Q/M list, and the Timer/Counter list. Since the entire area is shown for each of these memory areas, you can easily determine what locations are available and if any were improperly assigned.

The I/Q/M assignment list, as shown in the figure below, is a table in which the byte locations of each area are depicted as rows in the table. Each byte is divided into eight boxes, each of which indicates a bit location. The bits are numbered from the least-significant bit starting from the right, as seen in the table header. Each bit location that is used in your program, is marked by an 'x'. Furthermore, the column header also indicates whether the locations are accessed in the program as a byte (B), word (W), or doubleword (D) location. As seen in Figure 6-1a, the individual bits of byte-4 and byte-5 are each used; these two bytes are also accessed as a word location IW 4. The Timer/Counter list also shows uses a table format to show which timers and counters are used in the program.

Inputs, outputs, bit memory

	7	6	5	4	3	2	1	0	B	W	D
IB 0											
IB 1											
IB 2			X	X	X	X	X	X			
IB 3					X			X			
IB 4	X	X	X	X	X	X	X	X			
IB 5	X	X	X	X	X	X	X	X			
IB 6					X	X	X	X			

Figure 6-2a. I/Q/M list: Input Memory Area

Inputs, outputs, bit memory

	7	6	5	4	3	2	1	0	B	W	D
MB229											
MB230											
MB231											
MB232											
MB233											
MB234											

b) I/Q/M list: Bit Memory Area. MD 230 is used.

Program Structure

The program structure provides a visual map of how blocks are called within your S7 user program. In a tree-like structure, that can be expanded and collapsed, all of the nesting levels of block calls of the program can be viewed. This graphical view of the program can be a useful tool to quickly see what blocks are called from a given block as well as the nesting depth of a given call path. Furthermore, the program structure gives you the exact temporary local data requirements for each call path (i.e., OB calls).

Program Cross Reference Data

The program cross-reference report provides a detail listing of where memory addresses are used throughout the entire user program. For each address, this report will list the absolute address, its assigned symbolic address if it has been defined, the language used, whether the address is used in a read (R) or write (W) operation, and the networks and blocks in which the address is found. In the on-screen report, you can select an address, right click and navigate directly to the specific location where the address is used.

Unused Symbols List

The unused symbols list is an account of all symbol addresses that were defined, yet not used anywhere within the S7 program. As shown in the figure below, the table lists each symbol address, its corresponding absolute address, the data type, and a comment if it was defined. This report may reveal blocks that may have inadvertently been overlooked and not called anywhere in the program.

S7 Program(1) (Unused symbols) -- New_prj\SIMATIC 300(1)\CPU 316

Symbol	Address	Data type	Comment
LOOP A CONT	FB 10	FB10	Tank A Loop Control
LOOP B CONT	FB 20	FB20	Tank B Loop Control
Filter_4CH	FB 50	FB50	4-Channel Analog Input Filter with no block parameters
FILTER_8CH	FB 51	FB51	Analog Input Value Fileter 8-channel
Totalizer	FB 60	FB60	Totalizer without block parameters
Totalizer_Par	FB 61	FB61	Totalizer with block parameters
VALVE TANK A	Q 40.1	BOOL	Main Valve Tank A
VALVE TANK B	Q 40.7	BOOL	Main Valve Tank B

Figure 6-3. Program reference data: Unused Symbols report.

Addresses without Symbols

The Addresses without Symbols report is an account of all absolute addresses that were found within the program, yet are without assigned symbolic address. As shown in the figure below, the table lists each symbol address, its corresponding absolute address, the data type, and a comment if it was defined.

Determining the Cause of CPU STOP

When the S7 CPU makes a transition from a RUN state to STOP, the cause of the stop, whether hardware-related or program-related, can be ascertained by accessing the CPU's diagnostic buffer. There, you can determine the general nature and cause of the fault and should be able to determine whether it is resulting from problems in the hardware or in the program. If the cause is program-related you may then use the CPU stacks to further diagnose the problem and the specific cause.

Using the CPU Diagnostic Buffer

An S7 CPU logs up to one hundred or more significant event messages in its *diagnostic buffer*. Messages entered into the buffer include module faults, process wiring errors, system-related errors, program-related errors, CPU operating mode transitions, and user-defined events and messages. As a primary tool in diagnosing problems, the diagnostic buffer is the first check when a fault has resulted in a CPU STOP. If the CPU automatically switches to STOP, you can determine the cause by evaluating the last events leading up to the STOP. The diagnostic buffer behaves as a ring buffer for a maximum number of entries, which is CPU dependent. When the maximum number of entries is made, the next new event is placed at the top, the remaining entries are all pushed down by one position, and the oldest entry is deleted. The newest entry is always the first (top) entry in the buffer.

Diagnosing S7 Hardware Online

When the cause of a CPU STOP is determined to be hardware-related, you can obtain further details of the actual component in question by opening the Diagnose Hardware online utility. With the Simatic Manager open to the desired project, select **View ➢ Online** to open the online project window. With the station selected, right click and select Open Object. If the project is not on the PG/PC, then from the Simatic Manager menu select **PLC ➢ Display Accessible Nodes**. Select the appropriate CPU and then **PLC ➢ Diagnose Hardware**. A quick view of any fault modules will be displayed.

Diagnosing Program-Related Faults Using the CPU Stacks

When the cause of a CPU STOP is determined to be program-related, using the diagnostic buffer, you can obtain further details of the actual causes by examining the CPU stacks. For instance, the *B-stack* (short for block-stack), lists all of the blocks called in the program just prior to the STOP, and that did not complete execution. The *I-stack* (short for interrupt-stack), contains important data or statuses that were in effect at the time the CPU stopped. For example, accumulator contents, numbers of the open data blocks and their length, status word contents, number of the interrupted block, and the block where processing should resume. The stack information can only be accessed when the CPU is in STOP mode.

Comments on Working with Monitoring and Diagnostic Tools

The STEP 7 programming system offers many online tools and utilities that aid in managing the various components of your program, as well as for diagnosing problems in either the program or in the control system. In this chapter we will present several online monitoring and diagnostic tools, typically utilized in the runtime diagnosing of problems associated with the installation, startup, and maintenance of the control system.

Like in preceding chapters several brief tutorials have been presented here, to offer a quick overview of working with online monitoring and diagnostic operations. In the remainder of this chapter examples of monitoring and diagnostic operations such as monitoring program status with or without a project, forcing variables, diagnosing hardware, and using the CPU diagnostic buffer, are all presented in a step-by-step manner. The following checklist highlights some of the key points in working with the S7 monitoring and diagnostic tools.

Checklist: Working with Monitoring and Diagnostic Tools

- *Ensure that there is a direct connection between the CPU and programming system, using standard serial cable (and PC Adapter for PC) or an MPI subnet connection.*
- *From the Simatic Manager, use the Online Window toolbar button to access the online operations of a specific CPU with the project.*
- *From the Simatic Manager, use the Accessible Nodes toolbar button to access the online operations of a specific CPU without a project.*
- *Use the Debug Monitor tool in RUN mode to view the S7 program logic and data, with visual aids, as the program is being processed.*
- *Create Variable Tables (VATs) as a convenient tool and starting point for Monitoring/Modifying Variables and Forcing Variables.*
- *Use the Monitor/Modify tool to view and assign new values to memory and program variables as the control program is being processed.*
- *Use the Force Variables tool for overriding the actual status/values of memory and I/O variables as the control program is being processed.*
- *Use the CPU Diagnostic Buffer to quickly determine the cause of a stop by ascertaining what events lead up to the CPU entering the STOP mode.*
- *Use the Diagnose Hardware tool to quickly pinpoint errors or faults in the hardware of a station that has stopped.*
- *Use the CPU Stacks (e.g., B-Stack, I-Stack, L-Stack) to determine the causes of a CPU STOP, due to program-related faults.*
- *Generate the various program reference data and memory address usage reports to assist in diagnosing program-related problems.*

Monitoring Program Status with a Project

Basic Concept

The STEP 7 *Debug Monitor* is an online debug utility, available in the LAD/FBD/STL editor, when an online connection to a CPU is established and the CPU is in the RUN or RUN-P operating mode. Displaying a block in this mode allows you to view and to evaluate the code of a block as it is being processed. Combined with the Monitor/Modify Variables utility you are able to monitor and evaluate both your program and data online.

Essential Elements

An online connection must be established with the CPU to display a block in the monitor mode. With the connection in place and the block onscreen, simply enable the *Monitor* mode from the menu (**Debug ➢ Monitor**), or using the Monitor toolbar button (eyeglasses). Each LAD network is highlighted to show binary results along the network, as contacts open and close. Similar logic continuity is shown in FBD, and STL. In addition to showing the logic continuity, the digital contents of each program variable is also shown.

Application Tips

While in the Debug Monitor mode, the screen is only updated from the point of the cursor forward. If you need to switch between the language representations (i.e., LAD/FDB/STL), the status mode must first be turned off temporarily.

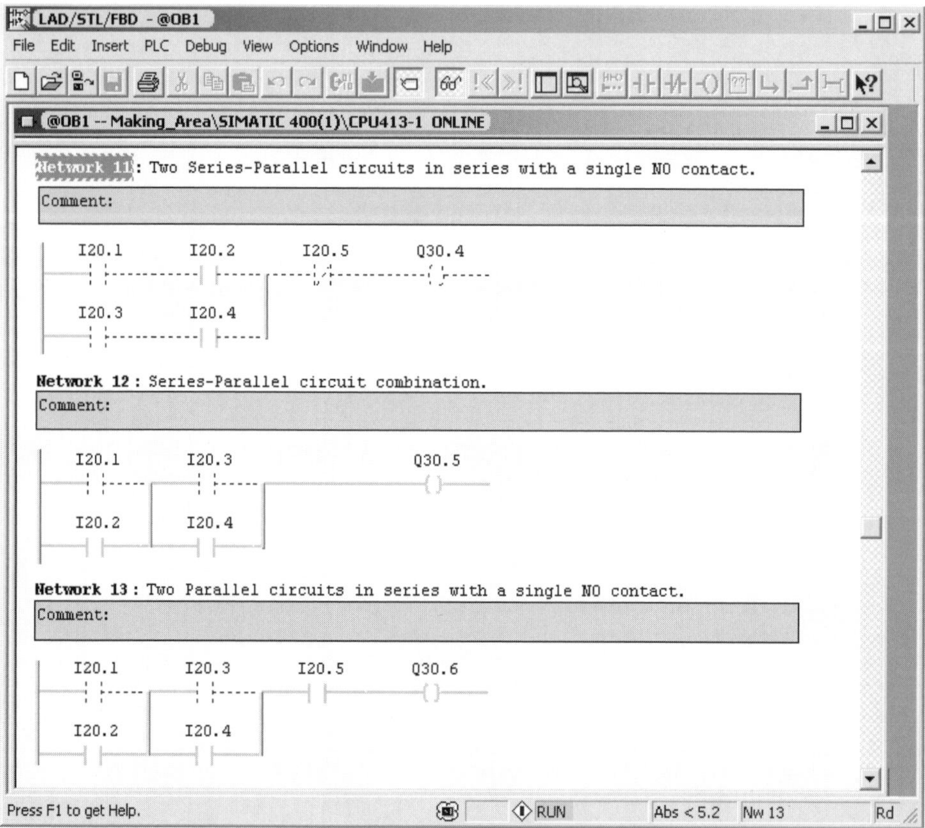

Figure 6-4. Block in Monitor mode. Status fulfilled = Green; Not fulfilled = Red Dash.

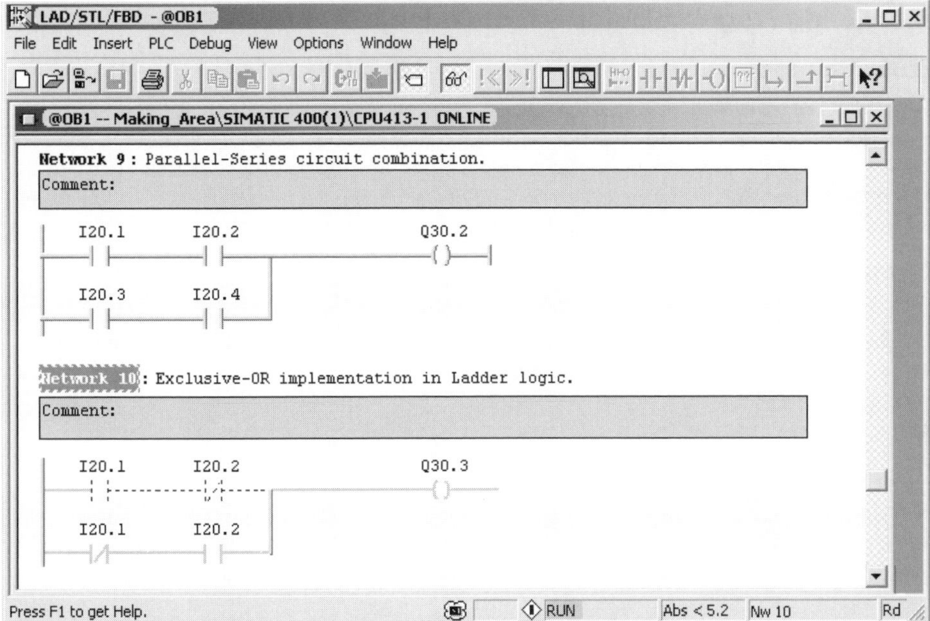

Figure 6-5. Block in Monitor Mode with cursor at Network 10; prior networks are not updated.

Quick Steps: Monitoring Program Status with a Project

STEP	ACTION
1	Launch the Simatic Manager application from the Windows **Start** button and open to the desired project and the S7 Program whose status you wish to monitor.
2	Select the *Blocks* folder of the Program, and then from the right pane double-click on the block you wish to open or right-click on the block and select **Open Object**.
3	From the STEP 7 menu, select **View** ➤ then select **LAD**, **FBD**, or **STL** as the language representation in which the block was created.
4	From the menu select **Debug** ➤ **Monitor** or press the Monitor toolbar icon (eye glasses) to view the block status, while processing (CPU in RUN or RUN-P mode).
5	Select the network header (e.g., **Network 11**), for the network you wish to evaluate first. Although the entire block is being processed, only networks below the selected network are updated — thereby minimizing the overhead due to status update.
6	If the program status does not appear to be updating, check the status bar of the editor window to verify that the CPU is in the RUN or RUN-P operating mode.
7	After the editor opens, maximize the window and use the **View** menu to hide the Program Elements browser (de-activate the **Overview**), to enlarge the viewing area.
8	For additional diagnostic aids while in the monitor mode select **View** ➤ **Detail**.
9	To open and monitor another block, select **File** ➤ **Open** from the menu.

Monitoring Program Status without a Project

Basic Concept

In some cases you will need to access and monitor the program in a CPU, without opening the associated project. It may be that the PG/PC that you currently have access to does not have a copy of the project or that you simply want to quickly go online without the delay of having to find and open the project. Any CPU program can be accessed without the project while using the *Accessible Nodes* utility.

Essential Elements

Accessing a CPU program without the project means that you are not accessing the offline program, but instead the program residing in the CPU. Either a direct MPI link to the CPU or an MPI or Profibus network must be in place since the Accessible Nodes utility only operates over these links (Accessible Nodes Utility is not possible via Industrial Ethernet). To view the program while it is processing, the CPU must be in the RUN or RUN-P operating mode.

Application Tips

Although you may gain access to any CPU program and data opened through the Accessible Nodes utility, this method should be used only for monitoring and not for editing a program. Editing CPU programs and data without the project may result in the loss or distorting of critical project information.

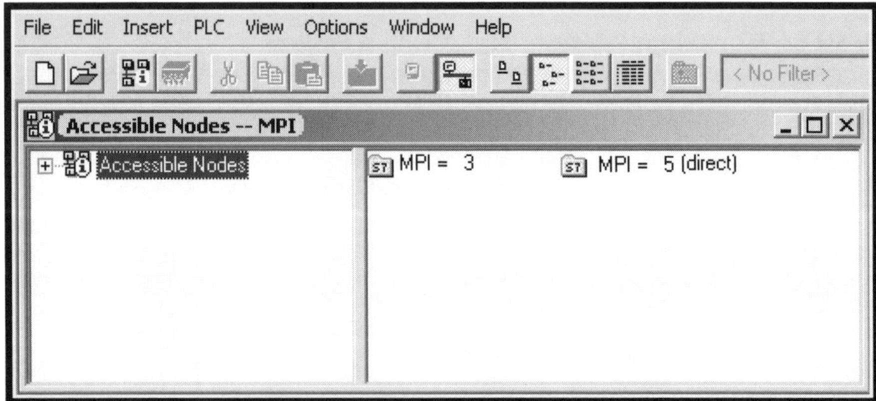

Figure 6-6. Displaying active MPI stations found using the *Accessible Nodes* utility.

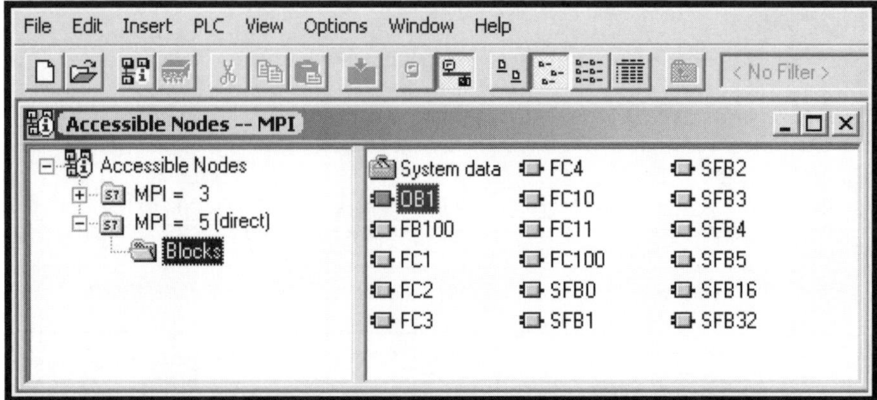

Figure 6-7. Accessing *online blocks folder* of MPI address = 5, without project.

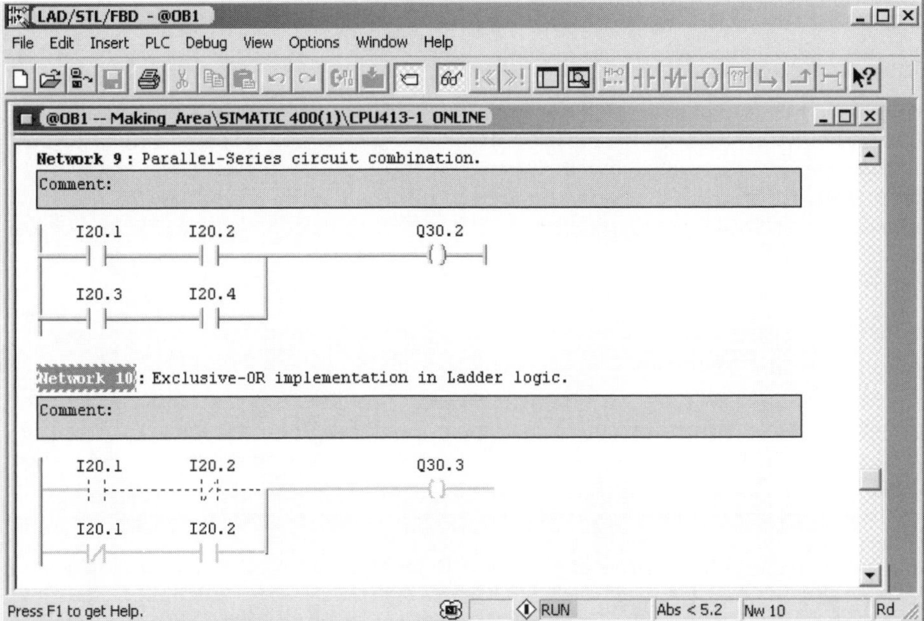

Figure 6-8. LAD/FBD/STL in Status mode, with the view set to LAD representation.

Quick Steps: Monitoring Program Status without a Project

STEP	ACTION
1	Launch the Simatic Manager application from the Windows **Start** button and from the menu, select **PLC ➢ Display Accessible Nodes**; or from the toolbar press the *Accessible Nodes* icon. Each station folder is identified by the station MPI address.
2	Select the Blocks folder of the desired station, then from the right pane double-click on the block you wish to open or right-click on the block and select **Open Object**.
3	From the menu, select **View** ➢ then select **LAD**, **FBD**, or **STL** as the language representation in which the block was created.
4	From the menu select **Debug ➢ Monitor** or press the Monitor toolbar icon (eye glasses) to view the block status, while processing (CPU in RUN or RUN-P mode).
5	Select the header (e.g., **Network 10**) for the network you wish to evaluate first. Although the entire block is being processed, only networks below the selected network are updated — thereby minimizing the overhead due to status update.
6	If the program status does not appear to be updating, check the status bar of the editor window to verify that the CPU is in the RUN or RUN-P operating mode.
7	After the editor opens, maximize the window and use the **View** menu to hide the Program Elements browser (de-activate the **Overview**), to enlarge the viewing area.
8	For additional diagnostic aids while in the monitor mode select **View ➢ Detail**.
9	To open and monitor another block, select **File ➢ Open** from the menu.

Creating and Editing Variable Tables (VATs)

Basic Concept

A variable table (VAT) is a table of specific memory addresses that can be created and used to assist you in monitoring, modifying, or forcing memory and process variables in your program. VATs are generated from the Simatic Manager, like other blocks, and are placed stored in the offline Blocks folder. Since VATs are not code or data blocks, they are not downloaded to the PLC. Once created, the VAT can be recalled to view and manipulate a particular variable set in the Monitor/Modify Variables or Force Variables utilities. Up to 255 VATs may be generated, numbered from VAT 1 to VAT 255.

Essential Elements

VATs may be comprised of any S7 memory locations, including the peripheral area, and timers and counters. Bit, byte, word, and double word locations from the Input (I), Output (Q), Bit memory (M), and data block memory (DB) areas may be used in the table. For example, bit Q 4.7, byte QB 4, word QW 4, and double word QD 4 are all valid entries. Only byte, word and double word areas of the Peripheral memory area are allowed (e.g., PIB 256, PQW 512). When defining each variable, you may use either its absolute or symbolic address.

For each VAT, you may define a *Trigger Point* and *Trigger Frequency* for managing how the variables are monitored and modified. The trigger point, defines the point at which the new values are assigned to variables being modified, and at which variables being monitored are updated. The trigger frequency defines how often variables you are modifying should be assigned the new values, and how often the variables you are monitoring should be updated. You may set the trigger point for the Beginning of Scan Cycle, End of Scan Cycle, or Transition to STOP. You may set the trigger frequency for only Once, or Every Cycle.

Application Tips

Using VATs can be a great monitoring and troubleshooting tool. Try to create tables for the sections of your machine or process where you anticipate problems. To enhance readability of each table, limit the number of entries to only those memory locations that are essential.

VAT_1 -- Making_Area\SIMATIC 400(1)\CPU413-1\S7 Program(1)

		Address	Symbol	Display format	Status value	Modify value
1		MW 30		DEC		
2		T 0		SIMATIC_TIME		
3		C 3		COUNTER		
4		I 32.0		BOOL		
5		I 32.1		BOOL		
6		I 32.2		BOOL		
7		I 32.3		BOOL		
8		I 32.4		BOOL		
9		I 32.5		BOOL		
10		Q 32.0		BOOL		
11		Q 32.1		BOOL		
12						

Figure 6-9. Variable Table (VAT) entry dialog. Enter absolute address, and display format.

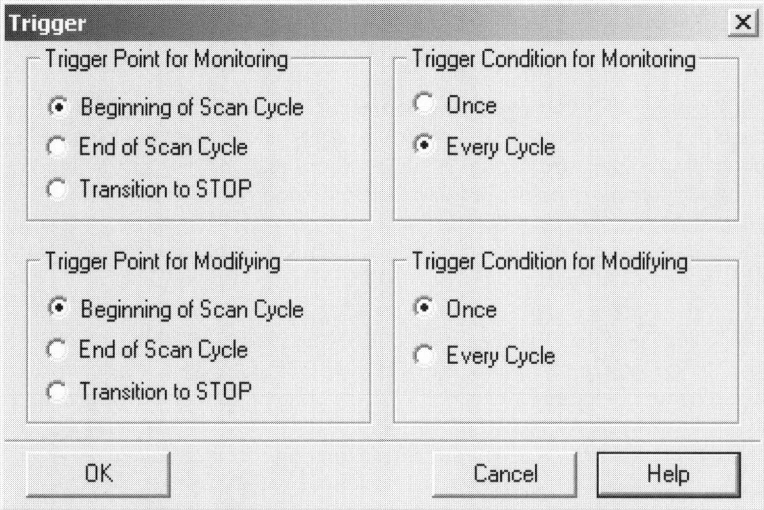

Figure 6-10. Default VAT trigger settings for Monitoring and Modifying.

Quick Steps: Creating and Editing Variable Tables (VATs)

STEP	ACTION
1	Open the Simatic Manager to the project and S7 Program you wish to use.
2	From the menu, **Insert S7 Block** ➢ **Variable Table** and accept the next available number (e.g., **VAT 1**). The number for each new VAT is automatically assigned Confirm the new VAT with **OK**.
3	Next, double-click on the new VAT to have it open in the variable table editor.
4	To add or remove a standard column from the variable table, select **View** from the menu and activate or deactivate the desired column (e.g., Symbol Comment).
5	Next, build the table with a group of variables you wish to monitor or modify; enter each **Address**, and right-click in the **Display Format** field to select the format in which you wish to view the data (e.g., BOOL, Decimal, Hex, Binary, or Counter).
6	To enter a range of addresses click in the address field, right-click and select **Insert Range of Variables** — then in the appropriate field, type the starting address, the number of locations, and the display format to use. Confirm your entries with **OK.**
7	To insert a comment line to describe a variable entry, select the entry and then from the menu or the right click, select **Insert** ➢ **Comment Line**. The comment is inserted above the variable row.
8	From the menu, select **Variable** ➢ **Trigger** to set the *trigger point* and *trigger frequency* for both monitoring and modifying the variables of the VAT.
9	From the menu, select **Table** ➢ **Save** to save entries under the VAT you opened; select **Table** ➢ **Save As** to save entries under another VAT number.

Monitoring and Modifying Variables

Basic Concept

The Monitor/Modify Variables utility allow you to both monitor and modify the defined constants and variables of your process and S7 program. Using this utility, you will be able to view and manipulate, in real time, the various S7 memory areas (e.g., bit memory, timers, counters, inputs, outputs, peripheral inputs, peripheral outputs, and data block variables), that you have defined in a variable table (VAT).

Essential Elements

To monitor or modify the variables you entered in a variable table (VAT), you must already have a physical online connection to the desired CPU. You must also establish a logical connection to the CPU. The logical connection is established using the **PLC** menu option **Connect To➢**. You may connect to the *Configured CPU,* which is the CPU associated with the S7 program in which the VAT was created; to a *Direct connect CPU,* which is to any CPU that you have a direct connection; or any *Accessible CPU*, which can be reached via the accessible nodes utility. Whenever a VAT is open with an online connection, you will see "Online" in the title bar. The status bar will display the operating states "RUN," "STOP," "Disconnected," or "Connected."

Application Tips

You may open as many VAT windows as required, all of which may be online simultaneously. Although each VAT you create is generally connected to the CPU of the associated program, it is possible to link each variable table with any CPU. For example, you might create a VAT in a CPU-independent program, which has variables that are common to several CPUs. You may then connect to any one of the CPU to and use the VAT.

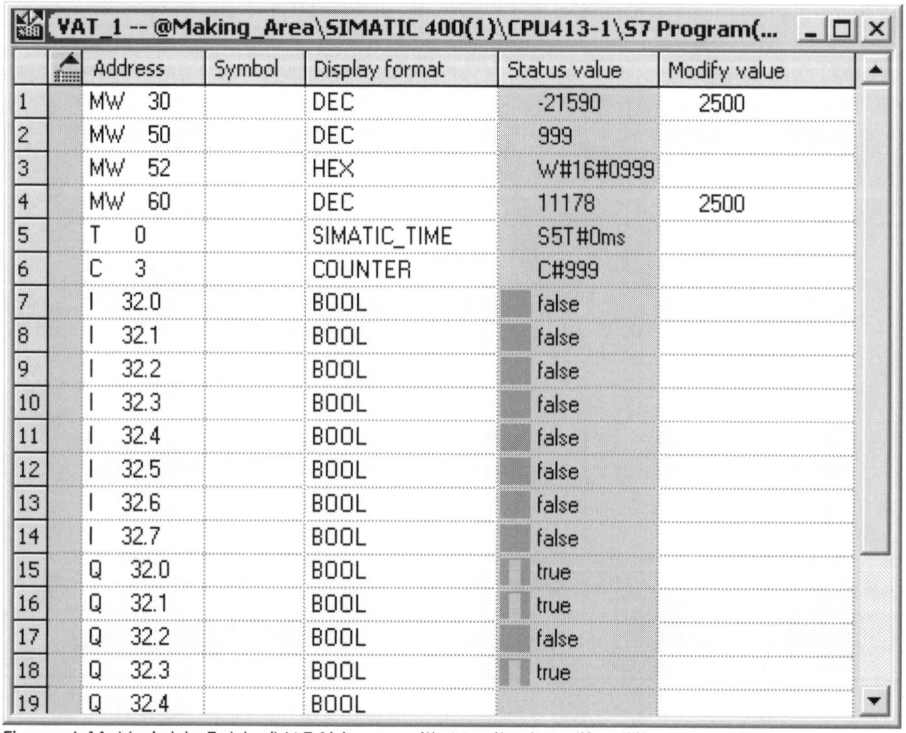

VAT_1 -- @Making_Area\SIMATIC 400(1)\CPU413-1\S7 Program(...

	Address	Symbol	Display format	Status value	Modify value
1	MW 30		DEC	-21590	2500
2	MW 50		DEC	999	
3	MW 52		HEX	W#16#0999	
4	MW 60		DEC	11178	2500
5	T 0		SIMATIC_TIME	S5T#0ms	
6	C 3		COUNTER	C#999	
7	I 32.0		BOOL	false	
8	I 32.1		BOOL	false	
9	I 32.2		BOOL	false	
10	I 32.3		BOOL	false	
11	I 32.4		BOOL	false	
12	I 32.5		BOOL	false	
13	I 32.6		BOOL	false	
14	I 32.7		BOOL	false	
15	Q 32.0		BOOL	true	
16	Q 32.1		BOOL	true	
17	Q 32.2		BOOL	false	
18	Q 32.3		BOOL	true	
19	Q 32.4		BOOL		

Figure 6-11. Variable Table (VAT 1) in use with Monitor/Modify utility. Direct CPU Connection and Monitor (plain eyeglasses) toolbar buttons are selected.

Quick Steps: Monitoring and Modifying Variables

STEP	ACTION
1	Open the Simatic Manager to the desired project and program; then from the Blocks folder double click on the desired variable table (VAT), to open the table in the Monitor/Modify utility.
2	Based on your actual situation, from the menu, select **PLC ➢ Connect To ➢ Configured CPU**; **Direct CPU**; or **Accessible CPU** to make an online connection to the desired CPU (e.g., the Configured CPU or any accessible CPU).
3	To start updating your monitored variables, from the toolbar, press the **Monitor** icon (eyeglasses); or form the menu select **Variable ➢ Monitor**.
4	To assign a new value to one or more of the variables of your VAT, select the row of the address and click in the **Modify** column and type in a new value; then from the menu, select **Variable ➢ Modify**; or from the toolbar, press the **Modify** icon.
5	From the menu, select **PLC ➢ Disconnect** to break the online connection between the CPU and the active variable table.
6	The *Configured CPU* is the CPU linked to the S7 program in which the VAT was created; the *Direct CPU,* would be a CPU to which the PG/PC is directly connected via the standard cable; an *Accessible CPU,* is any one of the project configured CPUs or other CPU that can be accessed.
	Note: Any variable modified while in the STOP or RUN mode will remain as set unless overwritten by the program when the CPU is in the RUN mode. If the statuses of inputs are modified at the beginning of the scan, the actual states are only overwritten until the next I/O update.

Forcing I/O and Memory Variables

Basic Concept

Forcing is a useful program and I/O diagnostic tool available in all S7-400s, and with limited scope in some S7-300s. This feature allows you to selectively override the normal program influence on memory variables of your S7 program, as well as the influence of the normal I/O update on the actual status of input and output devices. Using this utility, you will be able to force the desired statuses of I/O devices or the values of memory variables, until they are released by your command. With forcing, output wiring can be checked, and control logic can be fully tested without having to go out and manually open or close switch contacts or to wait until a temperature or speed setpoint is reached.

Essential Elements

The variables you wish to force are handled from a so-called "Force Values Window." Like with monitor/modify variables, to force variables, you can create the table of variables as needed, or you can use an existing variable table (VAT). A connection with the CPU is also required. The connection can be set to configure automatically, in the **Options** menu, or you can use the menu command **PLC ➢ Connect To➢**. You may connect to the *Configured CPU,* associated with the S7 program; a *Direct connect CPU*; or any *Accessible CPU*. In the S7-400, you may force bit memory (M), input (I), output (Q), peripheral input (PI), and peripheral output (PQ) memory. In the S7-300 you may force input (I) and output (Q) memory.

Application Tips

Before you start a Force job you must ensure that a force job is not already active in the CPU. To ensure safety, however, you should not stop a force job started by someone else, unless you are aware of the total impact on the control system. When you are done with a forcing job, you <u>must</u> explicitly terminate the job. A force job is <u>not</u> terminated by closing the application or by breaking the CPU connection; instead, forcing is <u>only</u> terminated by the menu command **Variable ➢ Stop**. Until a force job is stopped, the force values are still in the CPU and not deleted. An active force job is indicated by an LED on the CPU face plate and on the Force Values Window status bar.

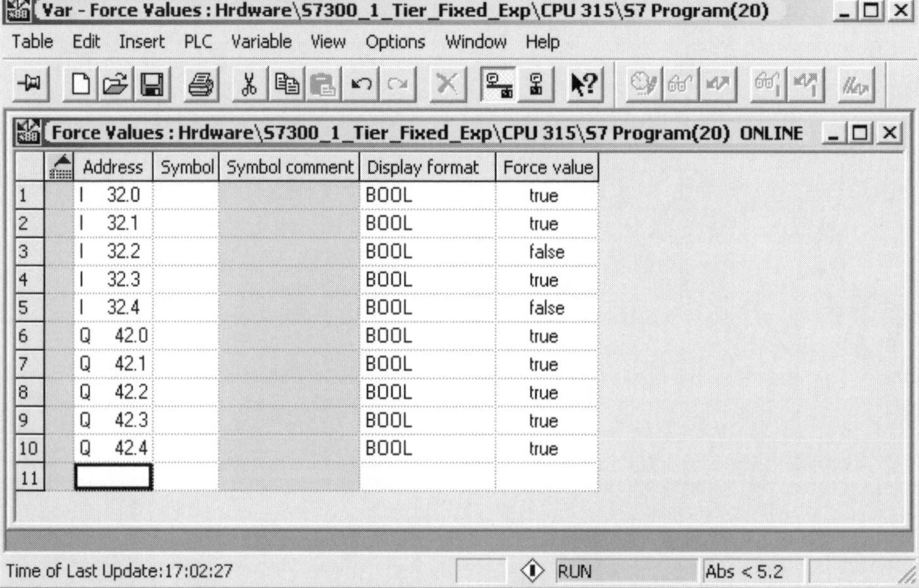

Figure 6-12. Force value table shown <u>normal</u>, to reflect that the table values are being edited and not yet "forced" in CPU.

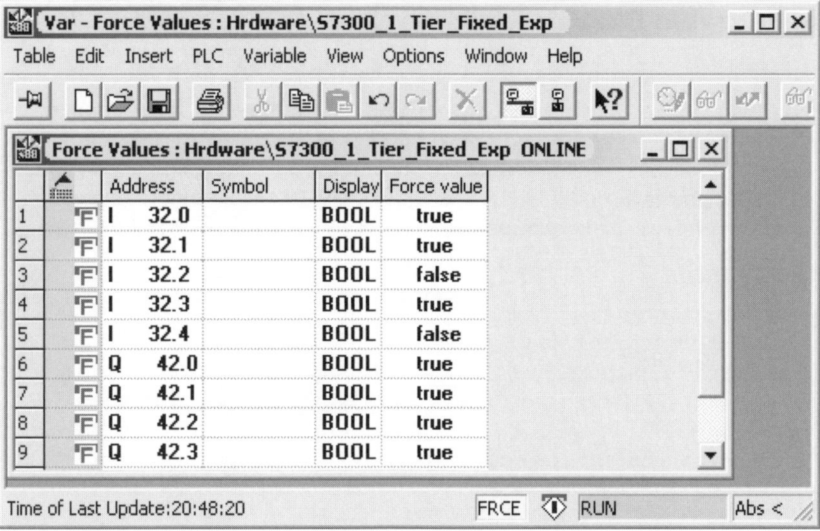

Figure 6-13. Force value table shown bold, to reflect values already "forced" in CPU.

Quick Steps: Forcing I/O and Memory Variables

STEP	ACTION
1	Open the Simatic Manager to your project; and select the CPU where you wish to initiate or work with "Forcing." A physical connection should already be established.
2	You may also establish an online connection with the desired CPU, from the menu, by selecting **PLC** ➢ **Connect To** ➢ **Configured CPU**; **Direct CPU**; or **Accessible CPU**.
3	With the CPU selected, from the menu or the right click, select **PLC** ➢ **Display Force Values**; this opens the **Force Value Window.** Forcing menu options are grayed out if forcing is not supported on the CPU.
4	If a force job is active already, it will be indicated in the Status bar as "FRCE." You must decide if the existing force job can be removed or replaced. (See Figure 6-12).
5	With the Force Value window open, you may enter the memory variables you wish to force in the table, or use an existing variable table (VAT).
6	To use an existing variable table (VAT), from the menu select **Insert** ➢ **Variable Table;** If there is no existing VAT for the force job, then enter each variable you wish to force in the **Address** column; then, in the **Force Value** column enter the values that you want to assign to each variable (Figure 6-11).
7	To initiate the force job, from the menu select **Variable** ➢ **Force**. If no force job is currently active, the variables are assigned the new force values.
8	To save a force window as a variable table, from the menu select **Table** ➢ **Save As**.
9	To stop a force job, select **Variable** ➢ **Stop Forcing**. This causes the forced job to be deleted from the CPU and the forced values to be released to normal behavior.
10	From the menu, select **PLC** ➢ **Disconnect** to break the online connection between the CPU and the active variable table.

Using the Hardware Diagnostics Utility

Basic Concept

Traditionally, diagnosing problems in the hardware has involved methods that include use of the programming system (PG/PC) and of course visual inspection of module indicators. While these methods are still viable, having a complete object-based configuration of each PLC station in the project and resident in the PLC itself, diagnosing hardware problems are drastically simplified and can be largely managed visually from the programming system.

Essential Elements

Diagnostic information obtained via the PG/PC, comes from a number of sources. SIMATIC S7 provides various diagnostic functions. Some of these functions are integrated on the CPU, others are provided by the modules (SMs, CPs, &FMs). System diagnostics detect, evaluate, and report errors that occur within a programmable controller. For this purpose, every CPU and every module with system diagnostics capability (for example, FM 354) has a diagnostic buffer in which detailed information on each diagnostic event is entered in the order it occurs.

Application Tips

The online diagnostics can also be opened from an online project window in the Simatic Manager. With the Simatic Manager open to the desired project, select **View ➢ Online** to open an online project window. With the station selected, right click and select Open Object, or double click on the station's Hardware object to open the Hardware Configuration online.

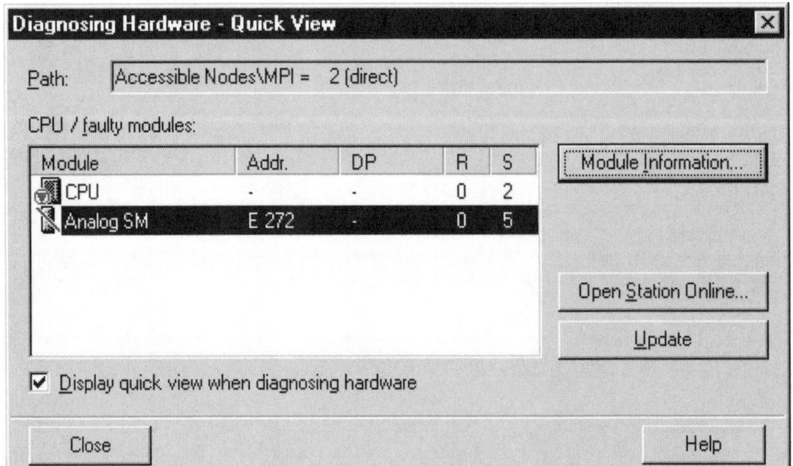

Figure 6-14. Using Hardware Dignostics in Quick View. Further details are available by using the Module Information button or Open Station Online button.

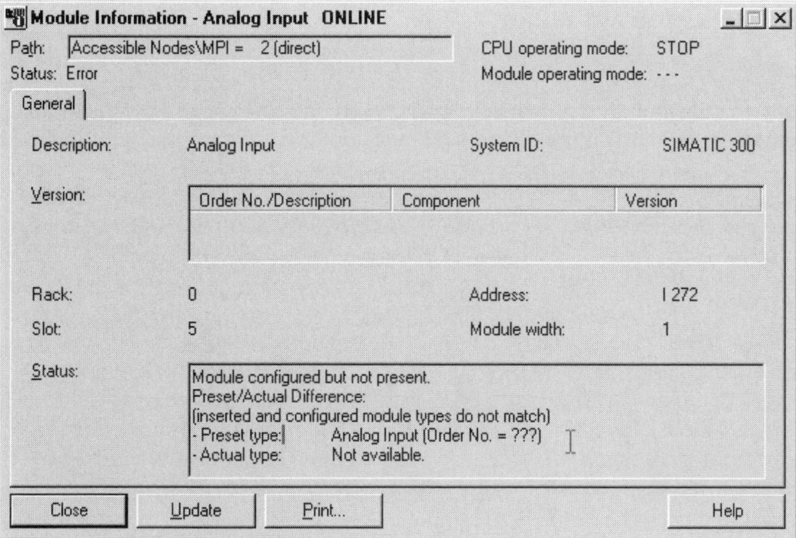

Figure 6-15. Module Information view.

Quick Steps: Using the Hardware Diagnostics Utility

STEP	ACTION
1	Open the Simatic Manager and the project to go online with the problem PLC, or without the project use Accessible Nodes to select and go online with the station.
2	From the menu, select **PLC ➤ Diagnostics/Settings ➤ Hardware Diagnostics** or with the CPU or MPI folder selected, right click and select **PLC ➤ Hardware Diagnostics**. The **Quick View** dialog will be displayed (unless the check box on the Quick View dialog has been deactivated. Along with the CPU, the *Faulty Modules* are listed.
3	Select a module that is indicating a fault and press the **Module Information** button, to obtain further diagnostic information on the selected module.
4	You can view brief diagnostic information for the module, in the **Status** box on the *General* properties tab. The available diagnostic details will depend on the module. CPUs, FMs, and CPs, for example may have additional diagnostic screens.
5	Next, you can press the **Open Station Online** button, to open the station's hardware configuration online, as a graphic diagnostic tool.
6	With the Hardware Configuration tool open online, the racks and any faulty modules will be shown, with visual status indicators; use the STEP 7 online help system for context-sensitive assistance with the visual error indicators.
7	In the configuration, you will be able to verify the module's address, and obtain specific information by opening the module's properties dialog.
8	Finally, you may also select the CPU from the Quick View dialog or from with the hardware configuration online, to then select Module Information. By viewing the diagnostic buffer, you may see the events leading up to the CPU going into STOP.

Using the CPU Diagnostic Buffer

Basic Concept

An S7 CPU logs up to one hundred or more significant event messages in its *diagnostic buffer*. Messages entered into the buffer includes module faults, process wiring errors, system-related errors, program-related errors, CPU operating mode transitions, and user-defined events and messages. As a primary tool in diagnosing problems, the diagnostic buffer is the fist check when a fault has resulted in a CPU STOP. If the CPU automatically switches to STOP, you can determine the cause by evaluating the last events leading up to the STOP.

Essential Elements

The **Events** list box of the diagnostic buffer contains a list of each diagnostic event that has occurred. Each event is listed with the event number (e.g., 1-100, where 100 is the oldest event); the Date/Time of the event; and the Event with description. In the lower window **Details of Event**, additional information is displayed for the currently selected event. This information provides a more detailed description, such as CPU Mode transition caused by the event, and reference to the location of the error in a block (e.g., block type, number, and relative address).

Application Tips

The information entered in the diagnostic buffer when a system diagnostic event occurs is identical to the start information transferred to the corresponding organization block.

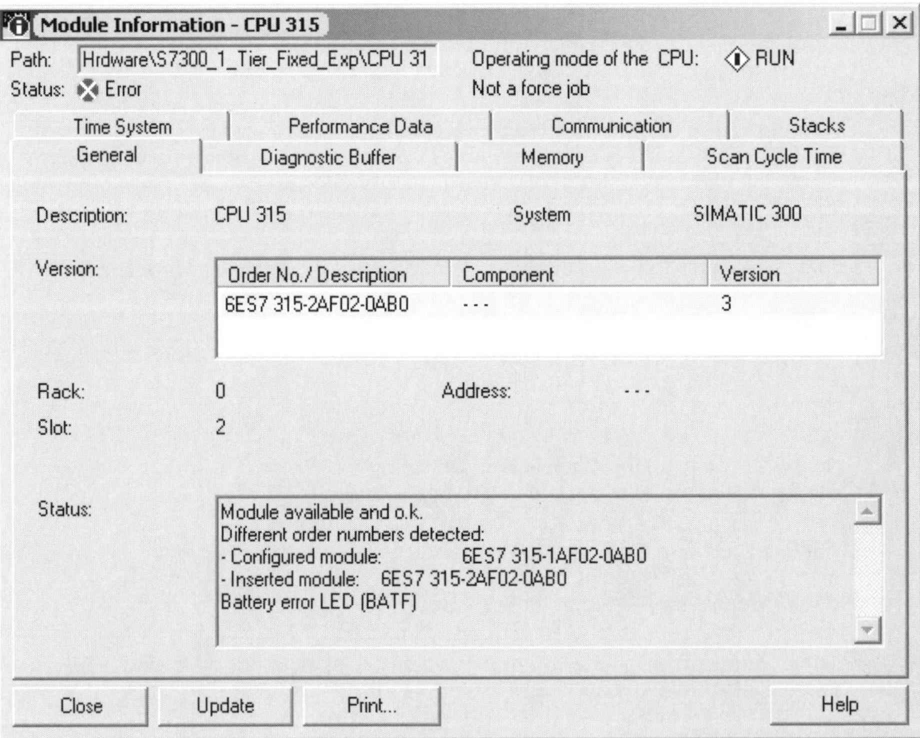

Figure 6-16. The CPU General Properties tab: Status box gives brief diagnostic information.

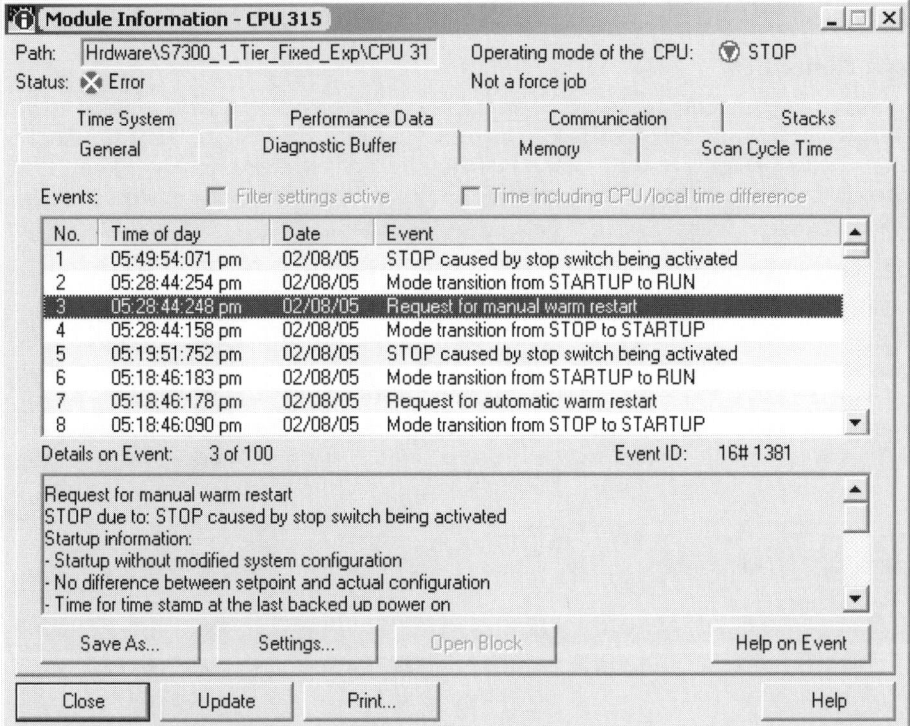

Figure 6-17. The CPU diagnostic buffer screen.

Quick Steps: Using the CPU Diagnostic Buffer

STEP	ACTION
1	With the Simatic Manager open to project and station, from the menu select **View ➢ Online**; or select **PLC ➢ Display Accessible Nodes** to go online without the project.
2	From the online project window (or the accessible nodes window), select the CPU module whose diagnostic buffer you wish to open; then right click and select **PLC ➢ Module Information**.
3	You can view brief diagnostic information for the CPU, in the **Status** box of the *General* properties tab. For example, a battery failure is indicated; or a difference in the configured CPU order number and the inserted order number may be shown.
4	Open the diagnostic buffer by selecting the *Diagnostic Buffer* tab.
5	The events leading up to the CPU stop are listed in the **Events** box, starting with the most recent event. As you step down through the list, a list of information details about the specific event are shown below in the **Detail on Event** box.
6	To get further assistance on an event, and possible causes and corrections for eliminating the error, press the **Help on Event** button.
7	If you make any corrections, while still in the diagnostic buffer, you can press the **Update** button to view changes in the CPUs ability to startup.

Using the CPU Stacks Dialog to Diagnose Program Faults

Basic Concept

When some fault has caused the CPU to transition to a STOP mode, the likely cause of the interruption can be generally determined by using the CPU's diagnostic buffer. If the exact cause is not immediately apparent yet it is determined that the interruption is program-related, then you can obtain further details of the actual causes by examining the CPU stacks.

Essential Elements

The *B-stack* (short for block-stack), lists the all of the blocks called in the program just prior to the STOP, but did not complete execution. The *I-stack* (short for interrupt-stack), contains important data or statuses that were in effect at the time the CPU stopped. For example, accumulator contents, numbers of the open data blocks and their length, status word contents, number of the interrupted block, and the block where processing should resume. The stack information can only be accessed when the CPU is in STOP mode.

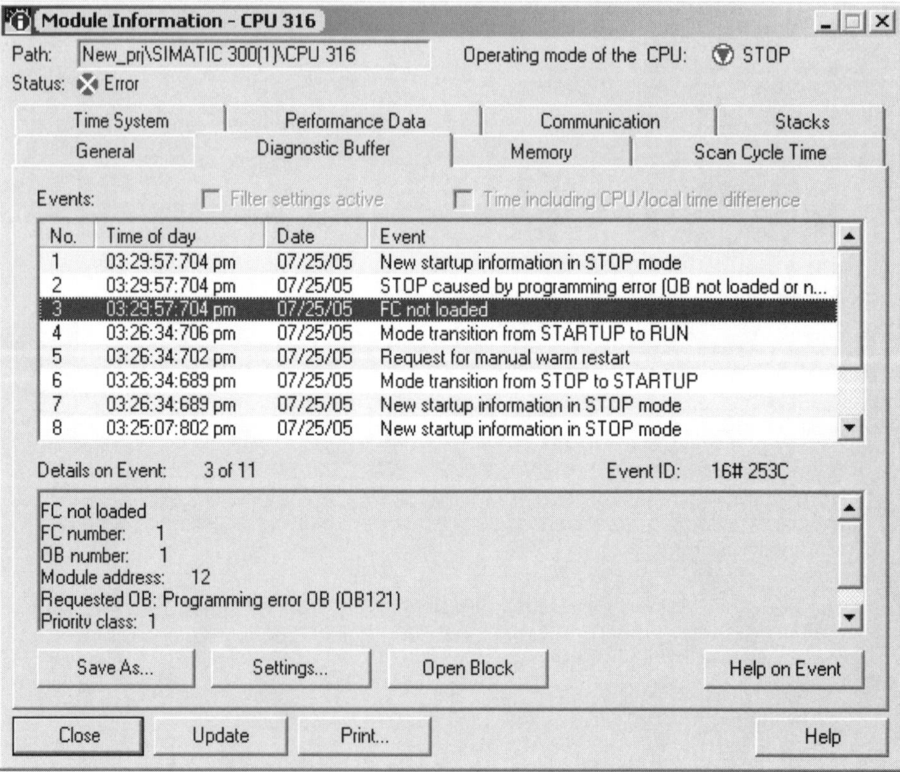

Figure 6-18. S7 CPU Diagnostic Buffer. Entries 2 and 3, indicate a fault in the program.

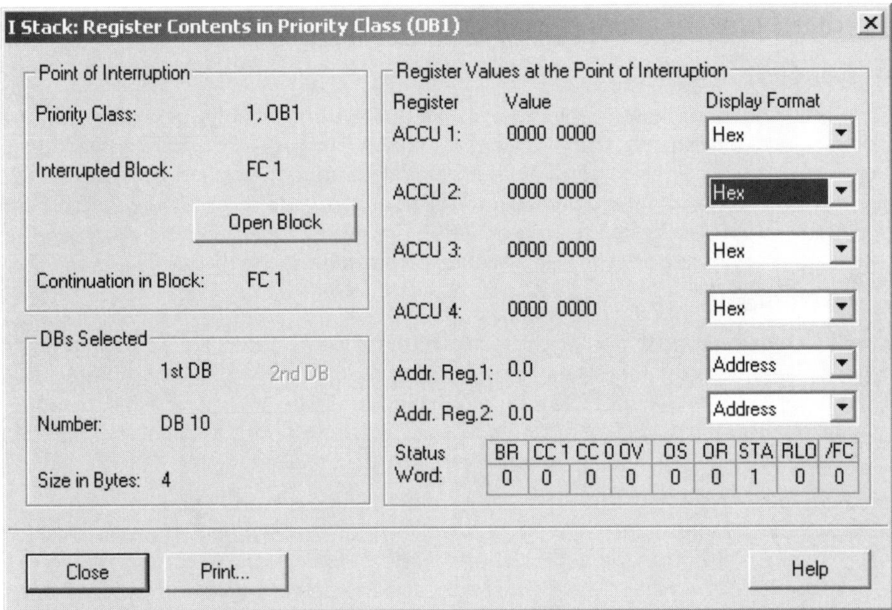

Figure 6-19. S7 CPU I-Stack dialog. This dialog can only be opened in the STOP mode.

Quick Steps: Using the CPU Stacks to Diagnose Program Faults

STEP	ACTION
1	With the Simatic Manager open to project and station, from the menu select **View** ➢ **Online**; or select **PLC** ➢ **Display Accessible Nodes** to go online without the project.
2	From the online project window (or the accessible nodes window), select the CPU module whose diagnostic buffer you wish to open; then right click and select **PLC** ➢ **Module Information**.
3	Select the *Stacks* tab, and the Block Stack will be displayed, with a listing of the complete block call sequence leading up to the STOP. The last executing block (e.g., FB22), the block that was interrupted, will be at the top of the stack.
4	Click the **I-Stack** button, at the bottom of the dialog, to display the interrupt stack as shown in Figure 6-19. Note that the dialog indicates the interrupted block as well as the block in which execution will resume when the fault is corrected.
5	Click the **Open** button, to launch the LAD/FBD/STL editor and open the interrupted block.
6	Click the **L-Stack** button, at the bottom of the dialog, to display the current contents of the local data stack (i.e., temporary local variables of the interrupted block).

Generating and Displaying Program Reference Data

Basic Concept

STEP 7 Program reference data provides useful information about your program that can be used during development and later when making modifications, or when troubleshooting problems. The main report elements of the program reference data include *cross-references, address assignments, program structure, unused symbols*, and *addresses without symbols*. As a developer, you will likely generate program reference reports throughout development and upon completion, as part of the documentation provided to the end user.

Essential Elements

The **cross-references report** is to list each address found in the program, the various locations where used, and how used. The **assignment list** is an overview of how S7 memory areas including timers, counters, bit memory, inputs, and outputs are being used. Input (I), output (Q), and bit memory (M) addresses are listed by byte, with individual bits, and shows whether program usage is by bit (X), byte (B), word (W), or double word (D).

The **program structure** describes the call path of all of the blocks in an S7 program. The program structure is represented as a tree, showing the blocks called from each OB, starting with OB 1. The structure shows the nesting depth of each call path as well as local memory usage for each block and the complete call path. The unused symbols report lists assigned symbols that are not used anywhere in the program. Finally, the **addresses without symbols** report list all addresses used in the program, but do not have an assigned symbol address.

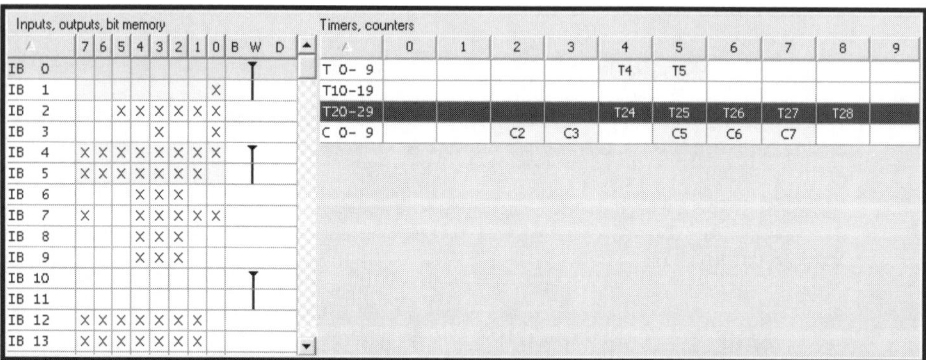

Inputs, outputs, bit memory

	7	6	5	4	3	2	1	0	B	W	D
IB 0											
IB 1								X			
IB 2			X	X	X	X	X	X			
IB 3					X			X			
IB 4	X	X	X	X	X	X	X	X			
IB 5	X	X	X	X	X	X	X				
IB 6				X	X	X					
IB 7	X			X	X	X	X	X			
IB 8				X	X	X					
IB 9				X	X	X					
IB 10											
IB 11											
IB 12	X	X	X	X	X	X	X				
IB 13	X	X	X	X	X	X	X				

Timers, counters

	0	1	2	3	4	5	6	7	8	9
T 0- 9					T4	T5				
T10-19										
T20-29					T24	T25	T26	T27	T28	
C 0- 9			C2	C3		C5	C6	C7		

Figure 6-20. STEP 7 Program Reference Data: Inputs/Outputs/Bit Memory (I/Q/M) Assignments.

Block(symbol), Instance DB(symbol)	Local data (in path)	Language	Location		Local data (for blocks)
S7 Program					
OB1 [maximum: 30]	[26]				[26]
FC1	[26]	LAD	NW	1	[0]
FC8	[26]	LAD	NW	2	[0]
FC7	[26]	LAD	NW	3	[0]
FC6	[26]	LAD	NW	4	[0]
DB6	[26]	LAD	NW	1	[0]
FB1, DB1	[30]	LAD	NW	5	[4]
DB6	[26]	LAD	NW	6	[0]
FC3	[26]	LAD	NW	7	[0]
FC9	[26]	LAD	NW	8	[0]
FC5	[26]	LAD	NW	9	[0]
FC4	[26]	LAD	NW	10	[0]
FC2	[26]	LAD	NW	11	[0]
FC10	[26]	LAD	NW	12	[0]
OB121	[0]				[0]

Figure 6-21. STEP 7 Program Reference Data: Program Structure Report; starting from OB 1.

Address	Block (symbol)	Type	Language	Location					Location				
C 2	FC100	R	STL	NW	31	Sta	10	/L	NW	31	Sta	12	/LC
C 3	FC100	R	STL	NW	31	Sta	25	/L	NW	31	Sta	27	/LC
C 5	FC10 (Box_Counters)	R	LAD	NW	1	/A			NW	1	/L		
C 6	FC10 (Box_Counters)	R	LAD	NW	2	/A			NW	2	/L		
C 7	FC10 (Box_Counters)	R	LAD	NW	3	/A			NW	3	/L		
FC 1	OB1	R	LAD	NW	1	/CC							
I 1.0	FB100	R	LAD	NW	1	/A							
I 2.0	FC100	R	STL	NW	24	Sta	6	/A					
I 2.1	FC100	R	STL	NW	19	Sta	1	/A	NW	22	Sta	1	/A
I 2.2	FC100	R	STL	NW	19	Sta	2	/A	NW	22	Sta	2	/A
I 2.3	FC100	R	STL	NW	20	Sta	2	/A	NW	22	Sta	5	/A
I 2.4	FC100	R	STL	NW	20	Sta	3	/A	NW	22	Sta	6	/A
I 2.5	FC100	R	STL	NW	21	Sta	2	/A	NW	22	Sta	9	/A
I 3.0	FC100	W	STL	NW	35	Sta	10	/=					
I 3.3	FC100	R	STL	NW	36	Sta	2	/A					
I 4.0	FC100	R	STL	NW	27	Sta	2	/A					
I 4.1	FC100	R	STL	NW	23	Sta	1	/A	NW	27	Sta	5	/A
I 4.2	FC100	R	STL	NW	26	Sta	2	/O	NW	29	Sta	25	/A
I 4.3	FC100	R	STL	NW	26	Sta	3	/O	NW	29	Sta	4	/A
I 4.4	FC100	R	STL	NW	31	Sta	2	/A	NW	33	Sta	1	/A
I 4.5	FC100	R	STL	NW	31	Sta	5	/A	NW	33	Sta	4	/A

Figure 6-22. STEP 7 Program Reference Data: Cross-References Report.

Quick Steps: Generating and Displaying Program Reference Data

STEP	ACTION
1	With the Simatic Manager open to the desired project and S7 program, select the offline blocks folder for which you wish to generate or display reference data.
2	With the Blocks folder selected, from the menu select **Options ➢ Reference Data ➢ Generate;** when **prompted,** to update or regenerate the reference data. Select the Re-generated option to delete existing reference data and to generate the reference data as new; select the Updated option to ensure data is generated for modified or new blocks, and that reference data for deleted blocks is removed.
3	To display reference data onscreen or to print reports, first select the desired Blocks folder, then right click and select **Reference Data ➢ Display**. If prompted by the *Customize* dialog to select a report to open, you may choose a report and then decide whether this dialog should always be opened when displaying reference data.
4	After the report you selected is displayed, from the menu select **Reference Data ➢ Print**, to have the report printed. You may need to adjust the Print Setup first.
5	Use the **View** menu or the appropriate toolbar icon to display other reports.
6	To filter the content data for any report that is currently open, select **View ➢ Filter** from the menu, then use the appropriate tab (i.e., Cross References, Assignments, Program Structure, or Unused Symbols) to select the elements you wish to have filtered out or shown in the report.

Step

7

Working with SIMATIC NET Networks

Objectives

- Introduce SIMATIC NET Networks
- Introduce Communications Services
- Introduce S7 Communication Processors
- Introduce PC Communication Processors
- Introduce Communications Drivers
- Introduce the NetPro Configuration Tool
- Create and Configure Subnets
- Configure S7 Communications Processors
- Configure Communications Connections
- Establish Remote PG/PC Connections
- Download Network Configurations

Introduction to SIMATIC NET Networks

SIMATIC NET represents a family of communications components and services, designed to cover a comprehensive range of networking requirements. With components that comply with international standards, SIMATIC NET provides open communications that supports a multi-vendor environment for PLCs, HMIs, computers, I/O devices, electric drives, and a variety of factory and process automation systems and devices. SIMATIC NET encompasses components such as communications processors (CPs) for PLCs and PCs, network media, attachment components, configuration and diagnostic software, and software drivers.

As shown in Table 7-1, the SIMATIC NET family includes MPI, Industrial Ethernet (IEEE 802.3/802.3u), and PROFIBUS (IEC 61 158/EN 50 170) networks. In addition to these networks, SIMATIC NET covers communications processors for point-to-point connections, including printers and other serial devices; and the AS-Interface (EN 50 295), a device level network for direct connection of field level sensors and actuators. The AS-Interface network and point-to-point communications topics are not covered in this book.

Networks and Subnets

In a manufacturing plant, a *network* connects devices such as PLCs, computers, human machine interfaces (HMIs), and other devices for the purposes of communication. Often, the range of a plant network spans one or more buildings, and is comprised of one or more subnets. In a subnet, all stations are connected to a common medium, of the same physical and operating characteristics, and communicate via the same protocol. A subnet may, however, involve two or more identical cable segments joined via repeaters. In your STEP 7 project, you will be able to work with MPI, PROFIBUS, Industrial Ethernet, or Point-to-point (PTP) subnets. Large projects, may involve multiple subnets to complete the network.

MPI (Multi-Point Interface)

In the S7 environment, the multi-point interface (MPI) serves as a low-performance network, supporting small amounts of data exchange between S7 PLCs, programming devices, human machine interface devices (e.g., operator panels), as well as other Simatic systems. Each S7 CPU has an integrated MPI interface, which allows it to connect as an MPI node without additional network modules. As a network, MPI supports up to 32 nodes, uses the same RS-485 and fiber transmission media and components as PROFIBUS, and typically operates at 187.5 Kbs. You can access all MPI nodes by placing a programming system (PG/PC) on the MPI subnet.

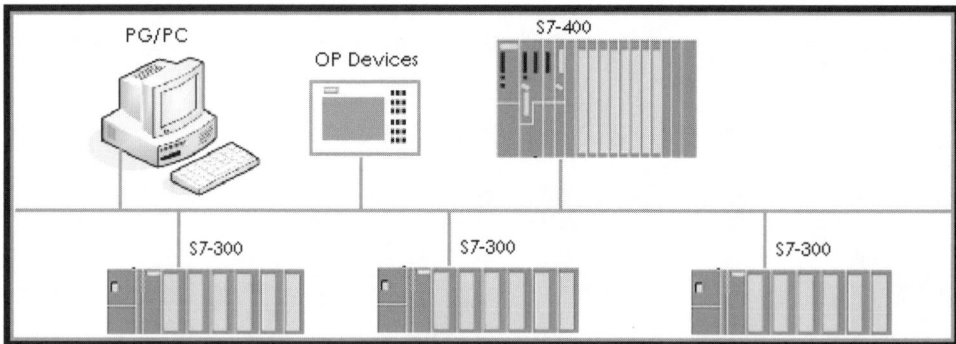

Figure 7-1. Typical MPI (Multi-Point Interface Subnet.

PROFIBUS

PROFIBUS, acronym for **PRO**ocess **FI**eld **BUS**, is a standard according to European standards IEC 61 158 and EN 50170 Vol.2. Comprised of three major protocol definitions, this standard protocol, serves networking of compliant devices at both the cell and device levels of the plant. The DP-component of this protocol supports high-speed distributed I/O connections to devices like ET-200, while the PA (Process Automation) component extends the DP protocol to include a transmission system that meets intrinsic safety requirements. The FMS (Fieldbus Message Specification) component of PROFIBUS supports client/server communications between compliant partners (e.g., S7-to-S7, S7-to-S5) in a multi-vendor environment.

The PROFIBUS network uses shielded twisted pair, glass or plastic fiber transmission media, supports 126 nodes, and transmission rates of up to 12 Mbaud. SIMATIC S7-300, S7-400, and PC stations require a communications processor to connect to a PROFIBUS subnet. CPUs that have an integrated PROFIBUS-DP interface (e.g., CPU 315-2 DP, CPU 416-2 DP), for establishing a station as a DP-Master or DP-Slave do not require additional modules. PROFIBUS CPs and the communications services they support are described later.

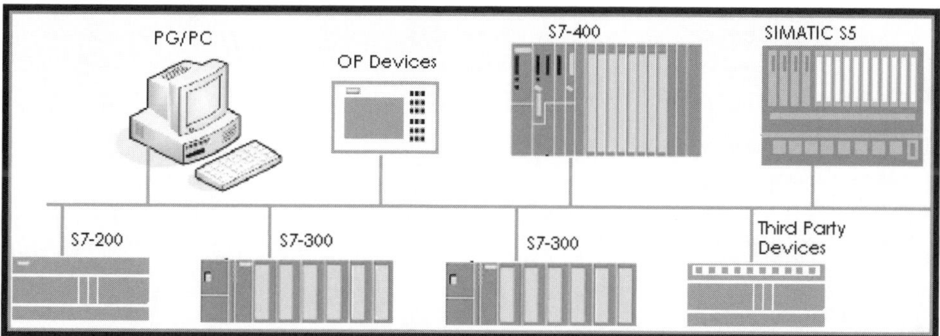

Figure 7-2. Typical PROFIBUS Subnet.

Industrial Ethernet

Industrial Ethernet serves the requirements of plant areas and cells, where manufacturing systems are monitored and coordinated —, and perhaps tied into plant information systems. Supporting both the ISO and TCP/IP transport protocols, Industrial Ethernet supports transmission of large data quantities over standardized local networks with access to worldwide networks according to IEEE 802.3/802.3u.

The Industrial Ethernet network uses double shielded coaxial cable, industrial twisted pair, glass, or plastic fiber transmission media, and supports over 1000 nodes at transmission rates of 10 Mbit/s or 100 Mbit/s. SIMATIC S7-300, S7-400, and PC workstations require the appropriate communications processor to connect to an Ethernet subnet. Industrial Ethernet CPs and the communications services they support are discussed later.

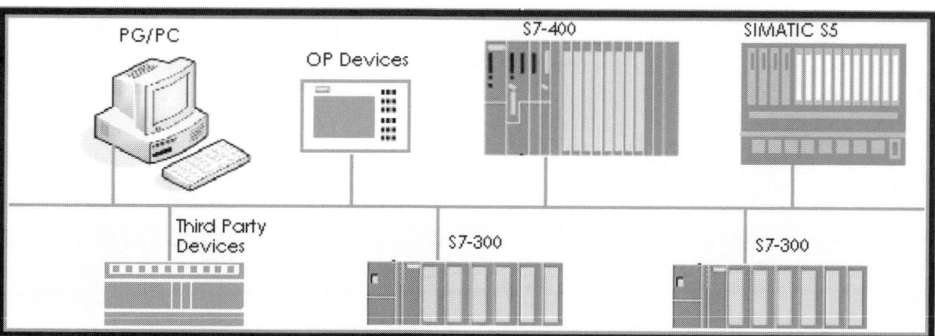

Figure 7-3. Typical Industrial Ethernet Subnet.

Table 7-1 Overview of MPI, PROFIBUS, and Ethernet Subnet Characteristics.

Characteristic	MPI	PROFIBUS	Ethernet
Standard	Siemens Procedure	EN 50170 Vol. 2	IEEE 802.3
Transmission Media - Electrical	Shielded 2-core	Shielded 2-core	Twisted Pair Shielded Coax
- Fiber Optic	Glass Fiber Plastic Fiber	Glass Fiber Plastic Fiber	Glass Fiber Plastic Fiber
Transmission Rates	19.2Kbits/187.5 Kbits 12.0 Mbits/s	9.6 Kbits/s to 12.0 Mbits/s	10 Mbits/s 100 Mbits/s
LAN Distances - Electrical	100 m (max.)	9.6 km (max.)	1.5 km (max.)
- Optical		90 km (max.)	200 km (max.)
WAN Distances	-	-	Worldwide via TCP/IP
Access Methods	Token Passing	Token Passing	CSMA/CD
Number of Nodes - Typical - Maximum	2-10 32	2-16 126	2-100 > 1000
Topology	Line	Line, Tree, Ring, Star, Redundant	Line, Tree, Ring, Star, Redundant
Automation Level	Cell/Field	Cell/Field	Cell/Management
Attachable Systems	SIMATIC S7/M7/C7 SIMATIC PG/PC SIMATIC HMI	SIMATIC S7/M7/C7 SIMATIC PG/PC SIMATIC HMI SIMATIC S5 Host Computer	SIMATIC S7/M7/C7 SIMATIC PG/PC SIMATIC HMI SIMATIC S5 Workstation
Communication Services	- PG/OP Functions - S7 Functions - Global Data	- PG/OP Functions - S7 Functions - PROFIBUS FDL - PROFIBUS FMS	- PG/OP Functions - S7 Functions - ISO Transport - ISO-on-TCP - TCP/IP - E-Mail

Note: S5 Compatible communication is possible via PROFIBUS FMS and PROFIBUS FDL, and on Industrial Ethernet using the ISO transport, ISO-on-TCP, and TCP/IP transport services.

Communications Connections and Services

The following discussions provide a brief overview of the communications with S7 controllers and with other communications partners. S7 communication with a partner always takes place via a connection. A *connection*, also called a "link," is a logical relationship established between two partners for the express purposes of a specific type of communication service. A *communication service*, as described later, represents a set of supported communications functions. The "Communicating with SIMATIC" manual provides a more detailed description of the Simatic NET communications connections and services.

Communications Connections

There are two types of connections, un-configured and configured. You determine the type of connection needed, based on its inherent characteristics, and on the requirements of your application. In addition to this, each CPU supports a fixed number of connections, which includes both configured and non-configured connections. In each CPU, one connection is reserved for a programming device and one for an operator panel (OP).

Un-Configured Connections

Un-configured connections are possible between S7 partners on MPI, whereby the connection is set up at runtime, when you call the related system function (SFC). The fact that the connection is set at runtime makes it a *dynamic connection*. With un-configured connections, a CPU can exchange small amounts of data with an external partner (e.g., S7 PLCs on MPI) or with an internal partner over the MPI backplane (e.g., S7-400 CPUs, or CPU-to-FM). Only one non-configured connection can exist between two partners.

An un-configured connection, because of its dynamic nature, might be used to make efficient use of the available link resources of a CPU. At least one connection resource must always be available in each partner, since the connection is established when the SFC is called in the S7 program. With this single link, a CPU may exchange data, in a sequential manner, with several partners. The following SFCs are used for non-configured connections.

Table 7-2. System Functions for Communication via Un-Configured Connections

SFC	NAME	Brief Description
SFC 65	X_SEND	Send Data to External S7 Partner
SFC 66	X_RCV	Receive Data from External Partner
SFC 67	X_GET	Read Data from External S7 Partner
SFC 68	X_PUT	Write Data to External S7 Partner
SFC 69	X_ABORT	Abort Connection to External S7 Partner
SFC 72	I_GET	Read Data from Internal S7 Partner
SFC 73	I_PUT	Write Data to Internal S7 Partner
SFC 74	I_ABORT	Abort Connection to Internal S7 Partner

Note: Internal partners share the MPI bus in a station (e.g., S7-400 Multi-CPU configuration)

Configured Connections

Configured connections support connections between S7 partners, and non-S7 partners (e.g., computers, Simatic S5, and other systems). Whereas an un-configured connection is dynamic, configured connections are static. In a *static connection*, the properties that establish the relationship between the partners is built-up at the start-up of the CPU and remains even when the CPU is switched to STOP. To use configured connections, you must define the connection in the CPU's connection table, using the NetPro configuration tool.

When defining configured connections, you may specify as many as required, within the limits of the available connection resources of the CPU. The connection type you choose will be based on the communications service you wish to implement (see Table 7-9). Multiple configured connections are possible between different partners or between the same partners. A Local ID is assigned in each partner for each connection you configure. This **local ID** is required as an input parameter to the communications blocks that you call in your program, to communicate over configured connections. Configured connections are always required when the communications involve communications processors (CPs).

Communications Services

Performance features are what characterizes a specific communications service. For example, a data service operates on a specific subnet (e.g., MPI, PROFIBUS, or Industrial Ethernet), and is based on a specific transmission procedure. User access to a communications service is via a particular interface. Access to a service in the S7 program, for instance, may require system blocks (SFCs or SFBs) or loadable user blocks (e.g., FCs, or FBs), available in the STEP 7 library. Access to a service in a PC station might involve an OPC Server, or C-functions. SIMATIC NET encompasses the following services, each of which requires that you configure a connection in the connection table of the partner stations.

Table 7-3. SIMATIC NET Communications Services.

Communications Services		
MPI	**PROFIBUS**	**Industrial Ethernet**
S7 Functions	S7 Functions	S7 Functions
Global Data Transfers	PROFIBUS DP	ISO Transport
	PROFIBUS FDL	ISO-on-TCP
	PROFIBUS FMS	TCP/IP
		UDP

S7 Functions

S7 Functions provides optimized services for communication among S7 systems (e.g., PLCs, operator panels (OPs), and HMIs). You must configure an S7 connection for the partner stations to use this service. The S7 services are supported on MPI, PROFIBUS, and Industrial Ethernet. With these service S7 PLCs may participate in send/receive, and direct memory access communications with other S7 PLCs or with PC stations.

In the S7 program, you may implement secure data transfers of up to 64 Kbytes per job, using the integrated blocks **B_SEND** (SFB 12) and **B_RCV** (SFB 13). You may implement high-speed unsecured data transfers (i.e., without acknowledgement), of up to 440 bytes, using the blocks **U_SEND** (SFB 8) and **U_RCV** (SFB 9). Using **GET** (SFB 14) and **PUT** (SFB 15) blocks, direct access to the memory in S7 PLCs is possible without programming in the remote partner. The call interface for these system blocks are included in the STEP 7 standard library of communications blocks so that you may incorporate them in your control program.

Table 7-4. Communications blocks for services over Configured S7 Connections.

S7-400	S7-300	Name	Brief Description
SFB8	FB8	USEND	Uncoordinated Send – Max. Length 440/160 bytes
SFB9	FB9	URCV	Uncoordinated Receive - Max. Length 440/160 bytes
SFB12	FB12	BSEND	Send data blocks - Max. Length 64/32 Kbytes
SFB13	FB13	BRCV	Receive data blocks - Max. Length 64/32 Kbytes
SFB14	FB14	GET	Read data in remote partner - Max Length 400/160 bytes
SFB15	FB15	PUT	Write data in remote partner - Max. Length 400/160 bytes
SFB19	NA	START	Execute a Warm Restart on remote partner
SFB20	NA	STOP	Switch remote partner to STOP mode
SFB21	NA	RESUME	Execute a Hot Restart on remote partner
SFB22	NA	STATUS	Query status of remote partner
SFB23	NA	USTATUS	Receive status from remote partner
SFC62	NA	CONTROL	Query status of a connection belonging to SFB instance
FC62	NA	C_CNTRL	Query status of a connection (in S7-300)

Note: The S7-300, depending on the PROFIBUS CP, may participate as a server only in S7 Communication functions. See these blocks in the SIMATIC NET CP Library folder in STEP 7.

PROFIBUS FMS

FMS services facilitate open communications with any partner that supports the transmission of structured data (FMS variables) in accordance with the Field bus Message Specification services of PROFIBUS. FMS supports the reading, writing, and reporting of named variables in a client-server relationship. In the client-server transaction, outside of defining variables, no further programming is required in the server. FMS communications is possible between S7 PLCs, with SIMATIC S5 PLCs (via CP 5431), and with PCs and other FMS systems.

In the S7 program, data transfer of up to 240 bytes are handled using the **READ** (FB 3) and **WRITE** (FB 6) blocks, over a configured *FMS connection*. You can find these loadable user blocks in the STEP 7 standard library of communication blocks. You need to define the FMS variables on the FMS server only for READ/WRITE services, and only on the client when using the REPORT service. Since these are FBs, you must create an instance DB as memory for each FB. FMS services in PC stations are implemented as C functions and in the OPC server.

Table 7-5. Communications blocks for services over Configured FMS Connections.

Name	S7-400	S7-300	Brief Description
IDENTIFY	FB2	FB2	Identify the remote partner for the user
READ	FB3	FB3	Read an FMS variable from specified remote partner
REPORT	FB4	FB4	Report an FMS variable to a remote partner
STATUS	FB5	FB5	Provide the status of a remote device on user request
WRITE	FB6	FB6	Write an FMS variable to a remote device

Note: See these standard blocks in the SIMATIC NET CP Library folder in STEP 7.

PROFIBUS FDL

This service allows communication on PROFIBUS, between partners that support the SDA *Send Data with Acknowledge (SDA)* and *Send Data with no Acknowledge (SDN)* features of the Field bus Data Link Services of PROFIBUS. With FDL services, S7 PLCs may participate in send/receive, broadcast, and multicast communications with the S5 115U/115H, S5 135U/155U (via the CP 5431 FMS/DP), or the S5 95U DP.

Use of this service requires that you configure an *FDL* connection in the partner stations. In the S7 program, FDL secure data transfers of up to 240 bytes are handled using the blocks **AG_SEND** (FC 5) and **AG_RECV** (FC 6). This is the so-called Send/Receive user interface, which corresponds to the PLC/PLC connection in the SIMATIC S5 PLCs. Data receipt is acknowledged by the remote partner, and can be confirmed in the sending station. These blocks, listed in the table below, are included in the STEP 7 standard library for communications blocks.

Table 7-6. Communications blocks for services over Configured FDL Connections.

Name	S7-400	S7-300	Brief Description
AG_SEND	FC5	FC5	Send data up to 240 bytes on configured connection
AG_RECV	FC6	FC6	Receive data up to 240 bytes on configured connection
AG_LSEND	FC50	FC50	Send data up to 8 Kbytes on configured connection
AG_LRECV	FC60	FC60	Receive data up to 8 Kbytes on configured connection
AG_LOCK	FC7	FC7	Lock data access by Fetch/Write from remote partner
AG_UNLOCK	FC8	FC8	Unlock data access by means of Fetch/Write

PROFIBUS DP

PROFIBUS DP is a part of the PROFIBUS protocol designed to facilitate open distributed I/O communications between compliant PROFIBUS DP-master and DP-slave devices. These compliant devices operate according to the European standard EN 50170 Vol. 2. The DP communications services operate on PROFIBUS subnets, and support both Mono-Master (single master) and Multi-Master (two or more) systems and their associated DP slaves.

As detailed in Chapter 3, you can configure the S7-300/S7-400 CPUs with integrated DP ports as DP-master of modular and compact slave devices (e.g., ET200 M, ET200 B). In these configurations, you will address the DP slaves in the same manner as you would the I/O in the central, local, and remote I/O racks. When the S7-300 is configured as an intelligent slave, using the CP 343-5, then the communication between a DP-master and DP-slaves involves the use of standard function calls **DP_SEND** (FC 1) and **DP_RECV** (FC 2) in the S7 programs of both the master and slaves. PROFIBUS DP services for PC stations are provided as C functions.

Table 7-7. Communications blocks for DP-Master/DP-Slaves in Intelligent Slave configuration

Name	FC	Brief Description
DP_SEND	FC 1	In **Master**, transfers data of a specified DP output area to the PROFIBUS CP for output to the distributed I/O.
		In **Slave**, transfers data of a specified DP data area on the CPU to the send buffer of the PROFIBUS CP for transmission to DP master.
DP_RECV	FC 2	In **Master**, receives processed data of the distributed I/O and status information in a specified DP input area.
		In **Slave**, takes data transferred by the DP master from the receive buffer of the PROFIBUS CP and enters it in a data area on the CPU.
DP_DIAG	FC 3	Queries diagnostic data on the DP master and DP slaves.
DP_CTRL	FC 4	Executes control functions in for the DP slave stations.

Note: See these standard blocks in the SIMATIC NET CP Library folder in STEP 7.

Global Data Communication (GD)

Global data communications is a cyclic data transfer service, integrated in every S7 CPU, and is possible on MPI subnets only. With this service, S7 PLCs may exchange small areas of input, output, data block, and bit memory. Since the data is transferred cyclically during the normal I/O update, no block programming is required. The response time of global data exchange is dependent on the program cycles of the partners involved in the exchange.

To implement global data transfers you must simply define the global data table, using the *NetPro Network Configuration Tool*. The table defines the CPUs involved, the memory areas sent, and memory areas used to receive data in each CPU. Global data exchange operates on the broadcast method, and the receiving partner does not acknowledge data receipt. If your particular application requires secure data exchange, then global data transfer is not the appropriate solution. In the S7-400, global data exchange can be triggered in the S7 program, using the system functions **GD_SND** (SFC 60) and **GD_RCV** (SFC 61).

ISO Transport

This service allows communication on Industrial Ethernet, between any partners that support data transmission in accordance with the ISO transport protocol. The ISO transport services supports S7-to-S7 data transfer, but also allows S7 PLCs to communicate with SIMATIC S5 PLCs that use the CP 1430 TF. To use this service you must configure an *ISO connection* in partner stations, using the NetPro Network Configuration Tool.

In the S7 program, you may initiate secure data transfers of up to 240 bytes, using **AG_SEND** (FC 5) and **AG_RECV** (FC 6) functions; or up to 8 Kbytes, using **AG_LSEND** (FC 50) and **AG_LRECV** (FC 60). In both cases, the remote partner acknowledges data receipt. S5 PLCs may use its FETCH/WRITE handling blocks to gain direct access to S7 memory. Table 7-8 lists the S7 communication blocks used for exchanging data over ISO transport connections.

ISO-on-TCP

This service allows communication on Industrial Ethernet, between any partners that supports the mapping of the ISO transport protocol onto the standard TCP/IP protocol, using the RFC 1006 extension. ISO-on-TCP services allow S7-to-S7 data transfer, but also allows S7 PLCs to communicate with S5 PLCs that use the CP 1430 TCP. In both cases, the remote partner acknowledges data receipt. To use this service you must configure an *ISO-on-TCP connection* in partner stations, using the NetPro Network Configuration Tool.

In the S7 program, you may initiate secure data transfers of up to 240 bytes, using **AG_SEND** (FC 5) and **AG_RECV** (FC 6) blocks; or up to 8 Kbytes, using **AG_LSEND** (FC 50) and **AG_LRECV** (FC 60). S5 PLCs may use its FETCH/WRITE blocks to gain direct access to S7 memory. Table 7-8 lists the S7 communication blocks used for exchanging data on ISO-on-TCP connections.

TCP/IP

The TCP service corresponds to the widely used TCP/IP protocol. With this service S7 PLCs may communicate with one another or with S5 PLCs (CP 1430 TCP), PCs, or any partner that supports the TCP/IP standard. Communication on TCP/IP networks uses data streaming without blocking the data into messages, so a transmitting partner does not receive an explicit acknowledgment for each job. Data security, however, is achieved by automatic repetition and block checking (CRC) at Layer 2. To use this service you must configure a *TCP connection* in the partner stations, using the NetPro Network Configuration Tool.

In the S7 program, you may use the **AG_SEND** (FC 5) and **AG_RECV** (FC 6) blocks to handle data exchange of up to 240 bytes; or the **AG_LSEND** (FC 50) and **AG_LRECV** (FC60) blocks, for up to 2 Kbytes. The TCP transport services for PC communication are provided as C functions within the framework of the Socket interface. Table 7-8 lists the S7 communication blocks used for exchanging data over TCP/IP connections.

Table 7-8. Communication blocks for ISO, ISO-on-TCP, TCP/IP, UDP, and E-Mail Connections.

Name	S7-400	S7-300	Brief Description
AG_SEND	FC5	FC5	Send data up to 240 bytes on configured connection
AG_RECV	FC6	FC6	Receive data up to 240 bytes on configured connection
AG_LSEND	FC50	FC50	Send data up to 8 Kbytes on configured connection
AG_LRECV	FC60	FC60	Receive data up to 8 Kbytes on configured connection
AG_LOCK	FC7	FC7	Lock data access by Fetch/Write from remote partner
AG_UNLOCK	FC8	FC8	Unlock data access by means of Fetch/Write

Note: See these blocks in the SIMATIC NET CP Library folder in STEP 7.

UDP

This service allows communication on Industrial Ethernet, between any partners that supports the standard UDP (User Datagram Protocol). The UDP protocol offers services for S7-to-S7, and S7-to-S5 communication for cross-network data transmission to SIMATIC S5 115U, 135U, and 155U stations that use the CP 1430 TCP. Data transfers using UDP is not acknowledged by the partner, and is therefore intended for simple data grams where acknowledgement is otherwise implemented, or the guarantee of correct data transfer is not required. To use this service you must configure a *UDP connection* in the partner stations, using the NetPro Network Configuration Tool. The connection, will operate on an exclusively TCP/IP network.

In the S7 program the unsecured data transfer of up to 240 bytes is handled, using the blocks **AG_SEND** (FC 5) and **AG_RECV** (FC 6) blocks; or up to 2 Kbytes, using the blocks **AG_LSEND** (FC 50) and **AG_LRECV** (FC60). The UDP transport services for PC communication are provided as C functions within the framework of the Socket interface. Table 7-8 lists the S7 communication blocks used for exchanging data over UDP connections.

Configured Connection Summary

Table 7-9 briefly describes the main characteristics of the various types of configured connections available between Simatic S7, Simatic S5, PC stations, and a wide variety of non-Simatic devices. These connections are configured in the NetPro Configuration tool.

Table 7-9. Possible Configured Connection Services with S7-300/S7-400.

Connection	Brief Description
S7	- Communication among S7 PLCs (S7-300 Server only) - Communication with PCs, and Operator Panels (OP) - Supported on MPI, PROFIBUS, and Industrial Ethernet subnets - Integrated in all S7/M7/C7 systems - Secure data transfer between stations using BSEND/BRCV SFBs - High-speed unsecured data transfer using the USEND/URCV SFBs - Acknowledgement of data receipt from remote partner
S7 Redundant	- Supported on PROFIBUS and Ethernet subnets only - S7 connections, restricted to use in S7 H-Systems
PROFIBUS FDL	- Supported on PROFIBUS subnets only - Field bus Data Link Layer, According to IEC 61 158 - Medium Volume Data Transfer up to 240 bytes - Error-free S7-to-S5 Data Exchange via FDL Services - Supports data exchange - Send/Receive Data Exchange with SDA Request - Broadcast and Multicast with SDN Request
PROFIBUS FMS	- Supported on PROFIBUS subnets only - PROFIBUS Variant, According to IEC 61 158 - Supports Data Transfer of structured FMS Variables - Supports Open Communication to all FMS Compliant Systems - Acknowledgement of data receipt in the remote partner - In accordance with European standard EN 50170 Vol.2 PROFIBUS
ISO Transport	- Supported on Industrial Ethernet Subnets only - Communication with ISO transport compliant systems (e.g. S7-S7, S7-S5) - Supports Send/Receive and Fetch/Write functions as used in S5 PLCs. - Acknowledgement of data receipt by the ISO Transport in the partner - ISO Transport (ISO 8073 class 4) corresponds to layer 4 of the ISO model
ISO-on-TCP	- Supported on Industrial Ethernet Subnets only - Fulfills the TCP/IP standard with the RFC 1006 extension - Communication with ISO-on-TCP compliant partners (e.g. S7-S7, S7-S5) - Supports Send/Receive and Fetch/Write functions as used in S5 PLCs. - Acknowledgement of Data receipt is confirmed by remote partner
TCP/IP	- Supported on Industrial Ethernet Subnets only - Complies with Transmission Control Protocol/Internet Protocol (TCP/IP) - S7-S7, S7-S5, and communication with other TCP/IP compliant partners - Supports Send/Receive and Fetch/Write functions as used in S5 PLCs. - PCs generally use the TCP/IP implementation of the operating system
UDP	- Supported on Industrial Ethernet Subnets only (via TCP/IP protocol) - Unsecured transfer of contiguous blocks of data between partners
E-Mail	- Supported on Industrial Ethernet Subnets via the TCP/IP protocol only - Enables process data transfer via e-mail IT services - Event-triggered e-mail delivery

PROFIBUS Communications Processors

In the SIMATIC NET world of communications, a PLC or PC node is connected and established on the network using a communications processor (CP). You will determine what CP is required, based primarily on the required communications services. We reviewed PROFIBUS communications services in the previous discussion. Here, we will first look at the communications processors needed to connect the S7-300 and S7-400 PLCs to PROFIBUS; following this, we will look at the CPs required to connect PC stations to PROFIBUS.

PROFIBUS CPs for S7-300 and S7-400 Stations

The CPs in the table below, allow you to establish the S7-300 or S7-400as a station on a PROFIBUS subnet. With the appropriate selection, the S7-300 may serve as a PROFIBUS DP-Master, as a DP-Slave, or as an FMS Server. As PROFIBUS nodes, all S7-300 or S7-400 CPs will support remote programming, via PG/OP services; S7-to-S7 communication, via S7 functions; and communication with S5 PLCs (CP 5431), via the FDL services. Given these standard services, available on all PROFIBUS CPs, you must select a CP according to the specific protocols and communication services needed in your application (e.g., FMS Master, DP-Master/DP-Slave).

PROFIBUS CPs for the S7-300 include the CP 342-5, the CP 342-5 FO, and the CP 343-5; and for the S7-400, the CP 443-5 Basic and the CP 443-5 Extended. Each short name listed in the CP Module column actually represents several CPs — each of whose full part number contains the numeric part of the short name (e.g., 342-5). These short names are used as CP folder names in the catalog of the *Hardware Configuration Tool*. The CPs found in a folder are of the same type and basic feature set. For example, the CPs in the CP 342-5 FO folder support direct fiber attachment, and incorporate the same basic communications services. The different part numbers in a CP folder may reflect the CP at different revisions, resulting from different firmware. Feature enhancements are briefly described when you select the part in the catalog.

Table 7-10. Overview of PROFIBUS Communications Processors (CPs) for S7-300/S7-400 PLCs.

CP Module	Brief Application Description	Comm. Services
CP 342-5 CP 342-5 FO	This CP may serve as DP-master or slave in the **S7-300** using copper or fiber attachments. As DP-master, the CP 342-5 supports connection of modular, compact, or intelligent slaves such as the S7-300 with a DP-slave port or via an installed CP342-5 serving as a DP-slave.	- PG/OP Comm. - S7 Comm. - S5 Comm. - PROFIBUS DP (M) - PROFIBUS DP (S)
CP 343-5	This CP is required to establish the **S7-300** as a PROFIBUS FMS master on PROFIBUS subnets. Communications services support communication with other PROFIBUS stations (e.g., S7, S5, 505, C7, HMIs, OPs, and PCs).	- PG/OP Comm. - S7 Comm. - S5 Comm. - PROFIBUS FMS
CP 443-5 Basic	This CP is required to establish the **S7-400** as a PROFIBUS master (FMS) on PROFIBUS subnets. Communications services also allow communication with other PROFIBUS stations (e.g., S7, S5, 505, C7, HMIs, OPs, and PCs).	- PG/OP Comm. - S7 Comm. - S5 Comm. - PROFIBUS FMS
CP 443-5 Extended	This CP is required to establish the **S7-400** as a DP-master having DP-slave connections to modular, compact, or intelligent slaves. Communications services support communication with other PROFIBUS stations (e.g., S7, S5, 505, C7, HMIs, OPs, and PCs).	- PG/OP Comm. - S7 Comm. - S5 Comm. - PROFIBUS DP (M)

Note: PROFIBUS DP (**M**) = DP-master; PROFIBUS DP (**S**) = DP-slave. S7-to-S5 Communication is via PROFIBUS FDL (layer-2) connection services, using the Send/Receive interface.

PROFIBUS CPs for PC and PG/PC Stations

SIMATIC NET classifies communications processors for PCs, as either a *HARDNET* CP or *SOFTNET* CP. The HARDNET CP design incorporates a dedicated microprocessor, which makes it suitable in applications with more stringent performance requirements. HARDNET CPs supports loadable firmware, and allows autonomous handling of multiple protocols onboard. SOFTNET CPs, do not incorporate a microprocessor and must rely on the resources of the PC for implementing the protocol software. They support single-protocol operation and suited to less stringent requirements. The PROFIBUS CPs in the table below, allow PC stations to serve in a variety of industrial applications.

The CP 5613 and 5614 are HARDNET CPs, and are typically used where several CPs are required in one PC, in high performance HMI systems, (e.g., WinCC), or in large supervisory/data acquisition systems (e.g., more than five S7-PLCs). SOFTNET CPs are typically used for STEP 7 programming systems, and in small supervisory/data acquisition systems (e.g., less than five S7-PLCs). In the catalog of the Hardware Configuration Tool, these modules are found under the *SIMATIC PC Station* in the *CP PROFIBUS* folder. Software drivers for operating PCs on PROFIBUS are described later in *PROFIBUS Communications Services and Software*.

Table 7-11. PROFIBUS Communications Processors for PC and Programming Stations.

CP Module	H/S	Brief Description	Comm. Services
CP 5511 (PCMCIA)	S	This PC network card is certified for use with all PROFIBUS SOFTNET drivers. The CP 5511 can serve as both DP-master and DP-slave, and is intended for use in non-critical/non-stringent applications (e.g., PG/PC, small HMI system).	- PG/OP Comm. - S7 Comm. - S5 Comm. - PROFIBUS DP (M) - PROFIBUS DP (S)
CP 5611 (PCI)	S	This PC network card is certified for use with all PROFIBUS SOFTNET drivers. The CP 5611 can serve as both DP-master and DP-slave, and is intended for use in non-critical/non-stringent applications (e.g., PG/PC, small HMI system).	- PG/OP Comm. - S7 Comm. - S5 Comm. - PROFIBUS DP (M) - PROFIBUS DP (S)
CP 5613 CP 5613 FO (PCI)	H	This PC network card is compatible with all CP 5412 A2 software. The CP 5614/CP 5614 FO can serve as DP-master, and is able to handle high-speed PROFIBUS applications (e.g., PC-Based Control, HMI, or redundant systems).	- PG/OP Comm. - S7 Comm. - S5 Comm. - PROFIBUS DP (M) - PROFIBUS FMS
CP 5614 CP 5614 FO (PCI)	H	This PC network card is compatible with all CP 5412 A2 software. The CP 5614/CP 5614 FO can serve as DP-master and DP-slave, and is able to handle high-speed PROFIBUS applications (e.g., PC-Based Control, HMIs, or redundant systems).	- PG/OP Comm. - S7 Comm. - S5 Comm. - PROFIBUS DP (M) - PROFIBUS DP (S) - PROFIBUS FMS

Note: **H** = HARDNET PC adapter; **S** = SOFTNET PC adapter; PROFIBUS DP (**M**) = DP-master; DP (**S**) = DP-slave. S5 Communication is via PROFIBUS FDL services, using Send/Receive Interface (Data exchange via standardized services provided as C-Functions).

PROFIBUS Software for PC and PG/PC Stations

Software drivers needed to operate PROFIBUS communications services on PCs and programming stations are available on the SIMATIC NET CD. This CD contains drivers for the different Windows operating systems, for both HARDNET and SOFTNET PC communications processors (i.e., network interface cards). As shown in the table below, DP-5613, FMS-5613, and S7-5613, are driver packages for both the CP 5613 and CP 5614 modules. SOFTNET-S7 and SOFTNET-DP are for SOFTNET adapters CP 5511 and CP 5611. Drivers for PG/OP services, needed by operator panels (OPs) and programming systems on PROFIBUS are contained in each of the listed driver packages shown with an asterisk. The same is true for PROFIBUS FDL drivers, which are required for S7-to-S5 compatible communication on PROFIBUS.

Table 7-12. PROFIBUS Communications Drivers for PC and Programming (PG/PC) Stations.

Software	Used with	Description
DP-5613	CP 5613/5614	Software drivers for DP master/slave functionality. *
FMS-5613	CP 5613/5614	Software drivers for PROFIBUS FMS functionality*
S7-5613	CP 5613/5614	Software drivers for S7 Functions*
SOFTNET-S7	CP 5511/5611	SOFTNET drivers for S7 Functions *
SOFTNET-DP	CP 5511/5611	SOFTNET drivers for DP master/DP slave functionality. *
SOFTNET-DP (S)	CP 5511/5611	SOFTNET drivers for DP slave functionality.
OPC Server	All PROFIBUS CPs	Contained in each package as interface to OPC capable Windows applications (e.g., Office, HMIs)

Notes: * Drivers for PG/OP and PROFIBUS FDL services are contained in this software package. PG/OP and FDL drivers are installed when the HARDNET adapters CP 5613/5614 are installed.

Ethernet Communications Processors

In the SIMATIC NET world of communications, a PLC or PC node is connected and established on the network using a communications processor (CP). You will determine what CP is required, based primarily on the required communications services. We reviewed Ethernet communications services in the previous discussion. Here, we will first look at the communications processors used to connect the S7-300 and S7-400 PLCs to Ethernet; following this, we will look at the CPs required to connect PC stations to Ethernet.

Ethernet CPs for S7-300 and S7-400 Stations

The CPs in the following table allow you to establish the S7-300 or S7-400as a station on an Ethernet subnet. As Industrial Ethernet nodes, all S7-300 or S7-400 CPs will support remote programming, S7-to-S7 communication, and communication with S5 PLCs (CP 1430), via the ISO or TCP services. Given these standard services, available on all Ethernet CPs, you must select a CP according to the specific protocols and communication services needed in your application (e.g., routing via gateways or IT services). The main features of a specific CP are highlighted when the part is selected in the STEP 7 *Hardware Configuration* catalog.

Ethernet CPs for the S7-300 include the CP 343-1, and the CP 343-1 IT; and for the S7-400, the CP 443-1 and the CP 443-1 IT. Each short name listed in the CP Module column actually represents several CPs — each of whose full part number contains the numeric part of the short name (e.g., 343-1). These short names are used as CP folder names in the catalog of the *Hardware Configuration Tool*. The CPs found in a folder are of the same type and basic feature set. For example, the CPs in the CP 343-1 IT folder all support internet capability and incorporate the same basic communications services. The different part numbers in a CP folder may reflect the CP at different revisions, resulting from different firmware. Feature enhancements are briefly described when you select the part in the catalog.

Table 7-13. Overview of Ethernet Communications Processors (CPs) for S7-300/S7-400 PLCs.

CP Module	Brief Application Description	Comm. Services
CP 343-1	This CP attaches the **S7-300** to Industrial Ethernet, with standard capabilities including remote programming between networks. Services of this CP also supports S7-300 communication with other devices (e.g., S7, S5, HMIs, PCs), via ISO and TCP/IP transport connections.	- PG/OP Comm. - S7 Comm. - S5 Comm. - ISO Transport - TCP/IP Transport
CP 343-1 IT	This CP attaches the **S7-300** to Industrial Ethernet, with standard and internet capabilities (e.g., Web browser access to data, Event E-mail). Services of this CP also supports S7-300 communication with other devices (e.g., S7, S5, HMIs, and PCs), via TCP/IP transport connections.	- PG/OP Comm. - S7 Comm. - S5 Comm. - ISO Transport - TCP/IP Transport
CP 443-1	This CP attaches the **S7-400** to Industrial Ethernet, with standard capabilities including remote programming between networks. Services of this CP also supports S7-400 communication with other devices (e.g., S7, S5, HMIs, PCs), via ISO and TCP/IP transport connections.	- PG/OP Comm. - S7 Comm. - S5 Comm. - ISO Transport - TCP/IP Transport
CP 443-1 IT	This CP attaches the **S7-400** to Industrial Ethernet, with standard and internet capabilities (e.g., Web browser access to data, Event E-mail). Services of this CP also supports S7-400 communication with other devices (e.g., S7, S5, HMIs, and PCs), via TCP/IP transport connections.	- PG/OP Comm. - S7 Comm. - S5 Comm. - ISO Transport - TCP/IP Transport

Note: S7-to-S5 communication is via the Send/Receive interface, using *ISO transport*, *ISO-on-TCP*, *TCP/IP transport*, or *UDP* connection services.

Ethernet CPs for PC and PG/PC Stations

SIMATIC NET classifies communications processors for PCs, as either a *HARDNET* CP or *SOFTNET* CP. The HARDNET CP design incorporates a dedicated microprocessor, which makes it suitable in applications with more stringent performance requirements. HARDNET CPs supports loadable firmware, and allows autonomous handling of multiple protocols onboard. SOFTNET CPs, do not incorporate a microprocessor and must rely on the resources of the PC for implementing the protocol software. They support single-protocol operation and are suited to less stringent requirements.

HARDNET CPs like the CP 1613, are recommended in cases where use of PC resources must be minimized. Example uses include redundant S7-H systems, where several CPs are needed in one PC, in high performance HMI applications (e.g., WinCC), and large (more than five S7-PLCs) supervisory/data acquisition systems. SOFTNET CPs, like the CP 1512 and the CP 1612, are typically used for STEP 7 programming systems, and in small supervisory/data acquisition systems (e.g., less than five S7-PLCs). These modules are found under the *SIMATIC PC Station* in the *CP Industrial Ethernet* folder of the hardware catalog. Software drivers for operating PCs on Ethernet are described later in *Ethernet Communications Services and Software*.

Table 7-14. Ethernet Communications Processors for PC and Programming (PG/PC) Stations.

CP Module	H/S	Brief Description	Services
CP 1512 (PCMCIA)	S	This PC network card is certified for use with all Industrial Ethernet SOFTNET drivers. The CP 1512 supports both ISO and TCP/IP protocols, and is intended for use in non-critical/non-stringent applications (e.g., PG/PC, small HMI system).	- PG/OP Comm. - S7 Functions - S5 Comm. - ISO and TCP/IP
CP 1612 (PCI)	S	This PC network card is certified for use with all Industrial Ethernet SOFTNET drivers. The CP 1612 supports both ISO and TCP/IP protocols, and is intended for use in non-critical/non-stringent applications (e.g., PG/PC, small HMI system).	- PG/OP Comm. - S7 Functions - S5 Comm. - ISO and TCP/IP
CP 1613 (PCI)	H	This PC network card is compatible with all CP 1413 protocol software. The CP 1613 supports ISO and TCP/IP protocols and is able to handle high-speed protocol processing for a large number of connections.	- PG/OP Comm. - S7 Comm. - S5 Comm. - ISO and TCP/IP - TF Protocol

Note: **H** = HARDNET PC network adapters; **S** = SOFTNET PC network adapters. S5 compatible communication is via ISO, TCP/IP, or UDP connections, using the Send/Receive interface.

Ethernet Software for PC and PG/PC Stations

Software drivers needed to operate Ethernet communications services on PCs and programming stations are available on the SIMATIC NET CD. This CD contains drivers for the various Windows operating systems, for both HARDNET and SOFTNET PC communications processors. As shown in the following table, S7-1613, TF-1613, and S7-Redconnect, are driver packages for the CP 1613. SOFTNET-S7 and SOFTNET-PG packages are used in conjunction with SOFTNET adapters CP 1512 and CP 1612.

PC drivers that support PG/OP services for programming systems on Industrial Ethernet are contained in each of the listed software drivers shown with an asterisk. The same is true for S5 compatible communications drivers, required for S7-to-S5 communication on ISO, ISO-on-TCP, TCP, and UDP connections.

Table 7-15. Ethernet Communications Drivers for PC and Programming (PG/PC) Stations.

Software	CPs Used With	Description
S7-1613	CP 1613	Software for S7 Functions.*
S7-Redconnect	CP 1613	Software for S7 Functions on redundant Ethernet. *
TF-1613	CP 1613	Software for Technological Functions/Manufacturing Message Specifications (TF/MMS services).*
PG-1613	CP 1613	Software for PG/OP Functions only.
SOFTNET PG	CP1512/CP1612	SOFTNET drivers for PG/OP Functions only.
SOFTNET-S7	CP1512/CP1612	SOFTNET drivers for S7 Functions. *
OPC Server	All Ethernet CPs	Interface to OPC capable Windows applications

Notes: * Drivers for PG/OP and S5 compatible communication (i.e., ISO, ISO-on-TCP, TCP, and UDP) are contained in these software packages; corresponding OPC server is also included.

The Network Configuration Tool

In STEP 7, creating a network configuration is to create a software model of the actual networked stations of an S7 project. The configuration includes the graphic arrangement of one or more subnets with their connected stations. PROFIBUS DP subnets will include both master and slave devices, where slave stations are shown connected to a maser station, and may be displayed or hidden from view. Configuration also involves module addressing and the setting of module parameters.

Configuring the Network

Creation of the network configuration requires the *Network Configuration Tool*, which may be started from the SIMATIC Manager or the Hardware Configuration tool. After a station is created, the configuration tool is launched by opening the Station folder and then double-clicking on the hardware object. With the hardware configuration tool, an object model of your actual hardware arrangement is developed, with each component of your hardware installation having a matching object in the hardware configuration. Once developed, a network configuration may be copied to other STEP 7 projects, modifying it as required.

The completed configuration can be checked for errors using a consistency check, compiled and then saved. The configuration is saved to the *System Data* object, which is placed in the offline *Blocks* folder. The System Data object may then be downloaded to the CPU, thereby providing the CPU with complete information of the network configuration. The CPU in turn transfers configuration parameters to the appropriate programmable modules.

Menu and Toolbar

Menu headings of the Network Configuration tool include *Network, Edit, Insert, PLC, View, Options, Window,* and *Help*. Network operations allow you to create, open, save, compile, and check the configuration for errors. Standard Edit operations including Cut, Copy, Delete, and Paste allow subnet and station objects to be edited as required. Standard online operations such as configuration upload and download, monitoring and diagnostic tools are supported by PLC operations. View and Window operations allow components of the configuration window to be displayed, hidden, or arranged to your convenience. The toolbar buttons, listed below, represent some of the most frequently used menu operations.

Table 7-16. NETPRO Network Configuration Toolbar Buttons.

Icon	Toolbar Function	Icon	Toolbar Function
	Open Station Offline Window		Download Network Configuration
	Open Station Online Window		Upload Network Configuration
	Save Configuration		View Network Catalog
	Save and Compile Configuration		Network Configuration Help
	Print Network Configuration		Insert New Connection
	Copy Selected Object		Change Connection Partner
	Paste Object from Clipboard		

Network Components Catalog Window

The network components catalog contains the various component objects used to create a configuration of your networking solution. When the configuration tool is opened, the catalog, which may be hidden from view, can be displayed by selecting **View ➢ Catalog** from the menu. If the window is docked, you may double-click above the words "Selection of the Network" to cause it to undock. If it is already un-docked, you may double-click on the title bar to cause it to dock. When undocked, you may resize or move the window to suit your convenience. You may also re-dock the catalog by dragging and dropping it onto the left-edge or the right-edge of the configuration window.

The major object containers of the catalog, *PROFIBUS DP*, *PROFIBUS PA*, *Stations*, and *Subnets*, are presented in a tree structure. PROFIBUS-DP and PROFIBUS-PA containers include objects for representing DP or PA slave devices. The Station containers contain objects for SIMATIC 300, SIMATIC 400, SIMATIC PC, SIMATIC S5, SIMATIC S7-400-H stations, PG/PC, and Other Stations (Non-Siemens stations). The Subnets folder contains objects for MPI, PROFIBUS, Industrial Ethernet, and Point-to-Point Subnets. Once you place a subnet is in the configuration, then compatible nodes may be placed in the window.

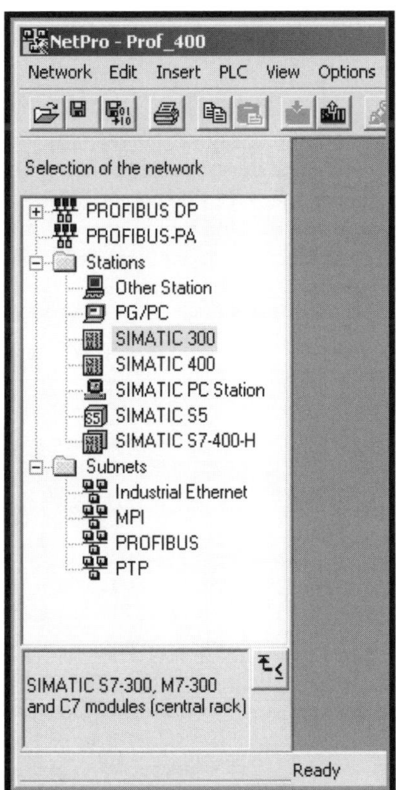

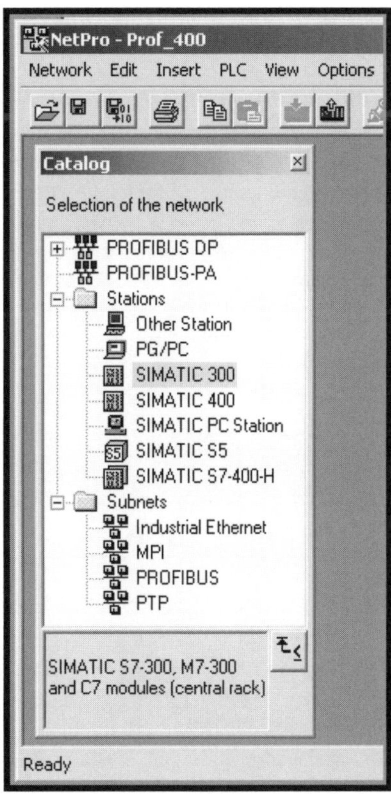

Figure 7-4. (a) Network Catalog Left-Docking. (b) Network Catalog Un-Docked window.

Configuration Window – Network Layout

As you develop your network configuration, the upper pane of the window will contain the subnets and stations that you define. If you have attached the nodes to a subnet interface, then they will appear attached to a specific subnet. Each connected S7 PLC station will contain an object for the programmable modules installed in the station (e.g., CPU, CP, or FM). Stations connected to MPI are attached via the MPI interface of a CPU module. Stations on PROFIBUS or Industrial Ethernet are attached to a communications processor object (e.g., CP 443-5, or CP 343-1). You can access and modify the properties of a subnet or any programmable module (e.g., station address), by double-clicking on the object.

During configuration, PLC station objects for the S7-300, S7-400, S7-400 H, may be dragged to the window. If the station's hardware is not yet configured, you can open the station in the hardware configuration tool by double clicking on the station name. You can return to the network configuration using the toolbar icon for Configuring Networks. PC workstations, PG/PC programming stations, and SIMATIC S5 stations may also be dragged and dropped into the station window. With these stations, you simply need to create and assign an appropriate network interface to the station object in order to attach it to a subnet.

PROFIBUS DP slaves are also shown as objects, based on the type of drop (e.g., modular, compact). As the configuration is developed, you can either hide or display the connection between each DP master and its associated DP slaves, from the View menu.

Configuration Window - Connection Table

When an S7-300/S7-400 CPU is selected in any of the stations of the upper pane of the network layout window, a list of its configured connections are displayed in a connection table in the lower pane of the configuration window. A new connection is added from the menu, by selecting **Insert ➢ New Connection**. For a given connection, the table columns list a connection ID in the local station, *Local ID*; a *Partner ID*, for the local station; the name of the remote partner, *Partner*; the connection *Type*, and the *Subnet* name. The Active Connection column indicates whether the local station is the *Active Connection Partner* in establishing the connection. Yes, indicates that the local node establishes the connection; 'No' indicates that the connection partner establishes the connection.

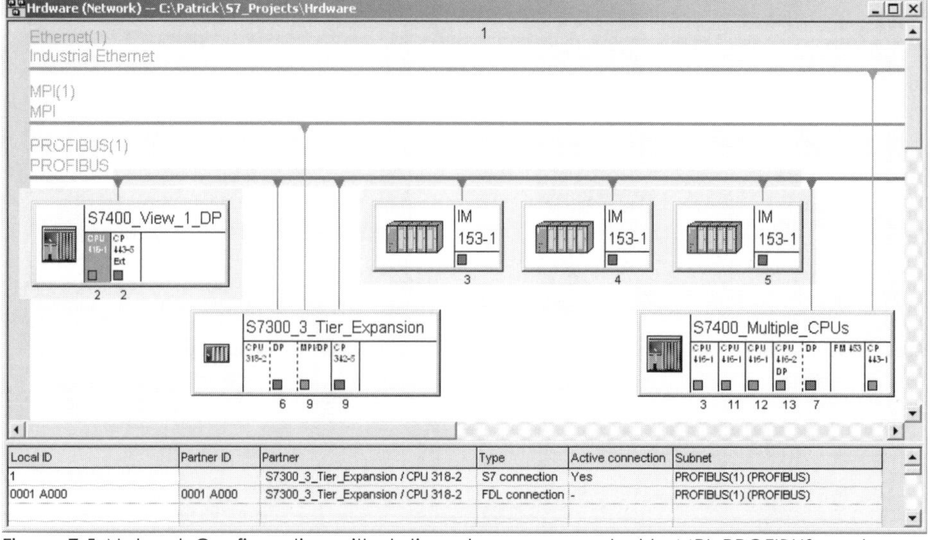

Figure 7-5. Network Configuration with stations shown connected to MPI, PROFIBUS, and Industrial Ethernet. PROFIBUS DP Master and associated modular DP Slaves highlighted.

Comments on Working with SIMATIC NET Networks

Network configuration tasks are accomplished using both the Hardware Configuration tool and the NetPro network configuration tool. Generally, at this stage, you will have configured each station's hardware (i.e., S7-300/S7-400 racks and modules) using the hardware configuration tool. You may or may not have inserted the various network communications processors in each station, assigned unique addresses, and attached the CPs to a subnet. If you did, however, you have already done much of the work and may only need to configure connections between communications partners.

This final chapter considers building the network configuration graphically from the top-down, using the NetPro Configuration Tool. As you may have already determined, the network is built from the bottom-up, as you install and configure the communications processors using the Hardware Configuration Tool. You may have already taken this approach. In either case, the network configuration is completed in the Network configuration tool, where you will need to configure connections between partners and download the results.

As a whole, establishing stations on a network can be accomplished in just a few short tasks. These tasks, some of which you may have already performed, are outlined below. Step-by-step examples of configuring MPI, PROFIBUS, and Industrial Ethernet networks are provided in the following pages.

Checklist: Working with SIMATIC NET Networks

- *Verify that "NCM for Industrial Ethernet" and "NCM for PROFIBUS" are installed (these packages are typically installed with the initial STEP 7 installation).*
- *Install the required protocol drivers in PC stations/programming systems.*
- *Insert the subnets (e.g., MPI, PROFIBUS, or Industrial Ethernet) required in your project, and configure the operating parameters of each subnet.*
- *Insert the required communications processor (e.g., Industrial Ethernet, or PROFIBUS) in each S7-300/S7-400 station; configure the address and operating parameters of the CP, and attach it to an existing subnet.*
- *Insert the required communications processor (e.g., MPI, PROFIBUS, or Industrial Ethernet) in each PC host or programming station; configure its address and operating parameters, and attach it to an existing subnet.*
- *Define required connections for S7 communications partners that will exchange data via configured connections (e.g., S7, FDL, ISO, ISO-on-TCP, TCP, or UDP).*
- *If global data transfers are required, then configure the Global Data table for the participating CPUs, which reside on an MPI subnet.*
- *Save and compile the network configuration after making changes, and perform consistency check for configuration errors prior to downloading.*
- *Download the network configuration to network stations.*
- *Use the Setting the PG/PC Interface utility, to select the appropriate ONLINE interface in your programming system (PG/PC).*

Building a Network Configuration Using NetPro

Basic Concept

The NetPro Configuration tool allows you to build and configure your MPI, PROFIBUS, and Industrial Ethernet networks in a graphical editor. Although it is possible to start creating the stations of your network using the SIMATIC Manager and the Hardware Configuration tool, more complex networks may be simpler to build by starting with NetPro. With NetPro, you may define the network stations and layout, and then finish up with the Hardware Configuration tool as you configure the racks and modules of each station.

Essential Elements

NetPro is similar to the Hardware Configuration tool in that it provides a component object catalog from which you may drag and drop network objects into the Network configuration window. In this fashion, you may build the framework of your network from the top, starting with the subnets (i.e., MPI, PROFIBUS, and Industrial Ethernet). After you have inserted one or more subnets, you may then insert the required network stations. Network stations include the S7-300, S7-400, S7-400 H, SIMATIC S5, PG/PC, PROFIBUS DP PA stations.

After you define the graphic layout of the network, you may then proceed to configure each station (i.e., insert modules and define parameters). When you double-click on a station, the station is opened in the Hardware configuration tool, where you may configure the central rack with power supply, CPU, and the required communications processor (CP). With the CP installed, you may then configure its operating parameters, its address, and finally attach the station to the subnet. As you attach each CP, the next available address is suggested.

Application Tips

As you insert and configure each communications processor, remember that each station must have a unique address. In addition, you will also want to manage the assignment of a unique MPI address to each CP installed in a rack and in the total subnet. Recall that in the S7-300, STEP 7 automatically assigns an MPI address to each FM or CP, by incrementing each address by one starting with the address assigned to the CPU.

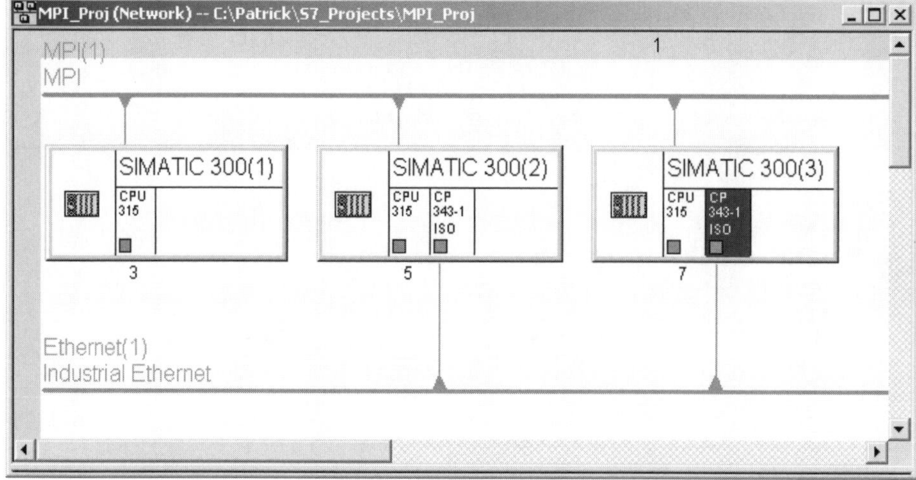

Figure 7-6. NetPro Graphic connection of S7-300 Stations on MPI and Industrial Ethernet.

Quick Steps: Building a Network Configuration Using NetPro

STEP	ACTION
1	Open an existing project or create a new project if necessary.
2	With the project window open, from the right pane, double click on the MPI subnet object or press the *Configure Network* icon, to open the NetPro configuration tool.
3	If the network objects catalog is not open in the window, select View ➤ Catalog.
4	From the catalog, expand the object folder for both the **Stations** and **Subnets**.
5	From the list of Subnets, double click on the desired subnet type (e.g., *PROFIBUS, MPI*, or *Industrial Ethernet*) to insert. Once the subnet object is inserted in the window, you may drag the object up or down to a new position, as desired.
6	To configure the operating properties of a subnet (e.g., **Transmission Rate**, **Highest Address**, and subnet **Profile** in the case of PROFIBUS), double click on the subnet object and select the *Network Settings* tab. The **subnet ID** and **Name**, are on the General tab, and may be modified; the subnet ID is generally unchanged.
7	From the Station folder, you may insert stations in the configuration window with a double click on the station type (e.g., SIMATIC 300, SIMATIC 400) or by dragging the object onto the screen. You may insert as many stations as required and may drag a station to any new position you choose.
8	Double click on a station object (e.g., on the Station name) to cause the station to open in the Hardware Configuration tool. Then, create a central rack, by inserting the required *Rack, CPU, Power Supply*, and the *Communications Processor*.
9	When you insert a CP *(e.g., PROFIBUS, Ethernet)*, the properties dialog will appear; *attach the CP to the subnet*, by selecting the subnet name — the next available address is automatically entered in the field. You may accept or alter the address.
10	Save the central rack configuration in order to have the results appear in the network configuration when you return to NetPro.
11	From the Network Configuration, you may alter any network address. For MPI, double-click on the CPU object; for PROFIBUS or Industrial Ethernet stations, double click on the associated communications processor (CP) object.
12	You can configure each station using the procedures of Step 8 through Step 10.
13	When all of the stations are completed, select **Network ➤ Save and Compile** to perform a consistency check, and to save the configuration to the *System Data*.

Downloading a Network Configuration Using NetPro

Basic Concept

Once you have completed your network configuration, you must then proceed to download to the individual stations. Like a station's hardware configuration, the network configuration is saved in system data blocks (SDBs), which are downloaded to the CPU of each station, which in turn transfers appropriate data to the communications processors. The SDBs associated with the network configuration include subnet properties, station addresses, configured connections, input/output addresses, and module parameters.

Essential Elements

You can download the network configuration for a station, using the **PLC ➢ Download** command, either from the Simatic Manager, or from the Hardware Configuration tool when configuring station hardware. By downloading the configuration from the Network Configuration Tool, you may download the entire configuration station by station to all stations on a subnet. You may also choose to download the network configuration of selected stations only. In addition, you can selectively download specific stations along with the associated connections and configurations for communications partners.

Application Tips

As you prepare to download the network configuration, consider that while it is possible to download the configuration over a subnet, that all stations must have been previously assigned unique station addresses. In addition, you will also want to ensure the assignment of a unique MPI address to each CPU, CP, and FM installed in a rack and in the total subnet. Recall that in the S7-300, STEP 7 automatically assigns an MPI address to each FM or CP, by incrementing each address by one starting with the address assigned to the CPU.

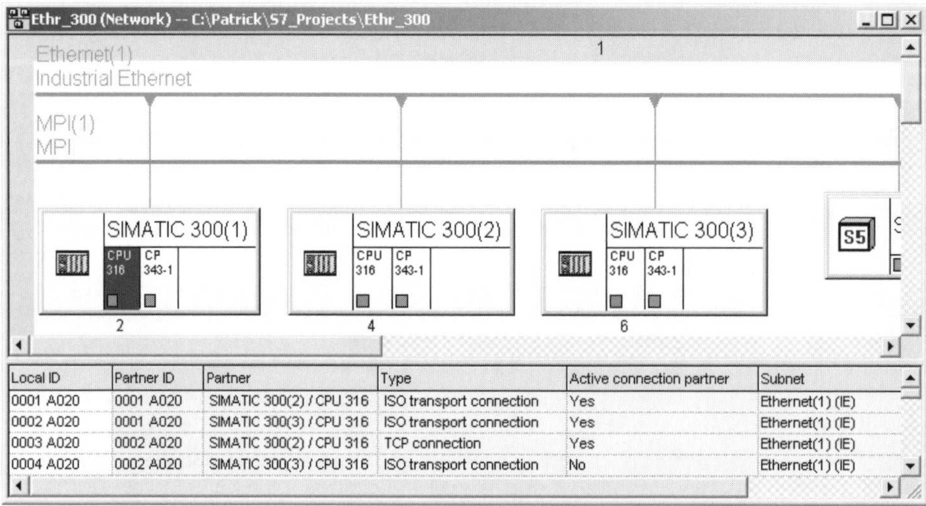

Local ID	Partner ID	Partner	Type	Active connection partner	Subnet
0001 A020	0001 A020	SIMATIC 300(2) / CPU 316	ISO transport connection	Yes	Ethernet(1) (IE)
0002 A020	0001 A020	SIMATIC 300(3) / CPU 316	ISO transport connection	Yes	Ethernet(1) (IE)
0003 A020	0002 A020	SIMATIC 300(2) / CPU 316	TCP connection	Yes	Ethernet(1) (IE)
0004 A020	0002 A020	SIMATIC 300(3) / CPU 316	ISO transport connection	No	Ethernet(1) (IE)

Figure 7-7. NetPro Graphic connection of S7-300 Stations on Industrial Ethernet.

Quick Steps: Downloading a Network Configuration Using NetPro

STEP	ACTION
1	When you are done with configuring your network, you must ensure that you have assigned unique addresses to each station (including stations whose addresses are set using switches.
2	Network connections must be completed for all stations that are communications partners in configured connections (e.g., S7, FDL, FMS, ISO, ISO-on-TCP, TCP/IP, etc.).
3	When all of the stations are completed, select **Network ➢ Save and Compile All** to perform a consistency check, and to save the configuration to the *System Data*.
4	When downloading for the first time, you must use a direct connection to each station's MPI port. Later after each CPU is assigned a unique MPI address, the configuration may be downloaded via the MPI subnet. After the initial download, it is also possible to download via PROFIBUS or Ethernet is subnets if they are in place.
5a	Download Stations on Subnet
	From the menu, select **PLC ➢ Download ➢ Stations on Subnet** to transfer the network configuration to all stations on the subnet — one-by-one. Connections, station addresses, I/O addresses, subnet properties, and module parameters are downloaded.
5b	Download Selected Stations
	From the menu, select **PLC ➢ Download ➢ Selected Stations** to transfer the network configuration of only the selected stations. Connections, I/O addresses, station addresses, subnet properties, and module parameters are all downloaded.
5c	Download Selected and Partner Stations
	From the menu, select **PLC ➢ Download ➢ Selected and Partner Stations** to transfer the network configuration of only the selected stations and its configured communications partners. Connections, I/O addresses, station addresses, subnet properties, and module parameters are all downloaded.
5d	Download Selected Connections
	From the menu, select **PLC ➢ Download ➢ Selected Connections** to download the selected connections of a station and to the partners involved in two-way communication. Only the selected connections are downloaded.
5e	Download Connections and Gateways
	From the menu, select **PLC ➢ Download ➢ Connections and Gateways** to transfer the configured connections to the selected module and its partner stations. Connections and gateway information are downloaded (in RUN-P or STOP modes).

Adding and Configuring an MPI Subnet

Basic Concept

An MPI subnet is automatically inserted into each newly created project — however, additional MPI subnets may be added as required to reflect each of the physical MPI subnets of your project. Once inserted, the operating parameters of the subnet must be configured as required, using the NetPro configuration tool.

Essential Elements

You may insert a new MPI subnet from the SIMATIC Manager or the network configuration tool. Once you insert the subnet, you must configure its *General* and *Network Settings* properties (See figure below). The General properties include a subnet **Name** and **Subnet ID.** Modifying the subnet ID is unnecessary, since this automatically derived ID (STEP 7 derived project number - subnet number) will always be unique. The Network Settings allow you to define the **transmission rate**, and the **highest station address (HSA)**. The standard transmission rate for MPI subnets is 187.5 Kbs, but this can be set to a maximum of 12Mbs to support the CPU 318-2 and the S7-400. The transmission rate, however, must be set for the slowest station.

Application Tips

The highest MPI address setting should be set to a value equal to or greater than the highest actual station address. A highest station address of 31 (HSA =31) is what is recommended in order to optimize the polling of network stations during normal, diagnostic, programming, or other network operations.

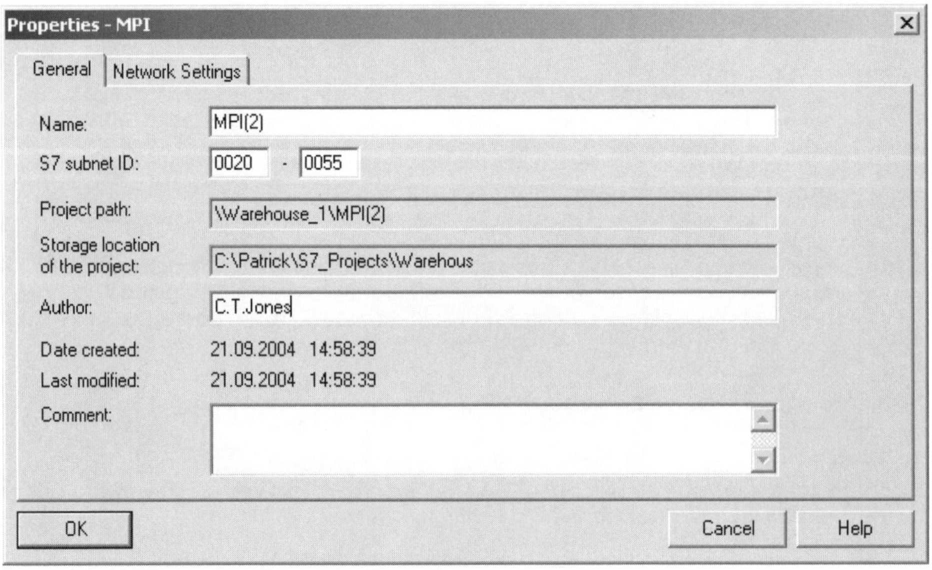

Figure 7-8. MPI Properties: General settings dialog.

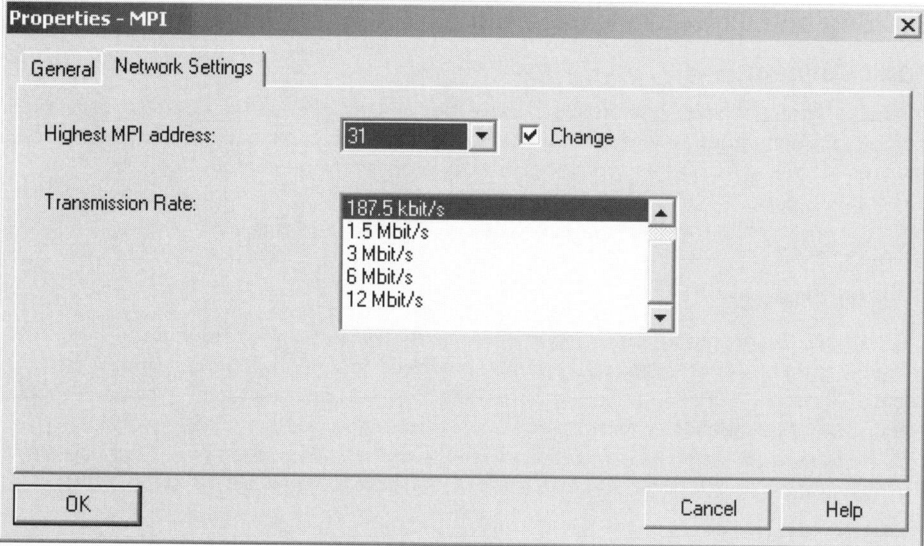

Figure 7-9. MPI Properties: Network Settings dialog.

Quick Steps: Adding and Configuring an MPI Subnet

STEP	ACTION
1	Open the SIMATIC Manager and the project where the MPI subnet is to be added.
2	Select the project folder, and then from the menu, select **Insert** ➢ **Subnet** ➢ **MPI**. This step is only required if you are adding an additional MPI subnet.
3	In the right pane of the project window, double click on any subnet object or press the *Configure Network* toolbar button to open the network configuration tool.
4	With the configuration tool open, select the MPI subnet object, double click on the subnet or right-click, and select **Object Properties** to open the properties dialog.
5	From the *General* tab, you may choose to keep the default **Name** of the MPI subnet [e.g., MPI (1)], or you may modify the name to reflect the subnet's function.
6	The **S7 Subnet ID,** represented by a project number separated by a subnet number, is automatically derived by STEP 7. You may modify the subnet ID, but this is unnecessary, since it will always be unique.
7	On the *General* tab, you may enter a department name in the **Author** field, or perhaps your name if you are defining the subnet properties; you may enter or modify the **Comment** lines to describe the subnet's function in the project.
8	From the *Network Settings* tab, activate the **Change** check box in order to modify the **Highest MPI Address**. For best performance set the actual highest address.
9	From the listed options, choose the **Transmission Rate** to use for the entire MPI subnet. This setting must not be higher than what is supported by the slowest station.
10	Select the **OK** button to accept the settings for the MPI subnet operating properties.

Building an MPI Network with Peer S7-300/S7-400 Stations

Basic Concept

The MPI port on each MPI-compatible device, though normally used for direct programming, can also be connected to form a low-performance network configuration. Communication services of the MPI network support local/remote programming functions for S7-300/400 CPUs, operator panels (OPs), observation stations (OS), and PCs. Data transmission services also support the automatic transfer of process variables between S7 CPUs and operator panels (OPs).

Essential Elements

Each S7 CPU has an integrated multipoint interface (MPI) port. Since the interface is integrated, no other cards are required. MPI stations are connected to the subnet in the same manner as PROFIBUS stations, and use the same media and attachment components. The choice of connection options, however, will depend on whether a copper or fiber optic (FO) media is used. Each S7 CPU will initially have a default MPI address of "2," but you will have to assign a unique MPI address to each station prior to connecting it to the subnet.

Application Tips

When assigning MPI station addresses, remember that an MPI subnet supports 32 stations, with addresses 0 -to- 31. In the S7-300, consider that CPs and FMs in each station are given addresses based on the CPU address. Each MPI device type has a factory default address, which you must modify if multiple devices of that type are to be connected. For a PG/PC programming station, "0" is the default address; the default address for an operator panel (OP), is "1"; and "2" is the default address of S7-300 and S7-400 H CPUs.

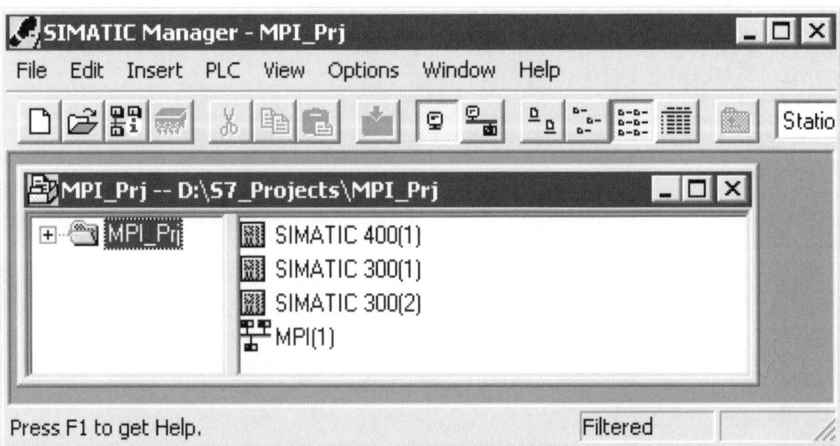

Figure 7-10. S7 Project with S7-300/S7-400 Stations, and an MPI subnet.

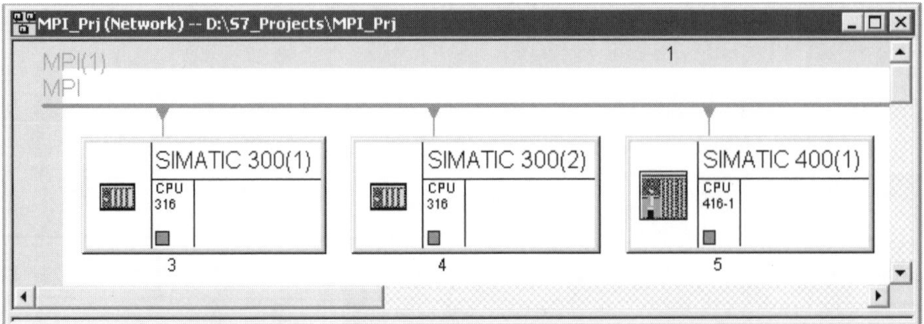

Figure 7-11. NetPro configuration tool, showing connection of S7-300/S7-400 stations on MPI.

Quick Steps: Building an MPI Network with Peer S7-300/S7-400 Stations

STEP	ACTION
1	Open the SIMATIC Manager to an existing project or create a new project using the **New** command.
2	With the project folder open in the SIMATIC Manager, from the menu select **Insert ➢ Station ➢ SIMATIC 300 Station**. Insert at least two additional S7-300 or S7-400 stations.
3	With the project folder selected, note in the right pane that STEP 7 has already generated an MPI subnet (created by default with each new project).
4	Double click on the subnet, or press the *Configure Network* toolbar button to open the Network Configuration tool. Your MPI network should appear, along with objects that represent the MPI subnet and the S7-300/S7-400 stations you created.
5	Double click on the MPI subnet object, to configure the subnet operating properties (e.g., **Transmission Rate**, **Highest MPI Address**) on the Network Settings tab. The subnet **Name** can be modified on the General tab (See the previous task, "*Adding and Configuring an MPI Subnet*").
6	Double click on the first station object (e.g., on the Station name) to open the station in the Hardware Configuration tool. Configure a central rack, inserting the required *Rack*, a *Power Supply*, and *CPU*.
7	Double click on the CPU object and when the dialog opens, press the properties button in the MPI Interface box.
8	When the MPI parameters dialog appears; *attach the CPU to the MPI subnet*, by selecting the correct subnet from the list; the next available address is automatically assigned when you select the subnet; you may also enter any valid address.
9	Save the central rack configuration and press the *Configure Network* icon on the toolbar to return to the network configuration. The station will appear connected to the MPI subnet when you return to the network configuration.
10	Repeat Step 6 through Step 9 to configure each remaining station.
11	When you are done with all stations, select **Network ➢ Save and Compile** to perform a consistency check for errors and to save the configuration to the *System Data*.

Configuring a Programming Station (PG/PC) on MPI

Basic Concept

If you will need to gain remote access to the S7 PLCs, connected to an MPI network, then you will have to configure and attach a PG/PC station to the subnet. As with other S7 stations, the PG/PC requires physical connection to the network and the logical configuration and attachment, in order to be known and recognized on the network.

Essential Elements

To establish a PC as a programming station on MPI, an MPI-compatible network interface card (e.g., CP 5511/CP 5611, or CP5412 A2)must be installed in the PC, and configured in Windows. If you have a, PG programming device (e.g., PG 720, PG 740), then it will already be equipped with the MPI network. The driver software for programming functions (PG/OP) over MPI is installed during the initial installation of STEP 7. With the physical connection in place, you only need to configure a logical interface connection for the PG/PC.

To configure your programming device, you must first insert a PG/PC object in the Simatic Manager or in the Network Configuration tool, and then configure and assign it an interface. An interface defines the subnet, and node address. When you assign the interface to one of the actual network cards on the PG/PC, then it will be adapted to the configured settings of the interface, including the operating properties of the subnet to which it is attached. If you modify the configured settings (e.g., the network transmission rate), the interface on your PG/PC is automatically adapted to match the new settings.

Application Tips

Since only one configured programming device (PG/PC) is allowed in a project, by creating several interfaces, which can be assigned to the PG/PC, you can conveniently switch interfaces. Several interfaces may be assigned to a PG/PC, however only one can be activated as the S7ONLINE interface at a given time. Without this function, you would have to call up the "Setting the PG/PC Interface" program and manually adapt the interface settings of your PG/PC to the configured settings.

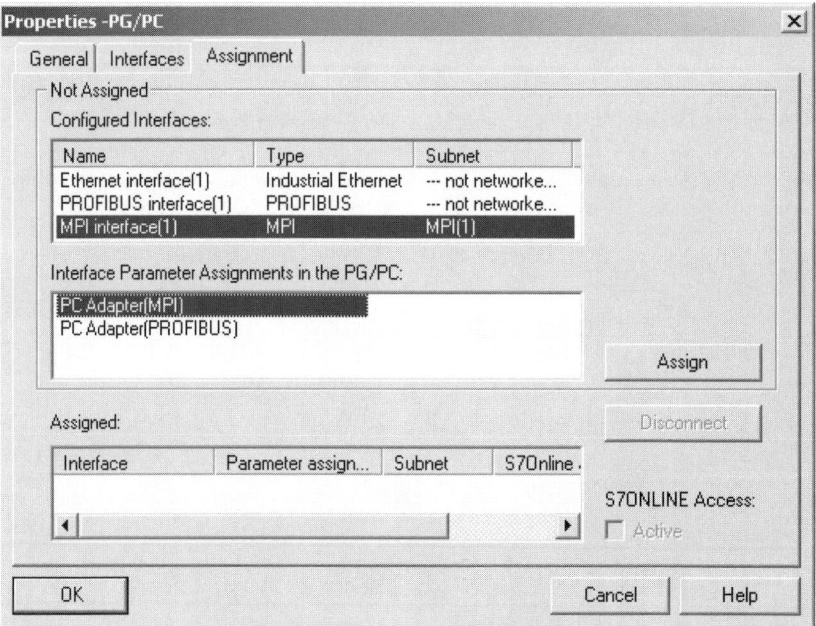

Figure 7-12. PG/PC Properties dialog: Interface Assignment tab.

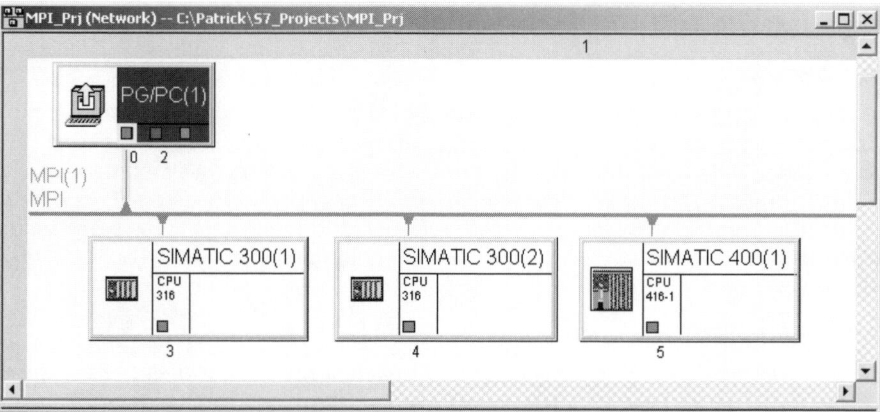

Figure 7-13. NetPro Graphic configuration of PG/PC assigned to a specific MPI interface.

Quick Steps: Configuring a Programming Station (PG/PC) on MPI

STEP	ACTION
1	Open the SIMATIC Manager, to the project where you wish to establish the PG/PC as a station on an MPI Subnet.
2	If the PG/PC object is not already installed, select the project folder, right click, and select **Insert New Object ➢ PG/PC**. Skip this step if the PG/PC is already installed.
3	With the project folder selected, from the right pane, double click on the MPI subnet object to open the NetPro graphical network configuration tool.
4	With the network configuration open, select the PG/PC object, right-click and select **Object Properties**; or double click on the PG/PC object to open the properties dialog.
5	From the *General* tab, you may modify the default name PG/PC (1), to a name you choose. You may also modify or enter a new **Author** of the interface parameters, and modify or enter a **Comment** describing the use of the programming system.
6	Select the *Interface* tab, to display existing interfaces or create a new MPI interface to which you wish to assign to the PG/PC.
7	Select the *Assignment* tab, to display the configured MPI interfaces and interface parameter sets (i.e., physically installed interfaces) previously defined in the PG/PC.
8	From the **Not Assigned** block, select the logical MPI interface you wish to assign to the physical interface in the PG/PC; then from the **Interface Parameter Assignment** window, select the physical interface of the PG/PC;
	Press the **Assign** button. The "Assign" button will be grayed out unless the configured interface you choose is networked and the selected interface parameter assignment matches the configured interface.
9	With a logical interface assigned to a physical MPI interface, you may then activate the **S7ONLINE** check box if you wish to activate this interface as the current means for establishing and the online connection to STEP 7.

Configuring Global Data Communications on MPI

Basic Concept

Global data (GD) communications is a simple cyclic data transfer service, integrated in all S7-300 and S7-400 CPUs. With this service, S7 PLCs may exchange input (I), output (Q), bit memory (M), and data block (DB) memory areas between CPUs on an MPI subnet (and via the S7-400 backplane). Global data transfer requires no programming since configured data is automatically transferred cyclically at the end of each CPU scan similar to the I/O update. The response time of global data exchange is therefore dependent on the CPU cycle of any two partners involved.

Essential Elements

For GD communication, you must configure a global data table, in the *NetPro Configuration Tool*. This table will define the CPUs involved, the memory areas sent, and the memory areas in which data are received. In the GD table, you may define up to fifteen CPUs, where each CPU is assigned to a column. As you define each CPU, the CPU and station identifier appear in the column header. In each row, in the cells beneath each CPU you will specify an address for the CPU to send or in which to receive data. A CPU is identified as the sender if the send address is preceded by the send character "➢". In a given row, a blank cell indicates the CPU in this column is not participating in the data exchange.

Application Tips

You may use Input (I) addresses as global data receive areas in the S7-300 and S7-400, if a corresponding module is not installed. In S7-400, you may handle global data transfer from you program, using SFCs. When the GD table is completed and compiled, each so-called global data row (GD row) receives a unique GD identifier (**GD ID**). You may use the GD ID as a parameter input to system functions (SFC), in the S7-400, when exchanging global data using SFC 60 (GD_SND) and SFC 61 (GD_RCV).

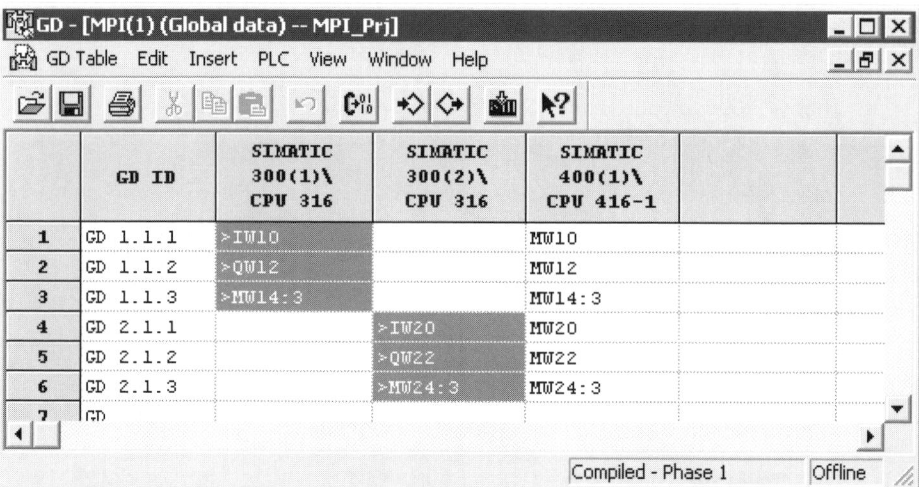

	GD ID	SIMATIC 300(1)\ CPU 316	SIMATIC 300(2)\ CPU 316	SIMATIC 400(1)\ CPU 416-1
1	GD 1.1.1	>IW10		MW10
2	GD 1.1.2	>QW12		MW12
3	GD 1.1.3	>MW14:3		MW14:3
4	GD 2.1.1		>IW20	MW20
5	GD 2.1.2		>QW22	MW22
6	GD 2.1.3		>MW24:3	MW24:3
7	GD			

Figure 7-14. Global Data Table, for configuring address for global data exchange.

Quick Steps: Configuring Global Data Communications on MPI

STEP	ACTION
1	Open the SIMATIC Manager and project where you wish to create a global data table.
2	From the menu select, **Options** ➢ **Configure Network**, or simply press the NetPro icon to open the NetPro network configuration tool.
3	When the network configuration window opens, select the MPI subnet object.
4	Verify that all stations that will participate in the global data table are attached to the MPI subnet. Use the following step to attach any stations not already attached.
5	Select the CPU object of the station, right click, select **Object Properties**; press the **Properties** button under the Interface box to open the MPI Interface dialog. Attach the station by selecting the subnet, set the address if required, and confirm with **OK**.
6	Select the MPI subnet, right-click and select **Define Global Data**.
7	Select the first column, right of the **GD ID** column, right-click and select **CPU...** to define the first column for a CPU that will participate in the global data exchange. When the project dialog opens, displaying the project, expand the station objects to display and select a CPU. Select a CPU and confirm the selection by pressing **OK**.
8	Repeat the procedure of defining a column for each CPU that will participate in the global data exchange.
9	Starting with the first GD row and first CPU, specify an address from which the CPU will send data, or in which it will receive data. To identify a send address, right-click and select **Sender**; all other addresses are marked **Receiver** by default.
10	When done, from the menu select **GD Table** ➢ **Save** to save the GD table.
11	From the menu, select **GD Table** ➢ **Compile** to generate a global data identifier (GD ID) for each set of global data (GD Row) and to compile the download data.

Adding and Configuring a PROFIBUS Subnet

Basic Concept

Connection of PROFIBUS stations on a common medium, comprised of one PROFIBUS segment or of several segments connected via repeaters, is considered a *subnet*. In a large plant or project, the complete PROFIBUS network may require multiple PROFIBUS subnets. In STEP 7, you will need to create a subnet object for each of your physical PROFIBUS subnets.

Essential Elements

You may insert a new PROFIBUS subnet from the SIMATIC Manager or the network configuration tool. Once you insert the subnet, you must configure its *General* and *Network Settings* properties (See figure below). The General properties include basic documentation such as **Name** and **Subnet ID.** Modifying the subnet ID is unnecessary, since this automatically derived ID (STEP 7 derived project number - subnet number) will always be unique. The Network Settings allow you to define PROFIBUS operating characteristics, including the **transmission rate**, the **highest station address (HSA)**, and the **bus parameters**.

PROFIBUS bus parameters are actually the timing values for the token bus. You may define the bus parameters, but since these values must be consistent for the entire network it is not recommended. Instead, STEP 7 offers predefined "**Profiles**," each of which optimizes the parameters for the makeup of a typical network. You can select a profile based on the makeup of your specific network.

Application Tips

If possible, you should create and configure nodes that you want to connect in a network in the same project. By placing the stations in the same project, STEP 7 is able to check your entries (e.g., addresses, connections, etc.) for consistency. In a large plant with several stations, it may be necessary to create a number of subnets. While you may manage these subnets in a single project, you may decide to handle the subnets in multiple projects. If stations in different projects need to be connected, you may assign the station to more than one subnet (i.e., in different projects).

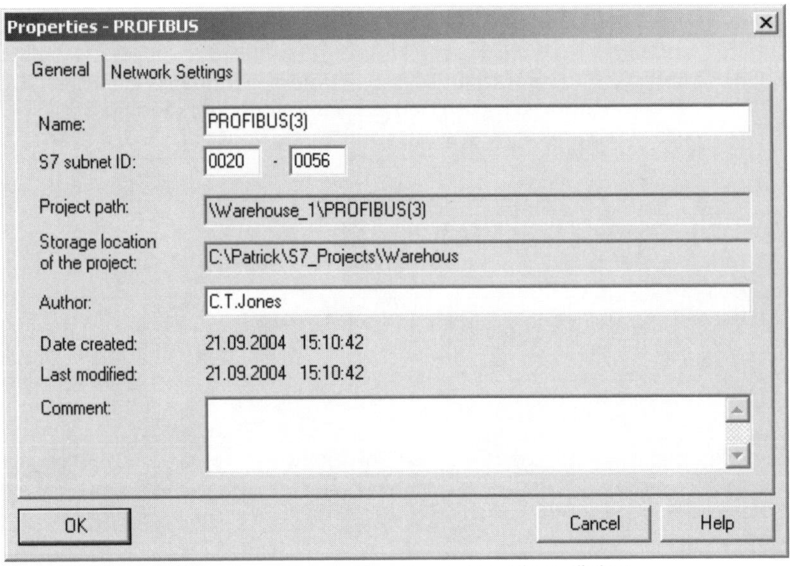

Figure 7-15. PROFIBUS subnet Properties: General settings dialog.

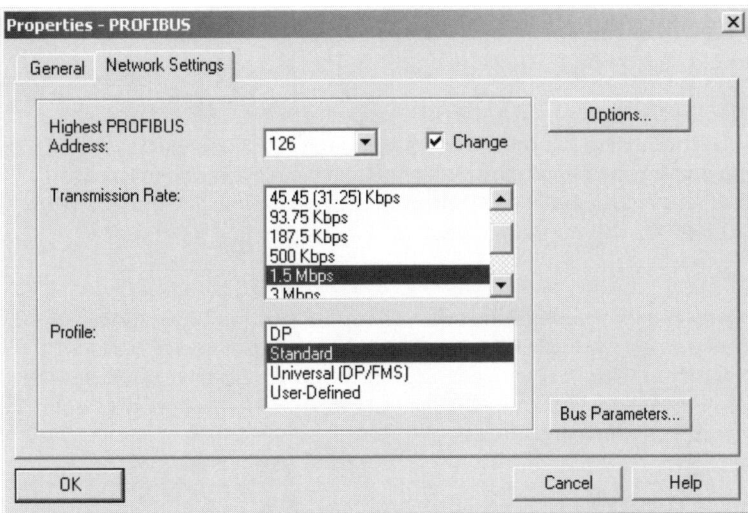

Figure 7-16. PROFIBUS subnet Properties: Network Settings dialog.

Quick Steps: Adding and Configuring a PROFIBUS Subnet

STEP	ACTION
1	With the SIMATIC Manager open to the project of the new PROFIBUS subnet, select the project folder, and then from the menu, select **Insert ➢ Subnet ➢ PROFIBUS**.
2	In the right pane of the project window, double click on any subnet object or press the *Configure Network* toolbar button to open the Network Configuration tool.
3	In the network configuration view, double click on the PROFIBUS subnet object or right-click on the object and select **Object Properties** to open the properties dialog.
4	From the *General* tab, you may keep or modify the default **Name** of the subnet, and the **S7 Subnet ID** (this STEP 7 derived ID number will always be unique).
5	Also on the *General* tab, you may enter a department name in the **Author** field, or perhaps your name if you are defining the subnet properties; you may enter or modify the **Comment** lines to describe the subnet's function in the project.
6	Select the *Network Settings* tab, and activate the **Change** check box to modify the **Highest PROFIBUS Address**. For best performance set the actual highest address.
7	Use the **Options** button, to specify further, the physical characteristics of the subnet; this will allow STEP 7 to determine the calculated bus parameters more accurately.
8	From listed options, choose the **Transmission Rate** for the entire PROFIBUS subnet. This setting must not be higher than what is supported by the slowest node.
9	From the listed options, select a **Profile** of the network configuration to determine predefined bus timing parameters based on network makeup. To set the bus-passing parameters manually, select the *User-Defined* profile, then the press **Bus Parameters** button.
10	Select **OK** to accept the settings for the PROFIBUS subnet operating properties.

Installing, Configuring, and Attaching an S7-300 PROFIBUS CP

Basic Concept

Attaching an S7-300 to a PROFIBUS subnet requires the use of a PROFIBUS communications processor (CP) selected according to application requirements. The communications processor must be physically inserted into a rack, included in the hardware configuration, and its interface parameters must be defined. Finally, attachment of the CP to one of the project's PROFIBUS subnets completes the connection of the station to the network.

Essential Elements

PROFIBUS networking options for the S7-300 include CPs that support the DP protocol (CP 342-5, CP 342-5 FO), or that support the FMS protocol (CP 343-5). CPs that support PROFIBUS DP may support both the DP-master and DP-slave functionality. This capability allows the S7-300 to act as a DP-master or as DP-slave. Newer PROFIBUS CPs may also support routing features, which allows communication between networks (e.g., Ethernet-to-PROFIBUS).

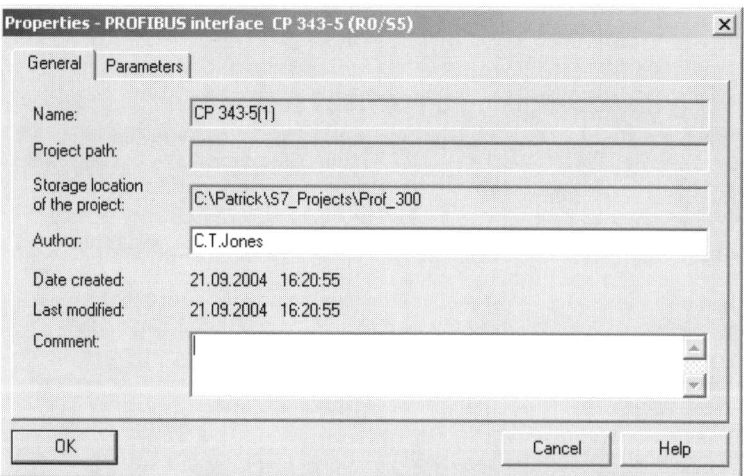

Figure 7-17. General Properties tab: CP 343-5 PROFIBUS Interface.

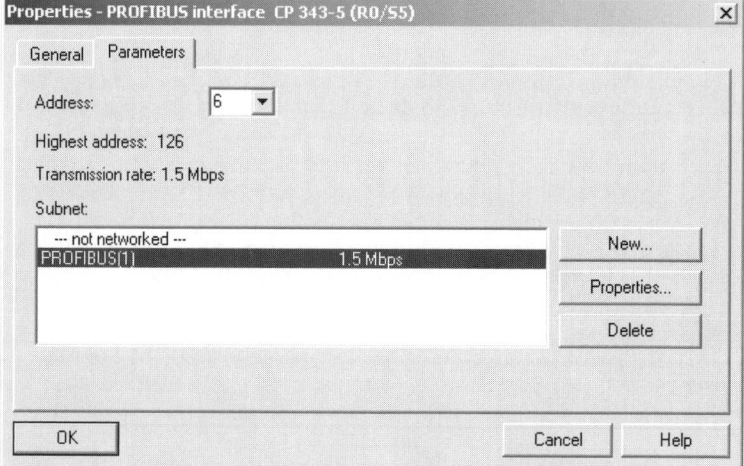

Figure 7-18. Parameters tab: CP 343-5 PROFIBUS Interface.

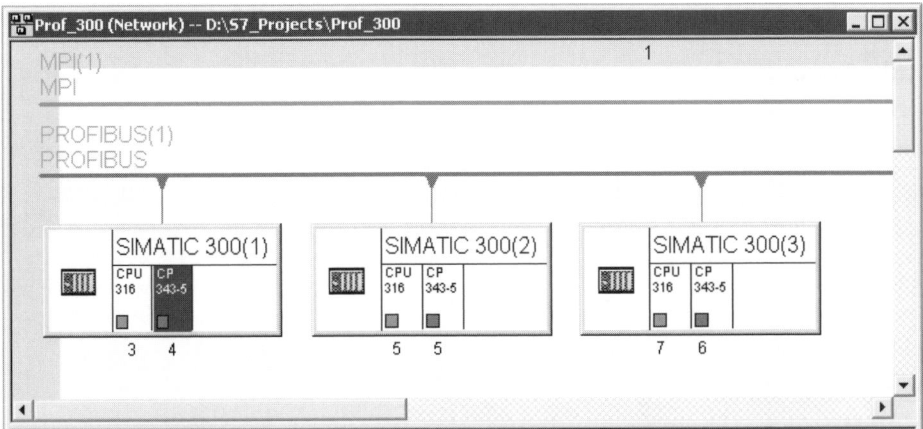

Figure 7-19. PROFIBUS Station addresses 4, 5, and 6, shown attracted to the subnet in NetPro.

Quick Steps: Installing, Configuring, and Attaching an S7-300 PROFIBUS CP

STEP	ACTION
1	Open the SIMATIC Manager and project where the PROFIBUS CP will be inserted, configured, and attached. If necessary, add and configure a PROFIBUS subnet as described in the previous task, "*Adding and Configuring a PROFIBUS Subnet*."
2	With the project folder expanded, select the station where you will install the CP, then from the right pane double click on the station hardware object to open the hardware configuration tool.
3	Open the **SIMATIC 300** catalog object to view the S7-300 component folders.
4	Open the **CP-300** folder, then under the **PROFIBUS** subfolder open the appropriate CP subfolder (CP-342-5, CP-342-5 FO, CP-343-5), to select a DP, DP over fiber, or FMS module. Select the desired CP and drag it to slot-4 or higher in the S7-300 rack.
5	If the interface properties dialog of the new CP does not open automatically, when inserted, then select the CP, right-click and select **Object Properties**; then press the Properties button to open the interface dialog.
6	From the *General* tab, you may modify the default name of the CP. You may also enter a new **Author** of the CP configuration, and enter a **Comment** on the CPU use.
7	From the *Parameters* tab, set the **Address** for this CP, as desired, or perform the next step to have the next available PROFIBUS address (1-126) automatically assigned.
8	From the **Subnet** list box, select the correct subnet to which this communications processor should be attached. Press the **OK** button to confirm your selections.
9	From the menu, select **Station** ➢ **Consistency Check** to check for errors.
10	From the menu bar, select **Station** ➢ **Save** to save the configuration. Use **Save** ➢ **Compile** to generate the *System Data* object that you will download to the CPU.

Building a Project with S7-300 Peer PROFIBUS Stations

Basic Concept

A PROFIBUS network of peer S7-300s is one in which multiple S7-300 stations are on the same subnet, have identical or similar characteristics, and have equal access to the network.

Essential Elements

Selecting and installing the appropriate PROFIBUS communications processor (CP) is all that is required to establish an S7-300 station as a network station. In this task, all of the stations will use the same communications processor; however, the exact module option will depend on whether a copper or fiber (FO) media is used, and what communications protocols must be supported in your application.

Application Tips

In cases where the central rack for the S7-300 stations will be identical or very similar, you should complete one station to include the rack, power supply, interface modules (IMs), and signal modules (SMs) that will be identical in all stations, and then use the Cut and Paste to duplicate the station. You might choose to duplicate the original station prior to inserting the communications processor. When you manually insert the CP in each station, STEP 7 will assign a unique address to each. Otherwise, you will have to modify each station address.

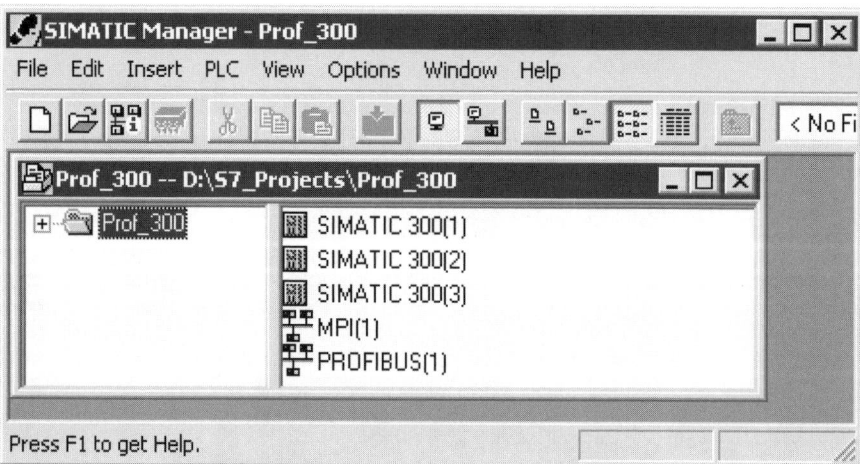

Figure 7-20. Project with S7-300 Peer stations, MPI and PROFIBUS subnets.

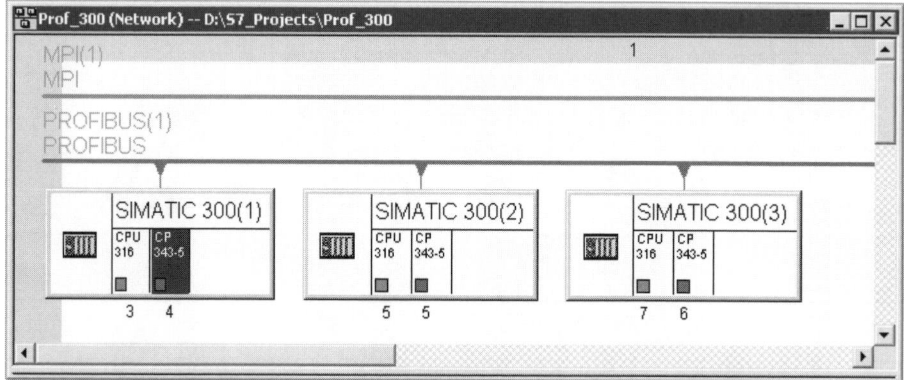

Figure 7-21. NetPro Graphic configuration of S7-300 Peer Stations on PROFIBUS.

Quick Steps: Building a Project with S7-300 Peer PROFIBUS Stations

STEP	ACTION
1	Start the SIMATIC Manager and Create a new Project using the **New** command.
2	When the New Project dialog appears, type in a project name in the **Name** field. Verify that the Storage Location Path is where you wish the project to be stored.
3	With the project folder open in the SIMATIC Manager, from the menu select **Insert** ➢ **Station** ➢ **SIMATIC 300 Station**. Insert at least two or more S7-300 stations.
4	With the project folder selected, from the menu select **Insert** ➢ **Subnet** ➢ **PROFIBUS**. The PROFIBUS subnet object will be inserted in the right pane of the project window.
5	Double click on the subnet or on the toolbar press the *Configure Network* button to open the Network Configuration tool. The network configuration window will open with objects that represent the PROFIBUS subnet, and the S7-300 stations you created.
6	Double click on the PROFIBUS subnet object, to define the subnet operating properties (e.g., **Highest PROFIBUS**, **Address**, **Transmission Rate**, and bus **Profile**) as described in the previous task "*Adding and Configuring a PROFIBUS Subnet.*")
7	Double click on the first station object (e.g., on the Station name) to open the station in the Hardware Configuration tool. Configure a central rack by inserting the required *Rack, Power Supply, CPU, and PROFIBUS communications processor (See Step 8)*.
8	When the CP is inserted, the properties dialog will open automatically; configure the properties (including a unique address) and attach the CP to the subnet as described in the previous task "*Installing, Configuring, and Attaching a CP.*"
9	Save the central rack configuration and press the *Configure Network* toolbar icon to return to the network configuration. The station will appear connected to the PROFIBUS subnet when you return to the network configuration.
10	Repeat Step 7 through Step 9 to configure the remaining stations.
11	From the menu, select **Network** ➢ **Check Consistency** to perform a check for errors.
12	From the menu bar, select **Network** ➢ **Save**, to save the network configuration. Use **Save and Compile** to generate the System Data that you will download to the CPU.

Installing, Configuring, and Attaching an S7-400 PROFIBUS CP

Basic Concept

Attaching an S7-400 to a PROFIBUS subnet requires the use of a PROFIBUS communications processor (CP), selected according to application requirements. You must insert the communications processor into a rack, include it in the hardware configuration, and define its interface parameters. Finally, attachment of the CP to one of the project's PROFIBUS subnets completes the connection of the station to the network.

Essential Elements

PROFIBUS networking options for the S7-400 include CPs that support the DP protocol (CP 443-5 Ext.), or that support the FMS protocol (CP 443-5 Basic). Newer PROFIBUS CPs may also support routing features, which allows communication between networks (e.g., Ethernet-to-PROFIBUS). PROFIBUS CPs must be installed in a rack that has the communications bus (C-Bus), [e.g., central rack (CR), or universal rack (UR)].

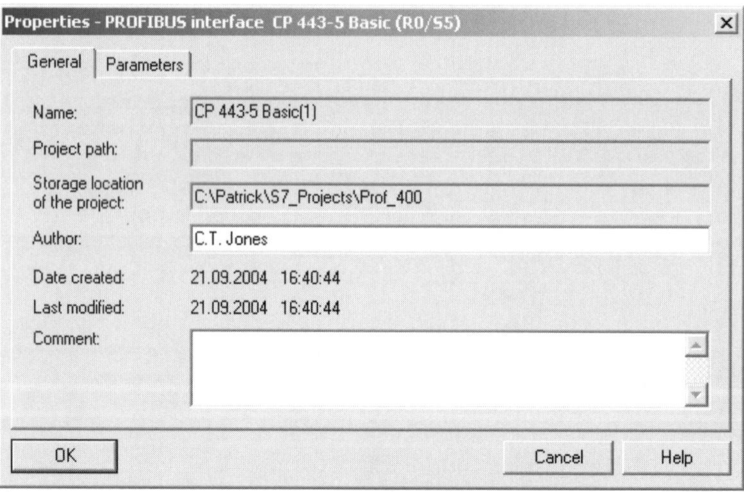

Figure 7-22. General tab: CP 443-5 Basic PROFIBUS Interface.

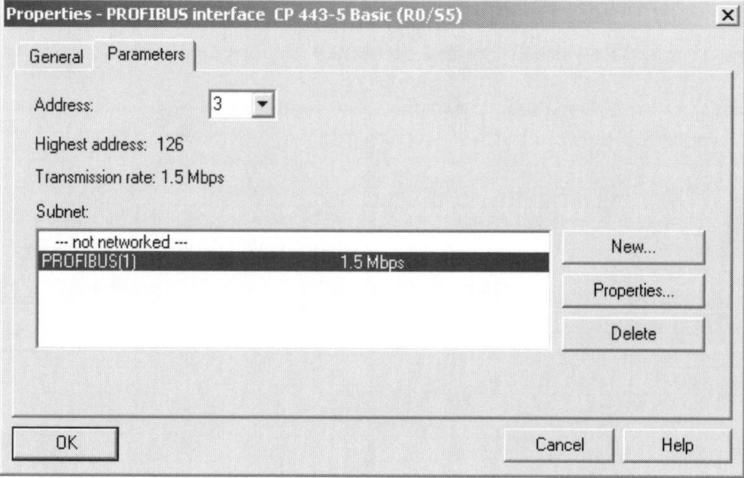

Figure 7-23. Parameters tab: CP 443-5 Basic PROFIBUS Interface.

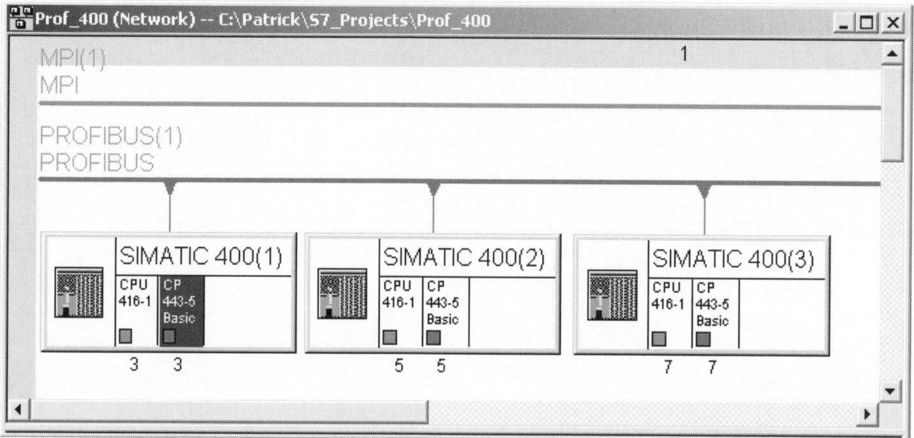

Figure 7-24. PROFIBUS Station addresses 3, 5, and 7, shown attracted to the subnet in NetPro.

Quick Steps: Installing, Configuring, and Attaching an S7-400 PROFIBUS CP

STEP	ACTION
1	Open the SIMATIC Manager and project where the PROFIBUS CP will be inserted, configured, and attached. If necessary, add and configure a PROFIBUS subnet as described in the previous task, "*Adding and Configuring a PROFIBUS Subnet*."
2	With the project folder expanded, select the station where you will install the CP, then from the right pane double click on the station hardware object to open the hardware configuration tool.
3	Open the **SIMATIC 400** catalog object to view the S7-300 component folders.
4	Open the **CP-400** folder, then under the **PROFIBUS** subfolder open the appropriate CP subfolder (CP-443-5 Basic, CP-443-5 Extended), to select from the FMS or DP modules. Select the desired CP and drag it to slot-4 or higher in the S7-400 rack.
5	If the interface properties dialog of the new CP does not open automatically, when inserted, then select the CP, right-click and select **Object Properties**; then press the Properties button to open the interface dialog.
6	From the *General* tab, you may modify the default name of the CP. You may also enter a new **Author** of the CP configuration, and enter a **Comment** on the CPU use.
7	From the *Parameters* tab, set the **Address** for this CP, as desired, or perform the next step to have the next available PROFIBUS address (1-126) automatically assigned.
8	From the **Subnet** list box, select the correct subnet to which this communications processor should be attached. Press the **OK** button to confirm your selections.
9	From the menu, select **Station** ➢ **Consistency Check** to check for errors.
10	From the menu bar, select **Station** ➢ **Save** to save the configuration. Use **Save** ➢ **Compile** to generate the *System Data* object that you will download to the CPU.

Building a Project with S7-400 Peer PROFIBUS Stations

Basic Concept

A PROFIBUS network of peer S7-400s is one in which multiple S7-400 stations are on the same subnet, have identical or similar characteristics, and have equal access to the network.

Essential Elements

Selecting and installing the appropriate PROFIBUS communications processor (CP) is all that is required to establish an S7-400 station as a network station. In this task, all of the stations will use the same communications processor; however, the exact module option will depend on whether a copper or fiber (FO) media is used, and what communications protocols must be supported in your application.

Application Tips

In cases where the central rack for the S7-400 stations will be identical or very similar, you should complete one station to include the rack, power supply, interface modules (IMs), and signal modules (SMs) that will be identical in all stations, and then use the Cut and Paste to duplicate the station. You might choose to duplicate the original station prior to inserting the communications processor. When you manually insert the CP in each station, STEP 7 will assign a unique address to each. Otherwise, you will have to modify each station address.

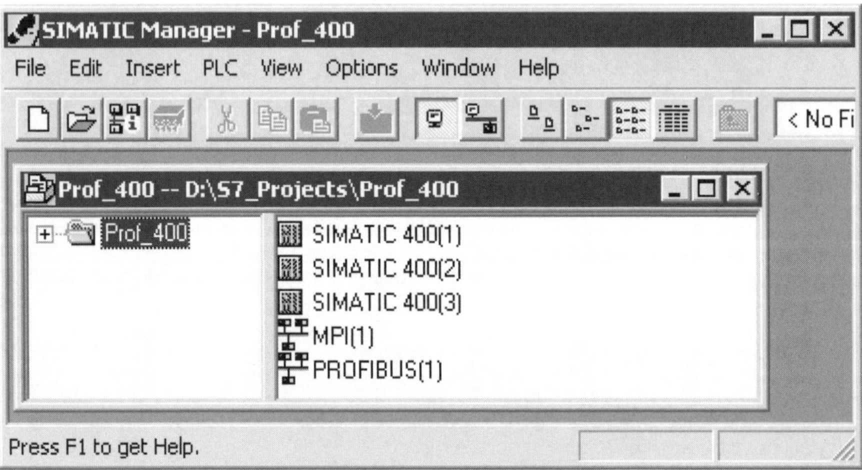

Figure 7-25. S7 Project with S7-400 stations, MPI and PROFIBUS subnets.

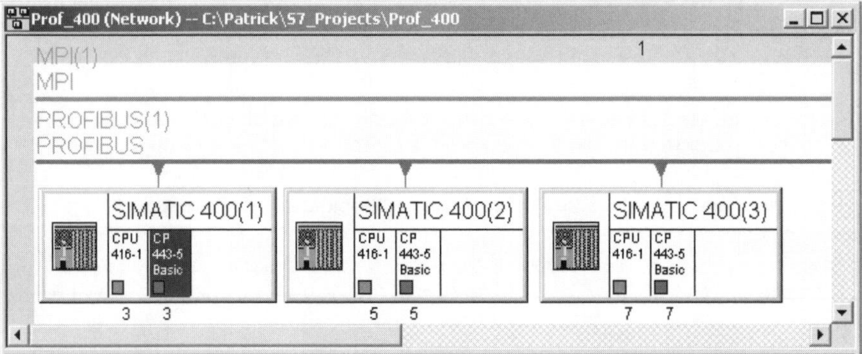

Figure 7-26. NetPro Graphic configuration of S7-400 Peer Stations on PROFIBUS.

Quick Steps: Building a Project with S7-400 Peer PROFIBUS Stations

STEP	ACTION
1	Start the SIMATIC Manager and Create a new Project using the **New** command.
2	When the New Project dialog appears, type in a project name in the **Name** field. Verify that the Storage Location Path is where you wish the project to be stored.
3	With the project folder open in the SIMATIC Manager, from the menu select **Insert** ➢ **Station** ➢ **SIMATIC 400 Station**. Insert at least two or more S7-400 stations.
4	With the project folder selected, from the menu select **Insert** ➢ **Subnet** ➢ **PROFIBUS**. The PROFIBUS subnet object will be inserted in the right pane of the project window.
5	Double click on the subnet or on the toolbar press the *Configure Network* button to open the Network Configuration tool. The network configuration window will open with objects that represent the PROFIBUS subnet, and the S7-400 stations you created.
6	Double click on the PROFIBUS subnet object, to define the subnet operating properties (e.g., **Highest PROFIBUS**, **Address**, **Transmission Rate**, and bus **Profile**) as described in the previous task "*Adding and Configuring a PROFIBUS Subnet.*")
7	Double click on the first station object (e.g., on the Station name) to open the station in the Hardware Configuration. Configure a central rack by inserting the required *Rack, Power Supply, CPU, and PROFIBUS communications processor (See Step 8)*.
8	When the CP is inserted, the properties dialog will open automatically; configure the properties (including a unique address) and attach the CP to the subnet as described in the previous task "*Installing, Configuring, and Attaching a CP.*"
9	Save the central rack configuration and press the *Configure Network* toolbar icon to return to the network configuration. The station will appear connected to the PROFIBUS subnet when you return to the network configuration.
10	Repeat Step 7 through Step 9 to configure the remaining stations.
11	From the menu, select **Network** ➢ **Check Consistency** to perform a check for errors.
12	From the menu bar, select **Network** ➢ **Save**, to save the network configuration. Use **Save and Compile** to generate the System Data that you will download to the CPU.

Configuring a Programming Station (PG/PC) on PROFIBUS

Basic Concept

If you will need to gain remote access to the S7 PLCs, connected to a PROFIBUS network, then you will have to configure and attach a PG/PC station to the subnet. As with other S7 stations, the PG/PC requires physical connection to the network and the logical configuration and attachment, in order to be known and recognized on the network.

Essential Elements

To establish a PC as a programming station on PROFIBUS, a PROFIBUS network card (e.g., CP 5511/CP 5611 or CP5412 A2) must be installed and configured in Windows, and the appropriate SOFTNET or HARDNET driver must be installed. If you have a, PG programming device (e.g., PG 720, PG 740), then it will already be equipped with the MPI network. The driver software for programming functions (PG/OP) over PROFIBUS is installed during the initial installation of STEP 7. With the physical connection in place, you only need to configure a logical interface connection for the PG/PC.

To configure your programming device, you must first insert a PG/PC object in the Simatic Manager or in the Network Configuration tool, and then configure and assign it an interface. An interface defines the subnet, and node address. When you assign the interface to one of the actual network cards on the PG/PC, then it will be adapted to the configured settings of the interface, including the operating properties of the subnet to which it is attached. If you modify the configured settings (e.g., the network transmission rate), the interface on your PG/PC is automatically adapted to match the new settings.

Application Tips

Since only one configured programming device (PG/PC) is allowed in a project, by creating several interfaces, which can be assigned to the PG/PC, you can conveniently switch interfaces. Several interfaces may be assigned to a PG/PC, however only one can be activated as the S7ONLINE interface at a given time. Without this function, you would have to call up the "Setting the PG/PC Interface" program and manually adapt the interface settings of your PG/PC to the configured settings.

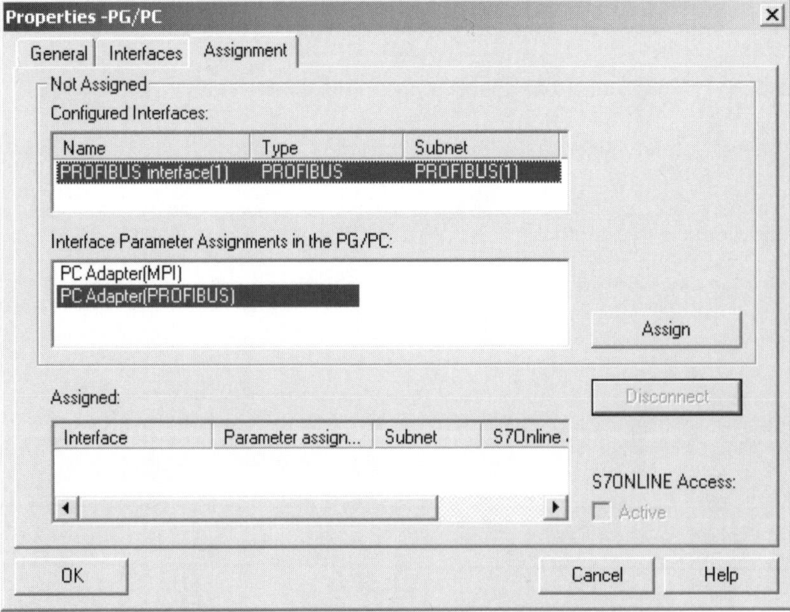

Figure 7-27. PROFIBUS Interface Properties: Interface Parameters dialog.

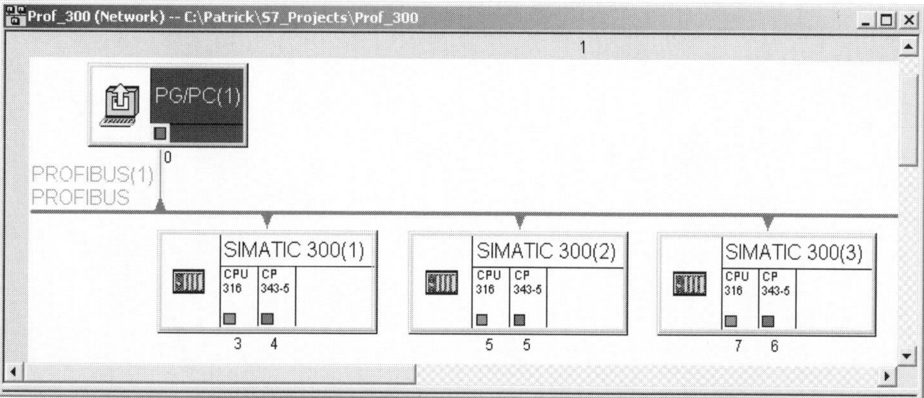

Figure 7-28. NetPro Graphic configuration of PG/PC and Peer S7-300 Stations on PROFIBUS.

Quick Steps: Configuring a Programming Station (PG/PC) on PROFIBUS

STEP	ACTION
1	Open the SIMATIC Manager, to the project where you wish to establish the PG/PC as a station on a PROFIBUS Subnet.
2	If the PG/PC object is not already installed, select the project folder, right click, and select **Insert New Object ➢ PG/PC**. Skip this step if the PG/PC is already installed.
3	With the project folder selected, from the right pane, double click on the PROFIBUS subnet object to open the NetPro graphical network configuration tool.
4	With the network configuration open, select the PG/PC object, right-click and select **Object Properties**; or double click on the PG/PC object to open the properties dialog.
5	From the *General* tab, you may modify the default name PG/PC (1), to a name you choose. You may also modify or enter a new **Author** of the interface parameters, and modify or enter a **Comment** describing the use of the programming system.
6	Select the *Interface* tab, to display existing interfaces or create a new MPI interface to which you wish to assign to the PG/PC.
7	Select the *Assignment* tab, to display the configured PROFIBUS interfaces and interface parameter sets (i.e., physically installed interfaces) previously defined in the PG/PC.
8	From the **Not Assigned** block, select the logical PROFIBUS interface you wish to assign to the physical interface in the PG/PC; then from the **Interface Parameter Assignment** window, select the physical PROFIBUS interface found in the PG/PC;
9	Press the **Assign** button. The "Assign" button will be grayed out unless the configured interface you choose is networked and the selected interface parameter assignment (i.e., PROFIBUS Network card) matches the configured interface.
	With a logical interface assigned to a physical PROFIBUS interface, you may then activate the **S7ONLINE** check box if you wish to activate this interface as the current means for establishing and the online connection to STEP 7.

Configuring PROFIBUS Communications Connections

Basic Concept

A *connection* is a logical channel established between two communication partners, for the purpose of data exchange using a specific communication service. Before a station can communicate over PROFIBUS using FDL, FMS, or S7 Functions, you must first configure a connection of that type between the station and one or more remote partners. In a given CPU, depending on compatibility, you may create connections of different types and to different partners. Connections for a CPU are created using the Network Configuration Tool, and are contained in what is referred to as the *CPU connection table*.

Essential Elements

On PROFIBUS, you may configure connections that facilitate **FDL**, **FMS**, or **S7** communication services. FDL communication is possible between S7 PLCs, S7 and S5 PLCs, or between S7 PLCs and other FDL compatible devices (e.g., PC stations). FMS communication is possible between S7 PLCs, between S7 and S5 PLCs, and between S7 and other FMS compatible devices (e.g., PC stations). Finally, S7 communication is possible between S7 PLCs, operator stations, and PCs. If you configure an S7 connection to a station in another project, you will have to specify a *connection name*. The connection name is used to link partners across project boundaries. After the connection is linked, you cannot edit the connection name.

Application Tips

To create connections for a station, you must have connected the intended partner stations to the subnet, using a communications processor that supports the connection types you intend to create. After defining the required connection table for the CPUs in your project, you should print the connections report as a reference tool for writing the communications functions in you program. Each S7 connection will generate a *connection ID*, which for some you will need as a block input parameter when programming the communication function.

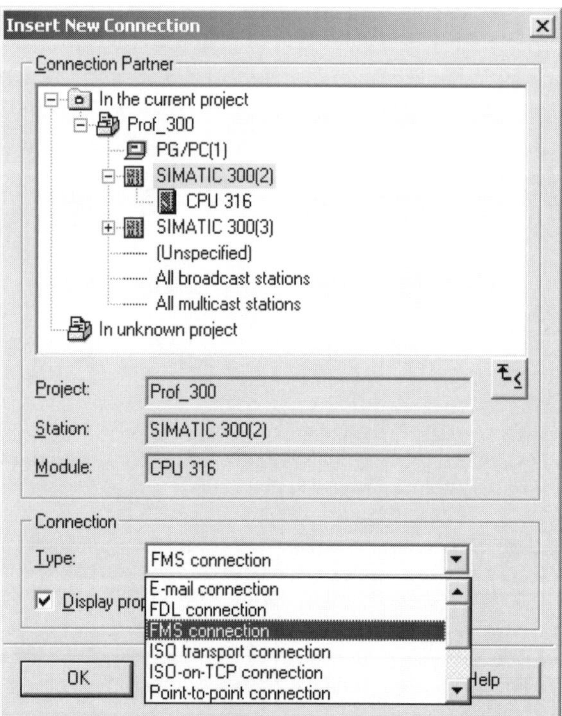

Figure 7-29. Dialog for new communications connections.

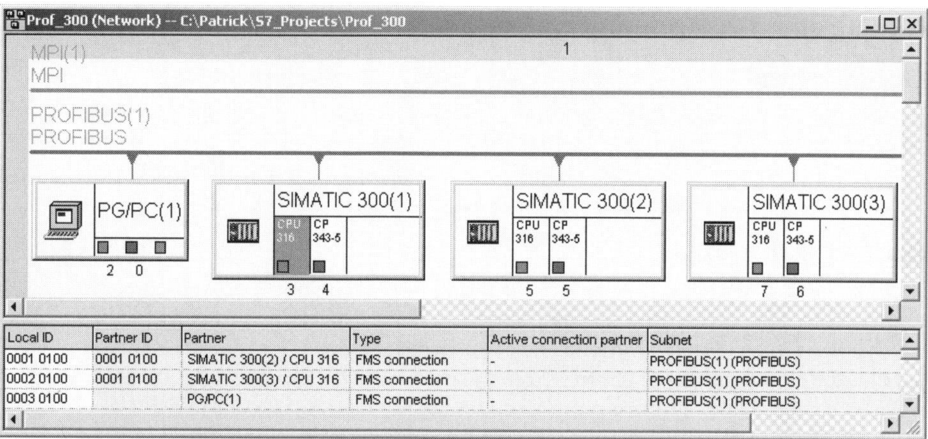

Figure 7-30. Network configuration tool, with *connection table* displayed for selected CPU.

Quick Steps: Configuring PROFIBUS Communications Connections

STEP	ACTION
1	Open the required project and double click on any subnet object to open the network configuration tool in order to create new communications connections.
2	With the network configuration in view, select the CPU module for which you wish to create connections. When the CPU is selected, its connection table is displayed.
3	With the desired CPU selected, you may double click on an empty row, or simply right click on an empty row and select **Insert New Connection.**
4	The New Connection dialog will display a project tree listing the stations of the project, other than the station you have selected, that you may define as the connection partner. You may click on a specific **S7-300 or S7-400** partner, or in the case of FDL, and FMS connections, you may also select *All Broadcast Stations*. FDL connections also support communication to *All Multicast Stations*.
5	Before selecting a connection type, activate the checkbox "**Display properties before inserting,**" if you wish to view or modify the derived connection parameters before the connection is inserted. De-activate the checkbox if you do not wish to view the assigned connection parameters before you save your entries.
6	Then, from the **Type** drop list, choose the connection type you wish to create (e.g., FDL, FMS, or S7 connection); and then press the **OK** button to enter the connection and view the connection properties.
7	Although the connection properties will vary depending on the connection type, you may get assistance from the online help on modifying the default parameters.
8	Confirm each connection entry with the **OK** button and STEP 7 will list the Local ID and Partner ID for the connection, the connection type, and whether the local station is Active or Passive in establishing the connection.
9	From the menu bar, select **Network ➢ Save**, to save the network configuration. Use **Save ➢ Compile** to generate the System Data that you will download to the CPU.

Adding and Configuring an Ethernet Subnet

Basic Concept

Connection of Ethernet nodes on a common medium, comprised of one or of several Ethernet segments connected via repeaters, is considered a *subnet*. In a large plant or project, the complete network may require several Ethernet subnets. In STEP 7, you will need to create a subnet object for each of your physical Ethernet subnets.

Essential Elements

You may insert a new Ethernet subnet from the SIMATIC Manager or the network configuration tool. Once you insert the subnet, you may configure its *General* properties (See figure below). The General properties include basic documentation such as **Name** and **Subnet ID.** STEP 7 derives the subnet name, but you change the name if necessary. STEP 7 also derives the subnet ID (project number - subnet number), which you may modify. This is unnecessary since this ID will always be unique. You do not need to set any operating parameters for an Ethernet subnet.

Application Tips

If possible, you should create and configure nodes that you want to connect in a network in the same project. By placing the stations in the same project, STEP 7 is able to check your entries (e.g., addresses, connections, etc.) for consistency. In a large plant with several stations, it may be necessary to create a number of subnets. While you may manage these subnets in a single project, you may decide to handle the subnets in multiple projects. If stations in different projects need to be connected, you may assign the station to more than one subnet (i.e., in different projects).

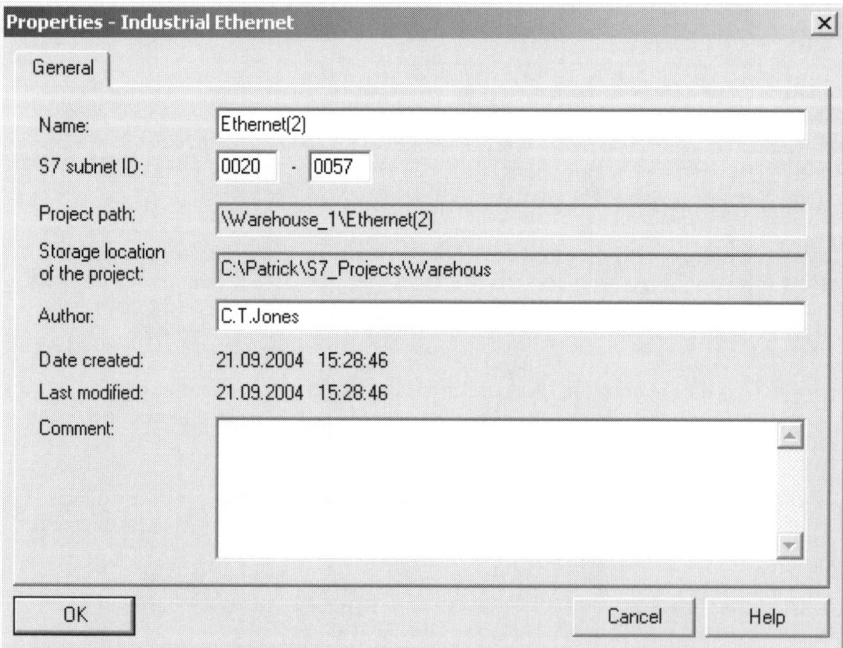

Figure 7-31. Industrial Ethernet subnet Properties: General settings dialog.

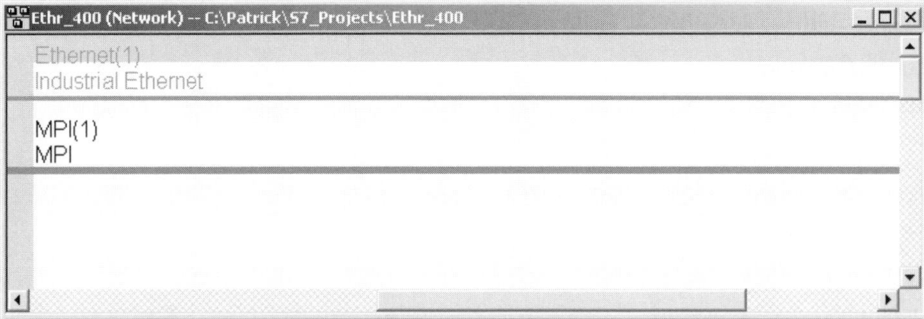

Figure 7-32. Newly inserted Ethernet Subnet, graphically shown in the Net Pro.

Quick Steps: Adding and Configuring an Ethernet Subnet

STEP	ACTION
1	Start and open the SIMATIC Manager to the project of the new Ethernet subnet, select the project folder, and then from the menu, select **Insert ➢ Subnet ➢ Industrial Ethernet**.
2	In the right pane of the project window, double click on any subnet object or press the *Configure Network* toolbar button to open the Network Configuration tool.
3	In the network configuration window, double click on the Ethernet subnet object or right-click on the object and select **Object Properties** to open the properties dialog.
4	From the *General* tab, you may keep or modify the default **Name** of the subnet, and the **S7 Subnet ID** (this STEP 7 derived ID number will always be unique).
5	Also on the *General* tab, you may enter a department name in the **Author** field, or perhaps your name if you are defining the subnet properties; you may enter or modify the **Comment** lines to describe the subnet's function in the project.
6	Select **OK** to accept the new Ethernet properties.

Installing, Configuring, and Attaching an S7-300 Ethernet CP

Basic Concept

Attaching an S7-300 to an Ethernet subnet requires the use of an Ethernet communications processor (CP) selected according to the application requirements. The CP must be physically inserted into a rack, included in the station's hardware configuration, and its interface parameters must be defined. Finally, you must attach the CP to one of the project's Ethernet subnets to complete the connection of the station to the network.

Essential Elements

Ethernet networking options for the S7-300 include CPs that support ISO transport connections only (CP 343-1 ISO); those that support TCP transport connections only (CP 343-1 TCP); CPs that support both ISO and TCP (CP 343-1); and CPs that support internet capabilities (CP-343-1 IT). Of these CP groups, newer CPs will also support routing features, which allows communication between networks (e.g., Ethernet-to-PROFIBUS).

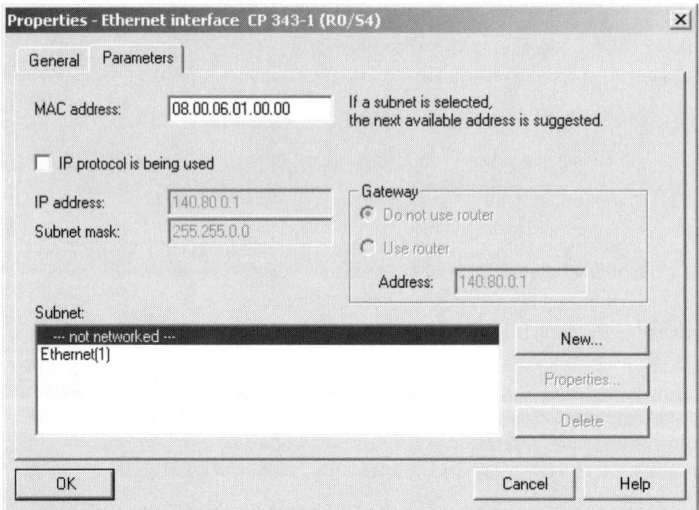

Figure 7-33. Parameters tab: CP 343-1Ethernet Interface.

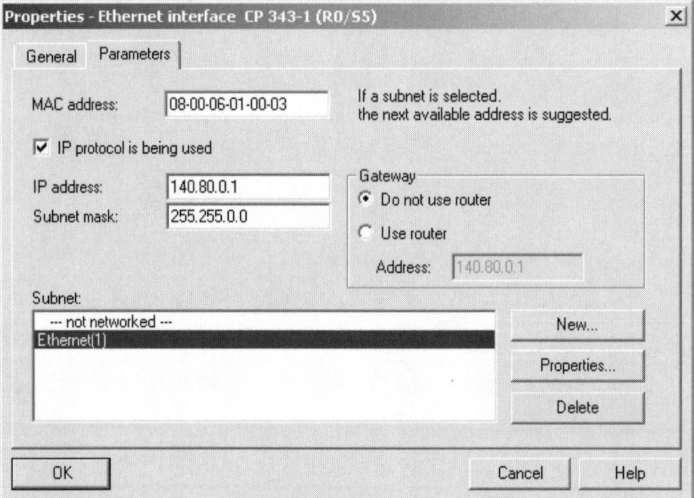

Figure 7-34. CP 343-1Ethernet Interface activated for IP protocol use.

Quick Steps: Installing, Configuring, and Attaching an S7-300 Ethernet CP

STEP	ACTION
1	Open the SIMATIC Manager and project where the Ethernet CP will be inserted, configured, and attached. If necessary, add and configure an Ethernet subnet as described in the previous task, "*Adding and Configuring an Ethernet Subnet*."
2	With the project folder expanded, select the station where you will install the CP, then from the right pane double click on the station hardware object to open the hardware configuration tool.
3	Open the **SIMATIC 300** catalog object to view the S7-300 component folders.
4	Open the **CP-300** folder, and then under the **Industrial Ethernet** subfolder open the appropriate CP subfolder (e.g., CP-343-1, CP-343-1 ISO, CP-341-1 TCP, CP-443-1 IT, or CP-341-PN). Select the desired CP and drag it to slot-4 or higher in the S7-300 rack.
5	If the interface properties dialog of the new CP does not open automatically, when inserted, then select the CP, right-click and select **Object Properties**; then press the Properties button to open the interface dialog.
6	From the *General* tab, you may modify the default name of the CP. You may also enter a new **Author** of the CP configuration, and enter a **Comment** on the CPU use.
7	From the *Parameters* tab, set the Ethernet MAC **Address** for this CP, or perform Step 8 to have the next available address automatically assigned. If no other stations are already connected, the default MAC address of 08.00.06.01.00.00 is displayed.
8	From the **Subnet** list box, select the correct subnet to which this communications processor should be attached. Press the **OK** button to confirm your selections.
9	If the CP supports the IP Protocol and TCP/IP will be used in this application, then activate the checkbox labeled **IP Protocol is being Used;** in the **IP Address** and **Subnet Mask** fields, enter the IP and subnet mask address provided for the CP, then accept your entries with the **OK** button.
10	From the menu, select **Station** ➢ **Consistency** to perform a check for errors.
11	From the menu bar, select **Station** ➢ **Save**, to save the configuration. Use **Save** ➢ **Compile** to generate the *System Data* that will be downloaded to the CPU.

Building a Project with S7-300 Peer Ethernet Stations

Basic Concept

An Ethernet network of peer S7-300s is one in which multiple S7-300 stations are on the same subnet, have identical or similar characteristics, and have equal access to the network.

Essential Elements

Selecting and installing the appropriate Ethernet communications processor (CP) is all that is required to establish an S7-300 station as a network station. In this task, all of the stations will use the same communications processor; however, the exact module option will depend on whether a copper or fiber (FO) media is used, and what communications protocols must be supported in your application.

Application Tips

In cases where the central rack for the S7-300 stations will be identical or very similar, you should complete one station to include the rack, power supply, interface modules (IMs), and signal modules (SMs) that will be identical in all stations, and then use the Cut and Paste to duplicate the station. You might choose to duplicate the original station prior to inserting the communications processor. When you manually insert the CP in each station, STEP 7 will assign a unique address to each. Otherwise, you will have to modify each station address.

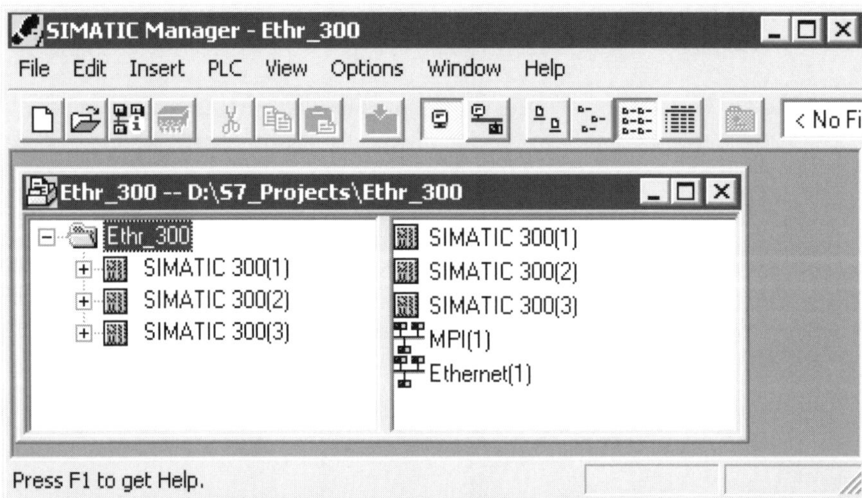

Figure 7-35. Project with S7-300 Peer stations, MPI and Industrial Ethernet subnets.

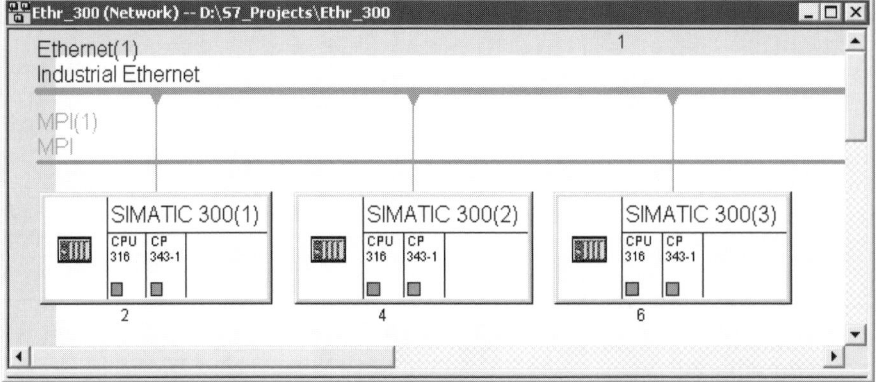

Figure 7-36. NetPro Graphic configuration of Peer S7-300 Stations on Industrial Ethernet.

Quick Steps: Building a Project with S7-300 Peer Ethernet Stations

STEP	ACTION
1	Start the SIMATIC Manager and Create a new Project using the **New** command.
2	When the New Project dialog appears, type in a project name in the **Name** field. Verify that the Storage Location Path is where you wish the project to be stored.
3	With the project folder open in the SIMATIC Manager, from the menu select **Insert ➢ Station ➢ SIMATIC 300 Station**. Insert at least two or more S7-300 stations.
4	With the project folder selected, from the menu select **Insert ➢ Subnet ➢ Industrial Ethernet**. The subnet object will be inserted in the right pane of the project window.
5	Double click on the subnet, or press the *Configure Network* toolbar button to open the Network Configuration tool. The network configuration window will open with objects that represent the Ethernet subnet, and the S7-300 stations you created.
6	Double click on the first station object (e.g., on the Station name) to open the station in the Hardware Configuration tool. Configure a central rack by inserting the required *Rack, Power Supply, CPU, and the Ethernet Communications Processor.*
7	When the CP is inserted, the properties dialog will open automatically; configure the properties (including a unique address) and attach the CP to the subnet as described in the previous task "*Installing, Configuring, and Attaching a CP.*"
8	Save the central rack configuration and press the *Configure Network* toolbar icon to return to the network configuration. The station will appear connected to the Industrial Ethernet subnet when you return to the network configuration.
9	Repeat Step 6 through Step 8 to configure the remaining stations.
10	From the menu, select **Network ➢ Check Consistency** to perform a check for errors.
11	From the menu bar, select **Network ➢ Save**, to save the network configuration. Use **Save and Compile** to generate the System Data that you will download to the CPU.

Installing, Configuring, and Attaching an S7-400 Ethernet CP

Basic Concept

Attaching an S7-400 to an Ethernet subnet requires the use of an Ethernet communications processor (CP) selected according to the application requirements. The CP must be physically inserted into a rack, included in the station's hardware configuration, and its interface parameters must be defined. Finally, you must attach the CP to one of the project's Ethernet subnets to complete the connection of the station to the network.

Essential Elements

Ethernet networking options for the S7-400 Ethernet includes CPs that support ISO transport connections (CP 443-1 ISO); that support TCP transport connections (CP 443-1 TCP); CPs that support both ISO and TCP connections (CP 443-1); and CPs that that support internet capabilities (CP-443-1 IT). Of these CP groups, newer CPs will also support routing features, which allows communication between networks (e.g., Ethernet-to-PROFIBUS).

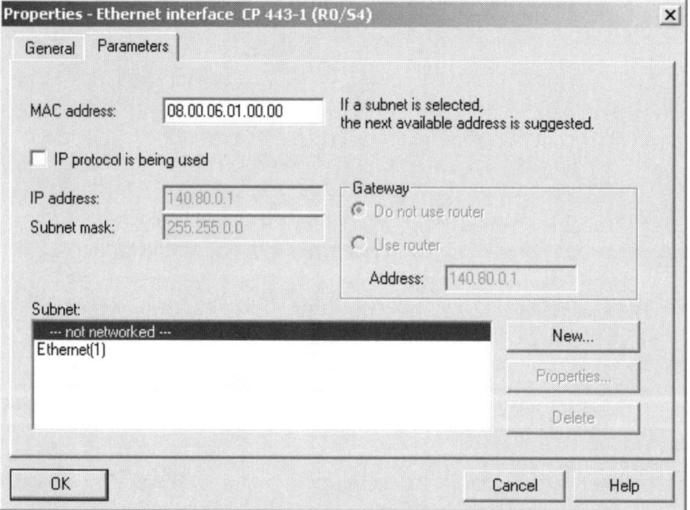

Figure 7-37. Parameters tab: CP 443-1Ethernet Interface.

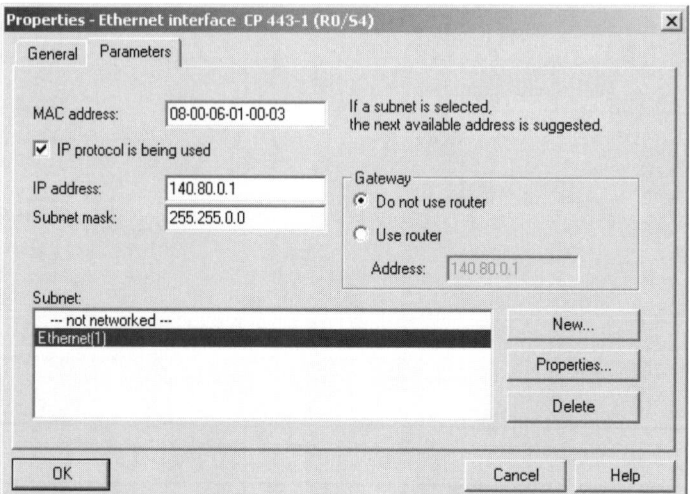

Figure 7-38. CP 443-1Ethernet Interface activated for IP protocol use.

Quick Steps: Installing, Configuring, and Attaching an S7-400 Ethernet CP

STEP	ACTION
1	Open the SIMATIC Manager and project where the Ethernet CP will be inserted, configured, and attached. If necessary, add and configure an Ethernet subnet as described in the previous task, "*Adding and Configuring an Ethernet Subnet*."
2	With the project folder expanded, select the station where you will install the CP, then from the right pane double click on the station hardware object to open the hardware configuration tool.
3	Open the **SIMATIC 400** catalog object to view the S7-300 component folders.
4	Open the **CP-400** folder, and then under the **Industrial Ethernet** subfolder open the appropriate CP subfolder (e.g., CP-443-1, CP-443-1 ISO, CP-443-1 TCP, CP-443-1 IT, or CP-443-1PN). Select the desired CP and drag it to slot-4 or higher in the S7-400 rack.
5	If the interface properties dialog of the new CP does not open automatically, when inserted, then select the CP, right-click and select **Object Properties**; then press the Properties button to open the Interface dialog.
6	From the *General* tab, you may modify the default name of the CP. You may also enter a new **Author** of the CP configuration, and enter a **Comment** on the CPU use.
7	From the *Parameters* tab, set the Ethernet MAC **Address** for this CP, or perform Step 8 to have the next available address automatically assigned. If no other stations are already connected, the default MAC address of 08.00.06.01.00.00 is displayed.
8	From the **Subnet** list box, select the correct subnet to which this communications processor should be attached. Press the **OK** button to confirm your selections.
9	If the CP supports the IP Protocol and TCP/IP will be used in this application, then activate the checkbox labeled **IP Protocol is being Used;** in the **IP Address** and **Subnet Mask** fields, enter the IP and subnet mask address provided for the CP, then accept your entries with the **OK** button.
10	From the menu, select **Station** ➢ **Consistency** to perform a check for errors.
11	From the menu bar, select **Station** ➢ **Save**, to save the configuration. Use **Save and Compile** to generate the *System Data* that you will download to the CPU.

Building a Project with S7-400 Peer Ethernet Stations

Basic Concept

An Ethernet network of peer S7-400s is one in which multiple S7-400 stations are on the same subnet, have identical or similar characteristics, and have equal access to the network.

Essential Elements

Selecting and installing the appropriate Ethernet communications processor (CP) is all that is required to establish an S7-400 station as a network station. In this task, all of the stations will use the same communications processor; however, the exact module option will depend on whether a copper or fiber (FO) media is used, and what communications protocols must be supported in your application.

Application Tips

In cases where the central rack for the S7-400 stations will be identical or very similar, you should complete one station to include the rack, power supply, interface modules (IMs), and signal modules (SMs) that will be identical in all stations, and then use the Cut and Paste to duplicate the station. You might choose to duplicate the original station prior to inserting the communications processor. When you manually insert the CP in each station, STEP 7 will assign a unique address to each. Otherwise, you will have to modify each station address.

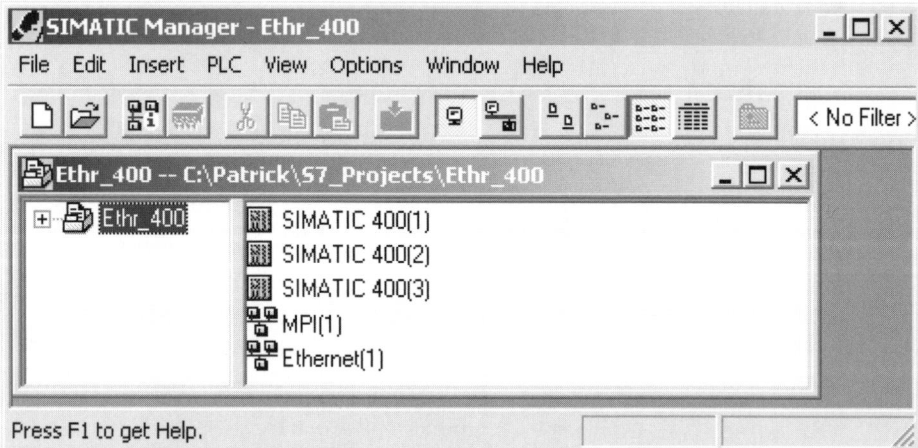

Figure 7-39. Project with S7-400 Peer stations, MPI and Industrial Ethernet subnets.

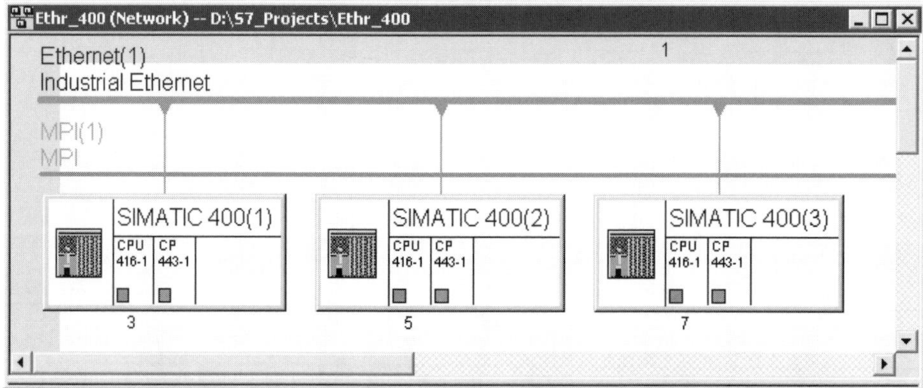

Figure 7-40. NetPro Graphic configuration of Peer S7-400 Stations on Industrial Ethernet.

Quick Steps: Building a Project with S7-400 Peer Ethernet Stations

STEP	ACTION
1	Start the SIMATIC Manager and Create a new Project using the **New** command.
2	When the New Project dialog appears, type in a project name in the **Name** field. Verify that the Storage Location Path is where you wish the project to be stored.
3	With the project folder open in the SIMATIC Manager, from the menu select **Insert ➢ Station ➢ SIMATIC 400 Station**. Insert at least two or more S7-400 stations.
4	With the project folder selected, from the menu select **Insert ➢ Subnet ➢ Industrial Ethernet**. The subnet object will be inserted in the right pane of the project window.
5	Double click on the subnet, or press the *Configure Network* toolbar button to open the Network Configuration tool. The network configuration window will open with objects that represent the Ethernet subnet, and the S7-300 stations you created.
6	Double click on the first station object (e.g., on the Station name) to open the station in the Hardware Configuration tool. Configure a central rack by inserting the required *Rack, Power Supply, CPU, and the Ethernet Communications Processor.*
7	When the CP is inserted, the properties dialog will open automatically; configure the properties (including a unique address) and attach the CP to the subnet as described in the previous task "*Installing, Configuring, and Attaching a CP.*"
8	Save the central rack configuration and press the *Configure Network* toolbar icon to return to the network configuration. The station will appear connected to the Industrial Ethernet subnet when you return to the network configuration.
9	Repeat Step 6 through Step 8 to configure the remaining stations.
10	From the menu, select **Network ➢ Check Consistency** to perform a check for errors.
11	From the menu bar, select **Network ➢ Save**, to save the network configuration. Use **Save and Compile** to generate the System Data that you will download to the CPU.

Configuring a Programming Station (PG/PC) on Ethernet

Basic Concept

If you will need to gain remote access to the S7 PLCs, connected to an Ethernet subnet, then you will have to configure and logically attach a PG/PC station to the network. As with other hardware stations in the subnet, there is the physical connection of the PG/PC to the network and the logical configuration using the STEP 7 *Hardware Configuration Tool.*

Essential Elements

To establish a PG/PC as a programming station on Industrial Ethernet, an Ethernet-compatible network interface card must be installed in the PC, the appropriate SOFTNET, or HARDNET driver must be installed on the PC, and the network interface card must be configured in Windows. With the physical connection in place, you need only to complete the logical connection of the PG/PC programming station.

Application Tips

As a rule, you should install the required Ethernet driver first, and then physically install the required Ethernet network interface card into your PC. In this way, the appropriate SOFTNET or HARDNET driver is in place when you attempt to install the network interface card in Windows. Recall, that the software drivers, for the programming services (PG/OP) over Ethernet, are installed as part of each Ethernet driver package.

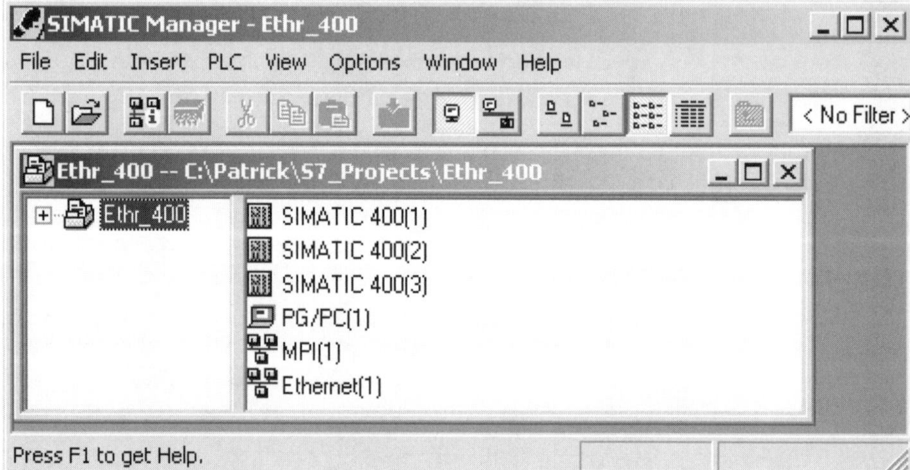

Figure 7-41. S7 Project with Peer S7-400 Stations, PG/PC station, MPI, and Ethernet subnets.

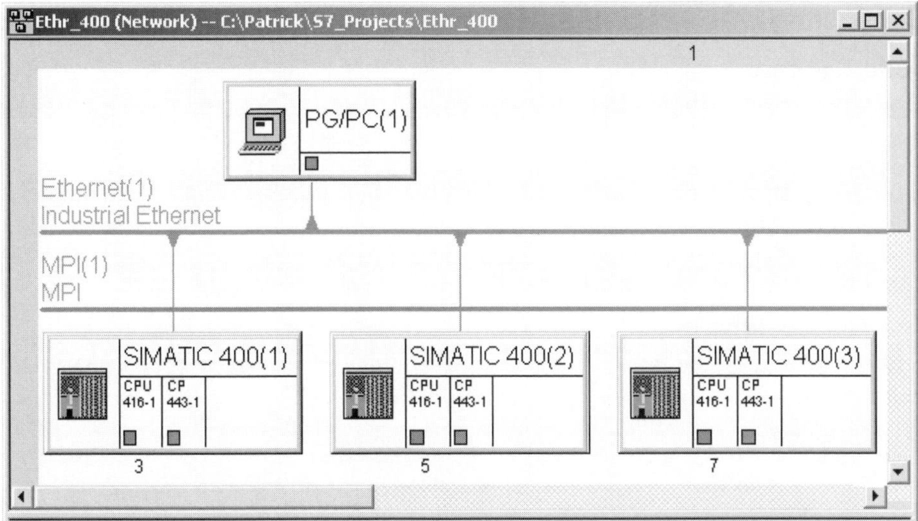

Figure 7-42. NetPro Graphic configuration of PG/PC and Peer S7-400 Stations on Ethernet.

Quick Steps: Configuring a Programming Station (PG/PC) on Ethernet

STEP	ACTION
1	Open the SIMATIC Manager, to the project where you wish to establish the PG/PC as a station on an Ethernet Subnet.
2	If the PG/PC object is not already installed, select the project folder, right click, and select **Insert New Object ➢ PG/PC**. Skip this step if the PG/PC is already installed.
3	With the project folder selected, from the right pane, double click on the Ethernet subnet object to open the NetPro graphical network configuration tool.
4	With the network configuration open, select the PG/PC object, right-click and select **Object Properties**; or double click on the PG/PC object to open the properties dialog.
5	From the *General* tab, you may modify the default name PG/PC (1), to a name you choose. You may also modify or enter a new **Author** of the interface parameters, and modify or enter a **Comment** describing the use of the programming system.
6	Select the *Interface* tab, to display existing interfaces or create a new Ethernet interface to which you wish to assign to the PG/PC.
7	Select the *Assignment* tab, to display the configured Ethernet interfaces and interface parameter sets (i.e., physically installed interfaces) previously defined in the PG/PC.
8	From the **Not Assigned** block, select the logical Ethernet interface you wish to assign to the physical interface in the PG/PC; then from the **Interface Parameter Assignment** window, select the physical Ethernet interface found in the PG/PC;
9	Press the **Assign** button. The "Assign" button will be grayed out unless the configured interface you choose is networked and the selected interface parameter assignment (i.e., Ethernet card) matches the configured interface.

Configuring Ethernet Communications Connections

Basic Concept

A *connection* is a logical channel established between two communication partners, for the purpose of data exchange using a specific communication service. Before a station can communicate over Ethernet using ISO, ISO-on-TCP, TCP/IP, UDP, or S7 communications services, you must first configure a connection of that type between the station and one or more partners. In a given CPU, depending on compatibility, you may create connections of different types and to different partners. Connections for a CPU are created in the Network Configuration Tool, and are contained in what is referred to as the *CPU connection table*.

Essential Elements

On Ethernet, you may configure connections to facilitate **ISO**, **ISO-on-TCP**, **TCP/IP**, **UDP**, **E-Mail**, or **S7 services**. ISO, ISO-on-TCP, TCP/IP, and UDP communications services are possible between S7 PLCs, S7 and S5 PLCs, or between S7 PLCs and PC stations with compatible ISO or TCP/IP protocols. Finally, S7 communication is possible between S7 PLCs, operator stations, and PCs. If you configure an S7 connection to a station in another project, you will have to specify a *connection name*. This reference is used to link partners across project boundaries. After the connection is linked, you cannot edit the connection name.

Application Tips

To create connections for a station, you must have connected the intended partner stations to the subnet, using a communications processor that supports the connection types you intend to create. After defining the required connection table for the CPUs in your project, you should print the connections report as a reference tool for writing the communications functions in you program. Each S7 connection will generate a *connection ID* that, for some, you will need as a block input parameter when programming the communication function.

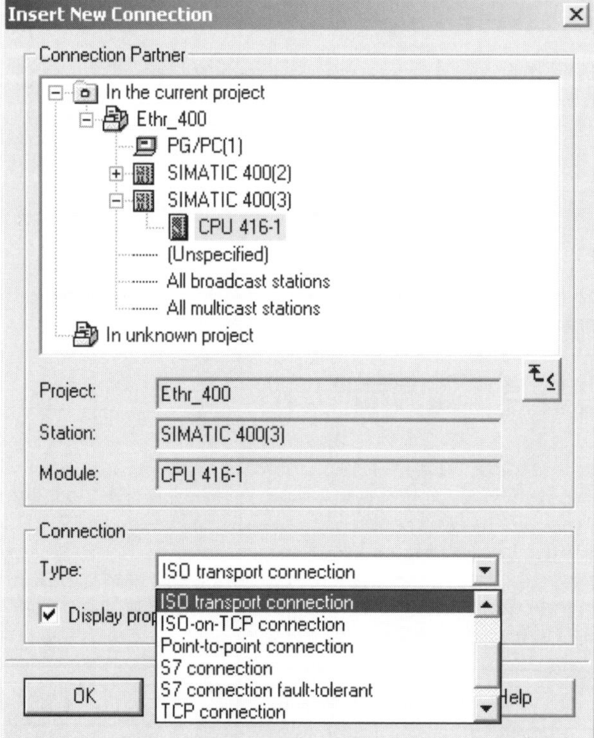

Figure 7-43. Dialog for new communications connections.

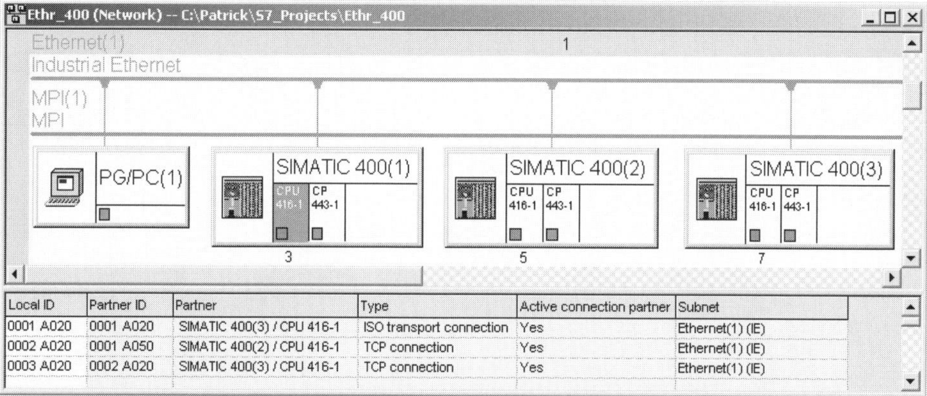

Figure 7-44. Network configuration tool, with *connection table* displayed for selected CPU.

Quick Steps: Configuring Ethernet Communications Connections

STEP	ACTION
1	Open the required project and double click on any subnet object to open the network configuration tool in order to create new communications connections.
2	With the network configuration in view, select the CPU module for which you wish to create connections. When the CPU is selected, its connection table is displayed.
3	With the desired CPU selected, you may double click on an empty row, or simply right click on an empty row and select **Insert New Connection.**
4	The New Connection dialog will display a project tree listing the stations of the project, other than the station you have selected, that you may define as the connection partner. You may click on a specific **S7-300 or S7-400** partner, or in the case of UDP connections, you may also select *All Broadcast Stations*.
5	Before selecting a connection type, activate the checkbox "**Display properties before inserting,**" if you wish to view or modify the derived connection parameters before the connection is inserted. De-activate the checkbox if you do not wish to view the assigned connection parameters before you save your entries.
6	Then, from the **Type** drop list, choose the connection type you wish to create (e.g., ISO, ISO-on-TCP, TCP, UDP, E-Mail, or S7 Connection); and then press the **OK** button to enter the connection and view the connection properties.
7	Although the connection properties will vary depending on the connection type, you may get assistance from the online help on modifying the default parameters.
8	Confirm each connection entry with the **OK** button and STEP 7 will list the Local ID and Partner ID for the connection, the connection type, and whether the local station is Active or Passive in establishing the connection.
9	From the menu bar, select **Network ➢ Save**, to save the network configuration. Use **Save ➢ Compile** to generate the System Data that you will download to the CPU.

Appendices

Contents

A Standard Library: Organization Blocks

B Standard Library: System Blocks (SFCs, SFBs)

C ASCII Charts

D Examples of Forcing/Modifying I/O and Memory Variables

E Condition Codes CC0 & CC1 as Results Bits

F Analog Input/Output Digital Representation

G Common Abbreviations and Acronyms

APPENDIX A: Standard Library - Organization Blocks

As described earlier in this book, you will be required to write the code for any Organization Blocks (OBs) you decide to include in your program. All of the OBs, however, are included in the Standard Library, provided with the STEP 7 Basic package. They are only provided in order to give you a template for the declaration area of each of the OBs. To include an OB in your program you must insert the block as you would with any other block. Remember, not all OBs are available in all CPUs – you must open *Module Information* for your CPU to determine which OBs are available with the CPU you are using.

Table A-1. Organization Blocks (OBs) list. Provided in STEP 7 Standard Library.

OB	Default Priority	Description
		Cyclical Processing
OB 1	1	Main Program Block
		Time-of-Day Interrupt Processing
OB 10	2	Time-of-Day Interrupt - 0
OB 11	2	Time-of-Day Interrupt - 1
OB 12	2	Time-of-Day Interrupt - 2
OB 13	2	Time-of-Day Interrupt - 3
OB 14	2	Time-of-Day Interrupt - 4
OB 15	2	Time-of-Day Interrupt - 5
OB 16	2	Time-of-Day Interrupt - 6
OB 17	2	Time-of-Day Interrupt - 7
		Time-Delay Interrupt Processing
OB 20	3	Time-Delay Interrupt - 0
OB 21	4	Time-Delay Interrupt - 1
OB 22	5	Time-Delay Interrupt - 2
OB 23	6	Time-Delay Interrupt - 3
		Cyclic Interval (Watchdog) Interrupt Processing
OB 30	7	Cyclic Interval Interrupt 0 - Default Interval = 5 sec
OB 31	8	Cyclic Interval Interrupt 1 - Default Interval = 2 sec
OB 32	9	Cyclic Interval Interrupt 2 - Default Interval = 1 sec
OB 33	10	Cyclic Interval Interrupt 3 - Default Interval = 500 msec.
OB 34	11	Cyclic Interval Interrupt 4 - Default Interval = 200 msec.
OB 35	12	Cyclic Interval Interrupt 5 - Default Interval = 100 msec.
OB 36	13	Cyclic Interval Interrupt 6 - Default Interval = 50 msec.
OB 37	14	Cyclic Interval Interrupt 7 - Default Interval = 20 msec.
OB 38	15	Cyclic Interval Interrupt 8 - Default Interval = 10 msec.

APPENDIX A: Standard Library - Organization Blocks (Cont.)

Table A-2. Organization blocks (OBs) list (continued).

OB	Default Priority	Description
		Hardware Interrupt Processing
OB 40	16	Hardware Interrupt 0
OB 41	17	Hardware Interrupt 1
OB 42	18	Hardware Interrupt 2
OB 43	19	Hardware Interrupt 3
OB 44	20	Hardware Interrupt 4
OB 45	21	Hardware Interrupt 5
OB 46	22	Hardware Interrupt 6
OB 47	23	Hardware Interrupt 7
		Multi-Computing Interrupt Processing
OB 60	25	Multi-Computing Interrupt
		Asynchronous Error Processing
OB 80	26, 28	Timing Error
OB 81	26, 28	Power Supply Fault
OB 82	26, 28	Diagnostic Interrupt
OB 83	26, 28	Insert/Remove Module Interrupt
OB 84	26, 28	CPU Hardware Fault
OB 85	26, 28	Priority Class Error
OB 86	26, 28	DP Error
OB 87	26, 28	Communications Error
		Background Processing
OB 90	29	Background Processing
		Start-up Processing
OB 100	27	Warm Restart
OB 101	27	Hot Restart (S7 400 only)
OB 102	27	Cold Restart
		Synchronous Error Processing
OB 121	27	Programming Error
OB 122	27	I/O Access Error

APPENDIX B: Standard Library - System Blocks (SFBs and SFCs)

System blocks, are another resource available to the STEP 7 program developer. These blocks which are integrated in the S7 operating system, may be called in your program. As part of the CPU, the actual code for these blocks is built into the operating system and do not use any of the user memory. However, to call these blocks as you develop your program, you will need the calling interface associated with each SFC or SFB. These interfaces are provided in the Standard Library supplied with STEP 7, under the program name "*System Function Blocks*." The block will appear in the *Blocks* folder of your program once used.

Calling an SFB or SFC in your program is the same as it would be for user-created FB or FC. Simply insert a new network or open an existing network and call the system block as required. If the block requires that actual parameters be supplied, you will need proper documentation that defines each parameter and the correct operand requirements. Remember, that SFBs are still FBs, and require an instance data block as its memory. You must create and download the required DB as part of the program.

Table A-3. System blocks, including SFBs and SFCs. Provided in STEP 7 Standard Library.

SFB/SFC	Abbreviation	Description
		IEC Counters and Timers
SFB 0	CTU	UP Counter
SFB 1	CTD	Down Counter
SFB 2	CTUD	UP/Down Counter
SFB 3	TP	Pulse Timer
SFB 4	TON	On Delay Timer
SFB 5	TOF	Off Delay Timer
		S7 Communications via Configured S7 Connections
SFB 8	USEND	Uncoordinated Send
SFB 9	URCV	Uncoordinated Receive
SFB 12	BSEND	Block-Oriented Send
SFB 13	BRCV	Block-Oriented Receive
SFB 14	GET	Read (Fetch) Data from Partner
SFB 15	PUT	Write Data to Partner
SFB 16	PRINT	Write Data to Printer
SFB 19	START	Initiate Complete Restart in Communications Partner
SFB 20	STOP	Initiate STOP in Set Communications Partner
SFB 21	RESUME	Initiate Restart in the Communications Partner
SFB 22	STATUS	Check Status of Communications Partner
SFB 23	USTATUS	Receive Status of Communications Partner
SFC 62	CONTROL	Check Communications status

APPENDIX B: Standard Library - System Blocks (SFBs and SFCs)

Table A-4. System blocks, including SFBs and SFCs (continued).

SFB/SFC	Abbreviation	Description
		Handling Integrated I/O Functions (for CPUs with Integrated I/O)
SFB 29	HS_COUNT	High-Speed Counter
SFB 30	FREQ_MES	Frequency Meter
SFB 38	HSC_A_B	Control "Counter A/B" (On CPU)
SFB 39	POS	Control "Positioning" (On CPU)
SFB 41	CONT_C	Continuous-Action Closed-Loop Control (On CPU)
SFB 42	CONT_S	Step-Action Closed-Loop Control (On CPU)
SFB 43	PULSEGEN	Generate Pulse (On CPU)
SFC 63 *	AB_CALL	Call Assembler Code Block
		Drum Sequencer
SFB 32	DRUM	Drum
		Generating Block-Related Messages
SFB 33	ALARM	Messages with Acknowledgement Display
SFB 34	ALARM_8	Messages without Embedded Values
SFB 35	ALARM_8P	Messages with Embedded Values
SFB 36	NOTIFY	Messages without Acknowledgement Display
SFB 37	AR_SEND	Send Archive Data
SFC 9	EN_MSG	Enable Messages (Block-Related, Symbol-Related, Group Status)
SFC 10	DIS_MSG	Disable Messages (Block-Related, Symbol-Related, Group Status)
SFC 17	ALARM_SQ	Generate Block-Related Message with ACK
SFC 18	ALARM_S	Generate Block-Related Message with Permanent ACK
SFC 19	ALARM_SC	Check ACK Status of Last ALARM_SQ Entering State Message
		CPU Clock/Run-Time Meter Operations
SFC 0	SET_CLK	Set System Clock
SFC 1	READ_CLK	Read System Clock
SFC 2	SET_RTM	Set Run-Time Meter
SFC 3	CTRL_RTM	Control Run-Time Meter Start/Stop
SFC 4	READ_RTM	Read Run-Time Meter
SFC 48	SNC_RTCB	Synchronize Slave Clocks
SFC 64	TIME_TCK	Read System Time
		H-System Operations
SFC 90	H_CTRL	Control Operation in H Systems

APPENDIX B: Standard Library - System Blocks (SFBs and SFCs)

Table A-4. System blocks, including SFBs and SFCs (continued).

SFB/SFC	Abbreviation	Description
		Copy Data Area/Block Operations
SFC 20	BLKMOV	Copy Variables
SFC 21	FILL	Initialize a Memory Area
SFC 22	CREAT_DB	Create a Data Block
SFC 23	DEL_DB	Delete a Data Block
SFC 24	TEST_DB	Test a Data Block
SFC 25	COMPRESS	Compress User Memory in CPU
SFC 44	REPL_VAL	Transfer Substitute Value to Accumulator 1
SFC 81	UBLKMOV	Un-Interruptible Block Move
		Program Control Operations
SFC 43	RE_TRIGR	Retrigger CPU Cycle Time Monitoring
SFC 46	STP	Switch the CPU to STOP Mode
SFC 47	WAIT	Delay User Program Execution
		Handling Interrupt Events
SFC 28	SET_TINT	Set a Time-of-Day Interrupt
SFC 29	CAN_TINT	Cancel a Time-of-Day Interrupt
SFC 30	ACT_TINT	Activate a Time-of-Day Interrupt
SFC 31	QRY_TINT	Query a Time-of-Day Interrupt
SFC 32	SRT_DINT	Start a Time-Delay Interrupt
SFC 33	CAN_DINT	Cancel a Time-Delay Interrupt
SFC 34	QRY_DINT	Query a Time-Delay Interrupt
SFC 35	MP_ALM	Trigger a Multiprocessor Interrupt
SFC 36	MSK_FLT	Mask Synchronous Faults
SFC 37	DMSK_FLT	Unmask Synchronous Faults
SFC 38	READ_ERR	Read Error Register
SFC 39	DIS_IRT	Disable New Interrupts and Asynchronous Errors
SFC 40	EN_IRT	Enable New Interrupts and Asynchronous Errors
SFC 41	DIS_AIRT	Disable Higher Priority Interrupts and Asynchronous Errors
SFC 42	EN_AIRT	Enable Higher Priority Interrupts and Asynchronous Errors
		System Diagnostic Operations
SFC 6	RD_SINFO	Read Start Information of an OB
SFC 51	RDSYSST	Read System Status List/Partial List
SFC 52	WR_USMSG	Write User-Defined Diagnostic Event to Diagnostic Buffer

APPENDIX B: Standard Library - System Blocks (SFBs and SFCs)

Table A-6. System blocks, including SFBs and SFCs (continued).

SFB/SFC	Abbreviation	Description
		Distributed I/O Operations
SFC 7	DP_PRAL	Trigger DP-master Hardware Interrupt
SFC 11	DPSYC_FR	Synchronize DP-slave Groups
SFC 12	D_ACT_DP	Deactivate/Activate DP-slaves
SFC 13	DPNRM_DG	Read DP-slave Diagnostic Data
SFC 14	DPRD_DAT	Read DP-slave Consistent Data
SFC 15	DPWR_DAT	Write DP-slave Consistent Data
		Process Image Update Operations
SFC 26	UPDAT_PI	Update Process Image of Inputs
SFC 27	UPDAT_PO	Update Process Image of Outputs
SFC 79	SET	Set a Range of Outputs
SFC 80	RSET	Reset a Range of Outputs
		Address/Access Module Operations
SFC 5	GADR_LGC	Get Logical Address of a Channel
SFC 49	LGC_GADR	Check Module Slot of a Specific Logical Address
SFC 50	RD_LGADR	Check Module Logical Addresses
		Data Record Transfer
SFC 54	RD_PARM	Read Defined Parameters
SFC 55	WR_PARM	Write Dynamic Parameters
SFC 56	WR_DPARM	Write Default Parameters
SFC 57	PARM_MOD	Assign Parameters to a Module
SFC 58	WR_REC	Write a Data Record
SFC 59	RD_REC	Read a Data Record
		S7 Global Data Communications
SFC 60	GD_SND	Send a Global Data
SFC 61	GD_RCV	Accept Received Global Data
		S7 Communications via Un-Configured Connections
SFC 65	X_SEND	Send Data to External Partner
SFC 66	X_RCV	Receive Data from Ext. Partner
SFC 67	X_GET	Read Data from External Partner
SFC 68	X_PUT	Write Data to External Partner
SFC 69	X_ABORT	Abort Connection to Ext. Partner
SFC 72	I_GET	Read Data from Internal Partner
SFC 73	I_PUT	Write Data to Internal Partner
SFC 74	I_ABORT	Abort Connection to Int. Partner

APPENDIX C: ASCII Character Chart

Table A-7. ASCII Characters 0-63, with Decimal, Binary, Octal, and Hex Equivalents.

CHAR	DEC	BINARY	OCT	HEX
NUL	0	00000000	000	00
SOH	1	00000001	001	01
STX	2	00000010	002	02
ETX	3	00000011	003	03
EOT	4	00000100	004	04
ENQ	5	00000101	005	05
ACK	6	00000110	006	06
BEL	7	00000111	007	07
BS	8	00001000	010	08
HT	9	00001001	011	09
LF	10	00001010	012	0A
VT	11	00001011	013	0B
FF	12	00001100	014	0C
CR	13	00001101	015	0D
SO	14	00001110	016	0E
SI	15	00001111	017	0F
DLE	16	00010000	020	10
DC1	17	00010001	021	11
DC2	18	00010010	022	12
DC3	19	00010011	023	13
DC4	20	00010100	024	14
NAK	21	00010101	025	15
SYNC	22	00010110	026	16
ETB	23	00010111	027	17
CAN	24	00011000	030	18
EM	25	00011001	031	19
SUB	26	00011010	032	1A
ESC	27	00011011	033	1B
FS	28	00011100	034	1C
GS	29	00011101	035	1D
RS	30	00011110	036	1E
US	31	00011111	037	1F

CHAR	DEC	BINARY	OCT	HEX
space	32	00100000	040	20
!	33	00100001	041	21
"	34	00100010	042	22
#	35	00100011	043	23
$	36	00100100	044	24
%	37	00100101	045	25
&	38	00100110	046	26
'	39	00100111	047	27
(	40	00101000	050	28
)	41	00101001	051	29
*****	42	00101010	052	2A
+	43	00101011	053	2B
,	44	00101100	054	2C
-	45	00101101	055	2D
.	46	00101110	056	2E
/	47	00101111	057	2F
0	48	00110000	060	30
1	49	00110001	061	31
2	50	00110010	062	32
3	51	00110011	063	33
4	52	00110100	064	34
5	53	00110101	065	35
6	54	00110110	066	36
7	55	00110111	067	37
8	56	00111000	070	38
9	57	00111001	071	39
:	58	00111010	072	3A
;	59	00111011	073	3B
<	60	00111100	074	3C
=	61	00111101	075	3D
>	62	00111110	076	3E
?	63	00111111	077	3F

APPENDIX C: ASCII Character Chart (Continued)

Table A-8. ASCII Characters 64-127, with Decimal, Binary, Octal, and Hex Equivalents (cont.).

CHAR	DEC	BINARY	OCT	HEX	CHAR	DEC	BINARY	OCT	HEX
@	64	01000000	100	40	`	96	01100000	140	60
A	65	01000001	101	41	a	97	01100001	141	61
B	66	01000010	102	42	b	98	01100010	142	62
C	67	01000011	103	43	c	99	01100011	143	63
D	68	01000100	104	44	d	100	01100100	144	64
E	69	01000101	105	45	e	101	01100101	145	65
F	70	01000110	106	46	f	102	01100110	146	66
G	71	01000111	107	47	g	103	01100111	147	67
H	72	01001000	110	48	h	104	01101000	150	68
I	73	01001001	111	49	i	105	01101001	151	69
J	74	01001010	112	4A	j	106	01101010	152	6A
K	75	01001011	113	4B	k	107	01101011	153	6B
L	76	01001100	114	4C	l	108	01101100	154	6C
M	77	01001101	115	4D	m	109	01101101	155	6D
N	78	01001110	116	4E	n	110	01101110	156	6E
O	79	01001111	117	4F	o	111	01101111	157	6F
P	80	01010000	120	50	p	112	01110000	160	70
Q	81	01010001	121	51	q	113	01110001	161	71
R	82	01010010	122	52	r	114	01110010	162	72
S	83	01010011	123	53	s	115	01110011	163	73
T	84	01010100	124	54	t	116	01110100	164	74
U	85	01010101	125	55	u	117	01110101	165	75
V	86	01010110	126	56	v	118	01110110	166	76
W	87	01010111	127	57	w	119	01110111	167	77
X	88	01011000	130	58	x	120	01111000	170	78
Y	89	01011001	131	59	y	121	01111001	171	79
Z	90	01011010	132	5A	z	122	01111010	172	7A
[	91	01011011	133	5B	{	123	01111011	173	7B
\	92	01011100	134	5C	|	124	01111100	174	7C
]	93	01011101	135	5D	}	125	01111101	175	7D
^	94	01011110	136	5E	~	126	01111110	176	7E
_	95	01011111	137	5F	DEL	127	01111111	177	7F

APPENDIX D: Examples of Modifying/Forcing I/O and Memory Variables

Table A-9. Typical I/O and Memory Addresses with Valid Force/Modify Values.

Possible Bit Addresses (I, Q, M, DBX)		
Address	**Force/Modify Value**	**Description**
I4.5	false	BOOL variable; FALSE also valid
M4.2	true	BOOL variable; TRUE also valid
Q11.6	0	Bit variable without binary identifier
DB5.DBX1.0	1	Bit variable without binary identifier
I4.4	2#1	Binary variable with binary identifier
M5.5	2#0	Binary variable with binary identifier
Possible Byte Addresses (IB, QB, MB, PQB, DBn.DBBn)		
Address	**Force/Modify Value**	**Description**
IB 10	2#11011001	BYTE variable as binary constant with binary identifier
MB 22	b#16#FF	BYTE variable as hex constant with hex identifier
MB 24	FF	BYTE variable as hex constant without hex identifier
QB 10	'y'	BYTE variable as CHAR constant
DB5.DBB 2	10	BYTE variable as positive INT constant
PQB 4	-12	BYTE variable as negative INT constant
Possible Word Addresses (IW, QW, MW, PQW, DBn.DBWn)		
Address	**Force/Modify Value**	**Description**
IW 3	2#0000111100001111	WORD variable as binary constant with binary identifier
MW 42	w#16#ABYN	WORD variable as hex constant with hex identifier
MW 44	ABYN	WORD variable as hex constant without hex identifier
QW 12	b#(20,44)	WORD variable as 2-byte unsigned decimal values
DB5.DBW 2	'yn'	WORD variable as CHAR constant
PQW 6	-10500	WORD variable as negative INT constant (-32,768, to 32,767)
MW 2	10500	WORD variable as positive INT constant (-32,768, to 32,767)
MW 4	S5T#10m30s	WORD variable as S5TIME value with S5TIME identifier
MW 8	c#175	WORD variable as COUNT value with Count identifier
MW 10	d#2005-07-04	WORD variable as DATE value without DATE identifier

APPENDIX D: Examples of Modifying/Forcing I/O and Memory Variables

Table A-10. Typical I/O and Memory Addresses with Valid Force/Modify Values (continued).

Possible Double Word Addresses (ID, QD, MD, PQD, DBn.DBDn)		
Address	**Force/Modify Value**	**Description**
ID 4	00011000111001010111010111010111	DWORD variable as binary constant without binary identifier
MD 2	10 E3	DWORD variable as REAL constant
MD 6	9	DWORD variable as hex value without identifier
QD 20	DW#16#abcdef10	DWORD variable as hex value with identifier
QD 24	ABCDEF10	DWORD variable as hex value without identifier
DB5.DBD 6	B#(12,34,56,78)	DWORD variable as 4-unsigned decimal numbers
PQD 4	'abcd'	DWORD variable as STRING constant
MD 20	L# -12	DWORD variable as DINT value with identifier
MD 24	L#12	DWORD variable as DINT value with identifier
MD 28	-123456789	DWORD variable as DINT value without identifier
MD 32	123456789	DWORD variable as DINT value without identifier
MD 36	T#12s345ms	DWORD variable as TIME value
MD 40	TOD#1:2:34.567	DWORD variable as Time-of-Day value
MD 44	p#e0.0	DWORD variable as POINTER value
Timer Memory Addresses (Tn)		
Address	**Force/Modify Value**	**Description**
T 10	0	Conversion to milliseconds (ms)
T 11	20	Conversion to ms
T 12	12345	Conversion to ms
T 13	S5TIME#10m30s	Timer value with S5TIME (or S5T) identifier
Counter Memory Addresses (Cn)		
Address	**Force/Modify Value**	**Description**
C 15	0	WORD variable as COUNT value without identifier
C 16	35	WORD variable as COUNT value without identifier
C 17	c#150	WORD variable as COUNT value with identifier

APPENDIX E: Condition Codes CC0 and CC1 as Result Bits

Table A-11. Digital Operations Results and Evaluation using Results Bit Equivalents.

Integer Operations	**CC0**	**CC1**	**OV**	**OS**	**Results Bit Equivalent**
Result < -32,768 for (ADD_I, and SUB_I)	0	1	1	1	>0 and OV and OS
Result < -32,768 for (MUL_I)	1	0	1	1	< 0 and OV and OS
Result = -32768 to -1	1	0	0	-	< 0
Result = 0	0	0	0	-	== 0
Result = +1 to +32,767	0	1	0	-	> 0
Result > +32,767 for (ADD_I, and SUB_I)	1	0	1	1	< 0 and OV and OS
Result > +32,767 for (MUL_I)	0	1	1	1	> 0 and OV and OS
Result = 32,768 for (DIV_I)	0	1	1	1	>0 and OV and OS
Division by Zero	1	1	1	1	
Real Operations	**CC0**	**CC1**	**OV**	**OS**	**Results Bit Equivalent**
+ Normalized	0	1	0	-	> 0
± De-normalized	0	0	1	1	==0 and OV and OS
± Zero	0	0	0	-	==0
- Normalized	1	0	0	-	<0 and NOT OV
+ infinite Result (Div by zero)	0	1	1	1	> 0 and OV and OS
– infinite Result (Div by zero)	1	0	1	1	< 0 and OV and OS
IN1 or IN2 is Invalid REAL	1	1	1	1	UO
Compare Operations	**CC0**	**CC1**	**OV**	**OS**	**Results Bit Equivalent**
Result IN2 = IN1	0	0	0	-	==0
Result IN2 < IN1	1	0	0	-	<0
Result IN2 > IN1	0	1	0	-	>0
IN1 or IN2 is Invalid REAL	1	1	1	1	UO
Conversion Operations (NEG_I/NEG_D)	**CC0**	**CC1**	**OV**	**OS**	**Results Bit Equivalent**
Result in Normal Positive Range	0	1	0	-	>0
Result Equal Zero	0	0	0	-	==0
Result in Normal Negative Range	1	0	0	-	<0
Result at Extreme Negative Range	1	0	1	1	< 0 and OV and OS
Word Logic Operations	**CC0**	**CC1**	**OV**	**OS**	**Results Bit Equivalent**
Result OUT Equal Zero	0	0	0	-	==0
Result OUT Not Equal Zero	0	1	0	-	>0, or <>0
Shift Operations	**CC0**	**CC1**	**OV**	**OS**	**Results Bit Equivalent**
Bit Shifted Out = 0	0	0	0	-	==0
Bit Shifted Out = 1	0	1	0	-	>0

Note: The Condition Code Bits CC0/CC1 are the same for DINT operations except that the number range is -2147483648 to +2,147,483,647.

APPENDIX F: Analog Input/Output Digital Representation

Table A-12. Analog Input/Output Representation for common Bipolar Measurement Ranges.

Range	± 500 mV	± 1 V	± 2.5 V	± 5 V	± 10 V	± 20 mA	
+ 32,767	**≥ 587.96**	**≥ 1.1760**	**≥ 2.9398**	**≥ 5.8796**	**≥ 11.759**	**≥ 23.5160**	OVERFLOW
+ 32,511 . . + 27,649	+ 587.94 . . + 500.02	+ 1.1750 . . + 1.0004	+ 2.9397 . . + 2.5001	+ 5.8794 . . + 5.0002	11.7589 . . 10.0004	23.5150 . . 20.0007	OVER-RANGE
+ 27,648	+ 500.00 **mV**	1.0000 **V**	+2.5000 **V**	5.000 **V**	+10.00 **V**	+20.00 **mA**	
+ 24,192	+437.50	+0.8750	+2.1875	+4.375	+8.75	+17.50	
+ 20,736	+375.00	+0.7500	+1.8750	+3.750	+7.50	+15.00	
+ 17,280	+312.50	+0.6250	+1.5625	+3.125	+6.25	+ 2.50	
+ 13,824	+250.00	+0.5000	+1.2500	+2.500	+5.00	+10.00	
+ 10,368	+187.50	-0.3750	+0.9375	+1.875	+3.75	+7.50	
+ 6,912	+125.00	+0.2500	+0.6250	+1.250	+2.50	+5.00	
+ 3,456	+62.50	+0.1250	+0.3125	+0.625	+1.25	+2.50	
0	0.00	0.0000	0.0000	0.000	0.00	0.00	NORMAL
- 3,456	-62.50	-0.1250	-0.3125	-0.625	-1.25	-2.50	
- 6,912	-125.00	-0.2500	-0.6250	-1.250	-2.50	-5.00	
- 10,368	-187.50	-0.3750	-0.9375	-1.875	-3.75	-7.50	
- 13,824	-250.00	-0.5000	-1.2500	-2.500	-5.00	-10.00	
- 17,280	-312.50	-0.6250	-1.5625	-3.125	-6.25	-12.50	
- 20,736	-375.00	-0.7500	-1.8750	-3.750	-7.50	-15.00	
- 24,192	-437.50	-0.8750	-2.1875	-4.375	-8.75	-17.50	
- 27,648	-500.00 **V**	-1.000 **V**	-2.5000 **V**	-5.00 **V**	-10.00 **V**	-20.00 **mA**	
- 27,649	-500.02 . . -587.96	-1.0004 . . -1.1750	-2.5001 . . -2.9398	-5.0002 . . -5.8796	-10.0004 . . -11.7590	-20.0007 . . -23.5160	OVER-RANGE
- 32,768	**≤ -588.98**	**≤ -1.176**	**≤ -2.935**	**≤ -5.880**	**≤ -11.76**	**≤ -23.517**	UNDERFLOW

APPENDIX G: Common Abbreviations and Acronyms

Table A-13. Common Abbreviations and Acronyms.

Abbreviation	Description
AI	Analog Input
AO	Analog Output
CFC	Continuous Function Chart
CP	Communications Processor Module
CPU	Central Processing Unit Module
DB	Data Block
DI	Digital Input
DO	Digital Output
DP	Distributed Periphery (or Distributed I/O)
EN	Enable Input Line
ENO	Enable Output Line
EPROM	Erasable Programmable Read Only Memory
FB	Function Block
FBD	Function Block Diagram language
FC	Function
FEPROM	Flash Erasable Programmable Read Only Memory
FM	Function Module
IM	Interface Module
LAD	Ladder Diagram Language
MPI	Multi-Point Interface
OB	Organization Block
OP	Operator Panel
PG/PC	Programming Device/PC Programming Device
PII	Process Image of Inputs
PIQ	Process Image of Outputs
RAM	Random Access Memory
RLO	Result of Logic Operation
SCL	Structured Control Language
SDB	System Data Block
SFB	System Function Block
SFC	System Function
SM	Signal Module
STL	Statement List Language
UDT	User Defined Type
VAT	Variable Table

Glossary

- A -

absolute address: A direct reference to a memory location, represented by a character identifier and a number. The term "Q 6.7" and "M 44.3" are absolute addresses. See symbolic address.

absolute addressing: Use of the direct references of memory locations like I 5.6, instead of substitute or name references such as "STOP_PB" to reference memory locations. See absolute address.

accessible nodes: A utility of the Simatic Manager that, when activated, determines what stations are configured as MPI nodes and can be accessed by the programming system.

accumulator (ACCU): Memory registers in the CPU that serves as buffers for load and transfer operations, as well as for comparisons, math, and conversion operations.

actual parameter: The address or value that replaces a formal parameter when a function block (FB) or function (FC) is called. For example, the formal parameter "STOP" may be replaced by the actual parameter "I 60.2." Also see *block parameters; formal parameter.*

address: 1) A reference number that identifies a unique memory bit, byte, word, or double word location. **2)** An identifying number designating a unique system entity or object.

address identifier: A single or double letter designation that precedes an address and indicates the specific memory area and size of the memory unit being referenced. The first letter indicates the area and always a bit if only one letter is present. A second letter 'B' will indicate byte, 'W' a word, and 'D' a double word.

AS-Interface (AS-I): A network intended to connect binary sensors and actuators at the lowest part of the field level of the plant networking hierarchy. The S7 300 connects to AS-I via the CP 342-2 communications processor.

asynchronous error OBs: The group of organization blocks called when one of the asynchronous errors occur.

asynchronous errors: Run-time errors that are not associated with or resultant from any particular aspect of the user program (e.g., power supply error, module inserted/removed). See *asynchronous error OBs.*

authorization: A right of use for the STEP 7 Basic and optional programming packages, stored on an authorization diskette. You must install required authorization to your hard drive before STEP 7 software or optional packages can run without interruption.

- B -

B stack: Block Stack. An S7 memory area (stack) that can be viewed to determine all non-terminated blocks called and the order in which they were called up to the block that erred and caused the CPU to stop. Also, see *I stack.*

background processing: A method of processing non-critical code (in OB90) in the slice of time that starts with the end of the actual CPU scan time and continues until the expiration of the most recent minimum CPU scan. If the actual execution cycle time is less than the user-specified minimum cycle time (S7 400), OB90 is processed for the remaining duration of the minimum cycle time. Then, the OB1 cycle is re-started.

bilateral communication: Refers to S7 communication between partners, in which there is a SFB on both the local and remote partner (e.g., "USEND" AND URCV"). See *unilateral communication.*

bit memory: The area of the system memory containing what may be referred to as the CPU's internal storage bits or software control relays. The letter 'M' precedes each bit memory address (e.g., M10.0 is a bit, MB 10 is a byte, MW 10 is a word, and MD 10 a double word.

block header: Each block has a header that contains attributes of the block, such as *Name*, *Family*, *Version*, and *Author*. The block header may be modified by opening the block properties. Select the block and then right click to select object properties.

blocks: Blocks make up the STEP 7 user program, and are distinguishable by function, structure, or purpose. STEP 7 blocks may be grouped as *logic code blocks* of types (FB, FC, OB, SFB, and SFC); *data blocks* of types (DB, SDB) and *user-defined data types* (UDT).

block stack (B-stack): See *B-stack*

block parameters: Refers to the named input/output channels that allow data to be passed into (input parameters) or output from (output parameters) a block function.

- C -

CC0 status bit: Condition code status bit of the CPU status word.

CC1 status bit: Condition code status bit of the CPU status word.

C-bus: Abbreviation for *Communication Bus*. Part of the backplane bus of the SIMATIC S7-300/M7-300 and S7-400/M7-400 programmable controllers. A separate communication bus increases data transfer rates between the CPU and programmable modules, as well as make programming access possible to all programmable modules (e.g., FMs, CPs), via a PG/PC connected to the CPU.

central rack (CR): For the S7300, the term refers to the main I/O rail (rack for S7400) that holds the CPU. The central rack may also contain signal modules, function modules, and communications processors. All other rails (or racks), connected to the central rack using interface modules, are referred to as expansion racks.

CFC: Abbreviation for Continuous Function Charts, an optional STEP 7 programming language used to link complex functions graphically. CFC links existing functions, many of which are available in standard functions libraries. CFC function libraries include logic, math, control, and data processing functions. Charts created in CFC are stored in the "Charts" folder, beneath the S7 program, before they are compiled to form the S7 blocks of the user program.

CFC charts: A special graphic source file, created using the Continuous Function Charts programming language. See *charts container*.

charts container: A folder or container used to store CFC "Charts" beneath the S7 program. See *CFC charts*.

clock memory: A system memory area comprised of individual bits that generate clock pulses that are usable in the control program by simply referencing the bit address. Eight clock memory bits are available, each representing a unique frequency from 0.5 hertz to 10 hertz.

code block: S7 program resource that allows program construction in modular sections of code. These sections are referred to as blocks. Logic block types include Organization blocks (OBs), Function blocks (FBs), and Functions (FCs). System Functions (SFCs) and System Function Blocks (SFBs), are code blocks that are part of the CPU operating system.

cold restart: An S7 start-up mode in which the organization block OB102 is processed if loaded; the process image (PII, PIQ), all timers, counters, and bit memory are all reset. Furthermore, Data Blocks generated by system functions are deleted from work memory.

communication bus (C-bus): See *C-bus*.

communications processor (CP): A class of S7 communications modules that include both those that provide network interfacing and point-to-point serial links between S7 PLCs or to other devices.

communications, bilateral: See *bilateral communication*.

communication, unilateral: See *unilateral communication*.

compressing memory: See *memory compression*.

configuration tables: The output of the STEP 7 hardware configuration tool, which has each I/O rail of an S7 300 or rack of an S7 400, and the modules installed in each slot, represented in the form of a table. The table includes rows for the module contained in each slot, and information associated with each module in columns.

configured connection: A communication link between S7 partners, in which the connection properties between two partners are preconfigured and fixed in a connection table. Connection types include: FDL, FMS, ISO, ISO-on-TCP, point-to-point, S7, and UDP.

connection table: A STEP 7 object that contains the preconfigured connections between the communications partners of a project.

consistency check: A utility that checks to see if the hardware configuration (*Hardware Configuration*) or if the network configuration (*NetPro*) is free of errors.

control logic: 1) The combination of conditions that must be satisfied to control a particular device or perform a particular function. **2)** Refers to the entire program logic or relay logic that will control a given machine or process.

controller rack (CR): A main S7 300 or S7 400 system component used to install the CPU. The controller rack may also contain signal modules, function modules, and communications processors.

counter memory (C): The area of the system memory containing locations used for software counters. Each counter word, when set, contains the user-defined preset value as a BCD value. The letter 'C' precedes each S7 counter address (e.g., C0, C1, C2, and C3 each is a unique counter.

counter, down: An S7 software counter instruction that decrements its count value by one for each off-to-on transition of its count-down input.

counter, up: S7 software counter instruction that increments its count value by one for each off-to-on transition of its count-up input.

counter, up-down: An S7 software counter instruction that combines the actions of both the up-counter and down-counter.

cyclic interrupt: S7 300 and S7 400 systems provide cyclic interrupt OBs to support periodic processing of certain portions of the user program, independent of the normal cyclic processing time of the CPU. Cyclic interrupt are available at intervals of 10 milliseconds to 5 seconds.

cyclic interrupt OBs: OB 30 to OB 38 are provided for servicing cyclic interrupts. Cyclic interrupt OBs available to a program, is CPU dependent.

- D -

Data Block (DB): A STEP 7 block that allows users to store data associated with the program and the process. There are two types of data blocks: the *shared data block* whose data is accessible by all logic code blocks, and the instance *data block,* whose data is associated with a specific function block (FB).

Data Block, instance: See *data block*.

Data Block, shared: See *data block*.

data type: Stipulates the characteristics of program data, with respect to its representation and the permissible range of values. Some example data types include BOOL, INTeger, REAL, S5_TIME, ARRAY, and STRING. In STEP 7, there are two categories of data — elementary, or complex.

data type, complex: Category of data types that reserve memory for variables that are comprised of two or more data element of elementary data type. Complex data types include DATE-AND-TIME, STRING, ARRAY, STRUCT, and UDT.

data type, elementary: Category of data types that reserve a single bit, byte, word, or double word of memory. Elementary data types include BOOL, BYTE, WORD, DWORD, INT, DINT, REAL, DATE, TIME, S5TIME, TIME-OF-DAY, and CHAR.

data type, user defined: A user-defined structure of any combination of elementary or complex data types, to be used as a template for creating data blocks of the same data structure or for declaring a frequently used structure as a data type.

declaration section: The area of a STEP 7 code block or data block, in which variables may be defined.

diagnostic address: An address that is assigned to a module's configuration in order to receive diagnostic information data. The diagnostic data can be checked from the user program using or from the STEP 7 monitor/modify data utility.

diagnostic buffer: A memory area of the CPU in which all diagnostic events are stored the order of occurrence. STEP 7 is used to view the contents of the buffer.

diagnostic entry: A single event entered into the S7 CPUs diagnostic buffer.

diagnostic interrupt: An interrupt used by a module with diagnostic capability, to report occurrence of a module fault to the CPU.

diagnostic message: A message that consists of a processed diagnostic event, which is then sent from the CPU to the display unit.

direct addressing: A method of addressing in which the memory location to be accessed or operated upon, is the address referenced by the instruction. The address can be absolute (e.g., I4.7) or symbolic (e.g., E-STOP).

direct I/O read: An instruction operation that uses the peripheral input (PI) address area to access an input module's data directly from the module or I/O bus, as opposed to normal reading from the process image of inputs (PII).

direct I/O write: An instruction operation that uses the peripheral output (PQ) address area to access an output module's data directly via the module or I/O bus, as opposed to normal writing to the process image of outputs (PIQ).

distributed I/O: Slave devices that are connected via a PROFIBUS-DP master. Also referred to as distributed periphery.

DP-master: An intelligent device that behaves in accordance to the Profibus DP standard EN 50170. A DP-master is able to send data to and request data from its slave devices.

DP-slave: A device that behaves in accordance to the Profibus DP standard EN 50170. A DP-master is able to send data to and request data from its slave devices.

dynamic connection: A connection that is set up at runtime when the active partner initiates the connection by calling the appropriate communications system function (SFC).

dynamic parameter: An S7 module parameter that can be set or modified via the STEP 7 program using a system block; contrasted to static parameters that are only set using STEP 7 configuration software.

- E -

error, asynchronous: Run-time errors that are associated with or resulting from some aspect of the user program (e.g., addressing error, I/O module access).

error, synchronous: Run-time errors that are not associated with or resulting from the user program (e.g., power supply fault).

error OBs: Organization blocks that allow users to write code to respond to system-related (asynchronous) or program-related (synchronous) errors. There is an associated OB for each error type. If the OB is loaded in the CPU as part of the user program, it will be called by the operating system if the associated error occurs.

expansion rack (ER): An I/O rack used to extend the S7 I/O system by holding additional modules. An expansion rack is connected to the central controller rack via an interface module placed in the central rack and in the expansion rack.

extended pulse timer: See *Pulse Timer Extended*.

- F -

FBD: Abbreviation for Function Block Diagram, one of the three language representations of the basic STEP 7 programming package. Control logic can be created using box instructions represented in Boolean gate logic format. The operations AND, OR, and NOT are combined with other box operations to construct simple and complex control logic. Also, see *LAD; STL*.

formal parameter: A placeholder for an actual parameter, used when creating a logic block that can be passed parameters. In FBs and FCs, formal parameters are defined by the user; in SFBs and SFCs, formal parameters are predefined. Also, see *actual parameter*.

Function (FC): According to the IEC 1131-3 standard, functions are logic blocks that do not reference an instance data block, meaning they do not have a 'memory'. A function allows passing of parameters in the user program, making it suitable for programming complex functions that are required frequently (e.g., calculations). Note: As there is no memory available, the calculated values must be processed immediately following the FC call.

Function Block (FB): According to the IEC 1131-3 standard, function blocks are logic blocks that reference an instance data block, meaning they have static data. A function block allows passing of parameters in the user program, making it suitable for programming complex reusable functions. Since FBs have a 'memory' in the form of the associated instance data block, its parameters are accessible at any time and any point in the user program.

Function Block Diagram (FBD): See *FBD*.

function module (FM): Any of the so-called intelligent I/O modules used to perform complex or time-critical tasks independent of the S7 300 or S7 400 CPU.

- G -

GD circle: The CPUs that participate in an exchange of global data packets form a circle. If there are other CPUs exchanging global data packets in a multipoint interface network, these CPUs form a second circle. A circle may be two-sided involving two CPUs, each of which can both send and receive a GD packet; it may be one-sided in which a single CPU sends a GD packet to several other CPUs. See *GD packets*.

GD communication: See *global data communication*.

GD packets: In global data communication, the small blocks of data transferred in global data communication; data that have the same receiver and sender are combined in a global data packet. The global data packet is sent in a frame. A global data packet is identified by a global data packet number. If the maximum length of a send global data packet is exceeded, a new global data circle is used.

global data: Information, such as input, output, bit memory, and data block memory areas, which may be exchanged between CPUs participating in global data communication.

global data communication: A data exchange method used by CPUs in an MPI network to exchange small amounts of data. See *global data*.

global symbol: A symbolic address known to the entire user program (i.e., all blocks), as opposed to a local symbol, defined in a block, and which is known only to the specific block. See *symbolic address*.

GRAPH: A STEP 7 optional programming package. Graph, also called S7-Graph, is a graphic language designed to implement sequential controls. The programming process includes creating a series of steps, defining the contents of each step, and defining the transitions from one step to the next. Also, see *HiGraph; CFC; SCL*.

group error: An LED indicator found on the front face-plate of some S7 300 modules, to indicate any internal or external error on the module.

- H -

hardware catalog: A component of the STEP 7 Hardware Configuration tool, that contains Simatic objects (e.g., S7 racks, and modules) used to create a software configuration that represents the various parts of the actual hardware configuration of a project.

hardware configuration: Definition of rack arrangement, modules, module slots, addressing, and initial operating parameters of the hardware components of a project. See *hardware configuration utility*.

hardware configuration utility: A STEP 7 software configuration utility used to define rack arrangement, modules, module slots, addressing, and initial operating parameters of the hardware components of a project. This tool also provides online hardware diagnostic capability of the actual components.

hardware interrupt: A signal ability of some S7 modules to trigger and alert to the CPU when some process or module event occurs.

HiGraph: An optional STEP 7 language, also called S7-Higraph, designed for implementing machines or processes that are considered "state machines." In such machines, only one state is possible at any given time. HiGraph allows programming of the blocks of your program as state graphs. This segments the plant into individual functional units that may each take on different states. Also, see *SCL; CFC; Graph*.

hot restart: An S7-400 start-up mode in which the organization block OB101 is processed if loaded; a hot start-up occurs following the return of power after a power loss while in the RUN mode; the S7-400 CPU resumes program execution at the point at which it was interrupted. Both a manual and automatic hot restart is possible if the user program was not modified in any way while in STOP mode.

H-station: A fault-tolerant S7 PLC station, consisting of at least two CPUs, one configured as master and one as standby.

- I -

I-stack: Interrupt Stack. An S7 memory area (stack) that can be viewed to show the content of the CPU registers (accumulators, address register, data block register, and status word) at the point of interruption and at the exact error that caused the CPU to stop. Also, see *B-stack*.

image table: See *process image*.

indirect addressing: An addressing method in which the desired absolute address is not used directly in an instruction operation, instead the address is contained in another memory location, used as a pointer to the actual address.

Industrial Ethernet: The Simatic NET network intended to connect S7 300, S7 400 and other systems at the management and cell levels of the plant networking hierarchy. Defined by the IEEE 802.3 standard, the network was formerly designated as SINEC H1.

input image table: See *input memory*.

Input (I) memory: Memory area that stores snapshot status of connected digital inputs. Each connected input has a bit in the input memory that corresponds to the configured byte address for the module. The address identifier "I' indicates an input memory address, when used in the program (e.g., I10.0 is a bit, IB 10 is a byte, IW 10 is a word, and ID 10 a double word. Also, see *process image inputs*.

instance: An "instance" is the call of a function block. If a Function is called four times in a STEP 7 program, then there are four instances. An instance data block is assigned to each call. See *data block*.

instance Data Block: See *data block*.

intelligent DP-slave: A DP-slave involving an intelligent device such as an S7-300 connected as a slave via its integrated DP port, or via a communications processor (CP), installed in an S7-300, which can serve as a DP-slave.

interface module (IM): Module components available in both the S7 300 and S7 400 systems, used to expand the I/O system by allowing connection of additional I/O expansion racks.

interrupt: An efficient method of requesting the immediate attention of a central processor to gain some type of service. Interrupts are usually classified as hardware (e.g., process signal, I/O module with interrupt lines) or software (e.g., timed interrupts).

interrupt stack: See *I-stack*.

I/O image table: See *process image*.

I/O rack: The mounting unit in which I/O modules are installed. In the S7-300 system, the rack is a standard DIN rail.

- K -

K-bus: Same as *C-bus;* The backplane communications bus of S7-300/S7400 PLCs.

Know_How_Protect: A STEP 7 keyword that allows a user-written code block to be compiled and locked from view. A block that has been protected may not be viewed, printed, or modified. Only the block header and declaration area will be displayed.

- L -

LAD: Abbreviation for Ladder, one of the three language representations of the basic STEP 7 programming package. Control logic can be created using relay-like instructions may be combined with box instructions to perform simple to complex control operations. Also, see *FBD; STL*.

ladder logic (LAD): See *LAD*

ladder network: See *ladder rung*.

ladder rung: A single Ladder diagram network, part of a complete program or block, that performs the desired control logic in part or in whole for a single device or function.

library: A STEP 7 object used for storing reusable programs or program components (i.e., block folder, source folder, CFC chart folder, symbol table). Users may create libraries from the Simatic Manager, and certain standard libraries are supplied with STEP 7 basic and optional packages

library, standard: A STEP 7 library that is provided with the basic package.

linear program: Refers to a STEP 7 user program designed such that the entire main program is coded in organization block 1 (OB1). See *partitioned program; structured program*.

load memory: S7 memory that holds the entire user program or objects that are downloadable to the CPU (e.g., code blocks, data blocks, and system configuration data). In some CPUs, load memory may also contain symbol, comments. See *work memory*.

local data: Temporary data stored in the local (L) stack and accessed exclusively by the executing block. The local data area is reserved in the block header of a block, when temporary (TEMP) variables are defined. Also, see *local stack*.

local data stack: See *local stack*.

local I/O: Input/output rack restricted to a short distance from the CPU connected using local I/O interface modules (IMs); contrasted to remote I/O, which may be located long distances from the CPU.

local data: Temporary data stored in the local (L) stack memory and used exclusively by the executing block for temporary results. The local variables defined in the block header determine the size requirements of the local data. See *local stack*.

local memory (L): See *local stack*.

local stack: Temporary memory locations made available by the CPU to each code block as it is called. The so-called L-stack, is released when a block terminates, and is available for use by the next called block. The block header reserves the L-stack length however the maximum length is CPU dependent.

local symbol: A name assigned to a block variable and known only to the block in which it is defined, as opposed to global symbols assigned in the Symbol editor, and that are known to the entire user program (i.e., all blocks). See *symbolic address*.

- M -

memory card: A plug-in RAM or Flash EPROM memory expansion card.

memory compression: An S7 memory reclamation operation that reclaims memory by eliminating gaps in the CPU's work memory and in the RAM load memory. Such gaps are the result of multiple block deletions and reloading.

message block: A system function block (SFB) or system function (SFC) incorporated in the S7 program in order to generate messages.

message table: Table that defines the text for messages and that assigns these messages to specific events.

message number: A unique number assigned to a message for identification and use purposes.

minimum scan cycle time: A CPU parameter that allows setting of a minimum time for processing the main program. If the actual processing time of the main program (OB1) takes less time tan the minimum specified time, then the CPU will wait until the specified minimum cycle time expires before starting the next cycle of OB1.

module parameter: A value that can be set on an S7 module, which will affect the module's behavior. Some modules have several parameters, some of which can be set via the control program.

module catalog: See *hardware catalog*.

MPI address: In an MPI network, every programmable module (e.g., CPUs, CPs, and FMs) must have its own unique MPI address assigned.

MPI network: A Multi-Point Interface network of up to 31 nodes, used to provide a programming network for S7 300 and S7 400 CPUs, function modules (FMs), communications processors (CPs), text displays and operator panels. The network also facilitates *global data communications* among CPUs.

multi-computing: An S7 400 feature that supports up to four CPUs installed in one rack. The total control task can be divided among the CPUs, each of which has its own program, govern its own I/O and can communicate with the other CPUs.

multi-computing interrupt: An S7 400 feature that allows any of four possible CPUs operating in a multiple processor configuration, to generate an interrupt to all of the other participants such that a synchronized response is possible. See *multi-computing interrupt OB; multi-computing*.

multi-computing interrupt OB: OB 60 is provided in S7 400, for servicing user-configured multiprocessing. See *multiprocessing interrupt; multiprocessing*.

- N -

nesting depth: Refers to the number of block calls that can be made in a horizontal direction, starting for example from the main organization block (OB1). Nesting-depth is CPU-dependent. The first OB represents nesting depth of one, hence six more block calls in a horizontal direction would be a nesting depth of seven.

NETPRO: A STEP 7 utility used to configure networks or point-to-point links for S7 projects, using graphic objects. The tool allows for adding of new subnets or point-to-point links, attachment of new stations (S5, S7, PG/PC, Ops, etc.), network and module properties to be set, and for configuration of communications connections.

network: 1) A number of nodes linked together by connecting cables for the purpose of communication. 2) Divides Ladder Logic (LAD) and Function Block Diagram (FBD) blocks into logic rungs or Statement List code into smaller segments.

network configuration: Creation or modification of point-to-point links or networks, for S7 projects. Stations may be added to the network configuration after or before they are created in the STEP 7 project. See *NETPRO*.

- O -

OB 1: Organization Block 1; the main code block of a STEP 7 program. Once downloaded to the CPU, OB1 processes cyclically. All other blocks are called directly or indirectly from OB1 or from another organization block.

object: Refers to an item in a STEP 7 folder, which can be opened and edited, automatically starting the appropriate application in the software, for example, block, source file, or station.

off-delay timer (S_OFFDT): A timer whose output is activated when the enabling signal goes TRUE, and starts timing when the enable goes FALSE. After the programmed timed delay expires, the timer coil de-energizes.

off-line blocks: Blocks contained in the offline blocks folder of a CPU program; the blocks stored in the project on your hard drive. Also, see *online blocks*.

on-delay timer (S_ODT): A timer that starts timing when the enable signal goes TRUE and continues to time unless the enable goes false or the timer is reset. The output is activated after the preset time has elapsed, and stays energized until the enable signal goes FALSE or the timer is reset.

on-line blocks: Blocks contained in the CPU; the blocks listed in the online blocks folder. Also, see *offline blocks*.

operating mode: The various continuous or transitional states of operation of an S7 CPU. Operating modes may be selected from the PG/PC or from the mode selector switch on the CPU front plate (e.g., RUN, RUN-P, STOP, HOLD, and STARTUP).

Organization Block (OB): A special category of STEP 7 code block that provides the user program with an interface to the CPU's operating system. Organization blocks allow processing of the main user program as well as organized program response to various categories of system related conditions. What OBs are available is CPU dependent.

output image table: See *output memory*.

Output (Q) memory: System memory area for storing the status of connected digital outputs. Each connected output has a bit in the output memory that corresponds exactly to the terminal to which the output is connected. The letter 'Q' precedes each output memory address, when used in the program (e.g., Q10.0 is a bit, QB 10 is a byte, QW 10 is a word, and QD 10 a double word. Also, see *process image outputs*.

- P -

P-bus: Abbreviation for *Peripheral Bus;* the S7 I/O data bus. Also, see *C-bus*.

parameter: 1) A variable input to or output from an S7 code block. 2) A variable that can be set on an S7 module to determine one or more aspects of the module's behavior. Also see *dynamic parameter; static parameter*.

partitioned program: A STEP 7 user program that in essence is a linear program, however the program is subdivided into blocks of code. The blocks are then called in sequence. See *linear program; structured program*.

Peripheral bus (P-bus): See P-bus.

Peripheral inputs (PI): An address area that allows direct read access to the data of an input module via the I/O bus. In other words, addressing a module using its PI address allows a direct or immediate input read. See *direct I/O read*.

peripheral outputs (PQ): An address area that allows direct write access to the data area of an output module via the I/O bus. In other words, addressing a module using its PQ address allows a direct or immediate output write. See *direct I/O write*.

PG: A Simatic designation for programming device. The SIMATIC PG is a personal computer with a rugged compact design, suitable for industrial conditions. The Simatic PG is completely equipped for programming the SIMATIC programmable logic controllers.

PG/PC: An abbreviation that refers to Simatic programming system (PG) or a PC (personal computer) programming system.

point-to-point link: Refers to the type of communication link used to connect devices with a serial interface, such as bar code readers, printers, or another controller, to an S7 300 or S7 400. The S7 300 uses the CP 340 for this type of link. S7 400 uses the CP 441-1 or -2.

priority: A method of assigning the order in which portions of code are executed based on level of importance. A program of higher priority can interrupt one of lower priority. Since all blocks are called either directly or indirectly from an OB, priority is assigned to S7 OBs.

priority class: A program hierarchy level into which a STEP 7 organization block can be placed. There are 28 groupings or priority classes to which OBs are assigned. Priority classes determine how OBs may interrupt other OBs. OBs assigned the same priority class do not interrupt one another but are processed in sequence. See *priority*.

process image: In the S7 world, the area of system memory that contains a snap-shot image of the status of digital inputs and digital outputs that is updated on each CPU scan. Also, see *process image inputs; process image outputs.*

process image inputs (PII): The system memory area that contains a snap-shot image of the input status bits (ON = 1, OFF = 0) as read from the connected digital input modules, at the beginning of each CPU cycle.

process image outputs (PIQ): The system memory area that contains an image of the output status bits as set by the control logic (ON = 1, OFF = 0) and are transferred to the connected digital output modules at the end of each CPU cycle.

PROFIBUS: Acronym for **PRO**cess **FI**eld **BUS**. Profibus is an internationally accepted network standard defined by DIN 19245, for the networking of field level devices.

project: The main STEP 7 object; the project allows users to develop, organize, and store all programs and data associated with an automation task. A project contains one or more PLC stations and programs, hardware configurations, and network configurations.

pulse timer (S_PULSE): A timer that starts to time when the trigger input (RLO) goes from low to high, and continues to time for the programmed duration, or until the trigger signal goes from high to low. The timer is reset when the reset input goes high while the timer is timing.

pulse timer extended (S_PEXT): An S7 timer that starts to time when the trigger input transitions from low to high, and continues to time until the programmed duration expires; regardless of any change at the trigger input before the timer times out.

- R -

real-time clock: A clock that indicates the passage of actual time, in contrast to a non-real-time set up by a computer program.

remote I/O: Input/output racks located at long distances from the CPU using remote I/O interface modules (IMs).

restart, complete: See *warm restart.*

restart, cold: See *cold restart.*

restart, hot: See *hot restart.*

restart, warm: See *warm restart.*

result of logic operation (RLO): The current result of a series of logic operations in the user program, which is then used to further process signals digitally. Certain instructions are either executed or not, depending on the last RLO.

retentive bit memory: Part of the bit memory area designated to retain its contents or signal states even under off-circuit conditions. Retentive bit memory by default starts at byte 0 of the (M) area and ends according to user designation.

retentive on-delay timer (ODTS): A timer that starts timing when the enable signal goes TRUE and continues to time even if the enable goes false. If the enable signal changes back to '1' before the timer expires, the timer will restart. The output is activated after the preset time has elapsed, and remains activated until the timer is reset.

rewiring function: A STEP 7 utility that allows the replacement of the addresses used in one or more blocks or of the entire program, with another set of addresses. For example, the inputs I 10.0 to I 10.7 may be replaced by I 20.0 to I 20.7. Addresses of inputs, outputs, timers, counters, bit memory, as well as Functions (FC) and Function blocks (FBs) may be rewired.

rising edge: The rising edge of a signal (i.e., the OFF-to-ON or 0-to-1transition).

rung: See *network*.

RUN mode: An S7 CPU operating mode in which the user program is processed and the process image of inputs and outputs are updated cyclically.

RUN-P mode: An S7 CPU operating mode in which the user program is processed and the process image is updated cyclically, and restricted access to the program is allowed (e.g., via the PG/PC).

run-time meter: A CPU function that counts hours. Run-time meters are used for tasks such as determining total CPU run-time or the running time of certain devices connected to the CPU.

- S -

S7 GRAPH: See *Graph*.

S7-HiGraph: See *HiGraph*.

S7-SCL: See *SCL*.

SCL: An optional STEP 7 programming package called "Structured Control Language." SCL is a text-based language, whose definition conforms generally to the IEC 1131-3 standard. PASCAL-like, SCL simplifies programming of loops and conditional branches, making it quite suitable for formula calculations, complex algorithms, and extensive data management. Also, see *HiGraph; CFC; Graph*.

segmented rack: A feature of the S7400 allowing two CPUs with a shared power supply to be independent of one another. The two CPUs may interchange data via the C- bus yet each has its own P-buss for I/O data.

signal module (SM): Any of the digital or analog I/O modules used to interface digital and analog signals to S7 300 or S7 400 systems.

Simatic Manager: The main software development tool running under the STEP 7 package; sometimes referred to as the Project Manager.

Simatic NET: The product designation or brand name for Siemens networks and network components. Formerly called SINEC.

source file: Part of a program, which is created with a graphic or text-oriented editor and is compiled into an executable block as part of the S7 user program. S7 source files are stored in the folder "Source Files" beneath the S7 program.

source file container: See *source file*.

standard blocks: Turnkey S7 blocks provided as part of a library. Standard blocks are delivered in a library as part of the STEP 7 basic or optional software, on CDs or other storage mediums.

start information: Every organization block reserves 20-bytes of local data that the operating system supplies as start information when the OB is started. The start information specifies the start event of the OB, the date and time the OB starts, errors that have occurred, and diagnostic events. For example, OB40, a hardware interrupt OB, contains the address of the module that generated the interrupt in its start information.

Start-up Mode: A CPU operational mode just prior to starting cyclical program execution. S7 startups are triggered by: 1) switching ON the main power supply, 2) switching the CPU mode selector from STOP to RUN or RUN-P, or 3) a communication START request issued from the PG/PC or from another CPU.

Start-up OBs: Organization Blocks that are called on one of the three types of S7 start-up modes: warm restart (OB 100); hot restart (OB 101); and cold restart (OB 102). Start-up OBs are called during just prior to starting the cyclical program (OB1).

Statement List (STL): See *STL*.

static connection: A connection in which the properties that establish the logical relationship between the partners is configured in a CPU connection table, and is built-up at the start-up of the CPU and remains even when the CPU is switched to STOP.

static parameter: An S7 module parameter that can only be set or modified using the STEP 7 software, contrasted to dynamic parameters that may be set via the STEP 7 user program. An example static is the input delay on a digital input module.

static variable: One of the variable types that can be defined in an S7 function block (FB). Static variables are variables that must be maintained from one call of the FB to the next. These variables are saved in the associated instance data block when the FB is terminated.

status bits: Binary indicator flags used by the CPU during binary operations (e.g., AND/OR logic), and set by the CPU during digital operations (e.g., compare, arithmetic). Status bits are available as bit instructions that may be combined with other logic operations.

STL: Abbreviation for *Statement List*, one of three language representations of the basic STEP 7 programming software, is a line oriented language with assembler-like instructions. Also, see *LAD; FBD*.

STOP Mode: A CPU operational mode in which the user program is not scanned and the modules are set to the default initial states specified in the CPU. The STOP mode is caused by: 1) switching the CPU mode selector to STOP; 2) executing SFC 46 STP function; 3) an unrecoverable error during program execution; or 4) a communication STP request from the programming unit or another CPU.

structure: A composite grouping of variables with different data types. An example where a structure may be required is a recipe, using data types (e.g., REAL, INT, TIME, and DATE). It could also contain an array.

structured program: Refers to a STEP 7 user program designed to take advantage of STEP 7 resources that allow programming of modular block functions that can be reused throughout the program to minimize code writing and redundant code development.

subnet: A subnet comprises all nodes in a network, which are connected together without gateways. A repeater may be included within a subnet.

symbol: See *symbolic address*. Also, see *global symbol, local symbol*.

symbolic address: A substitute reference to an absolute address or memory location, represented by a name. The term "STOP_PB" could be a symbolic address for the absolute address I 6.7". See *absolute address*.

symbolic addressing: Use of substitute or name references like "STOP_PB" to reference memory locations, instead of absolute addresses, like I 6.7". See *absolute address*.

symbols table: A STEP 7 program component that allows users to define symbols to represent absolute addresses. Symbol definition involves specifying the name (symbol), absolute address, data type, and an associated comment. Symbols defined here are considered global. See *global symbol*.

synchronous error OBs: The group of organization blocks called when one of the synchronous errors occur.

synchronous errors: Run-time errors that are associated with or resultant from a particular aspect of the user program (e.g., addressing error, error accessing I/O module). See *synchronous error OBs*.

system blocks: S7 blocks that are actually components of the operating system. Such blocks include system functions (SFCs), system function blocks (SFBs), and system data blocks (SDBs).

system data object: The compiled configuration data that is loaded to the CPU. The data contains configured hardware and network configuration and connection information.

System Function (SFC): An S7 Function, integrated in the operating system of an S7 CPU. SFCs are called from the user program just as any other Function. What SFCs are usable, is CPU-dependent.

System Function Block (SFB): An S7 Function Block, integrated in the operating system of an S7 CPU. SFBs are called from the user program just as any other Function Block. What SFBs are usable, is CPU-dependent.

- T -

temporary local data: Variables defined using the (temp) designation in the declaration table of a block. Variables designated as such are only available to the local block and only while the block is being processed. See *local stack*.

time base: A unit of time generated by the system clock and used by software timer instructions. Normal time bases are 0.01, 0.1, and 1.0 second.

time-delay interrupt: An S7 interrupt capability that causes an interrupt to the CPU after a specific amount of time has expired (e.g., every 100 ms.) S7 300 and S7 400 systems provide time-delay-interrupt OBs to allow creation of service routines to respond to user-configured time-delay interrupts. See *time delay interrupt OBs*.

time-delay interrupt OBs: OB 20 to OB 23 are provided for servicing time-delay interrupts configured by the user. The code blocks called from these OBs will support timed regulated processing of certain portions of the user program. Time-delay interrupt OBs available to a program, is CPU dependent. See *time-delay interrupt*.

time-of-day interrupt: S7 300 and S7 400 systems provide time-of-day interrupts to support processing of certain portions of the user program at a specific time of day either once only or periodically (e.g., hourly, daily, weekly, monthly, yearly, or every minute). See *time-of-day interrupt OBs*.

time-of-day interrupt OBs: OB 10 to OB 17 are provided for servicing user-configured time-of-day interrupts. Time-of-day interrupt OBs available to a program, is CPU dependent. See *time-of-day interrupt*.

timer memory (T): The area of the system memory containing locations used for software timers. Each timer word, when set, contains the timer preset value in BCD, and the user-defined time-base of the timer. The letter 'T' precedes each S7 timer address (e.g., T0, T1, T2, and T3 each is a unique timer.).

timer preset value: A programmed value that determines the number of time-base intervals to be counted in a software timer and, subsequently, the programmed time duration.

trailing edge: The falling edge of a signal (i.e., the ON-to-OFF or 1-to-0 transition).

tree structure: A graphical report showing how blocks are called in a given program.

trigger frequency: When using the Monitor/Modify variable utility, setting the trigger frequency enables you to determine whether selected variables are monitored or modified once or every time the *trigger point* is reached.

trigger point: A setting in the Monitor/Modify variable utility, that defines a point in the active user program (e.g., start of cycle, end of cycle, on RUN-to-STOP transition), that will determine when selected variables are to be monitored or modified.

- U -

UDT: A composite of user-defined elements of arbitrary data types, that may be used to declare local variables of a block, specific variables of a data block, or the variables of an entire data block. In essence, a UDT is a structure. Up to 65,535 UDTs from UDT 1 to UDT 65,535 may be defined. A UDT is global, and once defined may be reused repeatedly where the same data structure is required. See *structure*.

un-configured connection: A logical communication link between S7 partners, in which the connection is established dynamically at runtime when the active partner in the link initiates the service, instead of being defined explicitly as with configured connections.

unilateral communication: Refers to S7 communication between partners, in which there is a communications system function block (SFB) only on the local partner (e.g., "GET" SFB). See *bilateral communication*.

User Defined Type (UDT): See *UDT*.

User Memory: S7 memory that holds the control program and data. In S7 300/S7 400 systems, the user program is in two areas, load memory and work memory. See *load memory; work memory*.

- V -

variable: A factor that can be altered, measured, or controlled; quantity that can change in value.

variable declaration: The act of defining a variable for use in the user program. Declaration requires naming the variable, defining its data type (e.g., BOOL, INT, etc.).

variable declaration table: The part of a block in which its local variables are defined. In the case of code blocks, depending on the block type, these are the block parameters as well as the temporary or static data. In the case of data blocks, the declared variables are the data addresses.

variable table (VAT): User created table containing a group of variables to be displayed, monitored and perhaps modified or forced. Up to 255 variable tables from VAT 1 to VAT 255 may be created as resources for monitoring and debugging the control program.

VAT: See *variable table*.

- W -

warm restart: The new name for the *complete restart* mode of earlier STEP 7 versions. This restart mode is triggered whenever the CPU mode selector switch is moved from STOP to RUN, or upon power up (provided the CPU has battery backup). On a warm restart, the CPU and all modules are reset to their initial states; non-retentive data is erased; OB 100 is called; the process image of inputs (PII) is read; and the main program (OB1) is called for processing.

work memory: High-speed S7 CPU RAM, that contains the relevant portions of the user program; these are essentially the program code blocks and the user data that the CPU calls when processing the user program. See *load memory*.

Index

- A -

AG_LRECV 353, 355, 356

AG_LSEND 353, 355, 356

AG_RECV 353, 355, 356

AG_SEND 353, 355, 356

Accessible nodes utility 300, 330

Accessible node window 294, 295, 327

Actual parameters 190, 191, 228, 246, 254

Address, absolute 38, 39, 76, 196

Address, symbolic 38, 76, 77, 226, 288

Addresses
- analog I/O 134, 135, 180, 181
- digital I/O 126, 127, 172, 173

Address identifier 22

Address overview 60, 68, 69

Analog input
- addressing 134, 135, 180, 181
- module properties 128, 129, 174, 175
- signal parameters 130, 131, 176, 177

Analog output
- addressing 134, 135, 180, 181
- module properties 132, 133, 178, 179

Archiving a project
- library 50
- project 50

Array, data type 206, 207, 230

Asynchronous error OBs 115, 161,

Asynchronous errors 194

Authorization 12-15

AuthorsW 12, 14, 16

- B -

BRCV 352, 356

BSEND 352, 356

B-stack 326, 327, 342

Binary result 213, 222

Bit logic instructions 212
- programming 256, 258, 260

Bit memory 106, 152, 196, 197

Blocks 36
- documenting 238
- downloading 296, 314-315
- editing and saving 236
- folder 33, 38, 209
- generating 232
- header 228, 232
- parameters 206, 208
- stack 326, 327, 342
- types 188, 190, 191, 192
- uploading 296, 316-317
- window 211

BOOL, data type 200, 246, 254

BYTE, data type 200

- C -

C-bus 56, 57, 138, 384

CFC 7

Central rack configuration
- S7-300 78, 79
- S7-400 136, 137

Central processing unit 56, 58, 61

CHAR, data type 205

Charts 7

Clock memory 104, 105, 150, 151

Code block 186
- documenting 238
- editing and saving 236
- generating 232

Cold restart 102, 148, 305

Communications bus 56, 57, 384

Communications processors 2, 59, 357-36
- Ethernet 359-361
- Profibus 357-358

Compact DP slave
- configuring and attaching 88-89

Compare instructions 220
- programming 268-269

Comparing programs
- online/offline 296, 318
- path 1/path2 297, 320

Compressing memory 312

Configuration tables 62, 70, 126, 132

Configuring Compact DP Slaves

Configuring
- Compact DP Slaves 88
- Ethernet Connections 404
- Profibus Connections 390
- GD Communication 376
- Hardware Stations 70
- Intelligent DP Slaves 90-99
- Modular DP Slaves 86
- Modules for Multi-computing 182
- Multi-computing Operation 138
- S7-300 as DP-Master 84
- S7-300 Central Rack 78
- S7-300 Local Expansion 80
- S7-400 as DP-Master 140
- S7-400 Central Rack 136
- S7-400 Local Expansion 142
- S7-400 Remote Expansion 144
- Simatic Workspace 18
- Subnets 370, 378, 392
- CP 342-5 as a DP Slave 92
- CPU 315-2 DP as DP Slave 90
- BM 147/CPU as a DP Slave 94

Connections 350, 390, 404
- configured 350, 351,353, 355, 356
- un-configured 351, 413

Connection table 295, 351, 364, 390, 391

Connection types 390, 404

Continuous function chart 7

Conversion instructions 216-217
- programming 266-267

Counter instructions 214
- programming 262-263

Counter memory 198, 214, 417

Counter, down 214, 262, 410

Counter, up 214, 262, 268, 410

Counter, up-down 214

CPU properties 100
- access protection 112, 158, 297, 304
- cycle and clock memory 104, 150
- cyclic interrupts 118, 164, 189
- diagnostics and clock 110, 156
- general 100
- interrupt 114, 160
- local memory 108, 154
- performance data 302
- retentive memory 106, 152
- startup 102, 148
- time-of-day interrupts 116, 162

- D -

Data Block 188, 192
- accessing data elements 193, 288
- creating 228
- editing 230
- instance 192
- shared 192

Data type 200
- complex 206
- elementary 200
- user defined 208
- parameter 208

DATE, data type 204

DATE_AND_TIME, data type 207

Declaration section 190, 191

Diagnostic buffer 104, 110, 130, 150, 298

Diagnostic interrupt 122,124,130,168,176

Digital input
addressing 126, 172
module properties 120, 166

Digital output
addressing 126, 172
module properties 124, 170

DINT, data type 203

Direct connection 26, 294

Direct I/O read 197, 286

Direct I/O write 197, 286

Double integer 203

Downloading
blocks 296, 314
network configuration 295, 368

DP-master
S7-300 84
S7-400 140

DP-slave
compact 88
intelligent 90, 92, 94
modular 86
S7-31x-DP as intelligent 90
CP-342-5 as intelligent 92
BM 147/CPU as intelligent 94

DWORD, data type 202

Dynamic connection 351

- E -

ET-200 86

ET-200 B 88

ET-200 M

ET-200 S 86

edge evaluation operations
programming 260

Error OBs 189

Expansion rack 56, 142

Extended pulse timer 215

- F -

FB 188, 190

FBD 5, 202, 209, 234

FC 188, 191

FDL 353

FDL connection 353

Floating-point 204, 218, 272

FMS 353, 356

FMS connection 353, 356

Formal parameters 190, 191

Function 191
calling 191, 242, 246
call with formal parameters 246
call without formal parameters 242
programming 240, 244

Function Block 190
calling 190, 250, 254
call with formal parameters 254
call without formal parameters 250
programming 248, 252

Function Block Diagram 5, 212, 234

Function modules 59

- G -

GD communication 354

GD table 376

Global data communication 354
configuring 376

Global data, receive 376

Global data, send 376

Global symbol 226

- H -

Hardware catalog 61, 64

Hardware configuration 60, 66, 70

Hardware interrupts 114, 122, 160, 168

HiGraph 10

Hot restart 102, 148, 305

- I -

IM 360/IM 361 57, 80, 82

IM 365 57, 82

IM 460 57, 142, 144

IM 460/IM 461 57, 142

IM 461 57, 142, 144

IM 463 57, 144

I-stack 326, 327, 342

Industrial Ethernet 349

Input bit 196

Input byte 196, 199

Input double word 199

Input image 172, 196, 199

Input memory 126, 172, 196

Input word 134, 180, 196, 199

INT, data type 203

Instance 190

Installing authorizations 14

Instance Data Block 190, 192

Intelligent DP slave
- S7-31x-2 DP as 90
- CP-342-5 as 92
- BM 147/CPU as 94

Interface module 57

Integer 203

Integer arithmetic instructions 218
- programming 270-271

Interrupt 194-195
- OBs 189
- hardware 114, 160

Interrupt input properties 122, 168

Interrupt stack 342

ISO transport service 354

ISO-on-TCP service 355

- L -

LAD 5

LAD/FBD/STL Editor 209
- Navigating 234

Library 33
- archiving 50
- S7 standard 34

Load memory 303, 306

Local I/O expansion
- S7-300 80
- S7-300 Single-Tier 82
- S7-400 142

Local memory 108, 154

Local stack 108, 154

Local symbol 226

Local variables 190, 191

- M -

Memory
- addressing 196
- areas 196
- bit 197
- byte 197, 199
- compressing 312
- double word 197, 199
- word 197, 199

Minimum scan cycle time 104, 150

Modular DP slave
- configuring and attaching 86

Monitor/modify variables 334

Move instructions 224
- programming 286-287

MPI 59, 348

MPI address 366, 368, 370, 372

MPI network 370, 372, 374

Multi-computing
- configuring I/O modules 182
- interrupt OB 189
- S7-400 Central rack 138

Multi-point interface 59, 348

- N -

Nesting level 108, 154

NETPRO 362, 366, 368

Network 348

Network configuration 362, 364, 366

New Project Wizard 44

- O -

OB 1 189, 194
  programming 291-292

Object 33, 38

Object properties 39

Off-delay timer 215

Off-line blocks 236

Off-line window 36

On-delay timer 215, 264

On-line blocks 236

On-line window 36

Operating modes 296, 298, 304

Optional tools 6, 12, 16

Organization Blocks 146, 160, 188, 189, 194

Other Station 48, 365

Output bit 196

Output byte 196, 199

Output double word 199

Output image 172, 196, 199

Output memory 126, 172, 196

Output word 134, 180, 196, 199

- P -

P-bus 56, 57

PC adapter 26, 28

PC Station 358-359, 360-361

PG/PC system 2, 12, 26, 374, 388, 402

Parameter data type 208

Password Protection 112, 158

Partitioned program 186

Peripheral input 196, 197

Peripheral input byte 197

Peripheral input word 134, 180, 197

Peripheral memory 180, 197

Peripheral output 196, 197

Peripheral output byte 197

Peripheral output word 134, 180, 197

Power supply 2, 58

Priority 114, 160, 194-195

Priority class 108, 154, 194-195

Process image inputs 196

Process image outputs 196

PROFIBUS 349

Profibus DP 354

Program flow control instructions 221
  programming 276-279

Project 32, 38

Project window 32, 38

Pulse timer 215

Pulse timer extended 215

- R -

Rack 2, 56, 57, 80

REAL arithmetic instructions 218-219
  programming 272-275

REAL, data type 204, 206

Receiver interface modules 57

Remote I/O expansion
  S7-400 144

Removing authorizations 14

Resetting CPU memory 306-307

Restart
  cold 102, 148, 305
  hot 102, 148, 305
  warm 102, 148, 305

Result of logic operation 256, 257

Retentive bit memory 106, 152

Retentive on-delay timer 215

RLO 212, 213, 256, 260

RUN mode 304-305

RUN-P mode 304-305

- S -

S5TIME, data type 205

S7 300
  as DP-master 84
  as intelligent DP-slave 90
  basic components 56-59
  central rack configuration 78-79
  local expansion 80-83

S7 400
  as DP-master 140
  basic components 56-59
  central rack configuration 136-137
  local expansion 142-143
  multi-computing configuration 138-139
  remote expansion 144-145

S7 connection 351, 352, 356

S7 CFC 7

S7 Functions 352

S7-GRAPH 8

S7-HiGraph 10

S7 Online Connection 26

S7-Program 33

S7-SCL 9

SFB 188

SFC 188

Sample projects 42-43, 51

Segmented rack 57

Set-Reset operations 213
  programming 258-259

Setting the PG/PC interface 28

Shift-Rotate instructions 224
  programming 282-283

Signal modules 59

Simatic Manager 32, 35, 38, 40

Simatic NET 348

Simatic PC-Station 61, 358, 360

Simatic Station 33

Single-user environment 18

Source folder 33

Standard library 34

Start-up OBs 189, 194

Statement List 5, 234

Static connection 351

Static variable 190, 192

Station 33, 46, 48, 68, 70-74

Station configuration 70
  downloading 72
  uploading 74
  viewing 68

Station window 62

Status bit instructions 222
  programming 284-285

Status bits 222, 284

STEP 7 Authorization 14

STL 5, 234

STOP Mode 304, 306, 312, 314, 326, 342

STRING, data type 206

STRUCT, data type 207

Structure 207

Subnet 348
  Industrial Ethernet 349, 392
  MPI 348, 370
  Profibus 349, 378

Subnet ID 370-371, 378-379, 392-393

Symbol 226, 235

Symbolic addresses 196, 226
  assigning to I/O modules 76
  viewing and editing 226

Symbols table 226

Synchronous error OBs 189

Synchronous error interrupts 189, 194

System blocks 188

System data blocks 188

System data object 60

System Function 351

System Function Block 352

- T -

TCP/IP service 355

Temporary local data 190-191

TIME, data type 204

TIME_OF_DAY, data type 204

Time-delay interrupts 114,160,184,194

Time-delay interrupt OBs 114,160,189,194

Time-of-day interrupts 116, 162,194

Time-of-day interrupt OBs 116,162,189,194

Timer instructions 215
  programming 264-265

Timer memory 198

- U -

UDP service 355

UDT, data type 208

Un-configured connection 351

Universal rack 56, 57

Uploading
  blocks 316
  S7 program 316
  station configuration 74

User blocks 188

User Defined Type 208

- V -

VAT 332

Variable 192, 200-208

Variable declaration table 190-191

Variable table 332

- W -

Warm restart 102, 148, 305

Watchdog interrupt 118, 164

Watchdog interrupt OBs 118, 164

WORD, data type 200-201

Word logic instructions 223
  programming 280-281

Work memory 303, 306

Notes:

Notes:

Notes:

Notes:

Notes: